BIOINFORMATICS

BIOINFORMATICS

Prakash S. Lohar

Head, Department of Biotechnology,
Reader, Department of Zoology
M.G.S.M's Arts, Science and Commerce College,
Chopda, Jalgaon Dt.
Maharashtra

MJP PUBLISHERS

Cataloguing-in-Publication Data

Lohar, Prakash S. (1965 –   ).
    Bioinformatics/ by Prakash S. Lohar. –
Chennai : MJP Publishers, 2009
    xxvi, 580 p. ; 23 cm.
    Includes Glossary, References and Index.
    ISBN 978-81-8094-066-8 (pbk.)
    1. Information, Biology 2. Bioinformatics
I. Title.
    570.020 dc22 LOH MJP 063

ISBN 978-81-8094-066-8 **MJP PUBLISHERS**
© Publishers, 2009 47, Nallathambi Street
All rights reserved Triplicane
Printed and bound in India Chennai 600 005

Publisher : J.C. Pillai
Managing Editor : C. Sajeesh Kumar
Project Editor : P. Parvath Radha
Acquisitions Editor : C. Janarthanan
Editoral Team : B. Ramalakshmi, N. Pushpa Bharathi,
L. Mohanapriya, M. Gnanasoundari
Lissy John, N. Yamuna Devi
CIP Data : Prof. K. Hariharan, Librarian
RKM Vivekananda College, Chennai.

To

My wife, daughter and son
and
Late Shri Nitin Pawar

PREFACE

It was a refresher course at the Department of Zoology, University of Pune, specifically designed for teachers to understand the role of computers in the biological sciences. I was fortunate enough to become a part of that course and completed the same with outstanding grade. Since then it was in my mind to write a comprehensive book for biologists that can incorporate the basics of computers and the latest developments in the field of bioinformatics, which occurred in the last two decades.

Bioinformatics describes any use of computers to handle biological information. Actually, it is a multidisciplinary subject with immense scope in molecular biology, biotechnology, and pharmaceuticals. Genomics and proteomics are the heart of bioinformatics. Genomics means the study of complete (haploid) DNA content of an organism whereas proteomics deals with the study of the set of proteins produced (expressed) by an organism, tissue or cell, and the changes in protein expression patterns in different environments and conditions. The expressed proteins in a particular cell or tissue type, are obtained by identifying the proteins from cell extract using a combination of 2D gel electrophoresis, X-ray diffraction, and mass spectrometry.

This book is aimed at students and teachers who need to know the basics of genomics and proteomics, and for this purpose the book is divided into 10 chapters. The first chapter provides an overview of various definitions, origin, history and applications of bioinformatics. The computer generations, their types, operating systems, Internet, E-mail, World Wide Web and Web browsers, search tools and search engines, Web servers, and role of PERL and JAVA are deliberately discussed in the second chapter. Since molecular biology plays a central role in bioinformatics, its central dogma with structural details of DNA, RNA, and proteins are given in the third chapter. The fourth chapter on fundamental techniques in bioinformatics includes chromatography, gel electrophoresis, isoelectric focusing and two-dimensional (2D) gel electrophoresis, blotting techniques, DNA microarrays, automated gene sequencing technique, mass spectrometry, and X-ray crystallography. The fifth chapter explains the concept of biological database, its classification, and database retrieval with PubMed, Entrez, SRS, OMIM, ExPASy, and EMBL.

The sixth chapter is focused on the concept of sequence alignment, global and local alignment, database similarity searching software packages like BLAST and FASTA to determine sequence matches in particular DNA or protein sequences. The multiple sequence alignment and various software packages with their web URLs are discussed in chapter seven. Computational gene prediction and its different approaches including gene discovery using ESTs and cDNA are given in chapter eight. The overall development in the field of proteomics, including protein sequence databases, Swiss-Prot, PDB, its file formats, software for visualization of protein structure, its classification and interactions with applications are elaborated in the ninth chapter. Last but not the least, chapter ten highlights the concepts of gene, genome and genomics including the prime role of GenBank, DDBJ, TIGR, different categories of JCVI software, Kyoto Encyclopedia of Genes and Genomes, genomics of bacteriophage to HGP, gene therapy, drug designing and pharmacogenomics.

I feel grateful to Arther M. Lesk, A.M. Campbell, Altschul, S.F., Attwood, T.K. and Jin Xiong, who are in real sense the father figures in the textual field of bioinformatics and from whom I got inspired to complete this endeavour. A lot of up-to-date information in relation to various fields of bioinformatics is derived from web pages and hence it is my duty to express sincere thanks to all the service providers including wikimedia foundation who made extraordinary contribution in developing the software packages for proteomics and genomics and for their efforts in uploading the contents on the web pages. To prepare the manuscript for this title, I have extensively reviewed thousands of research papers from many international journals and have accessed Web linked library with several URLs. I am thankful to those authors who permitted me to incorporate their findings in this manuscript.

I am deeply indebted to honorable Dadasaheb Dr. S.G.Patil, Founder President, Bhaiyasaheb Advt. Sandeep S. Patil, President, and other respected members of Management Council and Governing Council of Mahatma Gandhi Shikshan Mandal for their continuous encouragement throughout my services. Thanks are also due to the Principal, Vice Principal, and teaching and non-teaching staff of my college for their moral support. There are so many well-wishers of mine including Prof. V.L. Maheshwari, Director, School of Life Sciences and Prof. S.B. Chincholkar, Director, BCUD, North Maharashtra University, Jalgaon, the Principal and Staff of College of Pharmacy of my institution, Dr. Vikas Gudve from Bhusawal, my relatives Shri. P.N.Lohar, Mrs. Vaishali Pawar, Shri. Sharad Nile, Shri. Rajesh Chavan, Dr. Swati Chavan, who provided moral support to complete this task. Special thanks to my wife Savita, daughter Shivani and son Kaushal for their support to complete this manuscript.

I express thanks to Mr. J.C. Pillai, Director, C. Sajeesh Kumar, Managing Editor, and other staff of MJP Publishers, Chennai, for their skill and patience in bringing out this book.

Prakash S. Lohar

CONTENTS

8. Computational Gene Prediction   319

10. Genomics — 431

OVERVIEW

Bioinformatics derives knowledge from computer analysis of biological data. It can consist of information stored in the genetic code, and also experimental results from various sources, patient statistics, and scientific literature. Research in bioinformatics includes methods for the development of storage, retrieval, and analysis of the data. Bioinformatics is a rapidly developing branch of biology and is highly interdisciplinary, using techniques and concepts from informatics, statistics, mathematics, chemistry, biochemistry, physics, and linguistics. It has many practical applications in different areas of biology and medicine.

Roughly, bioinformatics describes any use of computers to handle biological information. In practice the definition used by most people is narrower; bioinformatics to them is a synonym for "computational molecular biology"—the use of computers to characterize the molecular components of living things.

DEFINITIONS OF BIOINFORMATICS

Fredj Tekaia at the Pasteur Institute defines bioinformatics as the mathematical, statistical and computing methods that aim to solve biological problems using DNA and amino acid sequences and related information.

There are other fields, for example medical imaging/image analysis, which might be considered part of bioinformatics. There is also a whole other discipline of biologically inspired computation, genetic algorithms, artificial intelligence, and neural networks. Often these areas interact in strange ways. Neural networks, inspired by crude models of the functioning of nerve cells in the brain, are used to predict, surprisingly accurately, the secondary structures of proteins from their primary sequences. In common, all bioinformatics is the processing of large amounts of biologically derived information, whether DNA sequences or breast X-rays.

Richard Durbin, Head of Informatics at the Wellcome Trust Sanger Institute, expressed an interesting opinion: "I do not think all biological computing is bioinformatics, e.g. mathematical modelling is not bioinformatics, even when connected with biology-related problems. In my opinion, bioinformatics has something to do with management and the subsequent use of biological information, particularly genetic information".

The NIH Biomedical Information Science and Technology Initiative Consortium agreed on the following definitions of bioinformatics and computational biology recognizing that no definition could completely eliminate overlap with other activities or preclude variations in interpretation by different individuals and organizations.

The National Center for Biotechnology Information (NCBI 2001) defines bioinformatics as follows: "Bioinformatics is the field of science in which biology, computer science, and information technology merge into a single discipline. There are three important sub-disciplines within bioinformatics: the development of new algorithms and statistics with which to assess relationships among members of large data sets; the analysis and interpretation of various types of data including nucleotide and amino acid sequences, protein domains, and protein structures; and the development and implementation of tools that enable efficient access and management of different types of information."

Molecular bioinformatics is defined as conceptualizing biology in terms of molecules (in the sense of physical chemistry) and applying informatics techniques (derived from disciplines such as applied mathematics, computer science and statistics) to understand and organize the information associated with these molecules, on a large scale. In short, bioinformatics is a management information system for molecular biology and has many practical applications.

Some other definitions of bioinformatics include the following.

- ▣ Bioinformatics or computational biology is the use of mathematical and informational techniques, including statistics, to solve biological problems, usually by creating or using computer programs, mathematical models or both. One of the main areas of bioinformatics is the data mining and analysis of the data gathered by the various genome projects. Other areas are sequence alignment, protein structure prediction, systems biology, protein–protein interactions and virtual evolution.

▣ Bioinformatics is the science of developing computer databases and algorithms for the purpose of speeding up and enhancing biological research.

▣ As a discipline that builds upon computational biology, bioinformatics encompasses the development and application of data-analytical and theoretical methods, mathematical modelling and computational simulation techniques to the study of biological, behavioural, and social systems. As a discipline that builds upon the life, health, and medical sciences, bioinformatics supports medical informatics; gene mapping in pedigrees and population studies; functional-, structural-, and pharmacogenomics; proteomics, and dozens of other evolving -omics. As a discipline that builds upon the basic sciences, bioinformatics depends on a strong foundation of chemistry, biochemistry, biophysics, biology, genetics, and molecular biology, which allows interpretation of biological data in a meaningful context. As a discipline whose core is mathematics and statistics, bioinformatics applies these fields in ways that provide insight to make the vast, diverse, and complex life sciences data more understandable and useful, to uncover new biological insights, and to provide new perspectives to discern unifying principles. In short, bioinformaticists bring a multidisciplinary perspective to many of the critical problems facing the health-science profession today.

▣ Biologists using computers, or the other way around. Bioinformatics is more of a tool than a discipline.

▣ The application of computer technology to the management of biological information. Specifically, it is the science of developing computer databases and algorithms to facilitate and expedite biological research.

▣ Bioinformatics is a combination of Computer Science, Information Technology and Genetics to determine and analyse genetic information. (*Source*: BitsJournal.com)

▣ Bioinformatics is the application of computer technology to the management and analysis of biological data. The result is that computers are being used to gather, store, analyse and merge biological data.

Even though the three terms—bioinformatics, computational biology and bioinformation infrastructure—are often times used interchangeably, broadly, the three may be defined as follows:

1. Bioinformatics is the research, development, or application of computational tools and approaches for expanding the use of biological, medical, behavioural or health data, including those to acquire, store, organize, archive, analyse, or visualize such data. It refers to database-like activities, involving persistent sets of data that are maintained in a consistent state over essentially indefinite periods of time.

2. Computational Biology encompasses the use of algorithmic tools to facilitate biological analyses.

3. Bioinformation infrastructure comprises the entire collection of information management systems, analysis tools and communication networks supporting biology. Thus, the latter may be viewed as a computational scaffold of the former two.

Bioinformatics is currently defined as the study of information content and information flow in biological systems and processes. It has evolved to serve as the bridge between observations (data) in diverse biologically related disciplines and the derivations of understanding (information) about the systems or processes function, and subsequently the application (knowledge). A more pragmatic definition in the case of diseases is the understanding of dysfunction (diagnostics) and the subsequent applications of the knowledge for therapeutics and prognosis.

An issue that regularly pops up is the definition of bioinformatics, particularly from individuals who are trying to get into the field. Most lament that there are too many definitions of bioinformatics.

This should be taken as an indication of the ubiquity of the subject. Bioinformatics, as both an enabling and enabled technology, will be defined differently depending on the domain of the person who is giving the definition. A computer scientist will give one definition, a biologist another, a biotechnologist yet another, and an individual from a pharmaceutical company will provide another definition. Each definition is as good as the other. This is just the nature of the beast.

A BIOINFORMATICIST VERSUS A BIOINFORMATICIAN

Bioinformatics has become a mainstay of genomics, proteomics, and all other -omics (such as phenomics), that many information technology companies have entered the business or are considering entering the business, creating an IT (information technology) and BT (biotechnology) convergence.

A distinction has to be made between a bioinformaticist and a bioinformatician. A bioinformaticist is an expert who not only knows how to use bioinformatics tools, but also knows how to write interfaces for effective use of the tools. A bioinformatician, on the other hand, is a trained individual who only knows to use bioinformatics tools without a deeper understanding. Thus, a bioinformaticist is to -omics as a mechanical engineer is to an automobile. A bioinformatician is to -omics as a technician is to an automobile. It has been argued that the professional categorization of bioinformatician and bioinformaticist may lead to some confusion. It is undeniable this may be the case. For example, a mathematician is an expert. However, note that a physicist is an expert, so is a physician, except that they are in very different professions.

FIELDS RELATED TO BIOINFORMATICS

Bioinformatics has various applications in research, medicine, biotechnology, agriculture, etc. Following research fields has integral component of Bioinformatics:

Computational Biology The development and application of data-analytical and theoretical methods, mathematical modelling and computational simulation techniques to the study of biological, behavioural, and social systems.

Genomics Genomics is any attempt to analyse or compare the entire genetic complement of a species or species (plural). It is, of course possible to compare genomes by comparing more-or-less representative subsets of genes within genomes.

Proteomics Proteomics is the study of proteins—their location, structure and function. It is the identification, characterization and quantification of all proteins involved in a particular pathway, organelle, cell, tissue, organ or organism that can be studied in concert to provide accurate and comprehensive data about that system. Proteomics is the study of the function of all expressed proteins. The study of the proteome, called proteomics, now evokes not only all the proteins in any given cell, but also the set of all protein isoforms and modifications, the interactions between them, the structural description of proteins and their higher order complexes, and for that matter almost everything "post-genomic".

Pharmacogenomics   It is the application of genomic approaches and technologies to the identification of drug targets. In short, pharmacogenomics is using genetic information to predict whether a drug will help make a patient well or sick. It studies how genes influence the response of humans to drugs, from the population to the molecular level.

Pharmacogenetics   It is the study of how the actions of and reactions to drugs vary with the patient's genes. All individuals respond differently to drug treatments; some positively, others with little obvious change in their conditions and yet others with side effects or allergic reactions. Much of this variation is known to have a genetic basis. Pharmacogenetics is a subset of pharmacogenomics which uses genomic/bioinformatic methods to identify genomic correlates, for example, Single Nucleotide Polymorphisms (SNPs), characteristic of particular patient response profiles and use those markers to inform the administration and development of therapies. Strikingly such approaches have been used to "resurrect" drugs thought previously to be ineffective, but subsequently found to work within subset of patients or in optimizing the doses of chemotherapy for particular patients.

Cheminformatics   The mixing of those information resources (information technology and information management) to transform data into information and information into knowledge for the intended purpose of making better decisions faster in the arena of drug lead identification and optimization. Related terms of cheminformatics are chemi-informatics, chemometrics, computational chemistry, chemical informatics, chemical information management/science, and cheminformatics.

Chemical informatics   This deals with computer-assisted storage, retrieval and analysis of chemical information, from data to chemical knowledge whereas "chemoinformatics" (and the synonymous cheminformatics and chem-informatics) focus on drug design.

Chemometrics   The application of statistics to the analysis of chemical data (from organic, analytical or medicinal chemistry) and design of chemical experiments and simulations.

Structural genomics or structural bioinformatics   This refers to the analysis of macromolecular structure particularly proteins, using computational tools and theoretical frameworks. One of the goals of structural genomics is the extension of idea of genomics, to obtain accurate three-dimensional structural models for all known protein families, protein domains or protein folds. Structural alignment is a tool of structural genomics.

Comparative genomics   The study of human genetics by comparison with model organisms such as mice, the fruit fly, and the bacterium *E. coli.*

Biophysics   The British Biophysical Society defines biophysics as: "an interdisciplinary field which applies techniques from the physical sciences to understanding biological structure and function."

Biomedical informatics/Medical informatics   It is an emerging discipline that has been defined as the study, invention, and implementation of structures and algorithms to improve communication, understanding and management of medical information.

Mathematical biology   Mathematical biology also tackles biological problems, but the methods it uses to tackle them need not be numerical and need not be implemented in software or hardware. It includes things of theoretical interest that are not necessarily algorithmic, not necessarily molecular in nature, and are not necessarily useful in analysing collected data.

Computational chemistry   Computational chemistry is the branch of theoretical chemistry whose major goals are to create efficient computer programs that calculate the properties of molecules such as total energy, dipole moment, vibrational frequencies and to apply these programs to concrete chemical objects. It is also sometimes used to cover the areas of overlap between computer science and chemistry. It also includes, synthesis planning, database searching, combinatorial library manipulation, etc.

Functional genomics   Functional genomics is a field of molecular biology that is attempting to make use of the vast wealth of data produced by genome sequencing projects to describe genome function. Functional genomics uses high-throughput techniques like DNA microarrays, proteomics, metabolomics and mutation analysis to describe the function and interactions of genes.

Pharmacoinformatics   It concentrates on the aspects of bioinformatics dealing with drug discovery.

***In silico* ADME-Tox Prediction**   Drug discovery is a complex and risky treasure hunt to find the most efficacious molecule that does not have toxic effects but at the same time has the desired pharmacokinetic profile. The hunt starts when the researchers look for the binding affinity of the molecule to its target. Huge amount of research requires to be done to come out with a molecule, which has the reliable binding profile. Once the molecules have been identified, as per the traditional methodologies, the molecule is further subjected to optimization with the aim of improving efficacy. The molecules which show better binding are then evaluated for its toxicity and pharmacokinetic profiles. It is at this stage that most of the candidates fail in the race to become a successful drug.

Agroinformatics/Agricultural informatics   It concentrates on the aspects of bioinformatics dealing with plant genomes.

Systems biology   It is the coordinated study of biological systems by investigating the components of cellular networks and their interactions, by applying experimental high-throughput and whole-genome techniques, and integrating computational methods with experimental efforts.

OBJECTIVES AND SCOPE

Bioinformatics is the technology that uses computers for storage, retrieval, manipulation, and distribution of information obtained by analysing sequence data of biological macromolecules like DNA, RNA and proteins. This scientific endeavour helps in better understanding of a living cell and its functioning at the molecular level through the study of nucleotide sequence in the gene and amino acid sequence in the protein. Thus, the ultimate goal of bioinformatics is to enable the discovery of new biological insights as well as to create a global perspective from which unifying principles in biology can be discerned. Bioinformatics can be thought of as a central hub that unites several disciplines and methodologies (Figure 1.1).

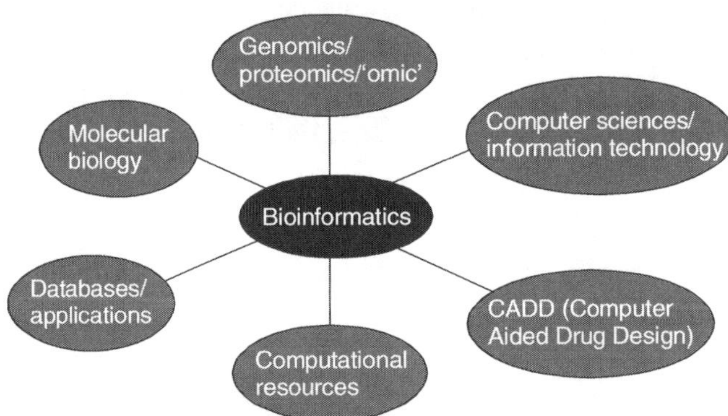

Figure 1.1   Interdisciplinary approach of bioinformatics

On the support side of the hub, information technology, information management, software applications, databases and computational resources all provide the infrastructure for bioinformatics. On the scientific side of the hub, bioinformatic methods are used extensively in molecular biology, genomics that deals with the study of analysis of sequence, structure and functions of genes, proteomics that involves prediction of sequence of amino acids, structure and functions of entire collection of proteins which are coded by the genome of an organism, other emerging areas (i.e., metabolomics, transcriptomics) and in Computer Aided Drug Design (CADD) research.

In other words bioinformatics involves:

1. Use of different computer software tools to create databases of genomics, proteomics, metabolomics, transcriptomics, and so on.
2. Use of experimental methods like X-ray crystallography and NMR spectroscopy for determination of three-dimensional structural aspects of genes and product of gene (mostly a protein molecule).
3. Development of algorithms for utilization and management of various databases in computer modelling.
4. Use of databases as well as analytical and computational methods such as numerical methods of analysis, simulation, visualization and modelling in the synthesis of better molecular species or a drug. Also, integrate this knowledge with genetic engineering or molecular engineering to synthesize desired product.

GENOME MAPPING AS A SOURCE OF BIOINFORMATICS

By 1981, 579 human genes had been mapped and mapping by *in situ* hybridization had become a standard method. Marvin Carruthers and Leory Hood made a huge leap in bioinformatics when they invented a method for automated DNA sequencing. In 1988, the Human Genome Organization (HGO) was founded. This is an international organization of scientists involved in Human Genome Project (HGP). In 1989, the first complete genome map of the bacteria *Haemophilus influenzae* was published.

The following year, the Human Genome Project was started. By 1991, a total of 1879 human genes had been mapped. In 1993, Genethon, a human genome research centre in France produced a physical map of the human genome. Three years later, Genethon published the final version of the human genetic map. This concluded the end of the first phase of the Human Genome Project.

Bioinformatics was fuelled by the need to create huge databases, such as GenBank, EMBL, and DNA Database of Japan to store and compare the DNA sequence data erupting from the human genome and other genome sequencing projects. Today, bioinformatics embraces protein structure analysis, gene and protein functional information, data from patients, pre-clinical and clinical trials, and the metabolic pathways of numerous species.

Origin of Bioinformatics/Biological Databases

The first bioinformatics/biological databases were constructed a few years after the first protein sequences began to become available. The first protein sequence reported was that of bovine insulin in 1956, consisting of 51 residues. Nearly a decade later, the first nucleic acid sequence was reported, that of yeast alanine tRNA with 77 bases. Just a year later, Dayhoff gathered all the available sequence data to create the first bioinformatic database. The Protein Data Bank followed in 1972 with a collection of ten X-ray crystallographic protein structures, and the Swiss-Prot protein sequence database began in 1987. A huge variety of divergent data resources of different types and sizes are now available either in the public domain or more recently from commercial third parties. All of the original databases were organized in a very simple way with data entries being stored in flat files, either one per entry, or as a single large text file. Each data entry in a file is then improved by providing it an index that allows convenient keyword searching for a given information.

Origin of Tools/Software

After the formation of the databases, tools became available to search sequence databases, at first in a very simple way, looking for keyword matches and short sequence words, and then more sophisticated pattern matching and alignment-based methods. The rapid but less rigorous BLAST algorithm has been the mainstay of sequence database searching since its introduction a decade ago, complemented by the more rigorous and slower FASTA and Smith–Waterman algorithms. Suites of analysis algorithms, written by leading academic researchers at Stanford, CA, Cambridge, UK and Madison, WI for their in-house projects, began to become more widely available for basic sequence analysis. These algorithms were typically single-function black boxes that took input and produced output in the form of formatted files. UNIX style commands were used to operate the algorithms, with some suites having hundreds of possible commands, each taking different command options and input formats. Since these early efforts, significant advances have been made in automating the collection of sequence information. Rapid innovation in biochemistry and instrumentation has brought us to the point where the entire genomic sequence of at least 30 organisms, mainly microbial pathogens, are known and projects to elucidate at least 150 more prokaryotic and eukaryotic genomes are currently under way. With new technologies one can directly examine the changes in expression levels of both mRNA and proteins in living cells, both in a disease state or following an external challenge. It is possible to identify patterns of response in cells that lead us to an understanding of the mechanism of action of an agent on a tissue. The volume of data arising from projects of this nature is unprecedented in the pharma industry, and will have a profound effect on the ways in which data are used and experiments performed in drug discovery and development projects. This is true not least because, with much of the available interesting data being in the hands of commercial genomics companies, pharmcos are unable to get exclusive access to many gene sequences or their expression profiles.

The competition between co-licensees of a genomic database is effectively a race to establish a mechanistic role or other utility for a gene in a disease state in order to secure a patent position on that gene. Much of this work is carried out by informatics tools. Despite the huge progress in sequencing and expression analysis technologies, and the corresponding magnitude of more data that is held in the public, private and commercial databases, the tools used for storage, retrieval, analysis and dissemination of data in bioinformatics are still very similar to the original systems gathered together by researchers 15–20 years ago. Many are simple extensions of the original academic systems, which have served the needs of both academic and commercial users for many years. These systems are now beginning to fall behind as they struggle to keep up with the pace of change in the pharma industry.

Databases are still gathered, organized, disseminated and searched using flat files. Relational databases are still few and far between, and object-relational or fully object-oriented systems are rare still in mainstream applications. Interfaces still rely on command lines, fat client interfaces, which must be installed on every desktop, or HTML/CGI forms. Whilst they were in the hands of bioinformatics specialists, pharmcos have been relatively undemanding of their tools. Now the problems have expanded to cover the mainstream discovery process, much more flexible and scalable solutions are needed to serve pharma-research and development informatics requirements.

A CHRONOLOGICAL HISTORY OF EVENTS IN BIOINFORMATICS

Listed below are some of the major events in bioinformatics over the last several decades. Most of the events in the list occurred long before the term, "bioinformatics", was coined.

1943   W. Astbury obtained the first X-ray diffraction pattern of DNA.

1951   Pauling and Corey proposed the structure for the alpha-helix and beta-sheet in a protein.

1953   Watson and Crick proposed the double-helix model for DNA-based X-ray crystallographic data obtained by Franklin and Wilkins.

1954   Perutz's group developed heavy atom methods to solve the phase problem in protein crystallography.

1955   The sequence of the first protein to be analysed, bovine insulin, announced by F. Sanger.

1958   The first integrated circuit was constructed by Jack Kilby at Texas Instruments.

The Advanced Research Projects Agency (ARPA) was formed in the U.S.

1962   Theory of molecular evolution given by Pauling's.

1965   Margaret Dayhoff's Atlas of Protein Sequences published.

1968   Packet-switching network protocols presented to ARPA.

1969   The ARPANET created by linking computers at Stanford, UCSB, The University of Utah and UCLA.

1970   The details of the Needleman–Wunsch algorithm for sequence comparison published.

1971   E-mail program invented by Ray Tomlinson (BBN).

1972   Protein Data Bank with collection of ten X-ray crystallographic protein structures (Bernstein *et al.*).

Paul Berg and his group created the first recombinant DNA molecule.

1973   The Brookhaven Protein DataBank announced.

1974   Vint Cerf and Robert Khan developed the concept of connecting networks of computers into an "internet" and developed the Transmission Control Protocol.

1975   Microsoft Corporation founded by Bill Gates and Paul Allen.

Two-dimensional electrophoresis, where separation of proteins on SDS polyacrylamide gel is combined with separation according to isoelectric points.

1976   The Unix-to-Unix Copy Protocol (UUCP) developed at Bell Labs.

E. M. Southern published the experimental details for the Southern Blot technique of specific sequences of DNA.

First genetic Engineering company Genentech founded.

1977   The full description of the Brookhaven PDB (http://www.pdb.bnl.gov) published.

Allan Maxam and Walter Gilbert (Harvard) and Frederick Sanger (U.K. Medical Research Council), reported methods for sequencing DNA and software to analyse it.

1978   The first Usenet connection established between Duke and the University of North Carolina at Chapel Hill.

1980   The first complete gene sequence for an organism (ϕX174) published. This viral DNA consists of 5,386 base pairs that code nine proteins.

Wüthrich *et al.* worked out multi-dimensional NMR for protein structure determination.

IntelliGenetics, Inc. founded in California. Their primary product is the programs for DNA and protein sequence analysis.

1981   The Smith–Waterman algorithm for sequence alignment published.

IBM introduces its Personal Computer to the market.

The concept of a sequence motif arised.

1982   Genetics Computer Group (GCG) created as a part of the University of Wisconsin of Wisconsin Biotechnology Center. The company's primary product is the Wisconsin Suite of molecular biology tools.

GenBank Release-3 made public.

Phage lambda genome sequenced. DNA contains 48,502 bp.

1983   The Compact Disk (CD) launched.

LANL (Los Alamos National Laboratory) and LLNL (Lawrence Livermore National Laboratory) begin production of DNA clone (cosmid) libraries representing single chromosomes.

DNA analysis becomes viable with the discovery of Polymerase Chain Reaction.

1984   Jon Postel's Domain Name System (DNS) placed online.

The Macintosh announced by Apple Computer.

1985   The FASTP/FASTN algorithm published.

Robert Sinsheimer holds meeting on human genome sequencing at University of California, Santa Cruz.

The first conference to assess the feasibility of the Human Genome Initiative.

1986   U.S. Department of Energy (DOE) announced Human Genome Initiative.

The term "Genomics" appeared for the first time to describe the scientific discipline of mapping, sequencing, and analysing genes. The term was coined by Thomas Roderick.

Amoco Technology Corporation acquires IntelliGenetics.

The Department of Medical Biochemistry of the University of Geneva and the European Molecular Biology Laboratory (EMBL) creates the Swiss-Prot database.

1987   The use of yeast artificial chromosomes (YAC) described.

The physical map of *E. coli* published.

Perl (Practical Extraction Report Language) released by Larry Wall.

1988   National Center for Biotechnology Information (NCBI) created at NIH/NLM.

EMBnet network for database distribution created.

The Human Genome Intiative started.

The FASTA algorithm for sequence comparison published by Pearson and Lupman.

1989   The Genetics Computer Group (GCG) becomes a private company. Oxford Molecular Group, Ltd.(OMG) founded.

1990   The BLAST program implemented.

Projects begun to mark gene sites on chromosome maps as sites of mRNA expression.

Research and development begun for efficient production of more stable, large-insert bacterial artificial chromosomes (BACs).

1991   The research institute in Geneva (CERN) announces the creation of the protocols that make-up the World Wide Web.

The creation and use of expressed sequence tags (ESTs) described.

1992   Low-resolution genetic linkage map of entire human genome published. Guidelines for data release and resource sharing announced by DOE and NIH.

1993   Sanger Centre established by Wellcome Trust and Medical Research Council, UK, to focus on mapping and sequencing the human genome.

CuraGen Corporation formed in New Haven, CT.

Affymetrix begun independent operations in Santa Clara, California.

International IMAGE Consortium established to coordinate efficient mapping and sequencing of gene-representing cDNAs.

1994   Netscape Communications Corporation founded and Navigator released.

Gene Logic is formed in Maryland.

The PRINTS database of protein motifs published.

Oxford Molecular Group acquires IntelliGenetics.

European Bioinformatics Institute EMBL, Hinxton, UK establishes the network called EMBnet for providing information of biological databases.

1995   The *Haemophilus influenzae* genome sequenced.

The high-resolution physical maps of chromosome 16 and chromosome 19 published.

The *Mycoplasma genitalium* genome sequenced.

Moderate-resolution maps of human chromosomes 3, 11, 12, and 22 maps published.

Physical map with over 15,000 STS markers published.

The term "proteome" introduced by Wilkins.

1996   The genome for *Saccharomyces cerevisiae* (Baker's yeast, 12.1 Mb) is sequenced.

Bacterium *Methanococcus jannaschii* genome sequenced.

Affymetrix produces the first commerical DNA chips.

1997   The genome for *E. coli* (4.7 Mbp) published.

NIH NCHGR becomes National Human Genome Research Institute (NHGRI).

Second large-scale sequencing strategy meeting held in Bermuda.

High-resolution physical maps of human chromosomes X and 7 completed. DOE-NIH task force on Genetic Testing releases final report and recommendations.

1998   The genomes for nematode worm *Caenorhabitis elegans* and Baker's yeast published.

The Swiss Institute of Bioinformatics established as a non-profit foundation.

Inpharmatica, a new genomics and bioinformatics company, established by University College London.

Gene Formatics, a company dedicated to the analysis and predication of protein structure and function established.

1999   Researchers in the Human Genome Project announced the complete sequencing of the DNA making up human chromosome 22.

2000   The genome for *Pseudomonas aeruginosa* (6.3 Mbp) published.

The crucifer weed *Arabidopsis thaliana* genome (100 Mb) sequenced.

The fruit fly *Drosophila melanogaster* genome (180 Mb) sequenced.

HGP leaders and President Clinton announced the completion of a "working draft" DNA sequence of the human genome.

Chromosome 21 genome, the smallest human chromosome completely sequenced.

DOE researchers announced completion of human chromosomes 5, 16, and 19 draft sequence.

2001   The human genome (3,000 Mbp) published. Human chromosome 20 finished.

2002   Structural Bioinformatics and GeneFormatics merge.

        The full genome sequence of the common house mouse (2.5 Gb) published.

        Draft sequence of mouse genome published.

        Draft sequence of *Fugu rubripes* published.

        Genome of fusion yeast *Schizosaccharomyces pombe* published.

2003   Human Genome Project completed.

        Chromosome 14—the fourth chromosome to be completely sequenced.

2004   The draft genome sequence of the brown rat, *Rattus norvegicus*, completed.

        Agilent Technologies Inc. comprised 41,534 oligonucleotide probes (genome microarrays) representing more than 41,000 mouse genes and transcripts.

        Proteomic profiling yields a trio of biomarkers for ovarian cancer.

        Georgia Institute of Technology (GIT) developed a method using nanoparticles for targeting and imaging cancer tumours *in vivo*.

2005   American Academy of Microbiology recommended a centralized genome annotation initiative.

        Novartis pharmaceuticals and Dana–Farber Cancer Institute published a report on targeted therapy for chronic myelogenous leukemia (CML).

2006   National Cancer Institute (NCI) constructed a gene expression database that comprises 18,927 genes.

        National Institute of General Medical Science (NIGMS), Models of Infectious Disease Agent Study (MIDAS) examined containment strategies for the influenza virus (H5N1).

        Neogenesis Pharmaceuticals Inc. created Automated Ligand Identification System (ALIS) to determine the affinities of individual compound from a chemical mixture.

2007   The International Crop Research Institute for the Semi-Arid Tropics (ICRISAT) and Department of Biotechnology, Government of India established a Center of Excellence in Genomics (CEG).

Dana–Faber Cancer Institute, Boston analysed both time and location of gene expression in *Caenorhabditis elegans* the worm, which has immense and growing popularity as a model for research and drug development.

Entire genetic blueprint of 2000 human and avian influenza viruses have been completed and the sequence data made available in a public database.

Researchers at University of Toronto successfully mapped all 70,000 nucleosomes present in yeast.

Invitrogen launched new-engineered stem cell line for disease research and drug discovery.

Researchers at University of Cornell discovered 300 new human genes with the help of supercomputer designed to root out genes that have been conserved over millions of years of evolution.

2008   Compugen Ltd. (NASDAQ: CGEN) and Roche signed agreement to identify genetic variation predictive of drug response in rheumatoid arthritis patients.

Oxford Genome Sciences and Amgen (AMGN) jointly discovered novel therapeutic antibodies in cancer.

FDA approves TOP2A (topoisomerase 2 alpha) gene FISH (Fluorescent *in situ* hybridization) pharmDx test that helps in assessing the risk of tumour recurrence and long-term survival for patients with relatively high-risk breast cancer.

Biomarker Research Initiatives in Mass Spectrometry (BRIMS) Center, Cambridge use quantitative proteomics in a clinical trial to monitor drug efficacy or as a diagnostic.

APPLICATION OF BIOINFORMATICS IN VARIOUS FIELDS

Molecular Medicine

The human genome will have profound effects on the fields of biomedical research and clinical medicine. Every disease has a genetic component.

This may be inherited (as is the case with an estimated 3000–4000 hereditary diseases including cystic fibrosis and Huntington's disease) or a result of the body's response to an environmental stress which causes alterations in the genome (e.g. cancers, heart disease, diabetes).

The completion of the human genome allows us to search the genes that are directly associated with different diseases and begin to understand the molecular basis of these diseases more clearly. This new knowledge of the molecular mechanisms of disease will enable better treatments, cures and even preventative tests to be developed.

Personalized Medicine

Clinical medicine will become more personalized with the development of the field of pharmacogenomics. This is the study of how an individual's genetic inheritance affects the body's response to drugs. At present, some drugs fail to make it to the market because a small percentage of the clinical patient population show adverse affects to a drug due to sequence variants in their DNA. As a result, potentially life-saving drugs never make it to the market place. Today, doctors have to use trial and error to find the best drug to treat a particular patient as those with the same clinical symptoms can show a wide range of responses to the same treatment. In the future, doctors will be able to analyse a patient's genetic profile and prescribe the best available drug therapy and dosage from the beginning.

Preventive Medicine

With the specific details of the genetic mechanisms of diseases being unravelled, the development of diagnostic tests to measure a persons susceptibility to different diseases may become a distinct reality. Preventive actions such as change of lifestyle or having treatment at the earliest possible stages when they are more likely to be successful could result in huge advances in our struggle to conquer disease.

Gene Therapy

In future, the potential for using genes themselves to treat disease may become a reality. Gene therapy is the approach used to treat, cure or even prevent disease by changing the expression of a person's genes. Currently, this field is in its infantile stage with clinical trials for many different types of cancer and other diseases ongoing.

Drug Development

At present all drugs on the market target only about 500 proteins. With an improved understanding of disease mechanisms and using computational tools to identify and validate new drug targets, more specific medicines that act on the cause, not merely the symptoms, of the disease can be developed. These highly specific drugs promise to have fewer side effects than many of today's medicines.

Microbial Genome Applications

Microorganisms are ubiquitous, that is they are found everywhere. They have been found surviving and thriving in extremes of heat, cold, radiation, salt, acidity and pressure. They are present in the environment, our bodies, the air, food and water. Traditionally, use has been made of a variety of microbial properties in the baking, brewing and food industries. The arrival of the complete genome sequences and their potential to provide a greater insight into the microbial world and its capacities could have broad- and far-reaching implications for environment, health, energy and industrial applications. For these reasons, in 1994, the U.S. Department of Energy initiated the MGP (Microbial Genome Project) to sequence genomes of bacteria useful in energy production, environmental clean-up, industrial processing and toxic waste reduction. By studying the genetic material of these organisms, scientists begin to understand these microbes at a very fundamental level and isolate the genes that give them their unique abilities to survive under extreme conditions.

Waste Clean-up

Deinococcus radiodurans is known as the world's toughest bacteria and it is the most radiation resistant organism known. Scientists are interested in this organism because of its potential usefulness in cleaning up waste sites that contain radiation and toxic chemicals.

Climate Change Studies

Increasing levels of carbon dioxide emission, mainly through the expanding use of fossil fuels for energy, are thought to contribute to global climate change. Recently, the Department of Energy (DOE), USA launched a program to decrease atmospheric carbon dioxide levels. One method of doing so is to study the genomes of microbes that use carbon dioxide as their sole carbon source.

Alternative Energy Sources

Scientists are studying the genome of the microbe *Chlorobium tepidum* that has an unusual capacity for generating energy from light.

Biotechnology

The archaeon *Archaeoglobus fulgidus* and the bacterium *Thermotoga maritima* have potential for practical applications in industry and government-funded environmental remediation. These microorganisms thrive in water temperatures above the boiling point and therefore may provide the DOE, the Department of Defence, and private companies with heat-stable enzymes suitable for use in industrial processes. Other industrially useful microbes include, *Corynebacterium glutamicum* which is of high industrial interest as a research object because it is used by the chemical industry for the biotechnological production of the amino acid lysine. The substance is employed as a source of protein in animal nutrition. Lysine is one of the essential amino acids in animal nutrition. Biotechnologically produced lysine is added to feed concentrates as a source of protein, and is an alternative to soybeans or meat and bonemeal.

Xanthomonas campestris pv. is grown commercially to produce the exopolysaccharide xanthan gum, which is used as a viscosifying and stabilizing agent in many industries. *Lactococcus lactis* is one of the most important microorganisms involved in the dairy industry, it is a non-pathogenic rod-shaped bacterium that is critical for manufacturing dairy products like buttermilk, yogurt and cheese. This bacterium, *Lactococcus lactis* is also used to prepare pickled vegetables, beer, wine, some breads and sausages and other fermented foods. Researchers anticipate that understanding the physiology and genetic make-up of this bacterium will prove invaluable for food manufacturers as well as the pharmaceutical industry, which is exploring the capacity of *L. lactis* to serve as a vehicle for delivering drugs.

Antibiotic Resistance

Scientists have been examining the genome of *Enterococcus faecalis*—a leading cause of bacterial infection among hospital patients. They have discovered a virulence region made up of a number of antibiotic-resistant genes that may contribute to the bacterium's transformation from harmless gut bacteria to a menacing invader. The discovery of the region, known as a pathogenicity island, could provide useful markers for detecting pathogenic strains and help to establish controls to prevent the spread of infection inwards.

Forensic Analysis of Microbes

Scientists used their genomic tools to help distinguish between the strain of *Bacillus anthracis* that was used in the summer of 2001 terrorist attack in Florida with that of closely related anthrax strains.

The Reality of Bioweapon Creation

Scientists have recently built the virus causing poliomyelitis using entirely artificial means. They did this using genomic data available on the Internet and materials from a mail-order chemical supply. The research was financed by the U.S. Department of Defence as part of a biowarfare response program to prove to the world the reality of bioweapons. The researchers also hope their work will discourage officials from ever relaxing programs of immunization. This project has been met with mixed feelings.

Evolutionary Studies

The sequencing of genomes from all three domains of life, eukaryotes, bacteria and archaea means that evolutionary studies can be performed in a quest to determine the tree of life and the last universal common ancestor.

Crop Improvement

Comparative genetics of the plant genomes has shown that the organization of their genes has remained more conserved over evolutionary time than was previously believed. These findings suggest that information obtained from the model crop systems can be used to suggest improvements to other food crops. At present the complete genomes of *Arabidopsis thaliana* (water cress) and *Oryza sativa* (rice) are available.

Insect Resistance

Genes from *Bacillus thuringiensis* that can control a number of serious pests have been successfully transferred to cotton, maize and potatoes. This new ability of the plants to resist insect attack means that the amount of insecticides being used can be reduced and hence the nutritional quality of the crops is increased.

Improve Nutritional Quality

Scientists have recently succeeded in transferring genes into rice to increase levels of vitamin A, iron and other micronutrients. This work could have a profound impact in reducing occurrences of blindness and anaemia caused by deficiencies in vitamin A and iron respectively. Scientists have inserted a gene from yeast into the tomato, and the result is a plant whose fruit stays longer on the vine and has an extended shelf life.

Development of Drought Resistance Varieties

Progress has been made in developing cereal varieties that have a greater tolerance for soil alkalinity, free aluminum and iron toxicities. These varieties will allow agriculture to succeed in poorer soil areas, thus adding more land to the global production base. Research is also in progress to produce crop varieties capable of tolerating reduced water conditions.

Veterinary Science

Sequencing projects of many farm animals including cows, pigs and sheep are now well under way in the hope that a better understanding of the biology of these organisms will have huge impacts for improving the production and health of livestock and ultimately have benefits for human nutrition.

Comparative Studies

Analysing and comparing the genetic material of different species is an important method for studying the functions of genes, the mechanisms of inherited diseases and species evolution. Bioinformatics tools can be used to make comparisons between the numbers, locations and biochemical functions of genes in different organisms. Organisms that are suitable for use in experimental research are termed model organisms. They have a number of properties that make them ideal for research purposes including short life spans, rapid reproduction, being easy to handle, inexpensive and they can be manipulated at the genetic level. An example of a human model organism is the mouse. Mouse and human are very closely related (>98%) and for the most part we see a one-to-one correspondence between genes in the two species. Manipulation of the mouse at the molecular level and genome comparisons between the two species can and is revealing detailed information on the functions of human genes, the evolutionary relationship between the two species and the molecular mechanisms of many human diseases.

REVIEW QUESTIONS

1. What are the different definitions of bioinformatics?
2. How can you distinguish between a bioinformaticist and a bioinformatician?
3. Explain the various objectives of bioinformatics.
4. Enlist the significant historical developments in the field of bioinformatics during the last five years.
5. "Bioinformatics has tremendous applications in various fields"—what are your views in support of this statement?
6. Explain the role of bioinformatics in the microbial genome project.
7. What is the contribution of bioinformatics in the field of biotechnology?
8. Explain the role of NCBI in origin of bioinformatics.
9. What is the significant achievement of European Molecular Biology Laboratory (EMBL)?
10. How can you differentiate computational biology from bioinformatics?
11. Write short notes on:
    i. Genomics
    ii. Proteomics
    iii. Swiss-Prot
    iv. Pharmacogenomics
    v. Molecular medicine
    vi. Landmarks in the Human Genome Project
    vii. Computational chemistry

The evolution of computers started from 16th century and resulted in the form that we see today. The present day computer, however, has also undergone a rapid change during the last fifty years. This period, during which the evolution of computer took place, can be divided into five distinct phases known as generations of computers. Each phase is distinguished from other on the basis of the type of switching circuits used.

FIRST-GENERATION COMPUTERS

These computers were large in size and writing programs on them was difficult. The first-generation computers used thermion valves and some of the computers of this generation were:

ENIAC It was the first electronic computer built in 1946 at the University of Pennsylvania and was named Electronic Numerical Integrator And Calculator (ENIAC). It was having the size of 30 feet × 50 feet and weight 30 tons, contained 18,000 vacuum tubes, 70,000 registers, 10,000 capacitors and required 150,000 watts of electricity.

UNIVAC Until 1951, electronic computers were the exclusive possessions of scientists, engineers, and the military. ENIAC's creators marketed a commercial version of their latest machine, known as UNIVAC (Universal Accounting Computer). The first of this kind of the computer was delivered to the Census Bureau of U.S. in June 1951. Unlike ENIAC, the UNIVAC processed each digit serially. But its much higher design speed permitted it to add two ten-digit numbers at a rate of almost 100,000 additions per second. This computer used IBM punched cards for input.

Following are the major limitations of first-generation computers:

- The operation speed was quite slow
- Power consumption was very high
- It required large space for installation
- The programming capability was quite low
- Arithmetic operations were done on bit-by-bit fixed point basis
- Program and data stored on separate memories
- Punched cards and papers were used as input and output devices respectively

SECOND-GENERATION COMPUTERS

First-generation computers were unreliable, largely because the vacuum tubes kept burning out. In 1947, John Bardeen *et al.* of Bell laboratories invented the transistors and this invention changed the way computers were built, leading to the second generation of modern computer technology. Unlike vacuum tubes, transistors are small, require very little power, and run cool. They have no filament and require no heating. The cost of manufacturing was also very low. Transistors were individually mounted in separate packages and interconnected on printed circuit boards by wires. Because second-generation computers were created with transistors instead of vacuum tubes, these computers were faster, smaller, and more reliable than first-generation computers.

In computers of this generation, memory was composed of small magnetic cores strung on wire within the computer and the memory capacity was about 100 kB. For secondary storage, magnetic disks were developed in addition to magnetic tapes. Second-generation were easier to program than first-generation computers. The reason was the development of high-level languages, which are much easier for people to understand and work than assembly languages. More sophisticated high-level languages such as COBOL (COmmon Business-Oriented Language) and FORTRAN (Formula Translator) came into common use during this time. Some of the second-generation computers are:

1. **IBM 1620**   Its size was smaller as compared to first-generation computers and mostly used for scientific purposes.
2. **IBM 1401**   Its size was small to medium and used for business applications.
3. **CDC 3600**   Its size was large and used for scientific purposes.

THIRD-GENERATION COMPUTERS

Though transistors were clearly an improvement over the vacuum tube, they still generated a great amount of heat, which damaged the sensitive parts inside the computer. Also transistors were individually mounted in separate package and interconnected on circuit boards by wires; this was complex, time consuming and error prone. In 1958, Jack St. Clair and Robert Noyce invented the first integrated circuit. Integrated Circuits (ICs) incorporate many transistors and electronic circuits on a single wafer or chip of silicon. ICs are sometimes called chips because of the way they are made.

In 1962, a new company built a plant in the "Silicon Valley" California. Digital Equipment Corporation (DEC) rocked the computer industry with the announcement of a revolutionary type of computer based on integrated circuits. Third-generation computers were IBM 360, ICL 1900, IBM 370, and VAX 750.

During this period, the concept of programming language was developed. Higher level language such as BASIC (Beginners All Purpose Symbolic Instruction Code) was developed with an additional facility of multiprogramming so that a computer runs a number of jobs simultaneously. Another feature of third-generation computers included the use of an operating system. An operating system serves two functions: First, it is a program that provides a buffer between the user and the machine. It helps the user to ask the computer to perform a high level task, and then the operating system translates the task into machine language instructions. Second, the operating system keeps the various parts of the computer running together smoothly. Other characteristic of third-generation computers included Video Display Unit (VDU) using cathode-ray tube was used as output device and keyboard was used as input device.

FOURTH-GENERATION COMPUTERS

After the invention of integrated circuits, the size of the computers decreased considerably. An engineer, associated with improvement of efficiency, speed and memory of the computer, learned to manufacture chips more easily, and connected the idea of grouping an assortment of functions on a single chip, creating a microelectronic "system" capable of performing various tasks required for a single job. This technology is known as Large-Scale Integration (LSI) that could fit hundreds of components onto one chip. In 1971, Intel created the first microprocessor containing a large-scale integrated circuit. The transistors on this single chip were capable of performing all of the functions of a computer's central processing unit. The reduced size, reduced cost, and increased speed of microprocessor lead to the creation of the first personal computers. The MITS (Micro Instrumentation Telemetry System) Altair, marketed in 1975, was the first commercially available microcomputer. Then in 1977, Apple Computer, Inc. became one of the leading forces in microcomputer market. In 1981, IBM personal computer with a microprocessor chip with the storage capacity of several GB made by Intel Corporation and a Microsoft operating system was released.

In these machines an Object-Oriented Programming (OOP) languages are used which encourages programmers to reuse code by maintaining libraries of code segments. Another fourth-generation development is the spread of high-speed computer networking that helps computers to communicate and share data. Within organization, Local Area Networks (LANs) connect several dozens or even hundred computers within a limited geographic area of either one building or several buildings near each other. Through Wide Area Networks (WANs) computers are connected globally.

It is not necessary that a computer operator must be a computer professional because computer manufacturers made the computers user-friendly by developing Graphical User Interface (GUI). A graphical user interface provides icons (pictures) and menus (lists of command choices) that an user can select with a mouse, keyboard or other sophisticated pointing devices. Microsoft has developed a GUI known as Windows that is popular IBM-compatible microprocessor.

FIFTH-GENERATION COMPUTERS

The improvement from one computer generation to the next takes years. It is also very difficult to be precise about the end of one generation and the beginning of the next generation. Also, the concept of computer "generations" has not been used extensively in recent years. This is because the technological innovations have come more gradually and less distinctly than changes in the past. Nevertheless, fifth-generation computers would be distinguished by their emphasis on artificial intelligence, focusing on such matter as machine reasoning and logic programming languages. The noticeable characteristic of fifth-generation computers is the ability to apply previously gained knowledge, draw conclusions and then execute an assigned work. The computer will, in short, simulate the human ability to reason.

Applications of fifth-generation computers are:

1. Intelligent robot that could "see" their environment (visual-input, e.g. a video camera) and could be programmed to carry out certain task without step-by-step instructions. The robot should be able to decide for itself how the task should be accomplished, based on the observations it made of its environment.
2. Intelligent systems that could control the route of missile and defence-system that could fend off attacks.
3. Word processors that could be controlled by means of speech recognition.
4. Programs that could translate documents from one language to another.

CLASSIFICATION OF COMPUTERS

There are many different ways in which computers can be classified. The classifications are:

- Based on the working principles or type of data they are capable of manipulating (analog computers, digital computers and hybrid computers).
- Based on the applications (special-purpose computers and general-purpose computers).
- Based on computing capabilities (microcomputer, minicomputers, mainframe computers and supercomputers).
- Based on number of users (single user and multi-user).

Based on the Working Principles

A computer system based on its working principle can be classified into the three types:

1. *Analog computers*   An analog computer is an electronic system that recognizes data as continuous measurements of physical quantities such as pressure, voltage or temperature along a continuous scale. The output of an analog computer will be usually in the form of dial gauge reading or graphs. The term "analog" means continuous and the value of an analog quality is often open to interpretation. For example, a processor attached to petrol pump, which converts the fuel flow measurements and displays the quantity and price.

Analog computers are highly suitable when the data are continuous. They are mainly used to solve differential, integral, simultaneous, polynomial, and partial differential equations. In addition, they are used to generate forcing functions, synthesize transfer functions and simulate feedback control systems. Analog computer is a powerful tool for designers. It is used for designing missiles and also in the development of new aircraft models. Though an analog computer can be used to solve many scientific and engineering problems, it has many disadvantages over digital computers. Some of the limitations of analog computers are as follows:

- Alpha numerical information cannot be processed in this system.
- Results are not accurate and are often open to interpretation by human beings.
- It cannot store a large amount of data or information since its memory capacity is limited.
- It is somewhat rigid because it is not so easy to understand the complexity of programming the analog system.

2. *Digital computers*   The word "digital" stands for discrete (step-by-step) and hence, digital computers can take only discrete values. A digital computer is a combination of electronic device, designed to manipulate physical quantities or information that are represented in a digital format. In other words, a digital computer represents data in terms of discrete numbers and processes data using the standard arithmetic operations. Unlike analog computers that deal with the measurement of physical quantities, digital computers directly count digits (or numbers), which represent letters, numerals or other special characters. In other words, digital computers are high speed, programmable electronic devices that perform mathematical calculations, compare values and store the results.

Advantages of digital computers

1. Digital systems can have many digits of precision as required, just by adding more switching circuits.
2. Digital computers are easy to design because the circuits used are switching circuits, where exact values of voltage or current are not important, but only the range (high or low) in which they fall is important.
3. Information storage of digital computers is fairly large.
4. The operations in digital system can be programmed and automated.
5. Digital circuits are less affected by spurious fluctuations in voltage (noise).

6. More digital circuits can be fabricated on an integrated circuit chips.

7. Digital computers have high reliability, fault tolerance and are portable.

8. They are well suited for solving scientific and engineering problems involving enormous amount of data with equal felicity.

3. *Hybrid computers* A computer that performs operations based on both analog and digital principles is known as hybrid computer. Many businesses, scientific and medical applications rely on the combination of analog and digital services. An example of hybrid computer is the ultrasonic digital scanner wherein the analog signals generated by the scanner, i.e., the continuous voltage are digitized (periodically measured and converted to numbers) and supplied to a small digital computer that monitors a patient's condition.

Based on the Applications

A majority of computers used in the world today are digital computers and based on the applications they can be classified as special-purpose computers or general-purpose computers.

General-purpose computers A computer that performs variety of processing tasks with almost same efficiency is called a general-purpose computer. All it requires is to load appropriate program for performing each particular task. For example, one has to play chess with computer, then chess-playing program is installed into computer's memory and computer is asked to execute the program. By loading another program into computer's memory, different task can be performed by the computer. Thus, a general-purpose computer

- performs a wide variety of processing task.
- performs any kind of job with equal efficiency, simply by changing application software.
- are in common use.

Special-purpose computers A computer that is built to perform a specific task; it can work according to the instructions provided for a specific function. This means that the programs are fixed for the machine. For example, a modern washing machine, which is a fully automatic machine containing a computer device that is programmed to accept the washing options which user chooses, and directs the machine to perform the assigned job accordingly. Similar examples of such automatic machines are modern microwave oven, modern vehicles, sophisticated embroidery machines, etc. that perform the specific task as the program being installed in them. Thus a special-purpose computer

- ▣ is fundamentally same as a general purpose.
- ▣ is programmed for a specific purpose.
- ▣ cannot perform other kind of task with equal efficiency.
- ▣ is an advent of cheap microprocessors which has lead to massive growth in usage/market penetration.
- ▣ is used in a variety of consumer products, microwaves, mobiles, washing machines, etc.

Based on the Functional Capabilities

General-purpose computers can be further classified into different categories depending upon the size, efficiency, memory and number of users. Broadly they are classified as microcomputers, minicomputers, mainframes and supercomputers.

Microcomputers   These are at the lowest end of the computer range in terms of speed and storage capacity. They are the smallest and cheapest of digital computers. They are called micro because of their miniature size and microprocessor as its CPU. They are used in a variety of applications ranging from business to engineering design. They are also used for entertainment, home appliances, commercial applications, office automations, and other personal uses. Microcomputers are designed to be used by one person at a time. Since microcomputers can be easily linked to large computers, they form a very important segment of the integrated information system. This category is further subdivided into (i) Personal computers and (ii) Home computers.

i.   Personal computers (PCs) are used by professionals, small business units and for office automation systems. Three categories of PCs are PC, PC-XT, and PC-AT.

ii. Home computers are used for entertainment, education, training, and for home management.

Minicomputers   These are more powerful, faster, and costlier than microcomputers. These are smaller and less powerful than mainframes, typically about the size and the shape of a wardrobe, mounted on a single tall rack. These are designed to support more than one user at a time. It possesses large storage capacity and operates at a higher speed. The minicomputer is used in multi-user system in which various users can work at the same time. This type of computer is generally used for processing large volume of data in an organization. They are also used as server in Local Area Networks (LAN). Minicomputers were characterized by short word lengths of 8 to 32 bits, limited hardware and software facilities. These have a wide variety of applications such as industrial control, where a small, dedicated computer that is permanently assigned to one application, is needed. In recent years, improvements in device technology have resulted in minicomputers that are comparable in performance to large second-generation computers and greatly exceed the performance of the first-generation computers.

DEC's PDP-1 was the first minicomputer and their PDP-11 was the most successful, closely followed by the VAX (Virtual Address Extensions). Another early minicomputer was the LINC developed by MIT in 1963. Some of the other minicomputers are IBM AS/400, the HP 9000 series, and Hallmark II of PCL.

Mainframes These are larger than minicomputers and microcomputers. Their significant features include operations at very high speed, very large storage capacity, and they can handle the workload of many users. They are generally used in centralized databases. Most of the world's business and financial information is stored in mainframes. They are also used as controlling nodes in Wide Area Networks (WANs). Other users include research organizations, large industries and governmental organizations. The term "mainframe" was originally used to refer to the metal framework (cabinet) that housed the computer's circuit, but today it refers to computer itself. Example of mainframe are DEC, ICL and IBM 3000 series.

Supercomputer It is the term used for the most powerful computer that operates at the fastest speed and the most expensive machine. They have higher processing speed compared to other computers. The speed of traditional computers is measured in terms of Million of Instructions Per Second (MIPS), whereas supercomputers are generally rated in billions of floating point operations per second. They have also multiprocessing technique. One of the ways in which supercomputers are built is by interconnecting hundreds of microprocessors in parallel. Supercomputers are mainly being used for weather forecasting, biomedical research, remote sensing, aircraft design, astro-physics and other branches of science and technology. PARAM 8000 is the example of first supercomputer, which is capable of doing 1 million mathematical calculations per second, was delivered by Centre for Development of Advance Computing (C-DAC) headed by Dr. Vijay P. Bhatkar. The Centre is situated in campus of University of Pune. Other examples of supercomputers are CRAY YMP, CRAY 2, NEC SX-3, and CRAY XMP from India.

Based on the Number of Users

A computer system may be classified on the basis of the number of users sharing the resources of the system. It can be classified as single user and multi-user.

Single user   The single user computer better known as a personal computer is one in which only one user can do the job at a time. Personal computers (PCs) are further classified as

Personal Computer (PC)

Personal Computer based on extended technology (PC-XT) and

Personal Computer based on advanced technology (PC-AT)

A single user computer system normally consists of the following configuration:

- Microprocessor chip
- Monitor
- Floppy drive and fixed disk drive
- Keyboard and mouse
- Printer and plotter

Multi-user   A multi-user is a special type of computer system that supports advanced computer configuration, mostly based on multiprocessing or parallel processing techniques. This type of computer is always controlled by an enhanced and complex Operating System (OS) on time-sharing basis, permitting more than two users to work at a time. Some examples for multi-user system are:

- The UNIX OS meant for multi-user, multitasking and time-sharing computer system.
- The VM (Virtual Machine) is another example for the multi-user computer system that makes a single real machine appears to be several real machines.

OPERATING SYSTEM

An operating system is a program that acts as an interface between user of a computer and computer hardware. The purpose of an operating system is to provide an environment in which a user may execute programs. The primary function of the operating system is the management and control of Computer Based Information System (CBIS) resources. This involves the allocation of hardware, programs, and resources to CBIS users in the wisest possible manner. Users have to be provided with pathways to access the hardware, software and data authorized for them to use. Thus, an operating system can be viewed as a resource manager or allocator.

Overall, the primary goal of an operating system is to make the computer system convenient to use with a secondary goal to use the computer hardware in an efficient manner. An operating system is an important part of almost every computer system. A computer system can be roughly divided into four components as listed in the Figure 2.1.

- The hardware (CPU, memory, I/O devices)
- The operating system
- The application programs (complier, interpreters, database management, video games, business programs)
- The users (people, machines or other computers)

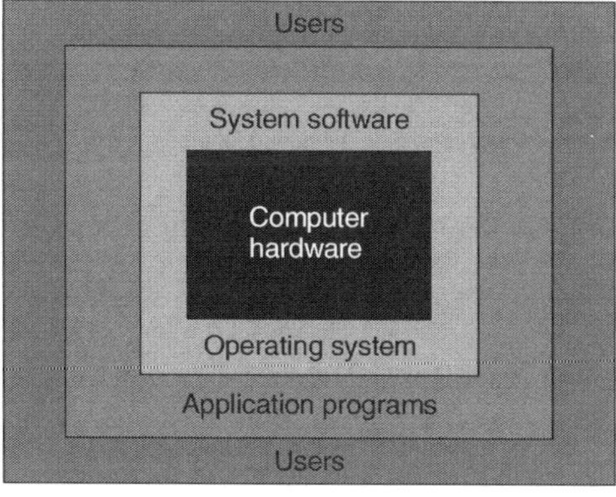

Figure 2.1 Components of a computer system

The basic resources of a computer system are provided by its hardware, software and data. An operating system provides the means for the proper use of these resources in the operation of the computer system and it works as a resource allocator. The resources like CPU, memory space, file storage space, I/O devices, and other peripherals are allocated to the programs and users can interact with these hardware through application of programs and operating system.

Services in Operating System

An operating system provides certain services to programs and to the user of these programs. There are three major categories of services:

1. *Information management* **(*IM*)**　It refers to a set of services used for sorting, retrieving, modifying or removing the information on various devices. These system services manages the organization of information in terms of directories and files, allocating and deallocating the sectors to various files, ensuring right people have access to the information and driving various devices. Some of the system in IM are listed below:

1. Create file
2. Create a directory
3. Open a file (for read, write or both)
4. Close a file
5. Read data from file to buffer
6. Write data from buffer to file
7. Move the file pointer
8. Read and return file's status
9. Create a link
10. Change working directory

2. *Process management* **(*PM*)**　In multi-user operating system, a number of users located at different locations run the same or different programs. A computer has only one CPU and it can execute only one instruction belonging to any one of all these programs. In such a case, the operating system has to keep track of all running programs, called as processes. Operating system has to schedule all processes and will have to dispatch them (i.e., run or execute) one after the other. While doing so, it gives an impression to user that he has the full control of the CPU. Some of the system calls in PM are as follows:

1. Create a child process similar to the parent
2. Wait for a child process to terminate
3. Terminate a process
4. Change the priority of a process
5. Block a process
6. Ready a process
7. Dispatch a process
8. Suspend a process

9. Resume a process
10. Delay a process

3. *Memory management* (*MM*)   The services provided under memory management are directed towards keeping track of memory and allocating/ deallocating it to various processes. The operating system keeps a list of free memory allocations. Before a program is loaded in the memory form—the disk, the MM consults this free list, allocates the memory to the process and updates the list of free memory. The memory allocated to process depends on the program size. The system calls in this category are as follows:

1. Allocate a chunk of memory to process.
2. Free chunk of memory form a process.

Disc Operating System

Microsoft developed the operating system referred to as MS-DOS or PC-DOS for the first IBM personal computer where MS-DOS is the acronym for Microsoft Disk Operating System, is an operating system with a command-line interface used on personal computers. The initial version of DOS 1.0 was released in August 1981. Later on Microsoft developed the higher version of DOS. Earlier, DOS, requiring the user to type in text-based commands using a keyboard. For example, to perform a task in DOS, one has to open a command prompt, click start, point to programs, point to accessories, and then click command prompt. To end your MS-DOS session, type exit in the command prompt Window at the blinking cursor, in the way like C:\exit. Users need to remember various commands and their syntax to work with the computer.

In early days, certain people who were computer literate used computers. The operating systems including MS-DOS used to have a set of cryptic commands. To perform an operation, the user needs to type command and were expected to remember all commands with their meanings. Later on the use of personal computers widely spreads over. The human–computer interface was a problem area in using computers. Attempts were made to replace commands by graphics. The interface that replaced cryptic commands by their graphical representation is called **Graphical User Interface** (GUI). Xerox Corporation first developed it for their xerox star computer. Apple computers used this interface in their Macintosh personal computers. It was popular among users. By 1990, Microsoft has a version of GUI known as Windows 3.0. This was more user friendly than DOS. Subsequently in the year 1995, Windows 95 and its upgraded version in the form of Windows 98 was released by Microsoft. Later on, Windows 98 was upgraded by Windows 2000, 2003, and so on. Its latest version is Windows Vista. The features of Windows 98 are as follows:

1. **Easier to use**   Windows 98 is a single user multitasking operating system. Navigation around the computer is easier in Windows 98. One can open a file by single click. Even use of multiple monitors with a single computer can be possible with this operating system. Thus one can increase the size of workspace. Installation of new hardware is easy because Windows 98 supports the Universal Serial Bus (USB) standards, allowing plugging in new hardware and using it immediately without restarting the computer. Windows 98 allows the access of digital cameras and other digital imaging devices.

2. **Faster**   Windows and programs open faster than ever before. By using maintenance wizard, speed and efficiency of the computer can be improved. One can use FAT 32 file system to store files more efficiently and save hard disk space.

3. **True web integration**   Using world wide web is easier and faster. The connection to web is simple. Webpages can be viewed in any window. You can make your favourite webpage. In Microsoft Outlook Express you can send e-mail and post messages to internet newsgroups. The conferences on internet can be arranged.

4. **More entertaining**   Windows 98 supports DVD and digital audio; one can play high quality digital movies and audio on the computer. The television broadcast can also be observed on computer.

The Windows 98 consist of program manager, file manager and control panel.

Program Manager

i. It provides user interface to start and stop application.
ii. It is used to organize various applications into different graphs.
iii. It is also used to end the MS-Window sections.

File Manager

i. It is a tool that helps to organize the user files and directories.
ii. It can also be used to traverse through the file system and to change drive
iii. It is also used to search, copy, move, create or delete files and directories.
iv. Application can be started directly from the file manager by clicking on an application icon.

Control Panel

i.   It can be used to choose or change schemes in the applications.

ii. It can also be used to select and display the background of the screen.

iii. One can add or remove fonts and view fonts on the screen.

iv. It can also be used to configure printers and other ports on the PC.

Windows NT   It is a fully-fledged operating system like windows. It is a portable operating system like UNIX. This ensures consistency of applications across a variety of platforms. The Intel X-86 family, silicon graphics MIPS RS 4000 RISC, alpha processor supports Windows NT. It is a multi-user and multitasking operating system. It has all the features of a modern operating system. It has virtual memory management, threads, symmetric multiprocessing and built in networking. It has modular and portable design.

Other features of Windows NT are as follows:

1. Windows NT is a multi-user, multitasking and multithreading operating system.

2. One can get a fast response even when multiple applications are running. Tasks are scheduled by the operating system by giving a time slice to each task and when its time slice is over, the next one is started. The previously running task is resumed when its turn comes again.

3. In multithreading, a task can be subdivided into smaller units called threads. Each thread has an independent flow of control within a task. Different threads can be scheduled and executed independently.

4. Symmetric multiprocessing will allow Windows NT to schedule various tasks on any CPU in a multiprocessor system.

5. Windows NT is built as 32-bit operating system. It supports virtual memory management and also supports file transfer, e-mail services and resource sharing on a network. Windows NT can interact with all networking systems like Novell's Netware. It uses New Technology File System (NTFS) in which file names can be 256 characters long. NTFS also implements fault tolerance, security and has support for very long files.

6. Windows NT is divided into different subsystems and layers. Its block diagram is shown in Figure 2.2. The lowest layer is Hardware Abstraction Layer (HAL); this layer provides all hardware specific functions and will be different on different hardware platforms. It handles interrupt, Direct Memory Access (DMA), memory management and multiprocessor synchronization.

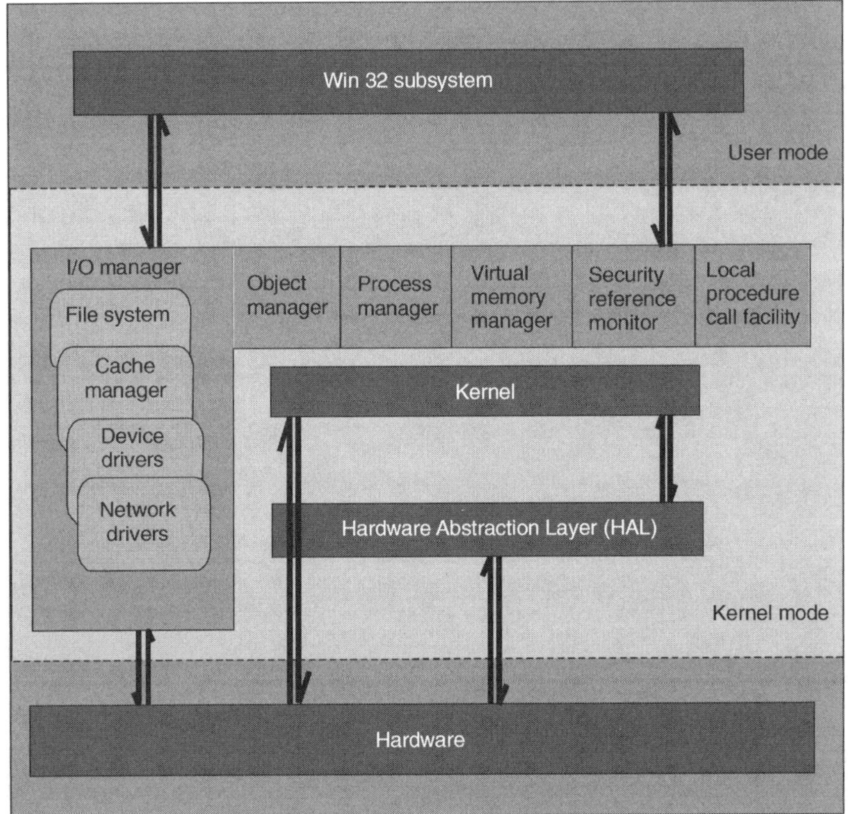

Figure 2.2   The block diagram of Windows NT

7. On the top of HAL layer is the Windows NT kernel that provides the basic operating system services like scheduling of threads and tasks. It supplies basic sets of objects like files, memory and processes for use of higher layers. On the top of kernel, there are six modules that together forms executive and are as listed below:

    i.   **Object manager** creates, manages and deletes objects, which are processes, threads, files, etc.

    ii. **Process manager** that creates, terminates, suspends and resumes processes, tasks, and threads.

    iii. **Virtual memory manager** manages the memory for processes. It allocates and frees memory. It also handles paging and memory protection from other processes. Each process access up to 2 GB of virtual memory.

iv. **Security reference monitor** enforces security policy and keeps track of files access rights, based on ownership and permissions for the user. It can be used to create alarm for the security violations.

v. **Local procedure call facility** is used to pass message between client systems and subsystems.

vi. **I/O subsystem** handles device drivers and passes data to and receives data from the devices of all subsystems. The file system is also handled by the I/O subsystem.

On the top of these subsystems is Win 32 layers that handles all the user interactions and decides the correct subsystem for each service requested by user application.

LINUX

It is a rapidly evolving operating system. It is similar to UNIX system. LINUX's development began in 1991. A student, Linus Torvalds, wrote kernel for 32-bit processor. In early days, Linux source code was available for free on internet hence so many users around the world have contributed in the development of Linux.

The Linux kernel is distributed under General Public License (GPL). Linux is not public domain implies that the authors have waived copyright over the software. The copyright over Linux is still held by the Code's various authors. Linux is free software. It is free in sense that people can copy it, modify it, and use it in any manner they want.

Features of Linux

1. Linux is multi-user, multitasking operating system with a full set of UNIX compatible tools.
2. Linux runs on a wide variety of platforms. It was developed exclusively on PC architecture.
3. It provides as much as functionality from limited resources. It can run on machine having 4 MB of RAM.
4. Linux present standard interfaces to both the programmer and user.
5. It supports a wide base of applications.
6. Linux is free software.
7. The file system in Linux obeys UNIX semantics.

8. The memory management system uses page sharing copy-on-write to minimize duplication of data shared by different processes. Page is loaded on demand when they are referenced first.

Components of Linux system   The Linux is composed of three main bodies of code as listed below and as shown in Figure 2.3.

i. **Kernel** maintains all significant abstractions of the operating system such as processes and virtual memory.

ii. **System libraries** define a standard set of functions through which applications can interact with kernel and implement much of the operating system functionality.

iii. **System utilities** are programs that perform individual, specialized management tasks. Some of utilities may be invoked just once to initialize and configure some aspects of system.

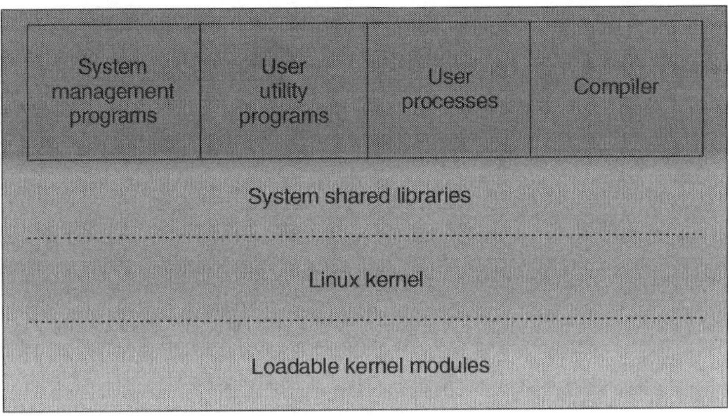

Figure 2.3   Components of Linux system

UNIX

It is a powerful, multi-user, multitasking operating system initially developed at AT & T Bell Laboratories in 1969 for use on minicomputers. UNIX is considered more portable, that is, less computer-specific, than other operating systems because it is written in C language. Newer versions of UNIX have been developed at the University of California at Berkeley and by AT & T. It is an interactive operating system for microcomputers, minicomputers and mainframes that is machine independent and supports multi-user processing, multitasking and networking.

This system was developed to help scientific researches to share data and programs while keeping other information as private. Many people can use UNIX simultaneously to perform same kind of task, or a single user can run many tasks on UNIX concurrently. This was developed to connect various machines together and is highly supportive of communication and networking. Originally it was designed for minicomputers. It has modified versions that suit other types of computers and readily customized. It can store and manage a large number of files. Earlier, UNIX is primarily used for workstations, minicomputers and inexpensive multi-user environments in small businesses, but its use in large business is growing because of its machine independence. Application programs that run under UNIX can be ported from one computer to run on a different machine with little modification.

INTERNET

The Internet (a network of networks) originated back in the 1960s when U.S. Department of Defence established the network to provide researchers and government officials with access to radio telescopes, weather analysis programs, supercomputers and specialized databases. From these origins for the exclusive use of government and educational institutes, the internet expanded. Over half of all U.S. Internet addresses belonging to people who got them through a private employer or a commercial access provider. More than 150,000 new users join the Internet each day and every 30 minutes, one big Internet (new) is added. The number of networks interconnected in the Internet doubles every year.

The most puzzling aspect of the Internet is that no one owns it and it has no formal management organization. To join the Internet an, existing network need only to a small registration fee and agree to certain standards based on TCP (Transmission Control Protocol)/IP (Internet Protocol). TCP/IP is necessary for communication over Internet. The value of the Internet lies precisely in its ability and inexpensively to connect so many diverse people from various places all over the world. Any one can log if he or she is having an Internet address and reach virtually every other computer available on network, regardless of location, computer type or an operating system.

Role of Modem in Internet

The Internet connection on the computer can be made available through telephone cable, that is, connected to computer via modem. Computers store digital data while telephone lines can transmit only analog data. Modems can transmit digital computer signals over telephone lines by converting them to analog form. Converting signal form to another is called modulation and recovering original signal is called demodulation. Modem may be external or internal that acts as modulator/demodulator through which data transfer activities are performed over the Internet. Recently, one connects to Internet by using Bluetooth technology, which involves wireless modem connected to the computer's serial port. The capacity of medium to transmit the data over the Internet is measured in bandwidth that can be of base band or broadband type. In earlier days, the Bharat Sanchar Nigam Limited (BSNL) provided the Internet services based voiceband in which the bandwidth of a standard telephone line is used and its typical speed are 9,600 to 56 kbps (can transmit data with maximum speed of 56 kilobits-per-second) but nowadays, BSNL is providing broadband Internet services through dial-up connections to colleges, universities, government offices, shopping malls, companies, so that one can access data at high speed. It is in the range of 264 Mbps to 30 Gbps.

Information Retrieval on Internet

Following are the three methods of accessing computers and locating the desired file on the Internet.

1. *Gophers* These are character-oriented approach tools for locating data on Internet that help the users to locate essentially all textual information stored on Internet servers through a series of easy-to-use hierarchical menus. It was originally developed at the University of Minnesota in 1991. It has thousands of Gophers server sites on the Internet throughout the world, with its own system of menus listing subject matter topics, local files, and other relevant sites. Each Gopher site may contain thousand listing within its various levels of menus. When the user initiates the Gopher software in search of a specific topic and selects a related item from a menu, the server will automatically transfer the appropriate file on the server or to the selected server wherever in the world it is located. In other words, Gopher is a software application that provides menu-based search and retrieval functions for a particular computer site.

2. *Archie* It is a tool for locating data on the Internet that performs keyword searches of an actual databank of document, software, and data files available for downloading from server around the world. Subjects like teleconferencing, telemedicine, teleshopping, etc., resulting in a list of sites contain files on the respective topics. Through a file transfer program, desired files can be downloaded.

3. *WAIS* (*Wide Area Information Servers*)   It is another way to locate the file around the world. A specific file can be located but it requires the name of database to be searched. With specific database names and keywords, WAIS searches for the keywords in all the files in those databases.

In addition to above mentioned Internet services, Telnet, FTP and the web are other commonly used services of Internet. Telnet is the Internet service is used that helps to run programs on remote computers. File transfer protocol (FTP) is an Internet service for transferring files. FTP allows to copy files on client's computer (downloading) as well as uploading the files from one computer to another computer on Internet. The service receiving the most attention today is the web, which is discussed separately in this chapter.

Internet is so vast that it is like an electronic cloud encircling the earth and providing the way of communication and easy to use offerings of information and products as well as research and entertainment all over the world. It will be appropriate to say that Internet is larger today compared to yesterday and it will be still larger tomorrow than today.

E-MAIL

As stated earlier, communication is one of the applications of Internet. Communicating via e-mail is by far the most common Internet activity. Electronic mail creates the communication using word processing techniques. The digital data is transmitted through a network to the terminal of the recipient. Internet-based e-mail has grown spectacularly from a mere-message carrier to a service that helps the user to exchange text, picture, movie clips and voicemail and most of it are free. E-mail, as it is available on the Internet is possibly the single biggest attraction for most of Internet subscribers-overweighing even the formidable information features and resources on the Internet.

One can communicate with any person in the world who has an Internet address of e-mail account with a system connected to the Internet. The person interested in doing so needs to access the Internet and an e-mail program. Two of the most widely used e-mail programs are Microsoft's Outlook Express and Netscape's Navigator. A typical e-mail message has three basic elements: header, message, and signature. The header appears first and typically includes the following information:

- ▣ **Subject**   A one-line description, used to present the topic of message. Subject lines typically are displayed when a person checks his or her mailbox.
- ▣ **Addresses**   Addresses of the persons sending, receiving, and optionally, anyone else who has to receive copies.
- ▣ **Attachment**   Many e-mail programs allow attaching files such as documents, pictures and worksheets. If a message has an attachment, the file name appears on the attachment line.

The letter or message comes in text. It is typically short and to the point. Finally, the signature line provides additional information about the sender. Typically, this information includes the sender's name, address, and telephone number. An e-mail message can be directed to a group of recipients, large or small so that each number of the group receives the same message virtually at the same time. It may be a sentence to communicate an idea or trigger an action or it may be an entire document of several pages containing graphical illustration.

Overall, electronic mails are designated to avoid telephone tag and to overcome the communication barriers created by time and distance. It is a service that transports messages from a sender to one or any number of receivers. Instant and unlimited distribution of messages, better time management and easy implementation of the program are additional advantages of e-mail facility.

WORLD WIDE WEB (WWW)

An individual linked to the Internet and to its most frequently used feature, the World Wide Web (WWW), can send and receive documents of all types including those containing sound, animation, video, and colourful graphics. The Internet was launched in 1969 when the United States funded a project that developed a national computer network called Advanced Research Project Agency Network (ARPANET). The web, also known as WWW and World Wide Web, was introduced in 1992 at the Center for European Nuclear Research (CERN) in Switzerland to allow the information sharing between internationally dispersed groups in the High Energy Physics community. Prior to the web, the Internet was "all text, no graphics", animation, sound, or video. The web is a multimedia interface to resources available on the Internet. From these research beginnings, the Internet and the web have evolved as tools for all of us to use. The web spread quickly and is now making a profound impact in the field of bioinformatics.

Web Browsers

World Wide Web is an information retrieval tool. The web is attractively easy to use, and lends itself to publishing or providing information, to anyone interested. Web is a series of interconnected documents stored in a digital format. The web browser is a special software that helps to access information displayed on the web using Universal Research Locator (URL) addresses. This software connects the desired person to remote computers, opens and transfer files, displays text and images, and interface to the Internet and Web documents. Netscape Navigator and Microsoft Internet Explorer are two well-known web browsers in addition to Lynx and Mosaic.

Universal Research Locators (URLs)

These are specific addresses or locations for browsers to connect to other resources. All URLs have at least two basic parts. The first part presents the protocol used to connect to the resource. Hypertext transfer protocol (http://) is by far the most common. The second part presents the domain name or the name of the server where the research is located. For example, http://www.ncbi.com is the URL of the website that provides information about the National Center for Biotechnology Information. In Figure 2.4 the two basic parts of URL address are distinguished, whereas the Figure 2.5 shows the computer is connected to specified URL with the help of Microsoft Internet Explorer as a web browser. Many URLs have additional parts specifying directory paths, file names and pointers like in the above URL, there may be additional features as http://www.ncbi.nlm.nih.gov

Figure 2.4 Basic parts of a URL

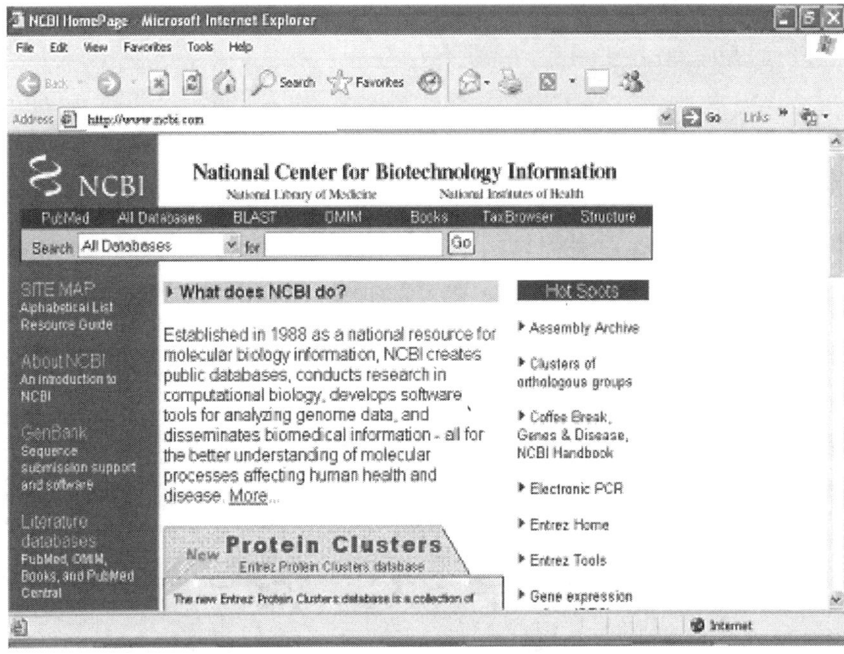

Figure 2.5   The Home page displayed as per URL

Netscape Navigator

It was developed in 1994 by Netscape communications corporation, California. It is the most popular web browser package for browsing information on the Internet. As per one estimate, about 80% of Internet users browse the web with Netscape. The current versions of the software include facilities such as Internet, e-mail, frames, real-time communication, audio and video support, and the latest technology to support creation of visually exciting and fully interactive pages (e.g. with Java applets). Web pages can also have links to special programs, called applets, written in a programming language called Java. These programs can be quickly downloaded and run by most browsers. Java applets are widely used to add interest and activity to a website presenting animation, displaying graphics, providing interactive games, and much more.

Internet Explorer

It was developed in 1995 by Microsoft Corporation Redmond, USA and was based on NCSA Mosaic. It is designed to work with PC-based operating system. It offers the familiar functionality of other hypermedia browsers, including support for frames, Java and ActiveX. Originally developed as Windows 95, NT product, current versions also run on Windows XP and Windows Vista.

Lynx

It was developed in the Academic Computing Services at the University of Kansas, USA, as a part of an effort to build a campus-wide information system. It runs on UNIX or VMS operating systems, providing a text-only interface via low-cost, dumb display devices. It is probably the most widely used text-only browser on the Internet. Where a comparatively slow dial-up connections in use, Lynx is more efficient than graphical browsers that use large amount of computer resources.

Mosaic

It was developed at National Center for Supercomputing Applications (NCSA), University of Illinois, USA. As a multimedia system designed for X-Windows, Apple Mac and Microsoft Windows platforms, it provided a single, user-friendly interface to diverse protocols. Data formats, and information servers are available on the Internet. Prior to Netscape Navigator, it was the Mosaic that undoubtedly increased the popularity of the web.

Search Tools

Once the Internet is connected, the web shows massive collection of interrelated webpages. It is very difficult to locate the precise information needed. Fortunately, a number of search tools have been developed and are available for us on certain websites. These are basically two types: indexes and search engines.

1. *Indexes* These are also called web directories, organized by categories such as an arts and humanities, computers and Internet, business and economy, entertainment, news, science, sports, and so on. Each category is further organized into subcategories. Using the browser, one can select a category and continue to select subcategories until the search has been narrowed and a list of relevant document appears. By selecting a document, an appropriate link can be made and the desired information can be displayed. One of the best-known and most widely used indexes is Yahoo! (Figure 2.6). Other indexes such as Google, Sify, and Rediffmail are also in routine use.

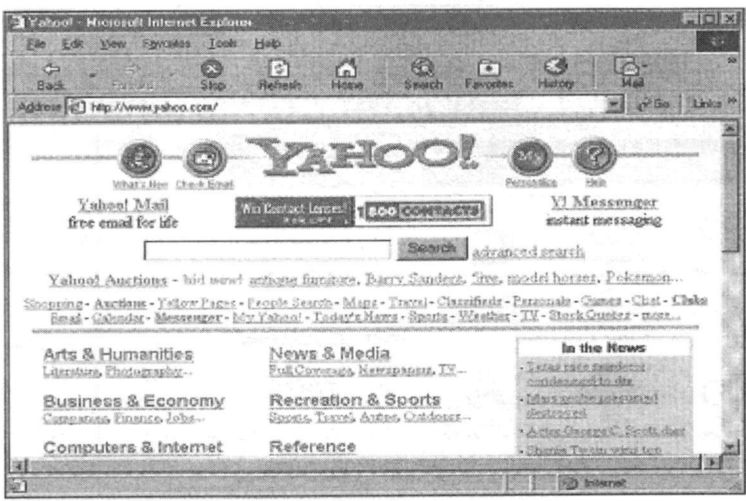

Figure 2.6   Index as Search tool: Yahoo

2. *Search engines* These are also known as Webcrawlers and Web Spiders. Information is not organized by major categories, rather, search engines are organized like a database and search through these engines is done by entering keywords and phrases. Special programs called agents, spiders or bots maintain these databases. They automatically search for new information on the web and update the databases. HotBot (Figure 2.7), WebCrawler, and Alta Vista are widely used search engines.

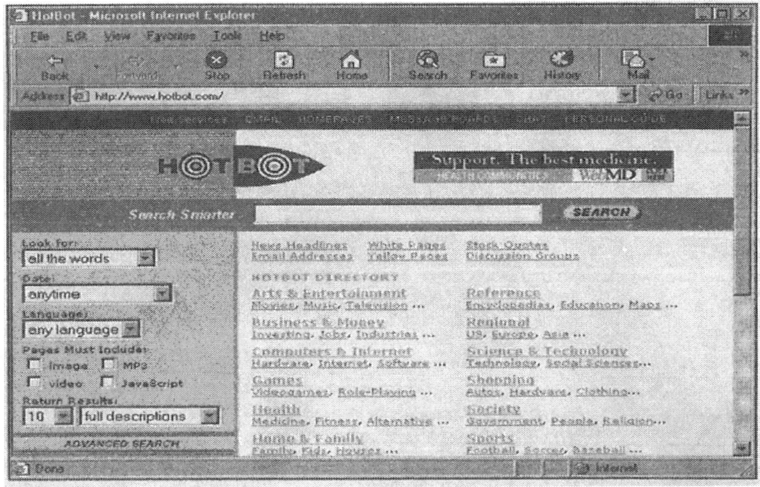

Figure 2.7   Search engine: HotBot

WEB SERVERS

Hypertext Transfer Protocol (HTTP)

Once the browser has connected to a website on the Internet, a document file is sent to the computer. This document that browsers display exploits hypertext and hypermedia techniques to make web browsing and publishing extremely easy. Hypertext documents contain text with embedded links (hence known as **hyperlinks**) to other documents. Typically, the first page of a website is referred as its **Home Page** that presents information about the site along with references and hyperlinks. Hyperlinks are usually distinguished by being highlighted in some way, either using a colour different from main body of the text, or by being boxed, etc. Selecting a highlighted link calls up the linked document, regardless of its locations, whether on the same server, or on a server in a different country.

Hypertext Markup Language (HTML)

It is a standard markup language used to write the hypertext documents. Markup is the process of taking ordinary text and adding extra symbols, such as editor's proof reading symbols. Each of the symbols used for markup in HTML is a command that tells the browser how to display the text. HTML is a computer language that is related to the computer programming languages like BASIC, C and PASCAL. It has its own syntax and rules for communication. HTML indicates which part of the document is the title, which is sub-heading, which is author name and so on. In short, web browser interprets the HTML commands and displays the document as a webpage.

The markup instructions allow the web author to render tags to the document. A tag is a unit of markup. It is a set of symbol defined in HTML to have special meaning. Tags start with a less than sign (<) followed by a keyword, and conclude with the greater than sign (>). These symbols are known as angle brackets. For example, the tags are:

```
<HTML>

<TITLE>

</B>
```

Every tag in HTML has a meaning. For example, in above case, <B> is used to render phrases in bold type. Similarly, < FONT> symbol is used to customize the size and colour of the font, <HR> symbol is used to insert the horizontal rulers, and <IMG> symbol is used to insert the image.

Each of these modes is switched off with the relevant </> symbol as shown in above example.

Every webpage consists of few tags that define the page as whole. For example, a simple webpage can be shown as follows:

```
<HTML>

<HEAD>

<TITLE> A Hello World Example </TITLE>

</HEAD>

<BODY>

Hello World!

</BODY>

</HTML>
```

The above HTML code can be typed in using text editor. The browser (Netscape Navigator or Internet Explorer) will display document Hello world! on the computer screen. The basic structure of the HTML is shown in the Figure 2.8.

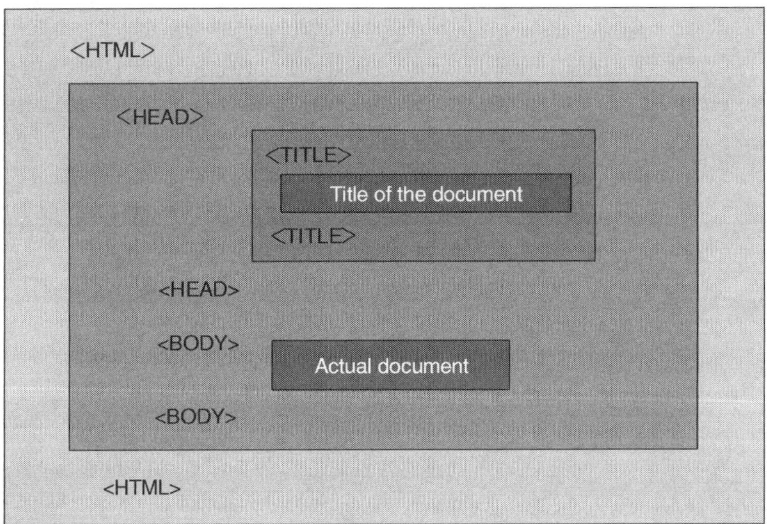

Figure 2.8   Basic structure of the HTML document

In short, HTML is designed primarily for formatting the text. It is basically a typesetting language that specifies the shape to the text, the colour, where to put it, and how to make it. It's not much different from the most other typesetting languages, except that it doesn't have as many typesetting options as most simple *What You See Is What You Get* (WYSIWYG) editors, such as Microsoft Word. Following is one such HTML program written to obtain the "exact output" for the code with font specification as mentioned in brackets.

```
<html>

<title> introduction </title>

<body>

<h2> <b> Terms in Bioinformatics </B> </h2>

<dl>

<dt> Bioinformatics

<dd> <b> Bioinformatics </b> is the science of developing
computer databases and algorithms for the purpose of speeding up
and enhancing biological research.

<dt> Genomics

<dd> <b> Genomics </b>is any attempt to analyse or compare the
entire genetic complement of a species.

<dt> Proteomics

<dd> The study of proteins with respect to their location,
structure and function is known as <b> Proteomics </b>

</d>

</body>

</html>
```

The output of the program will be as follows:

Terms in Bioinformatics

Bioinformatics

    Bioinformatics is the science of developing computer databases and algorithms for purpose of speeding up and enhancing bilogical research

Genomics

    Genomics is any attempt to analyze or compare the entire genetic complement of a species.

Proteomics

    The stydy of proteins with respect to their location ,structure and function is known as **Proteiomics**.

Advantages of HTML   HTML carries several advantages as mentioned below:

1. For creating HTML document, what you need is only text editor. No special software is required. For example, in Windows, one can use notepad for typing HTML document.
2. HTML document can be created on any hardware platform using any text editor. One can work on the website even when the person is away from the usual computer.
3. If something is not working, then finding an error is easy in HTML.
4. HTML will not cost anything for its use. It does not require any expensive licences to buy or no upgrades to purchase.
5. One can work independently and need not to rely on vendor or program.
6. Learning HTML is simple than any other programming language.

Disadvantages of HTML   Inspite of several advantages, HTML has certain limitations, such as:

1. HTML is not a programming language in true sense.
2. Any simple calculation can not be done in HTML.
3. It cannot be used to display even the date.
4. Interactive webpages cannot be built by HTML.
5. The webpages developed in HTML cannot behave like an application and such webpages do not have their own interface.
6. A hyperlink is provided in HTML, but to achieve that a trip to server at each step is needed.

Extensible Markup Language [XML]

It is a general-purpose specification for creating custom markup languages. It is classified as an extensible language because it allows its users to define their own elements. Its primary purpose is to facilitate the sharing of structured data across different information systems, particularly via the Internet, and it is used both to encode documents and to serialize the data. In the latter context, it is comparable with other text-based serialization languages such as JSON and YAML.

It started as a simplified subset of the Standard Generalized Markup Language (SGML), and is designed to be relatively human legible. By adding semantic constraints, application languages can be implemented in XML.

These include XHTML, RSS, MathML, GraphML, Scalable Vector Graphics, MusicXML, and thousands of others. Moreover, XML is sometimes used as the specification language for such application languages. XML is recommended by the World Wide Web Consortium (W3C). It is a fee-free open standard. The W3C recommendation specifies both the lexical grammar and the requirements for parsing.

XML Syntax for Well-formed Documents

As long as only well-formedness is required, XML is a generic framework for storing any amount of text or any data whose structure can be represented as a tree. The only indispensable syntactical requirement is that the document has exactly one **root element** (alternatively called the **document element**). This means that the text must be enclosed between a root start-tag and a corresponding end-tag. The following is a well-formed XML document:

```
<book>This is a book.... </book>
```

The root element can be preceded by an optional **XML declaration**. This element states the version of XML that is in use (normally 1.0); it may also contain information about character encoding and external dependencies.

```
<xml version= "1.0" encoding= "UTF-8">
```

The specification requires that processors of XML support the pan-unicode character encodings UTF-8 and UTF-16 (UTF-32 is not mandatory). The use of more limited encodings, such as those based on ISO/IEC 8859, is acknowledged and is widely used and supported.

Comments can be placed anywhere in the tree, including in the text. XML comments start with <!— and end with —>. Two dashes (—) may not appear anywhere in the text of the comment.

```
<!— This is a comment. —>
```

In any meaningful application, additional markup is used to structure the contents of the XML document. The text enclosed by the root tags may contain an arbitrary number of XML elements. The basic syntax for one **element** is:

```
<name attribute = "value">PrakashX 02586222354... txt me</
name>
```

The two instances of »name« are referred to as the **start-tag** and **end-tag**, respectively. Here, »content« is some text which may again contain

XML elements. So, a generic XML document contains a tree-based data structure. Here is an example of a structured XML document:

```
<recipe name="Tea" prep_time="5 mins" cook_time="5 mins">
    <title>Tea</title>
    <ingredient amount="1" unit="cup">Milk</ingredient>
     <ingredient  amount="1"  unit="teaspoon">Tea  powder</
ingredient>
    <ingredient amount="1" unit="teaspoon">Sugar</ingredient>
    <instructions>
     <step>Mix all ingredients together in tea pot.</step>
     <step>Place the tea pot in microwave.</step>
     <step>Set microwave at high power for 5 minutes.</step>
     </instructions>
    </recipe>
```

Attribute values must always be quoted, using single or double quotes; and each attribute name should appear only once in any element.

XML requires that elements to be properly nested—elements may never overlap. For example, the code below is not well-formed XML, because the **title** and **author** elements overlap:

```
<!— WRONG! NOT WELL-FORMED XML! —>

<title>Book on Biotechnology<author>Prakash Lohar<title>Another
Book  on  Biotechnology<author>Ignacimuthu</title></author></
title></author>

<!— Correct: Well-formed XML. —>

<title>Book on Biotechnology </title> <author> Prakash Lohar
</author> <title>Another Book on Biotechnology </title> <author>
Ignacimuthu </author>
```

Alternatively,

```
<title>Book on Biotechnology </title> <author> Prakash Lohar
<title>Another  Book  on  Biotechnology  <author>  Ignacimuthu </
author></title></author>
```

XML provides special syntax for representing an element with empty content. Instead of writing a start-tag followed immediately by an end-tag, a document may contain an empty-element tag. An empty-element tag resembles a start-tag but contains a slash just before the closing angle bracket. The following three examples are equivalent in XML:

```
<foo></foo>

<foo />

<foo/>
```

An empty-element may contain attributes:

```
<info author="Prakash Lohar" genre="Cell & Molecular Biology"
date="2007-Jan-01" />
```

Advantages of XML

1. It is text-based.
2. It supports Unicode, allowing almost any information in any written human language to be communicated.
3. It can represent common computer science data structures: records, lists and trees.
4. Its self-documenting format describes structure and field names as well as specific values.
5. The strict syntax and parsing requirements make the necessary parsing algorithms extremely simple, efficient, and consistent.
6. XML is heavily used as a format for document storage and processing, both online and offline.
7. It is based on international standards.
8. It can be updated incrementally.
9. It allows validation using schema languages such as XSD and Schematron, which makes effective unit-testing, firewalls, acceptance testing, contractual specification and software construction easier.
10. The hierarchical structure is suitable for most (but not all) types of documents.
11. It is platform-independent, thus relatively immune to changes in technology.
12. Forward and backward compatibility are relatively easy to maintain despite changes in DTD or Schema.
13. Its predecessor, SGML, has been in use since 1986, so there is extensive experience and software available.

Disadvantages of XML

1. XML syntax is redundant or large relative to binary representations of similar data, especially with tabular data.

2. The redundancy may affect application efficiency through higher storage, transmission and processing costs.

3. XML syntax is verbose, especially for human readers, relative to other alternative "text-based" data transmission formats.

4. The hierarchical model for representation is limited in comparison to an object-oriented graph.

5. Expressing overlapping (non-hierarchical) node relationships requires extra effort.

6. XML namespaces are problematic to use and namespace support can be difficult to implement correctly in an XML parser.

7. XML is commonly depicted as "self-documenting" but this depiction ignores critical ambiguities.

8. The distinction between content and attributes in XML seems unnatural to some and makes designing XML data structures harder.

JAVA AND JAVASCRIPT—BRINGING LIFE TO HTML

Java is an object-oriented programming language developed by Sun Microsystems, Mountain View, CA. Whatever we see in the form of updated stock information appearing "live" on browser window and accessing the webpage containing animated icons instead of static ones and much more are possible with Java.

Java is a full-fledge programming language tailored for network computing; it included hundreds of its own objects, including objects for creating user interfaces that appear in Java applets (in web browsers) or stand-alone Java applications. Instead of execution on the HTTP server, a Java application actually is downloaded and executed by the web browser. Java can be used for two types of program: Applets that are embedded in the webpages with the <APPLET> tag, and full-scale applications, that can be used with Java Interpreter. When a webpage containing Java applet is accessed, the entire application is downloaded to the browser. The browser then executes the code. In order to do this, the browser must include a Java interpreter.

In contrast, JavaScript is a cross-platform and object-based scripting language. It is a programming language that can be written directly into HTML pages to allow interaction between the webpage and the viewer. JavaScript was a new scripting language that can be executed within a browser's environment and not on the server. On December 4, 1995 Brendan Eich of Netscape developed JavaScript. It's Level 1.0 version, originally called LiveScript to the world was jointly developed by Netscape and Sun Microsystems.

Java and JavaScript share some common syntax; they were developed independently of each other and for different audiences. JavaScript language resembles Java but does not have Java's static typing and strong type checking. Java is class-based programming language designed for fast execution and type safety. Type safety means, for instance, that one can't cast a Java integer into an object reference or access private memory by corrupting Java byte-codes. Java's class-based model means that programs consist exclusively of classes and their methods. Java's class inheritance and strong typing generally require tightly coupled object hierarchies. These requirements make Java programming more complex than JavaScript.

In contrast, JavaScript descends in spirit from a line of smaller, dynamically typed language such as HyerTalk and dBASE. These scripting languages offer programming tools to a much wider audience because of their easier syntax, specialized built-in functionally, and minimal requirements for object creation. Some of the points of differences between Java and JavaScript are given in Table 2.1.

Table 2.1   Differences between Java and Javascript

Java	JavaScript
• Compiled byte-codes downloaded from server, executed on client.	• Interpreted (not compiled) by client.
• Class-base—Objects are divided into classes and instances with all inheritance through the class hierarchy. Classes and instance cannot have properties or methods added dynamically.	• Object-oriented—No distinction between types of objects. Inheritance is through the prototype mechanism, and properties and methods can be added to any object dynamically.
• Applets distinct from HTML (accessed from HTML pages).	• Code integrated with, and embedded in HTML.
• Variable data types must be declared (static typing).	• Variable data types not declared (dynamic typing).

Java treats all elements of the programs as object. An object can be a variable, a subroutine, or application itself. The idea behind object-oriented languages is that an object can include both data and code, for example, a "number" object, would include the value of the number and the code needed to display it. Following is an example of a short Java applet. This program simply displays the text Good Morning in the larger text on the browser's screen.

The following is an example of a simple Java applet.

```
01:        import browser, Applet;
02:            import awt. Graphics;
03:            class Good Morning extends Applet  {
04:            public void unit () {
05:                 resize (150, 25);
06:            }
07:            public void paint (Graphics g)  {
08:                g.drawString ('Good Morning!', 50, 25);
09:            }
10:            }
```

Example 1 Write the following code in TextPad and save it with extension.html and then open the file in browser to view the output (Figure 2.9).

```
import javax.swing.*;

import javax.swing.event.*;

import java.awt.*;

import java.awt.event.*;

import java.io.*;

import java.lang.*;

public class Test extends JApplet

{

    String capital;

    public  Test()

    {

        JLabel l;

            capital=JOptionPane.showInputDialog("What is
the Capital of India?: ");

    }

    public void paint(Graphics g)

    {

            if(capital.equalsIgnoreCase("Delhi"))

        {
```

```
        g.setColor(Color.blue);
        g.setFont(new Font("Cosmic new sans",Font.BOLD,15));
        g.drawString("Good Day...",50,50);
            }
    else
    g.drawString("Sorry You are Wrong...",50,50);
  }

 }
```

Use the following HTML code for executing the above program.

```
<HTML>
<Body Bgcolor="gray">
<TITLE>
 BioInformatics
</TITLE>
<Applet code=Test.class height=200 width=300>
</Body>
</HTML>
```

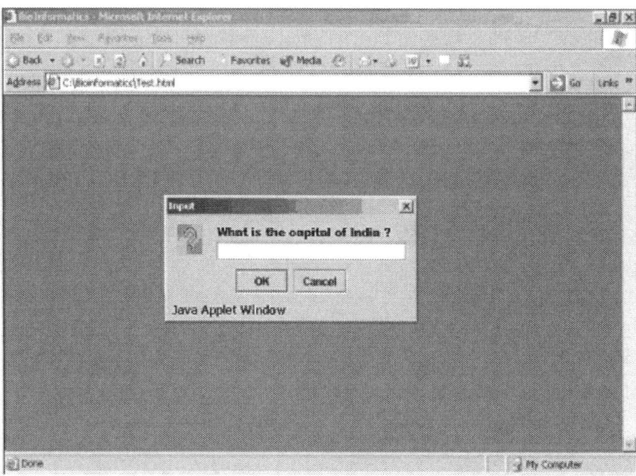

Figure 2.9 The output of the above code displayed in the browser

As mentioned earlier JavaScript (originally called LiveScript) is a scripting language which is actually based on Java's syntax, but it is a different language with different uses. The basic differences are summarized as follows:

▣ Instead of creating and compiling a Java Applet, one can embed JavaScript commands directly within HTML webpage.

▣ JavaScript uses simpler variable types and type checking than Java.

▣ JavaScript is executed by an interpreter within the browser. Unlike Java, it can access information on the webpage, such as links or the contents of forms.

To use JavaScript, embed it in the HTML of page using the <SCRIPT> tag. Following is the simple page that includes a very short JavaScript.

```
01:    <HTML>
02:    <HEAD><TITLE>Simple JavaScript Output</TITLE>
03:    </HEAD>
04:    <BODY>
05:    <SCRIPT LANGUAGE= "JavaScript">
06:    document.write (This is the output of the script.<BR>")
07:    </SCRIPT>
08:    Here's the body of the www page.
09:    </Body>
10:    </HTML>
```

Example 2 Write the following code in notepad or any other JavaScript compliant tool and save it with extention.htm. Open the file in browser to view the output (Figure 2.10).

```
<html>
    <head><title>Example1</title>
    <SCRIPT>
    <! - -
    var answer;
    var response;
    answer = prompt ("How many chromosomes are present in
    human genome?", " " );
    answer = answer.toUpperCase ()
    response = (answer = = 'numerical')? "Correct" : Incorrect"
    alert (response)
    function xyz ( )
```

```
{
ans =confirm ("Do you wish to exit?");
if (ans =1){document.write("Have a nice day");
}
window.close ( );
}
}
// - ->
</SCRIPT>
</head>
<body>
<form>
<input type = "button" value= "click me" onclick=xyz
</form>
</body>
</html>
```

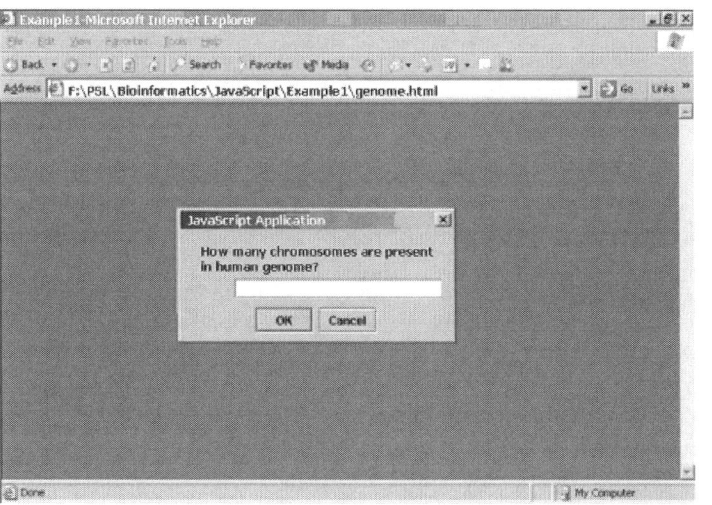

Figure 2.10   The output of above code displayed in the web browser

COMMON GATEWAY INTERFACE (CGI)

CGI programming is a writing application that acts as interface or gateway programs between the client browser, web server, and a traditional programming application. CGI programming involves designing and writing programs that receive their starting commands from a webpage that uses an HTML form to initiate the CGI program. CGI program can generate all the HTTP response headers required for sending data back to the client/browser by calling itself a non-parsed header CGI program. The way data is sent back and forth across the Internet is one of the most unique aspects of CGI programming.

PERL PROGRAM

For working in bioinformatics, one should develop expertise in using tools available on the web. Some skill, in writing simple scripts in a language like PERL (Practical Extraction Report Language) provides an essential extension to the basic facilities of the operating system. Perl is rapidly becoming one of the most popular scripting languages anywhere because it really does satisfy most of the needs of the person interested in bioinformatics. It's free, works on any platform, and run as soon as the person type in it.

In fact, Perl was designed originally for working with text, generating reports, and manipulating files. It does all these things fairly well and fairly easy. In addition, Perl has a lovely data structure called the associative array that helps the person to manipulate the databases. Since programming in computer science is similar to bricklaying in construction of buildings. Both activities are creative; one is an art and the other is craft. Here an attempt is made to execute simple Perl programs for those who are basically from the field of biology and who have some basic knowledge of computer science. This program doesn't have any CGI in it—it's just straight Perl. The entire steps for execution of Perl program are as follows:

1. Install Perl program with all its components (it includes the folders Bin, e.g. HTML, Lib, Site and Windows Installer Package).
2. Double click on TextPad icon present on the desktop.
3. TextPad [Document 1] will appear on the computer screen.
4. Open the configure menu of TextPad. Drop down to preferences. Activate the box of "allow multiple file on command lines" and then add Perl program in its tools by selecting the file "Perl" from the drive where it is installed.
5. Now the tool menu contains "Perl" ready to execute the program as the person type in it.
6. Type the following code in the TextPad editor and then save it to a specific file named Sunny.txt as shown in Figure 2.11.

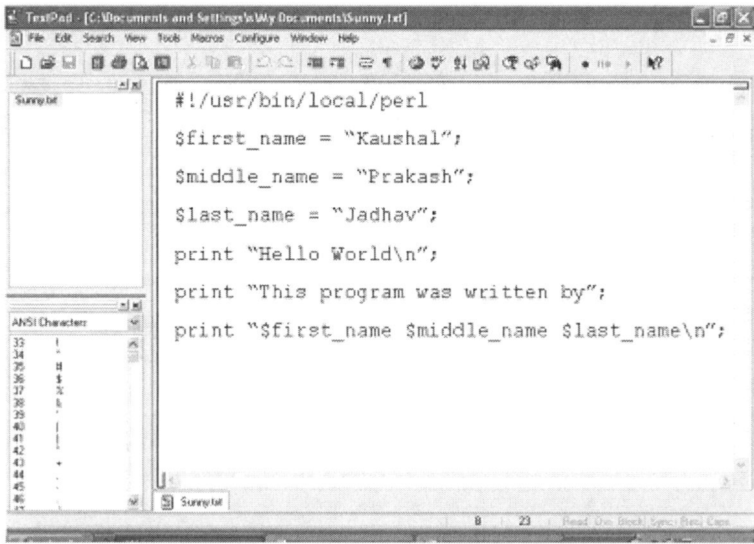

Figure 2.11 Personalizing the Hello World example with Perl

The first line of this file tells the computer where the Perl interpreter is located. If it is located somewhere else, this line should be changed. The sixth and seventh lines of the program tell the computer to print on the screen Hello World and the name of program writer respectively.

7. For command results, open the tool menu and drop down to Perl. On the computer screen following matter with the footnote "tool completed successfully" will be displayed.

Hello World

This program was written by Kaushal Prakash Jadhav

Perl was designed by Larry Wall as a tool for writing programs in the UNIX environment and is continually being updated and maintained by him.

Perl provides the best of several worlds. For instance:

- ▣ Perl has the power and flexibility of a high-level programming language such as C. In fact, many of the features of the language are borrowed from C.

▣ Like shell script languages, Perl does not require a special compiler and linker to turn the programs you write into working code. Instead, all you have to do is write the program and tell Perl to run it. This means that Perl is ideal for producing quick solutions to small programming problems, or for creating prototypes to test potential solutions for larger problems.

Example 3 Write a Perl program to convert miles to kilometers

The first character of the line in the program `#!/usr/local/bin/perl` is the **comment character**, #. When the Perl interpreter sees the #, it ignores the rest of that line.

```
1: #!/usr/local/bin/perl

2:

3: print ("Enter the distance to be converted:\n");

4: $originaldist = <STDIN>;

5: chop ($originaldist);

6: $miles = $originaldist * 0.6214;

7: $kilometers = $originaldist * 1.609;

8: print ($originaldist, " kilometers = ",
    $miles,

9: " miles\n");

10: print ($originaldist, " miles = ", $kilometers,

11: " kilometers\n");
OUTPUT
$ program_1
Enter the distance to be converted:10
10 kilometers = 6.2139999999999995 miles
10 miles = 16.09 kilometers
```

Perl is a very powerful tool and its strength at character-string handling makes it suitable for sequence analysis tasks in biology. Let's see another very simple Perl program that translates a nucleotide sequence into an amino acid sequence according to standard genetics code present on DNA molecule.

```
# ! / usr /local /bin /perl

# translate.pl. - - translate nucleic acid sequence to protein
sequence
```

```perl
# according to standard genetic code
# set up table of genetic code
% standardgeneticcode  = (
"ttt"=> "Phe",   "tct" => "Ser",  "tat" => "Tyr",  "tgt" => "Cys",
"ttc"=> "Phe",   "tcc" => "Ser",  "tac" => "Tyr",  "tgc" => "Cys",
"tta"=> "Leu",   "tca" => "Ser",  "taa" => "TER",  "tga" => "TER",
"ttg"=> "Leu",   "tcg" => "Ser",  "tag" => "TER",  "tgg" => "Trp",
"ctt"=> "Leu",   "cct" => "Pro",  "cat" => "His",  "cgt" => "Arg",
"ctc"=> "Leu",   "ccc" => "Pro",  "cac" => "His",  "cgc" => "Arg",
"cta"=> "Leu",   "cca" => "Pro",  "caa" => "Gln",  "cga" => "Arg",
"ctg"=> "Leu",   "ccg" => "Pro",  "cag" => "Gln",  "cgg" => "Arg",
"att"=> "Ile",   "act" => "Thr",  "aat" => "Asn",  "agt" => "Ser",
"atc"=> "Ile",   "acc" => "Thr",  "aac" => "Asn",  "agc" => "Ser",
"ata"=> "Ile",   "aca" => "Thr",  "aaa" => "Lys",  "aga" => "Arg",
"atg"=> "Met",   "acg" => "Thr",  "aag" => "Lys",  "agg" => "Arg",
"gtt"=> "Val",   "gct" => "Ala",  "gat" => "Asp",  "ggt" => "Gly",
"gtc"=> "Val",   "gcc" => "Ala",  "gac" => "Asp",  "ggc" => "Gly",
"gta"=> "Val",   "gca" => "Ala",  "gaa" => "Glu",  "gga" => "Gly",
"gtg"=> "Val",   "gcg" => "Ala",  "gag" => "Glu",  "ggg" =>  "Gly"
    );
# process data input
while ($ line = <DATA>) {                # read in the line of input
  print  "$ line";                       # transcribe to output
  chop ();                               # remove    end-of-line
                                         # character

  @triplets = unpack("a3" ×
(length($line)/3), $line);                # pullout    successive
                                         # triplets

  for each $codon (@triplets) {          # loop over triplets
    print "$standard
geneticcode{$codon}";                     # print out translation of
                                         # each

      }                                  # end loop on triplets
print "\n\n";                            # skip line on output
```

```
}                                       # end loop on input lines
# what follows is input data
__ End__
atggcacagtctcggtttacctag
```

(*Source* : Arthur, M.Lesk. *Introduction to Bioinformatics.* (2005)).

Running this program using the Perl tool in the TextPad, the output will be as follows:

```
MetAlaGlnSerArgPheThrTER
```

The same program entitled 'translate' can be written in Java

```java
//Translate
import java.io.*;
import javax.swing.*;
import java.awt.*;
import java.awt.event.*;
import java.lang.*;
class Translate
{
        public static void main(String[]args)
        {
                TranslateFrame ob=new TranslateFrame();
                ob.show();
        }
}
class TranslateFrame extends JFrame
{
        public TranslateFrame()
        {
                setSize(500,500);
                setLocation(200,10);
                setTitle("Translate Technique");
            Container cp=getContentPane();
            cp.add(new TranslatePanel());
```

```java
        }
}
class TranslatePanel extends JPanel implements ActionListener
{
        JTextField t1;
        JButton b1,b2;
        String s1,s2,temp;
String s4[]={«Phe»,»Phe»,» Leu»,»Leu»,»Leu»,»Leu»,»Leu»,»Leu»,
        »Ile»,»Ile»,»Ile», »Met»,»Val»,»Val»,»Val»,»Val»
        ,»Ser»,»Ser»,»Ser», »Ser»,»Pro»,»Pro»,»Pro»,
        »Pro»,»Thr»,»Thr», »Thr»,»Thr», »Ala»,»Ala»,»Ala»,»Ala»,
        "Tyr","Tyr","TER","TER", "His","His"," Gln",
        "Gln","Asn","Asn", "Lys","Lys","Asp","Asp","Glu","Glu",
        "Cys","Cys","TER","Trp","Arg","Arg","Arg", "Arg","Ser",
        "Ser","Arg","Arg", "Gly","Gly","Gly","Gly"};
String s3[]= {"ttt","ttc","tta","ttg","ctt","ctc","cta",
        "ctg","att","atc","ata","atg","gtt","gtc","gta","gtg",
        "tct","tcc","tca","tcg","cct","ccc","cca","ccg",
        "act","acc","aca","acg","gct","gcc","gca","gcg",
        "tat","tac","taa","tag","cat","cac","caa","cag","aat","aac",
        "aaa","aag","gat","gac","gaa","gag","tgt","tgc","tga",
        "tgg","cgt","cgc","cga", "cgg","agt","agc",
        "aga","agg","ggt","ggc","gga","ggg"  };
//String s5=
"PhePheLeuLeuLeuLeuLeuLeuIleIleIleMetValValValValSerSerSerSer
ProProProProThrThrThrThrAlaAlaAlaAlaTyrTyrTERTER
HisHisGlnGlnAsnAsnLysLysAspAspGluGluCysCys
TERTrpArgArgArgArgSerSerArgGlyGlyGlyGly";
//String s6=
"tttttcttattgcttctcctactgattatcataatggttgtcgtagtgtc
ttcctcatcgcctcccccaccgactaccacaacggctgccgcagcgtattactaatagca
tcaccaacagaataacaaaaaggatgacgaagagtgttgctgatggcgtcgccgacgga
gtagcagaaggggtggcggaggg";
        public TranslatePanel()
        {
JLabel l1=new s6= JLabel("Enter Sequence  :"); add(l1);
t1=new JTextField(30); add(t1);
```

```java
    b1=new JButton("Show Result");
    add(b1);
    b1.addActionListener(this);
    s1="atgcatccctttaat";
     }
    public void actionPerformed(ActionEvent e)
    {
          if(s1.length()>0)
            s1=t1.getText();
        repaint();
    }
    public void paintComponent(Graphics g)
    {
          super.paintComponent(g);
      setBackground(Color.red);
      int flag=0;
    int x=50;
    g.setColor(Color.white);
    g.setFont(new  Font("verdana",Font.BOLD,17));
    g.drawString(s1,50,170);
    for(int  i=0;i<s1.length();i=i+3)
            {
            for(int  j=0;j<s3.length;j++)//
            {
if(s1.charAt(i)==s3[j].charAt(0)&& s1.charAt(i+1)==s3[j].charAt(1)
&&  s1.charAt(i+2)==s3[j].charAt(2))
            {
                g.drawString(s4[j],x,199);
                x=x+50;
                flag=1;
                }
            }
    }
```

```
    if(flag==0)

    g.drawString(„Sorry Wrong sequence No Match Found",x,199);
}

}
```

Output of the above program   After running the above written program in TextPad with Run Java Application, the window with the title "Translate Technique" appears. After entering the particular sequence of genetic code in enter sequence box, press the tab "Show Result"; the entered sequence of the genetic code is translated into amino acid sequence that is shown below.

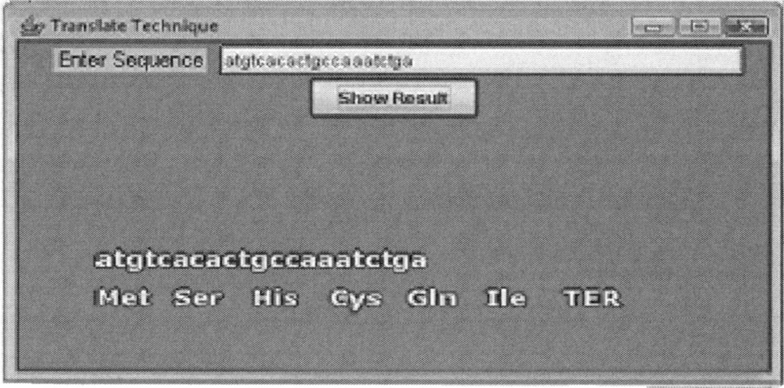

ROLE OF COMPUTERS IN BIOINFORMATICS

Since bioinformatics is the research, development, or application of computational tools and approaches for expanding the use of biological, medical, behavioural or health data, including those to acquire, store, organize, archive, analyse, or visualize such data. And computational biology is the development and application of data-analytical and theoretical methods, mathematical modelling and computational simulation techniques to study the biological behavioural and social systems. Computers play indispensable role in the development of bioinformatics. The bioinformatic tools are software programs that are designed for extracting the meaningful information from the mass of molecular biology/biological databases and to carry out sequence or structural analysis.

Factors that must be taken into consideration when designing bioinformatics tools, software and programs are:

- The end user (the biologist) may not be a frequent user of computer technology.

  ▣ These software tools must be made available over the internet given the global distribution of the scientific research community.

Major Categories of Bioinformatics Tools

There are both standard and customized products to meet the requirements of particular projects. There are data-mining software that retrieve data from genomic sequence databases and also visualization tools to analyse and retrieve information from proteomic databases. These can be classified as homology and similarity tools, protein functional analysis tools, sequence analysis tools and miscellaneous tools.

Here is a brief description of a few of these, everyday bioinformatics is done with sequence search programs like BLAST, sequence analysis programs, like the EMBOSS and Staden packages, structure prediction programs like THREADER or PHD or molecular imaging/modelling programs like RasMol and WHATIF.

Homology and similarity tools   Homologous sequences are sequences that are related by divergence from a common ancestor. Thus the degree of similarity between two sequences can be measured while their homology is a case of being either true or false. This set of tools can be used to identify similarities between novel query sequences of unknown structure and function and database sequences whose structure and function have been elucidated.

Protein function analysis   This group of programs allow to compare protein sequence to the secondary (or derived) protein databases that contain information on motifs, signatures and protein domains. Highly significant hits against these different pattern databases allow us to approximate the biochemical function of query protein.

Structural analysis   This set of tools allows to compare structures with the known structure databases. The function of a protein is more directly a consequence of its structure rather than its sequence with structural homologs tending to share functions. The determination of a protein's 2D/3D structure is crucial in the study of its function.

Sequence analysis   This set of tools allows us to carry out further, more detailed analysis on a query sequence including evolutionary analysis, identification of mutations, hydropathy regions, CpG islands and compositional biases. The identification of these and other biological properties are all clues that aid the search to elucidate the specific function of query sequence.

Some examples of bioinformatics tools are listed below.

BLAST BLAST (Basic Local Alignment Search Tool) comes under the category of homology and similarity tools. It is a set of search programs designed for the Windows platform and is used to perform fast similarity searches regardless of whether the query is for protein or DNA. Comparison of nucleotide sequences in a database can be performed. Also a protein database can be searched to find a match against the queried protein sequence. NCBI has also introduced the new quering system to BLAST (Q BLAST) that allows users to retrieve results at their convenience and format their results multiple times with different formatting options.

Depending on the type of sequences to compare, there are different programs:

- **BLASTP** compares an amino acid query sequence against a protein sequence database
- **BLASTN** compares a nucleotide query sequence against a nucleotide sequence database
- **BLASTX** compares a nucleotide query sequence translated in all reading frames against a protein sequence database
- **TBLASTN** compares a protein query sequence against a nucleotide sequence database dynamically translated in all reading frames
- **TBLASTX** compares the six-frame translations of a nucleotide query sequence against the six-frame translations of a nucleotide sequence database.

FASTA FASTA (FAST homology search All sequences) is an alignment program for protein sequences created by Pearsin and Lipman in 1988. The program is one of the many heuristic algorithms proposed to speed up sequence comparison. The basic idea is to add a fast pre-screen step to locate the highly matching segments between two sequences, and then extend these matching segments to local alignments using more rigorous algorithms such as Smith–Waterman.

EMBOSS EMBOSS (European Molecular Biology Open Software Suite) is a software-analysis package. It can work with data in a range of formats and also retrieve sequence data transparently from the web. Extensive libraries are also provided with this package, allowing other scientists to release their software as open source. It provides a set of sequence-analysis programs, and also supports all UNIX platforms.

ClustalW   It is a fully automated sequence alignment tool for DNA and protein sequences. It returns the best match over a total length of input sequences, be it a protein or a nucleic acid.

RasMol   It is a powerful research tool to display the structure of DNA, proteins, and smaller molecules. Protein Explorer, a derivative of RasMol, is an easier to use program.

Prospect   PROSPECT (PROtein Structure Prediction and Evaluation Computer Toolkit) is a protein-structure prediction system that employs a computational technique called protein threading to construct a protein's 3D model.

PatternHunter   PatternHunter, based on Java, can identify all approximate repeats in a complete genome in a short time using little memory on a desktop computer. Its features are its advanced patented algorithm and data structures, and the Java language was used to create it. The Java language version of PatternHunter is just 40 KB, only 1% the size of BLAST, while offering a large portion of its functionality.

COPIA   COPIA (COnsensus Pattern Identification and Analysis) is a protein-structure analysis tool for discovering motifs (conserved regions) in a family of protein sequences. Such motifs can be then used to determine membership to the family for new protein sequences, predict secondary and tertiary structure and function of proteins and study evolution history of the sequences.

Application of Programs in Bioinformatics

JAVA in bioinformatics   Since research centres are scattered all around the globe ranging from private to academic settings, and a range of hardware and OSs are being used, Java is emerging as a key-player in bioinformatics. Physiome Sciences' computer-based biological simulation technologies and Bioinformatics Solutions' Pattern Hunter are two examples of the growing adoption of Java in bioinformatics.

BioJava   The BioJava Project is dedicated to providing Java tools for processing biological data, which includes objects for manipulating sequences, dynamic, programming, file parsers, simple statistical routines, etc.

Perl in bioinformatics   Perl is an acronym, short for Practical Extraction and Report Language. String manipulation, regular expression matching, file parsing, data format, inter conversion, etc. are the common text-processing tasks performed in bioinformatics. Perl excels in such tasks and is being used by many developers. Yet, there are no standard modules designed in Perl specifically for the field of bioinformatics. However, developers have designed several of their own individual modules for the purpose, which have become quite popular and are coordinated by the BioPerl project.

BioPerl   The BioPerl project is an international association of developers of Perl tools for bioinformatics and provides an online resource for modules, scripts and web links for developers of Perl-based software. BioPerl is a collection of Perl modules that facilitate the development of Perlscripts for bioinformatics applications. As such, it does not include ready-to-use programs, in the sense that many commercial packages and free web-based interfaces do (e.g. Entrez, SRS). On the other hand, BioPerl does provide reusable-Perl modules that facilitate writing Perlscripts for sequence manipulation, accessing of databases using a range of data formats and execution and parsing of the results of various molecular biology programs including BLAST, ClustalW, T-Coffee, GenScan, ESTScan and HMMER. Consequently, BioPerl enables developing scripts that can analyse large quantities of sequence data in ways that are typically difficult or impossible with web-based systems.

In order to take advantage of BioPerl, the user needs a basic understanding of the Perl-programming language including an understanding of how to use Perl references, modules, objects and methods. If these concepts are unfamiliar the user is referred to any of the various introductory or intermediate books on Perl. For newcomers and people who want to quickly evaluate whether the package is worth using in the first place, there is a very simple module which allows easy access to a small number of BioPerl's functionality in an easy-to-use manner. The Bio::Perl (http://www.bioperl.org/wiki/Module:Bio::Perl) module provides some simple access functions.

BioXML   A part of the BioPerl project, this is a resource to gather XML documentation, DTDs and XML aware tools for biology in one location.

BioCorba   Interface objects have facilitated interoperability between BioPerl and other Perl packages such as Ensembl and the Annotation Workbench. However, interoperability between BioPerl and packages written in other languages requires additional support software. Corba is one such framework for inter-language support, and the BioCorba project is currently implementing a Corba interface for BioPerl. With BioCorba, objects written within BioPerl will be able to communicate with objects written in BioPython and BioJava (see the next subsection). For more information, see the BioCorba Project website at http://biocorba.org/. The BioPerl-BioCorba server and client bindings are available in the Bioperl-corba-server and Bioperl-corba-client Bioperl CVS repositories respectively. (see http://cvs.bioperl.org/for more information).

Ensembl   Ensembl is an ambitious automated-genome-annotation project at EBI. Much of Ensembl's code is based on BioPerl, and Ensembl developers, in turn, have contributed significant pieces of code to BioPerl. In particular, Ensembl developers have largely contributed the BioPerl code for automated sequence annotation. Describing Ensembl and its capabilities is far beyond the scope of this tutorial. The interested reader is referred to the Ensembl website at http://www.ensembl.org/.

Bioperl-db   BioPerl-db is a relatively new project intended to transfer some of Ensembl's capability of integrating bioperl syntax with a stand-alone Mysql database (http://www.mysql.com/) to the BioPerl code-base. More details on bioperl-db can be found in the BioPerl-db CVS directory at http://cvs.bioperl.org/. It is worth mentioning that most of the BioPerl objects mentioned above map directly to tables in the BioPerl-db schema. Therefore object data such as sequences, their features, and annotations can be easily loaded into the databases.

BioPython and BioJava   BioPython and BioJava are open-source projects with very similar goals to BioPerl. However their code is implemented in Python and Java, respectively. With the development of interface objects and BioCorba, it is possible to write Java or Python objects that can be accessed by a BioPerlscript, or to call BioPerl objects from Java or Python code. Since BioPython and BioJava are more recent projects than BioPerl, most effort to date has been to port BioPerl functionality to BioPython and BioJava rather than the other way around. However, in the future, some bioinformatics tasks may prove to be more effectively implemented in Java or Python in which case being able to call them from within BioPerl will become more important. For more information, go to the BioJava http://biojava.org/and biopython http://biopython.org/websites.

REVIEW QUESTIONS

1. Describe five generations of computer with their specifications.
2. What are different criteria for classification of computer? Explain the type of computers based on their functional capabilities.
3. Define operating system and explain the services in operating system.
4. What are point of differences between DOS and Windows operating systems?

5. Define Linux. Describe its components and special features.
6. Comment on Internet and information retrieval on Internet.
7. What are three basic elements of e-mail?
8. How does Internet help in communication?
9. What are search tools? Explain indexes and search engines used in Internet.
10. Explain the concept of World Wide Web and describe various web browsers.
11. How would you comment on Hypertext Transfer Protocol and Hypertext Markup Language?
12. What is Universal Research Locator? What are the components of URL?
13. Write the simple HTML program and enlist its advantages and disadvantages.
14. Define Extensible Markup Language (XML). How would you write the syntax for well formed XML document?
15. What are your views on the statement, that Java and JavaScript bring the life to HTML?
16. How would you compare Java and JavaScript?
17. Write the simple programs written in Java and JavaScript.
18. What is Perl? Write a Perl program for the code "Hello World".
19. Give brief idea about major categories of bioinformatics tools present on Internet.
20. Write short notes on:
    i. General-purpose computers
    ii. Graphical User Interface
    iii. Modem in Internet
    iv. Home Page
    v. WebCrawler
    vi. Java applet
    vii. Common Gateway Interface
    viii. RasMol
    ix. Ensembl
    x. BioPython
    xi. EMBOSS
    xii. BioCorba

3

BIOLOGICAL MACROMOLECULES

The molecule is the smallest entity of a chemical compound. There are four major classes of biological macromolecules in the cell—nucleic acids, proteins, carbohydrates and lipids—essential for the integrity of cell and essential for its sustainability and perpetuation. Carbohydrates are linear as well as branched chain polymers and they are involved in structural, energy storage and cell–cell communication. Biological membranes (lipids) are macromolecules, but are not polymers and they are involved in energy storage functions. From the point of view of bioinformatics, nucleic acids and proteins are most important biomolecules. Nucleic acids are responsible for storage, expression and transmission of genetic information. Proteins are polymers of amino acids and they carry out wide range of biological and biochemical activities, structural as well as functional.

As stated earlier, bioinformatics is the technology that uses computers for storage, retrieval, manipulation, and distribution of information obtained by analysing sequence data of biological macromolecules like DNA, RNA and proteins. It is essential to understand the structural features, vis a vis the structural and functional constitutes, is of significant to understand their functional characteristics. Each organism carries the blueprint of potential development and activity in the form of genetic material, DNA or in some viruses, RNA. Nucleic acids—DNA and RNA (except tRNA) are long thread-like macromolecules playing central role in all transmission of hereditary characters. They store information in the form of genetic code, replicate, transcribe and translate into proteins as per the central dogma of molecular biology.

DEOXYRIBONUCLEIC ACID (DNA)

CHEMICAL STRUCTURE OF DNA

It is the deoxyribonucleic acid (DNA) that carries genetic information of all living information, except RNA viruses. The basic building block of this macromolecule is nucleotide. Each nucleotide is composed of (1) a five-carbon containing (pentose) sugar, deoxyribose, (2) a phosphoric acid (biologically a phosphate), and (3) a cyclic nitrogen-containing compound called a base, either a pyrimidine or purine, as shown in Figure 3.1.

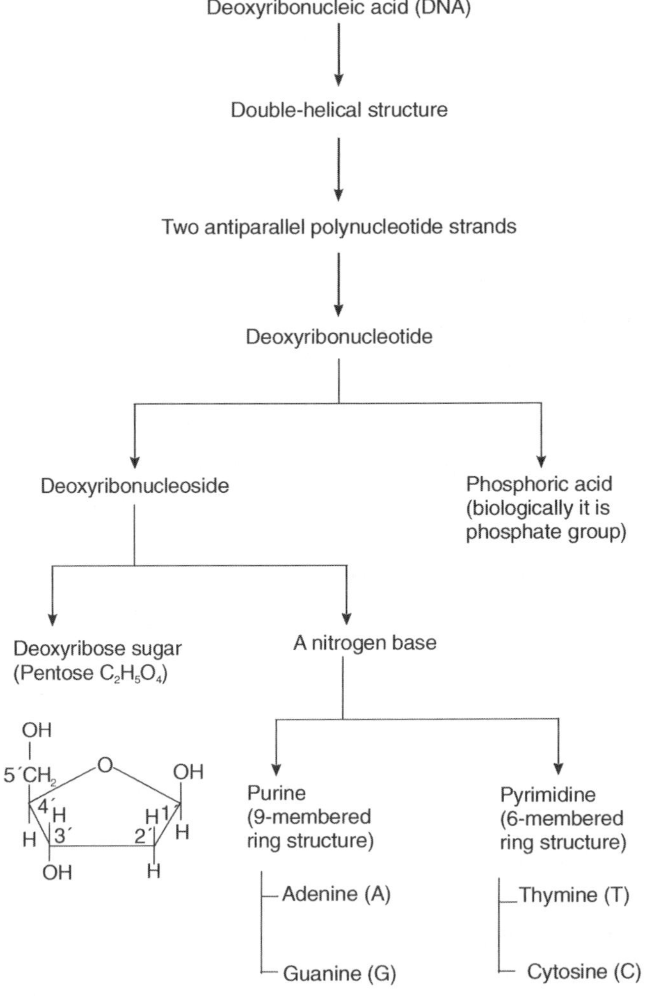

Figure 3.1   The building blocks of deoxyribonucleic acid (DNA)

The pentose sugar in DNA lacks the oxygen on the carbon 2 '(pronounced as two prime) position, hence called 2-deoxy-D-ribose or simply deoxyribose. The four nitrogen bases found in DNA are adenine (A), guanine (G), thymine (T), and cytosine (C). The A and G are purines and T and C are pyrimidines. The nucleoside in DNA is called deoxynucleoside that is a complex of deoxyribose sugar and a nitrogen base, where the purines with their N-9 position always attach to carbon atom 1 ' of deoxyribose sugar and pyrimidines with their N-1 attach to C-1 ' of deoxyribose sugar. While the deoxyribonucleotide consists of a deoxyribose sugar, a nitrogen base, and a phosphate group, where a phosphoric acid or phosphate group is attached to carbon 5 ' of the deoxyribose sugar that is already joined with any one of

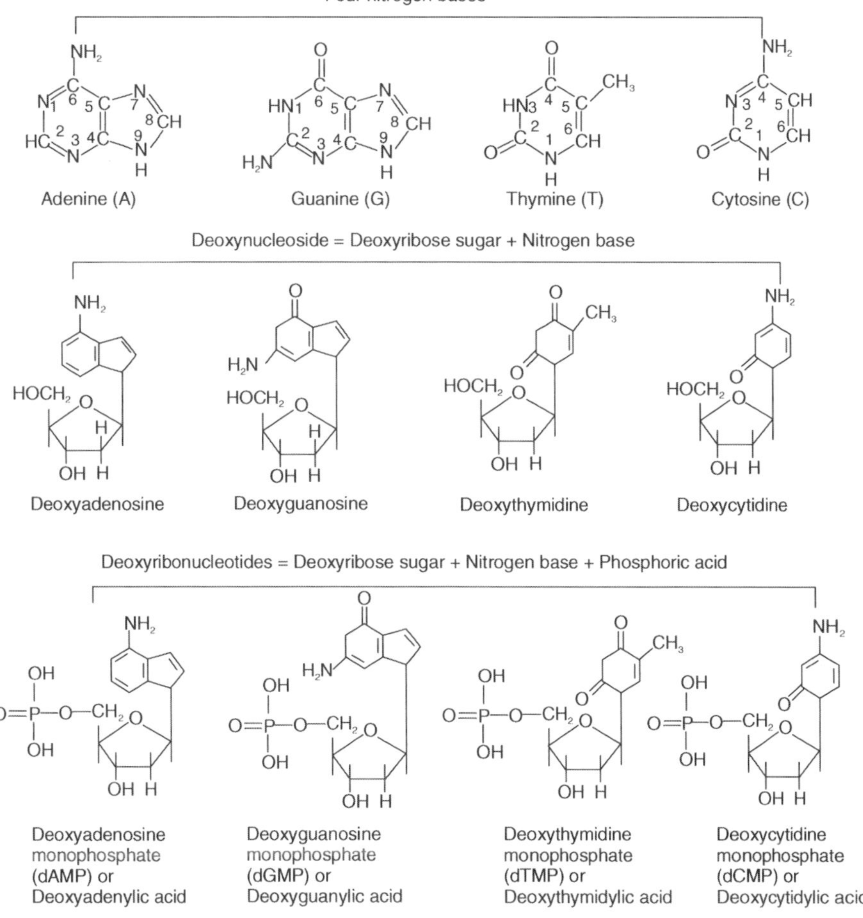

Figure 3.2 Chemical structures of nitrogen bases, deoxyribonucleosides, and deoxyribonucleotides of DNA

nitrogen bases. Four types of nitrogen bases, deoxynucleosides, and deoxyribonucleotides with their terminologies and abbreviations are mentioned in Figure 3.2. Each deoxyribonucleotide is a polar structure, at one edge where the phosphate is located, is called 5′end, while the other edge is called 3′end.

The phosphoric acid –OH groups are both ionized at physiological pH since one of the –OH groups has a pKa of around 2 and the other is of around 7. This means they are highly hydrophilic, whereas the nitrogen bases are almost water insoluble and DNA is strongly negatively charged.

The deoxyribonucleotides are known to be covalently linked to one another to form a linear polymer or strand, with a backbone composed of alternating sugar and phosphate groups joined by 3′, 5′-phosphodiester bonds (Figure 3.3). The nitrogen bases were thought to project from the backbone like a column of stacked shelves. Since each of the stacked nucleotides in a strand has polarity, the same direction is attributed to the entire strand. One end is 3′ end, and the other is the 5′ end.

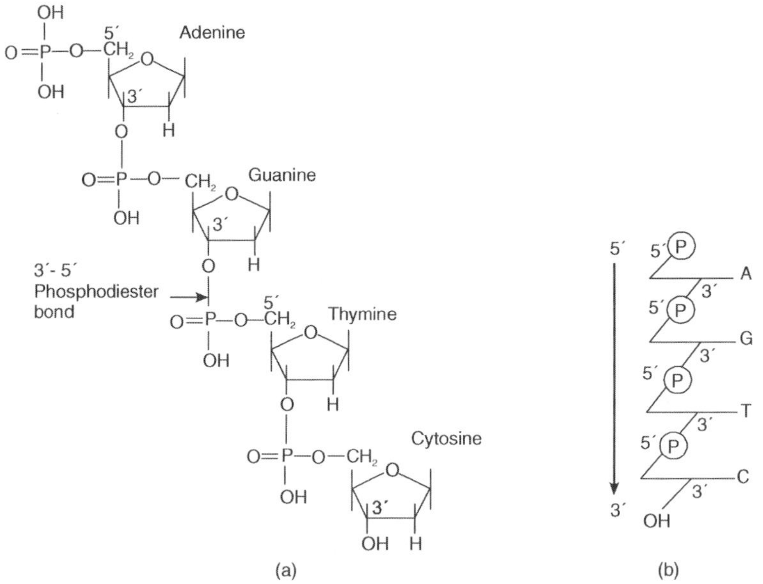

Figure 3.3   (a) Extended structure of a polynucleotide strand of DNA, and (b) Shorthand method for representation of a polynucleotide strand showing polarity

Chargaff's Rule for Base Composition in DNA

As per earlier concept, DNA was considered as a simple repeating tetranucleotide (e.g. —ATGCATGCATGC—). In 1950, Erwin Chargaff discovered the equivalence rule that suggested that despite wide compositional variation exhibited by different types of DNA, the total number of purines is equal to the total amount of pyrimidines (A + G = T + C); the amount of adenine is equal to the amount of thymine (A = T) and the amount of guanine is equal to the amount of cytosine (G = C). Chargaff analysed base composition that was accomplished by hydrolysing the bases from their attached sugars, separating the bases in hydrolysate by paper chromatography. His equivalence rule has been found to apply almost universally in different organisms.

However, he found that the ratios of four bases were quite variable from one type of organism to other. For example, DNA isolated from higher plants and animals was rich adenine and thymine (A : T) and relatively poor in guanine and cytosine (G : C). In case of human AT/GC ratio was 1.40 : 1, indicating higher percentage of A and T, whereas DNA isolated from microorganisms (viruses, bacteria, and lower animals and plants) was generally rich in guanine and cytosine and relatively poor in adenine and thymine. For example, AT/GC ratio of DNA of *Mycobacterium tuberculosis* was 0.60 : 1. Based on these observations, Chargaff discovered the following base composition rules in DNA:

$$(A) = (T), \quad (G) = (C), \text{ and } \quad (A) + (T) \neq (G) + (C)$$

These differences in base compositions give the specificity and individuality from one organism to another. It reflects the phylogenic, evolutionary and taxonomical relationship between different organisms.

X-ray Crystallography of DNA

The structure of DNA was deciphered only after many types of experimental evidence and theoretical considerations were combined. The crucial evidence was obtained by X-ray crystallography, the process that requires tremendous skill. Some chemical substances, when they are isolated and purified, can be made to form crystals. The positions of the atoms in a crystalline substance can be inferred from the pattern of diffraction of X-rays passed through it (Figure 3.4). The English chemist Rosalind Franklin and biophysicist Maurice Wilkins succeeded to obtain the X-ray diffraction photographs of samples containing very uniformly oriented DNA fibres. Such studies demonstrated that DNA was a helical structure with a diameter of 20 Å and one complete turn of helix is about 34 Å long. These findings provided significant information for Watson and Crick to propose the model of DNA.

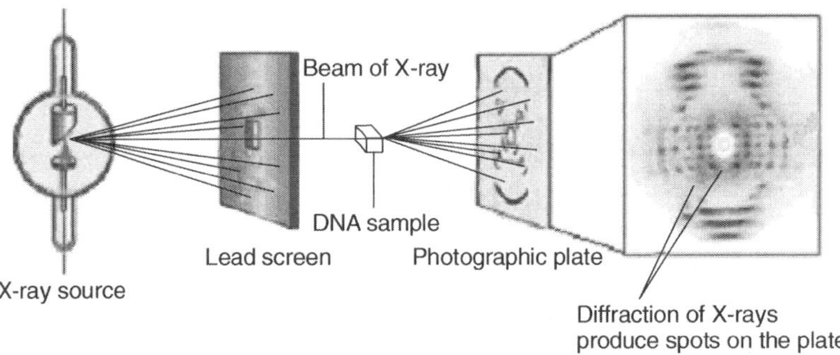

Figure 3.4   The X-ray crystallography technique revealed the helical nature of DNA

Watson and Crick Proposal for DNA Model

Taking into consideration of Chargaff's chemical data, Wilkins and Franklin's X-ray diffraction data, and inference drawn from model building, in 1953, J.D. Watson and F.H.C. Crick proposed that DNA exists as a double helix in which the two polynucleotide strands are coiled about one another in a spiral (Figure 3.5a) with uniform diameter of 20 Å. Each polynucleotide chain consists of a sequence of deoxyribonucleotides linked together by 3′-5′ phosphodiester bonds. Both the polynucleotide strands are held together in their helical configuration by hydrogen bonding between bases in opposite strands, the resulting base-pairing being stacked between the two chains perpendicular to the axis of the molecule like the steps in a spiral staircase. The double-helical structure of DNA is right-handed (that is, it twists to right, as do the threads on the most screws).

The base pairing is specific—adenine is always paired with thymine and guanine is always paired with cytosine. Thus, all base pairs consist of one purine and one pyrimidine. In their most common structural configurations, adenine and thymine form two hydrogen bonds (A = T), and guanine and cytosine forms three hydrogen bonds (G ≡ C). Since both the strands of DNA are complementary to each other, if the sequence of bases in one strand of a DNA double helix is known, the sequence of bases in the other strand is also known because of the specific base pairing. This property makes the DNA uniquely suited to store and transmit genetic information. The base pairs in DNA are stacked 3.4 Å apart with 10 base pairs per turn (each base pair rotates 36° with respect to the adjacent pair) of the double helix. Thus, one complete turn of the DNA helix has the length of 34 Å. The DNA double helix shows two external grooves, a deep-wide **major groove** and a shallow-narrow **minor groove**. Proteins that bind to the DNA often fit into these grooves.

The sugar–phosphate backbone of two polynucleotide strands run antiparallel to each other, i.e., they have opposite chemical polarity (Figure 3.5b). As one moves unidirectional along a DNA double helix, the phosphodiester bonds in one strand go from a 3′ carbon of one nucleotide to a 5′ carbon of the adjacent nucleotide, whereas those in the complementary strand go from a 5′ carbon to a 3′ carbon. This opposite polarity of the complementary strands is very significant in considering the mechanism of replication of DNA.

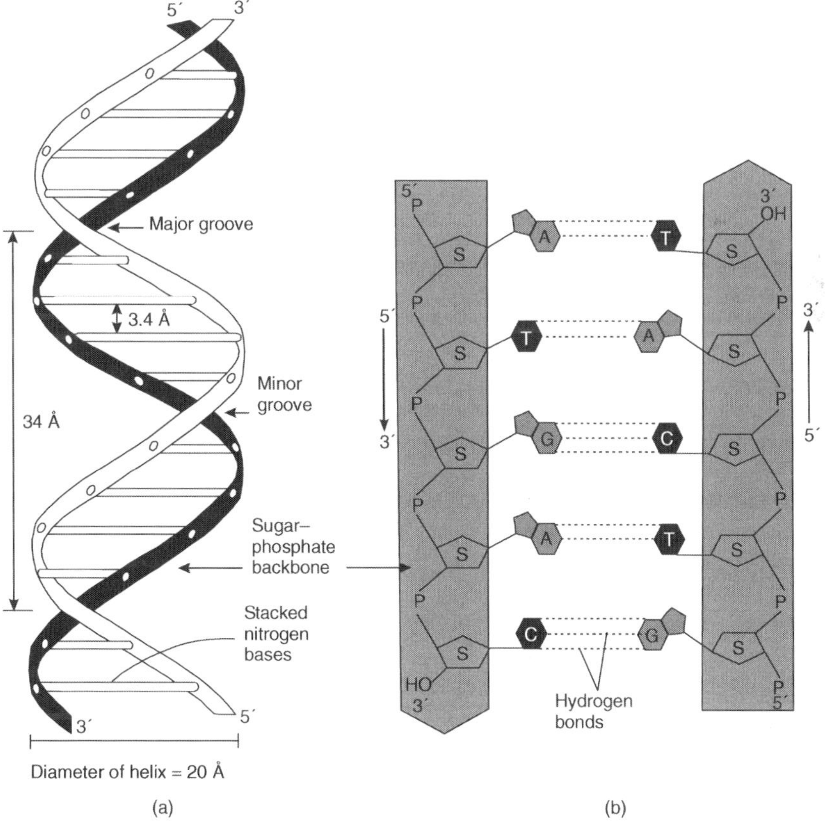

Figure 3.5   (a) Watson–Crick double helical model of the DNA molecule, and (b) Molecular structure of DNA showing sugar–phosphate backbones of polynucleotide strands with their antiparallel nature.

Alternate Forms of DNA Helix

The structure of DNA is not monotonous. The DNA model proposed by Watson and Crick was based on the X-ray diffraction data in which the fibres of DNA had been prepared when "wet" or fully hydrated. It is thought to represent the majority of DNA molecules present in living cells with the B-form or B-DNA (right-handed double helix). Further experiments revealed that DNA is much more polymorphic (indicating presence of alternate forms of DNA) than expected. When the DNA fibres prepared under lower humidity conditions and processed for X-ray crystallography, they had a more crystalline structure and generated more sharply defined X-ray diffracted patterns. The molecular arrangement of DNA in the dried fibres is somewhat different from the hydrated DNA fibres and it is referred to as the A-form of DNA. In 1978, Alexander Rich studied the X-ray diffraction pattern of GC-rich DNA and found that the strands of this synthetic molecule twisted in a counterclockwise spiral indicating its left-handed nature. Because of its zigzag conformation of the sugar–phosphate backbone, this new DNA structure is referred as Z-DNA. The comparative account of A-, B-, and Z-DNAs are mentioned in Table 3.1. In addition to above-mentioned types, C-DNA and D-DNA are also alternate forms of right-handed DNA helix that are shown in Figure 3.6.

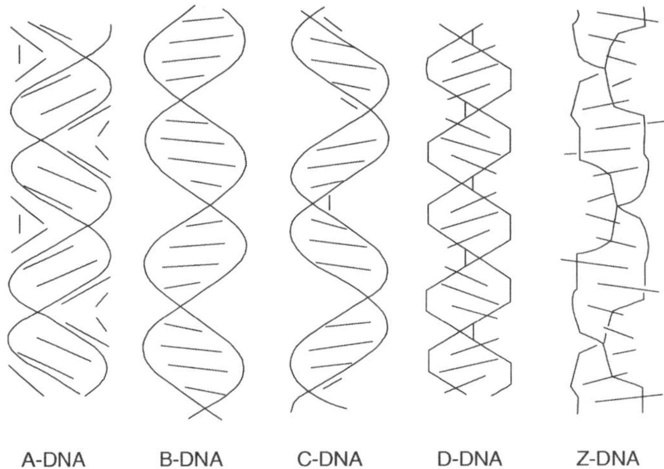

Figure 3.6   Five different molecular forms of DNA double helix. The A-, B-, C- and D-DNA are alternate DNA conformations for right-handed DNA helix and Z-DNA is polymorphic form of left-handed DNA.

Table 3.1 Comparison of some morphological parameters of three major alternate forms of DNA double helix

Parameter	A-DNA	B-DNA	Z-DNA
Overall proportion	Short and broad	Longer and thin	Elongated and slim
Helix sense	Right-handed	Right-handed	Left-handed
Residue per turn	11	10.5	22.6
Distance between adjacent bases	2.4 Å	3.4 Å	3.7 Å
Rotation per residue	32.7°	36°	–9°, –15°
Diameter of helix	25.5 Å	23.7 Å	18.4 Å
Major groove	Narrow, deep	Wide, deep	Flattened
Minor groove	Wide, shallow	Narrow, deep	Narrow, deep

DNA Replication

Every living organism has the ability to duplicate itself either by asexual or sexual reproduction, every living cell duplicates itself by cell division, similarly the genetic material duplicates by replication. As a carrier of genetic information, DNA performs heterocatalytic or autocatalytic functions. In a heterocatalytic role, DNA directs the synthesis of chemical molecules other than itself, e.g. the synthesis of RNA, proteins, etc., whereas in an autocatalytic activity, DNA directs the synthesis of DNA itself. The formulation of the structure of DNA by Watson and Crick in 1953 was accompanied by a proposal for its "self-duplication". Each strand of DNA double helix can serve as template for synthesis of new strand to form duplicated DNA. Watson and Crick proposed that replication of DNA involves breakdown of weak hydrogen bonds that holds the duplex together, the process is followed by a rotation and separation of both the polynucleotide strands (Figure 3.7), much like the separation of two halves of a zipper. Because of the complementary nature of both strands of DNA, each separated strand acts as a template as it contains the information required for the synthesis of other strand. Each base of template strand attracts a complementary nucleotide available within the cell in the presence of a polymerizing enzyme called **DNA polymerase** that holds the nucleotide in a position to continue the synthesis of a new strand. Finally, replicas (genetically identical with each other) of double-helical molecules are formed.

Semiconservative DNA replication   According to the proposal made by Watson and Crick for DNA duplication, once the DNA replication is initiated, both the old strands of the duplex serve as a template that directs the synthesis of new complementary strand. As a result, each daughter duplex should be hybrid in nature since it consists of one complete strand from parental duplex and one complete strand that has been newly synthesized. Thus, each daughter DNA duplex retains half of the parental DNA and the replication of this type is said to be semiconservative (Figure 3.8). When these two hybrid duplexes replicate themselves, there would be a formation of four duplexes, two of which would contain a single strand derived from the original chromosome and two of which contain totally new DNA strands.

Semidiscontinu ous DNA synthesis   During DNA replication, none of three prokaryotic DNA polymerases (I, II, and III) is able to construct daughter DNA strand in the $3' \rightarrow 5'$ direction. The opposing chain directions ($5' \rightarrow 3'$ and $3' \rightarrow 5'$) of the parental duplex mean that the two daughter strands being synthesized at each replicating fork must also run in opposite directions. Therefore, the overall direction must be $5' \rightarrow 3'$ for one daughter strand and $3' \rightarrow 5'$ for the other daughter strand. DNA polymerase can add nucleotide precursors to DNA only in $5' \rightarrow 3'$ direction, since the chemical reaction catalysed by three enzymes allow a nucleotide triphosphate to react only with the free $3'$–OH end of growing polynucleotide strand. Consequently, one of the newly synthesized strands grow towards the replication fork where the DNA strands are being separated, while the other strand grows away from the fork (Figure 3.9). Apparently, the strand that assembled in continuous fashion is known as leading strand. In contrast, the strand that grows away from the replication fork is synthesized discontinuously in the form of small pieces of DNA (approximately 1000–2000 nucleotide long in *E.coli* and about 100–200 nucleotide long in eukaryote). These small pieces or segments of DNA are often called "Okazaki fragments" named after R.Okazaki of Nagoya University, Japan, who first identified them. The strand being extended in overall $3' \rightarrow 5'$ direction grows by the synthesis of Okazaki fragments (synthesized $5' \rightarrow 3'$) is referred to as lagging strand because the initiation of each fragment must wait for parental strands to separate and expose additional templates. Subsequently, the Okazaki fragments are joined to form a continuous strand by the enzyme called DNA ligase. Since one strand is synthesized continuously and other strand discontinuously, replication is said to be semi-discontinous.

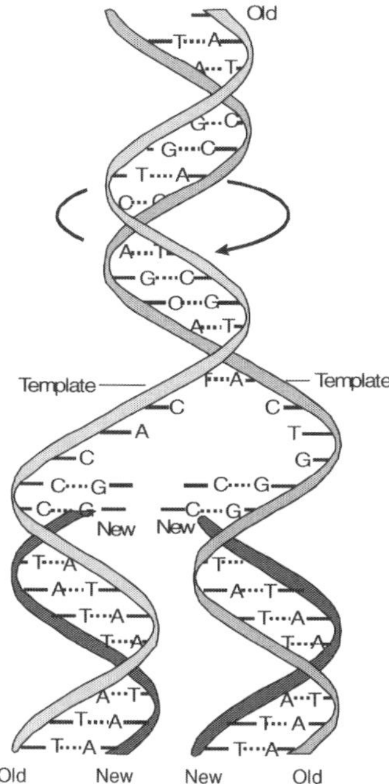

Figure 3.7   Watson and Crick proposal for replication of DNA double helix

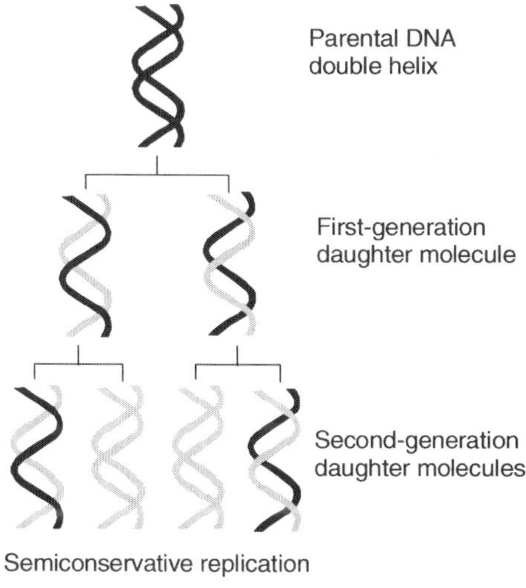

Figure 3.8   Semiconservative type of DNA replication

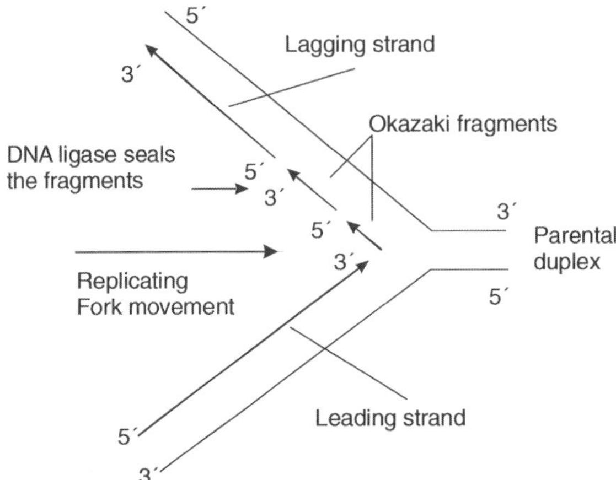

Figure 3.9   Semi-discontinuous synthesis of daughter DNA duplexes

THE CENTRAL DOGMA

Crick and his co-workers proposed the central dogma of molecular biology for the transfer of genetic information from genes to protein that occurs in two steps. The first step is known as transcription in which the messenger RNA (ribonucleic acid) is synthesized as a complementary copy of one of the two strands of DNA. The mRNA retains the same information as the gene itself. In this process, there is not much difference between the coded language (script) of DNA and RNA, hence it is a transcription due to the fact that a genetic code on DNA is in the form of ATGC and codons on the mRNA are in the form of AUGC, thus T in DNA is replaced by U (uracil) in RNA. The enzyme called DNA-dependant RNA polymerase or simply RNA polymerase catalyses this process. The second process is referred to as translation, which is a highly complex process that requires the participation of dozens of different components, including ribosomes containing ribosomal RNA (rRNA) and third major class of RNA such as transfer RNA (tRNA). This step involves the change of codons of mRNA to the particular sequence of amino acids in a polypeptide chain to form a protein. The process is called translation because it involves a complete translation of the language of codons on mRNA to amino acids in a protein that forms a phenotype. Thus, messenger RNA (mRNA) is intermediate between a gene and its polypeptide (Figure 3.10). Central dogma in simple words is a one-way transfer of genetic information from DNA to RNA, from RNA to protein and protein does not code for protein, RNA or DNA. Thus once the information is passed in protein it cannot be reversed.

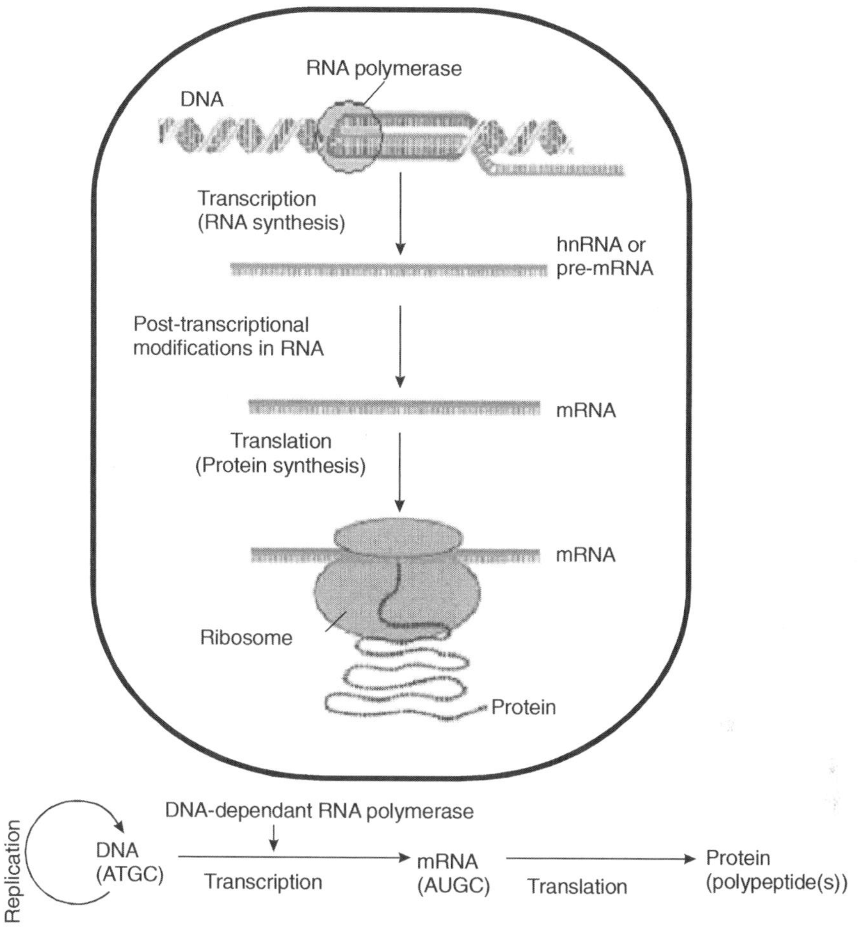

Figure 3.10 Central dogma of molecular biology showing one-way transfer of genetic information from DNA to protein

Modification of Central Dogma due to RNA Viruses

In 1962, Temin reported the presence of an enzyme called RNA-dependant DNA polymerase or reverse transcriptase in RNA viruses such as tobacco mosaic virus, influenza virus, poliovirus that have RNA as genetic material instead of DNA. This enzyme can synthesize DNA duplex from a single-stranded RNA template. Baltimore (1970) also independently reported the activity of this enzyme in certain RNA tumour viruses. Their findings gave rise to the new concept of "reverse transcription" or Teminism suggesting that flow of information is always not unidirectional from DNA to RNA but it can be from RNA to DNA (Figure 3.11).

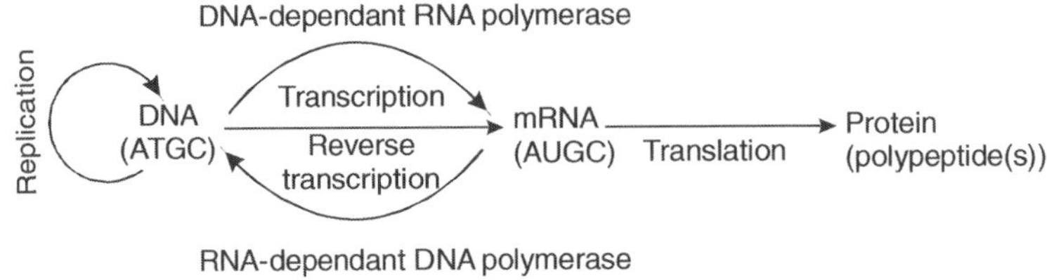

Figure 3.11   Modified central dogma showing reverse transcription

RIBONUCLEIC ACID (RNA)

CHEMICAL NATURE OF RNA

Ribonucleic acid (RNA) is a key intermediate between the genotype and its expression in the synthesis of protein. Chemically, the non-genetic RNA is a polynucleotide similar to DNA, but it differs from DNA in the following ways:

- RNA is usually single stranded, i.e., it has only one polynucleotide strand where adjacent ribonucleotides are linked by phosphodiester bonds. At some places the exposed nitrogen bases are complementary hence this single-stranded molecule may become folded to form complex shapes due to internal base pairing.

- Each ribonucleotide consists of ribose (a pentose sugar with the molecular formula $C_5H_{10}O_5$), rather than deoxyribose found in DNA, a nitrogen base and phosphoric acid.

- The nitrogen bases in RNA are purines (adenine and guanine) and pyrimidines (cytosine and uracil). The base uracil (U) is similar to thymine but lacks the methyl ($-CH_3$) group. The complementary base pairing rules are same as in DNA, except that adenine pairs with uracil.

- RNA molecules do not usually have complementary base ratios. The amount of adenine does not often equal the amount of uracil, and the amounts of guanine and cytosine also usually differ from each other.

Transcription—The Basic Process of DNA-directed RNA Synthesis

In eukaryotes, the most of the cell's DNA is confined to nucleus and during transcription there is flow of this genetic information from nucleus to the cytoplasm where the actual synthesis of protein takes place. To understand this mechanism, Crick and his colleagues proposed the **messenger hypothesis** suggesting that messenger RNA (mRNA) forms as a complementary copy of one strand of DNA of a particular gene. This mRNA molecule is then responsible for transfer of genetic information from nucleus to the cytoplasm of the cell and acts as a template for the synthesis of protein. Each gene consists of a definite number of nucleotides arranged in a specific order in DNA that codes for the particular protein is expressed as a sequence in mRNA. This hypothesis found correct since it has been tested for several times for genes that code for proteins. Similarly to understand the relationship between a specific nucleotide sequence in DNA and a specific amino acid sequence in a protein, Crick proposed the **adapter hypothesis,** indicating that there must be an adapter molecule that can bind a specific amino acid with one region and recognize the sequence of nucleotides with another region. Later on, these adapter molecules called transfer RNA (tRNA) were identified. They identify the codons on mRNA and accordingly carry specific amino acid to translate the coded language of DNA into the language of proteins. Thus, tRNA adapters line up themselves for the short span of time on mRNA and insert the amino acids in a proper sequence to elongate the polypeptide chain, constituting the process called translation. This hypothesis has been confirmed on the basis of actual observations of the expression of thousands of genes during which tRNA molecule acts as the intermediary between the sequence of nucleotides on mRNA and the sequence of amino acids in a protein.

Transcription is much simpler than translation since it involves the synthesis of one nucleic acid from the other nucleic acid. And it is accomplished by a single enzyme that works in conjunction with a variety of accessory proteins. The enzymes responsible for transcription in both prokaryotic and eukaryotic cells are called DNA-dependant RNA polymerases, or simply RNA polymerases. In addition to enzymes, the process requires a DNA template for complementary base pairing, the appropriate ribonucleoside triphosphates (ATP, GTP, CTP, and UTP) to act as substrate and the promoter as the site on the DNA to which RNA polymerase binds prior to initiate transcription. The synthesis of RNA from DNA takes place in three steps:

1. **Initiation** It is the step in which RNA polymerase binds the promoter region on DNA and   begins the synthesis of RNA.
2. **Elongation** In this step the RNA strand is extended using appropriate ribonucleotides and the majority of RNA synthesis takes place.

3.   **Termination**   During this step elongation of RNA, transcription ceases and nascent RNA (transcript) dissociate from the template.

Three Major Classes of RNAs

All three major types of RNAs, viz., messenger RNA (mRNA), transfer RNA (tRNA), and ribosomal RNA (rRNA) are derived from precursor RNA that are considerably longer than "mature" RNA product. The initial RNA molecule (called primary transcript or pre-RNA) copied from the template strand of DNA undergoes series of modifications to form a mature transcript that may no longer remain as an exact copy of the DNA. However, bacterial mRNAs are rarely processed. Usually in bacteria mRNA is transcribed on one end and its translation is initiated from other end before the transcription is completed. Eukaryotic pre-mRNA is also called heterogeneous nuclear RNA (hnRNA) because it represents a diverse nucleotide sequence with large molecular weight and it is found only in the nucleus of eukaryotic cell. RNA processing requires a variety of small RNAs (90 to 300 nucleotides long) and their associated proteins. These RNAs are called small nuclear RNAs (snRNAs) due to their small size and location in the nucleoplasm. There are more than a dozen different, low-molecular weight snRNAs that are rich in uracil residues.

Ribosomal RNA (rRNA)   It was explained in chapter 2, that the eukaryotic genes located in nucleolar organizer (rDNA genes) that encodes ribosomal RNA is a part of the moderately repeated fraction of the genome. Eukaryotic cells contain millions of ribosomes, each of which contains several molecules of rRNA associated with dozens of ribosomal proteins. Ribosomal RNA (rRNA), stable or insoluble RNA constitutes the largest part (up to 80%) of the total cellular RNA. To furnish the cell with such a large number of rRNA, the DNA sequences encoding rRNA called rDNA sequences are normally repeated hundreds of times. In interphase, the clusters of rDNA are gathered together as a part of one or more irregular-shaped nucleoli, which serve as ribosome-producing organelles. Nucleoli disappear during mitosis and then reappear in the nuclei of the daughter cells. The regions of a chromosome that contain rDNA are known as nucleolar organizers.

Eukaryotic cells consist of four types of rRNA molecules namely, 28S, 18S, 5.8S and 5S. The 28S, 5.8S and 5S rRNAs occur in 60S ribosomal (large) subunit, while 18S rRNA is present in 40S ribosomal (small) subunit. The large and small subunits together form a eukaryotic ribosome (80S). The prokaryotic cells consist of three kinds of rRNA molecules, viz., 23S, 16S, and 5S rRNA. The 23S and 5S rRNAs occur in 50S large ribosomal subunit and 16S rRNA occurs in 30S small ribosomal subunit, together they form 70S ribosome in prokaryotes. The most remarkable property of rRNA is a high degree of internal base pairing (the single-stranded RNA fold back on itself) to give an elaborate three-dimensional structure. There appears to be remarkable homologies in the general shape of analogous rRNAs from prokaryotes and eukaryotes. For example, the small-subunit rRNA from the prokaryotes, eukaryotic cytoplasm, and eukaryotic organelles have very different nucleotide sequences, yet their three-dimensional base-paired structures are quite similar (Figure 3.12).

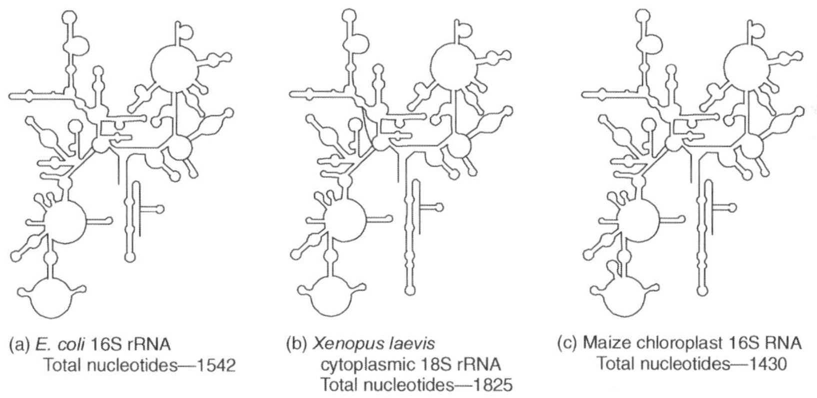

Figure 3.12 Three-dimensional structures of small-subunit rRNAs from (a) *E.coli,* (b) *Xenopus laevis,* and (c) Maize chloroplast

Transfer RNA (tRNA) These are small RNA molecules containing about 70 to 90 ribonucleotides, they are also known as soluble RNA (sRNA) playing significant role in transfer of amino acid from the "amino acid pool" to the site of protein synthesis. Transfer RNA (tRNA) or soluble RNA constitutes about 15% of the total cellular RNA. Eukaryotic cells are estimated to have approximately 60 different species of transfer RNA, each encoded by a DNA sequence that is repeated a number of times within the genome. The degree of repetition is species-specific, for example, there are about 300 tRNA genes in yeast cell, about 850 in fruit flies, and humans have about 1,300 genes. Genes responsible for tRNA synthesis are located within small clusters scattered around the genome. A single cluster typically contains multiple copies of different tRNA genes. According to Crick's adapter hypothesis, the translation of mRNA into proteins requires a molecule that links the information in mRNA codons with specific amino acids in proteins and that molecule is a tRNA. For each of the 20 amino acids, there is at least one specific type ("species") of tRNA molecule. Its molecular structure helps itself to relate three functions such as (i) it carries activated amino acid, (ii) it associates with mRNA molecules, and (iii) it interacts with ribosomes.

Robert Holley (1965) and his colleagues worked out the complete sequence of alanine tRNA molecule in yeast. Later on about 100 different "species" of tRNA are sequenced. All tRNA molecules have a generalized two-dimensional cloverleaf structure (Figure 3.13a) that has several unique properties including

i. presence of DNA-like double-helical segments due to internal pairing of complementary bases

ii. presence of number of unusual nucleotides like pseudouridine ψ or psi, inosine, dihydroxyuridine (DHU)

iii. presence of guanine residue at 5′-terminal end and unpaired CCA sequence at 3′ end that is called amino acid attachment site

iv. the amino acid stem that consists of seven paired bases

v. the T-stem or T-loop or TψC loop contains seven unpaired bases and five paired bases. This loop helps in binding of tRNA molecule to the ribosome

vi. the D-stem or DHU loop or simply D-loop is composed of three or four base pairs. It assists in binding of aminoacyl synthetase

vii. the anticodon stem or loop consists of seven unpaired bases, the 3rd, 4th and 5th of which from 3′ end of the molecule constitute the anticodon that is complementary to the codon of mRNA

viii. an extra small arm with variable number of nucleotides that may be absent in some tRNA molecules.

In 1973, S.H. Kim proposed the most acceptable three-dimensional structure of phenylalanine tRNA molecule found in yeast (Figure 3.13b). According to Kim, each tRNA molecule has a conformation in the shape of letter "L" with a thickness of 20 Å that is maintained by complementary base pairing within its own sequence. This conformation of tRNA molecule allows it to combine specifically with binding sites on the ribosomes. The L-shape model can be easily derived from two-dimensional cloverleaf model. At the 3′ end of each tRNA molecule is a site to which its specific, activated amino acid binds covalently. The anticodon loop consists of three nucleotides that are complementary to triplet of nucleotides on mRNA and specify the amino acid. For example, a codon on mRNA is AUG (coding for methionine), the corresponding anticodon on tRNA will be UAC. The codons and anticodons are antiparallel to each other.

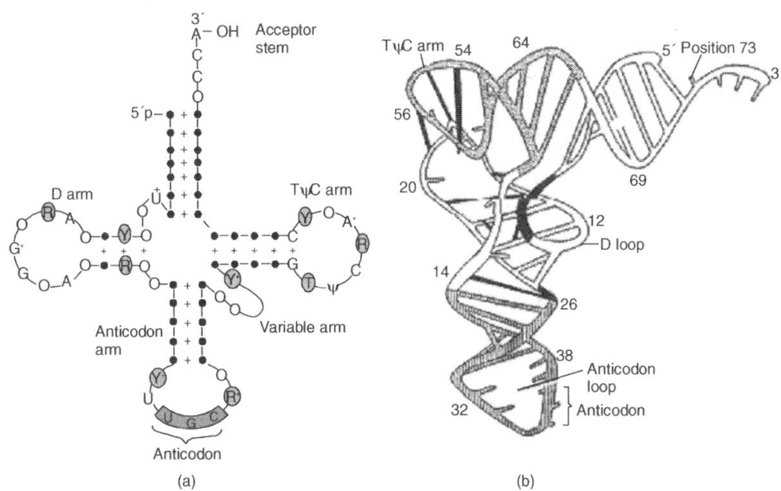

Figure 3.13 (a) Generalized two-dimensional cloverleaf structure of tRNA, (b) Diagrammatic representation of three-dimensional model of yeast tRNA^{Phe}

Messenger RNA (mRNA) The eukaryotic cell consists largely of rRNA, tRNA, and snRNA. In addition, there is a group represented by RNAs of diverse (heterogeneous) nucleotide sequence with molecular weights up to 80S, or 50,000 nucleotides that are found only in nucleus. These RNAs are referred to as heterogeneous nuclear RNAs (hnRNAs). Based on the experimental observations, the relationship between hnRNA in the nucleus and messenger RNA (mRNA) in the cytoplasm was interpreted. These studies revealed that the much larger hnRNA molecule is processed (post-transcriptionally modified) to form small mRNA molecule and even though mRNA and their hnRNA precursors constitute only a small percentage (about 5%) of total RNAs of most eukaryotic cell, they constitute a high percentage of RNA that is being synthesized by that cell at any given moment.

GENETIC CODE

How do transcription and translation produce specific and functional protein products? This was a major challenge faced by the molecular biologists. Once the structure of DNA had been described, the investigators directed their efforts to find out the relation between sequence of amino acids in a polypeptide and the sequence of nucleotides in the gene. It was widely assumed that the information for protein synthesis in the nucleotide sequence of the DNA was present as a type of genetic code. The coded message of DNA is called **cryptogram**. This genetic code specifies which amino acids will be used to build a protein. During transcription, the genetic information of DNA is handed over to the complementary mRNA molecule as a series of sequential, non-overlapping three-letter "words" called codons. The sequence of codons (triplets of nucleotides) on mRNA along the chain specifies the sequence of particular amino acid in a polypeptide chain. Our present knowledge of genetic code is based on the experimental outputs obtained by several investigators.

As per Gamow's model, there must be at least 3 (triplet) successive nucleotides for the specification of 20 different amino acids. Thus, there are 64 codons ($4^3 = 4 \times 4 \times 4 = 64$ combinations as shown in Table 3.2) possible that are more than the number of amino acids. Obviously then each amino acid will have more than one code indicating the degeneracy of codes. The triplet nature of the code was soon verified by number of genetic researchers. Francis Crick reasoning on how the genetic information encoded in 4-letter language of nucleic acids could be translated into the 20-letter language of proteins reveals that a small nucleic acid, tRNA act as an adaptor that "translates" the nucleotide sequence of mRNA into amino acid sequence of a polypeptide.

Table 3.2   The genetic code. Out of sixty-four possible codons, 61 codons specify amino acids, AUG acts as initiation codon and codes for methionine, while UAA, UAG, and UGA are non-sense codons that stop the process of protein synthesis. Single letter abbreviation for each amino acid is given in brackets.

Second letter in code

First letter in code	U	C	A	G	Third letter in code
U	UUU UUC ⌉ Phe (F) UUA UUG ⌉ Leu (L)	UUC UCC UCA UCG ⌉ Ser (S)	UAU UAC ⌉ Tyr (Y) UAA UAG ⌉ Termination codons	UGU UGC ⌉ Cys (C) UGA - Termination codon UGG - Trp (W)	U C A G
C	CUU CUC CUA CUG ⌉ Leu (L)	CCU CCC CCA CCG ⌉ Pro (P)	CAU CAC ⌉ His (H) CAA CAG ⌉ Gln (Q)	CGU CGC CGA CGG ⌉ Arg (R)	U C A G
A	AUU AUC AUA ⌉ Ileu (I) AUG - Met (M) Initiation codon	ACU ACC ACA ACG ⌉ Thr (T)	AAU AAC ⌉ Asn (N) AAA AAG ⌉ Lys (K)	AGU AGC ⌉ Ser (S) AGA AGG ⌉ Arg (R)	U C A G
G	GUU GUC GUA GUG ⌉ Val (V)	GCU GCC GCA GCG ⌉ Ala (A)	GAU GAC ⌉ Asp (D) GAA GAG ⌉ Glu (E)	GGU GGC GGA GGG ⌉ Gly (G)	U C A G

Mechanism of Translation

The RNA-directed protein biosynthesis is the most complex biochemical process in which the message of mRNA is decoded with the help of tRNA molecules that bring the activated amino acids to the site provided by ribosomes. Prokaryotic and eukaryotic ribosomes do not produce just one kind of protein. They can use any mRNA and all species are charged of tRNAs, and thus can be used to make many different polypeptide molecules.

1. *Activation of amino acids and formation of aminoacyl-tRNA* It is the initial step in protein biosynthesis in which amino acids are brought to the ribosome by their appropriate adapter tRNA molecules. In the cytoplasm, each of 20 amino acids are present in an inactive state and each of them before of its attachment to their specific tRNA is activated by a specific activating enzyme called aminoacyl-tRNA synthetase in the presence of ATP. Thus each amino acid has its appropriate synthetase enzyme, and made up of single or more polypeptides and all of them require Mg^{++} ions for their activity. Aminoacyl-tRNA synthetase catalyses the following two-step reaction:

$$AA + ATP \xrightleftharpoons{\text{Aminoacyl-tRNA synthetase}} AA{\sim}AMP + PPi$$

$$\text{Aminoacyl-adenylate complex} \qquad \text{Pyrophosphate}$$

$$AA{\sim}AMP + tRNA \xrightleftharpoons{\text{Aminoacyl-tRNA synthetase}} \text{Aminoacyl-tRNA} + AMP$$

It is critically important process during protein synthesis in which each tRNA molecule is attached to the "correct" (cognate) amino acid. Each amino acid covalently attaches to the 3′ end of its cognate tRNA in two-step reactions that are mediated by the enzyme aminoacyl-tRNA synthatase. As a result AMP (Adenosine monophosphate) and enzyme are released and a final product aminoacyl-tRNA is formed that moves towards the site of protein synthesis, i.e., ribosomes.

2. *Formation of initiation complex* It is of vital importance that a ribosome begins its translation of an mRNA exactly at the correct point. The mRNAs have 5′ and 3′ untranslated regions and coding regions lies in between. Initiation of protein synthesis involves the formation of complex between the small subunit of ribosomes, initiator tRNA and its cognate mRNA codon.

Translation of mRNA begins with the formation of a initiation complex, which consists of a charged tRNA bearing first amino acid of the polypeptide chain and a small ribosomal (30S) subunit, both bound to a complementary ribosome-recognition sequence on mRNA (Figure 3.14). This sequence of bases on mRNA is known as the Shine-Dalgarno sequence, which is complementary to a section of 16S rRNA present in the small ribosomal subunit. This sequence is "upstream" (towards the 5′end) of the actual start codon that begins translation. In the presence of initiation factors, mRNA, fMet-tRNA^{fMet}, and GTP, a complex of these with a 30S subunit is formed and IF3 is released. The charged tRNA, bearing anticodon UAC bringing first N-formyl methionine, is positioned at the P-site on the subunit with its anticodon paired with AUG codon. Once IF3 is released, the complex then associates with a 50S subunit. The event is accompanied by the hydrolysis of GTP and release of GDP and Pi and of IF1 and IF2. As a consequence, a complete 70S ribosome is positioned on mRNA in P-site with the donor fMet-tRNA^{fMet} anticodon base paired with initiation AUG codon and A-site vacant, awaiting for the second charged tRNA to deliver specific amino acid. This brings the completion of initiation process.

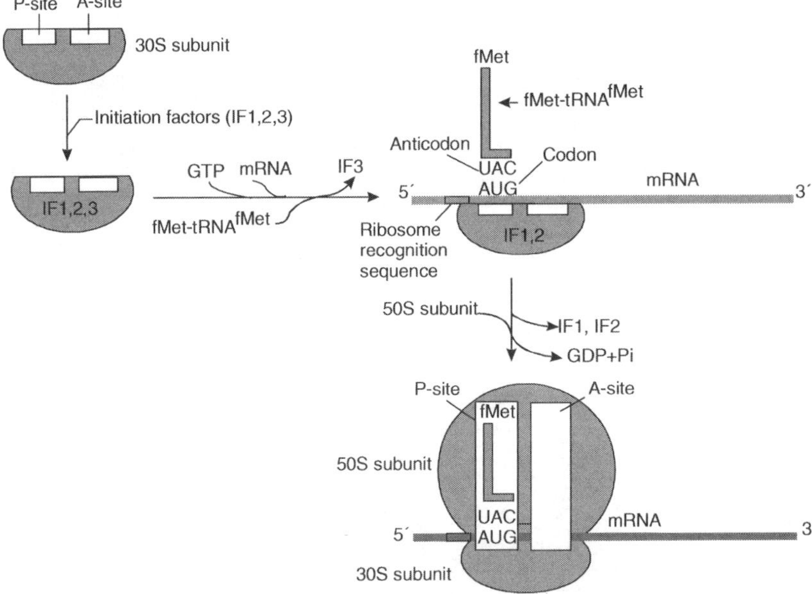

Figure 3.14   Steps in the formation of initiation complex during protein synthesis in prokaryotes

As compared to prokaryotes, in eukaryotes the process of protein synthesis follows the same general pattern so far its initiation complex is concerned except for the number of initiation factors involved. In eukaryotic cells (e.g. reticulocytes) about ten initiation factors have been identified and are referred to as eIF1, eIF2, eIF2B, eIF3, eIF4A, eIF4B, eIF4E, eIF4G, eIF5, and eIF6 (a prefix 'e' indicates the eukaryotic origin for each initiation factor).

3. *Mechanism of elongation of polypeptide*   Once the initiation complex is formed, the initiator fMet-tRNA^{fMet} occupies the P-site and A-site remains vacant (Figure 3.15a). Only in the initiation process does the P-site accept tRNA charged with an amino acid, in this case N-formylmethionine, while all subsequent aminoacyl-tRNAs enter the A-site. The second aminoacyl-tRNA is complexed with the elongation factor EF-Tu, carrying a molecule of GTP bound to it. The EF-Tu ensures the correct hydrogen bonding between codon of mRNA and anticodon of tRNA and helps in proper positioning of tRNA in A-site. This process requires energy that is provided by the hydrolysis of GTP. Once this activity is performed, the EF-Tu dissociates from the ribosome, and in the cytoplasm it is subsequently regenerated to its active form by another elongation factor, the EF-Ts. At this moment, both sites of ribosome are occupied by tRNAs, each of which carries amino acid. As shown in the Figure 3.15(b), the P-site is filled by fMet-tRNA^{fMet} and the A-site is occupied by Phe-tRNA^{Phe}.

The aminoacyl groups on the two tRNA molecules on the P-site and A-site are in the vicinity of ribosomal enzyme, peptidyl transferase, which transfers the first amino acid (N-formylmethionine) to the free amino (NH$_2$) group of the second amino acid (in this case it is phenylalanine). Thus, the first amino acid is placed on the top of the second amino acid by forming a peptide bond (–CO–NH–) between the carboxyl group of the first amino acid and the amino group of the second amino acid, resulting in a formation of a dipeptide. Now the A-site filled with dipeptide-carrying tRNA and P-site has discharged tRNA^{fMet} (Figure 3.15c).

Once dipeptide is formed, for continuous synthesis of polypeptide the next charged tRNA must enter the A-site of ribosome and to achieve this objective the ribosome moves one codon further along the mRNA. This process is called translocation of ribosome in which discharged tRNA^{fMet} is ejected from E-site, the third exit site on 50S subunit and later on this tRNA is to be re-used. The dipeptide-carrying tRNA is shifted from A-site to P-site constituting the transpeptidation reaction. The movement of the ribosome relative to mRNA requires an elongation factor called EF-G also called translocase and hydrolysis of GTP. As a result, the P-site is occupied by dipeptidyl-tRNA and A-site is vacated, which to be filled by third charged tRNA bringing the amino acid to elongate the polypeptide chain (Figure 3.15d).

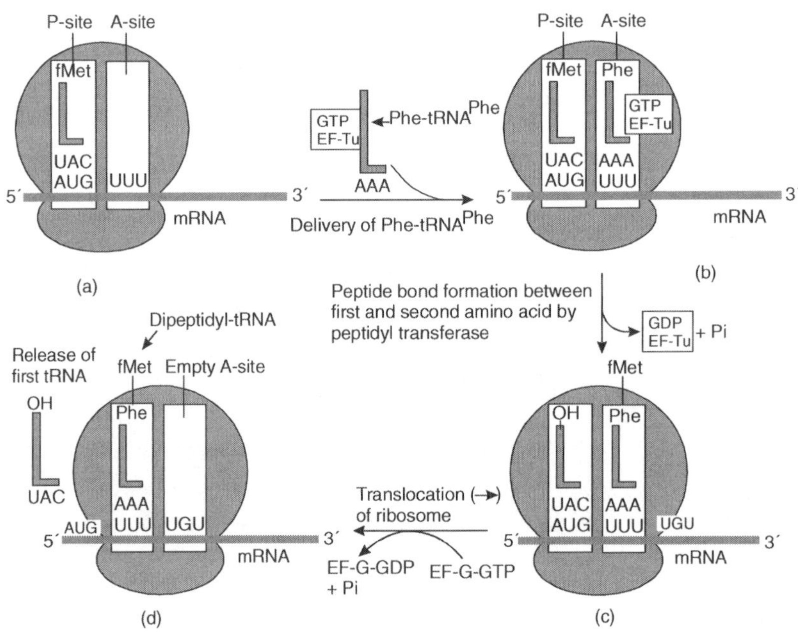

Figure 3.15   Stages in the process of elongation of polypeptide. (a) Binding of fMet-tRNA^{fMet} to P-site, (b) Entry of second charged tRNA at A-site, (c) Peptide bond formation between first and second amino acid, and (d) translocation of ribosome.

The elongation of polypeptide continues as ribosomes move exactly three nucleotides along mRNA molecule in 5′ to 3′ direction. The peptidyl-tRNA molecule in P-site transfers its burden of growing polypeptide to next succeeding tRNA, followed by the translocation of ribosome, release of discharged tRNA, and the filling of A-site by incoming new tRNA with activated amino acid and correct anticodon complementary to codon of mRNA. In prokaryotes, each cycle requires one twentieth of a second under optimal conditions. Thus an average size protein of 400 amino acids requires only 20 seconds for its synthesis.

4. *Termination of protein synthesis*   Polypeptide chain elongates as ribosome moves codon-by-codon on mRNA and it continues until it encounters any one of the three termination or stop codons for which no

tRNA with anticodons complementary to these codons exists; these are UAG, UAA, and UGA. When the ribosome reaches one of the stop codons, specific cytoplasmic protein-releasing factors (RFs), bind to A-site on ribosome. The RF1 identifies termination codon UAA and UAG, while the RF2 recognizes codon UGA, whereas RF3 seems to promote the action of RF1 and RF2. Overall, releasing factors cause release of the nascent polypeptide from tRNA by altering the peptidyl transferase such that it hydrolyses the ester bond between the –COOH of the protein and the –OH of the 3′ terminal nucleotide of tRNA. Finally ribosome detaches from mRNA and dissociates into subunits that are reused for the next round of initiation.

Post-translational modification of protein   Many proteins are covalently modified, either before their detachment from ribosome or after their synthesis has been completed. Since most of the modifications in polypeptide takes place after the translation, they are referred to as post-translation modifications that mainly include trimming or proteolysis, phosphorylational (addition of phosphate groups to proteins in presence of protein kinases), glycosylation (addition of "sugar coating" to proteins by the enzymes present in ER and the Golgi apparatus), and hydroxylation (Figure 3.16).

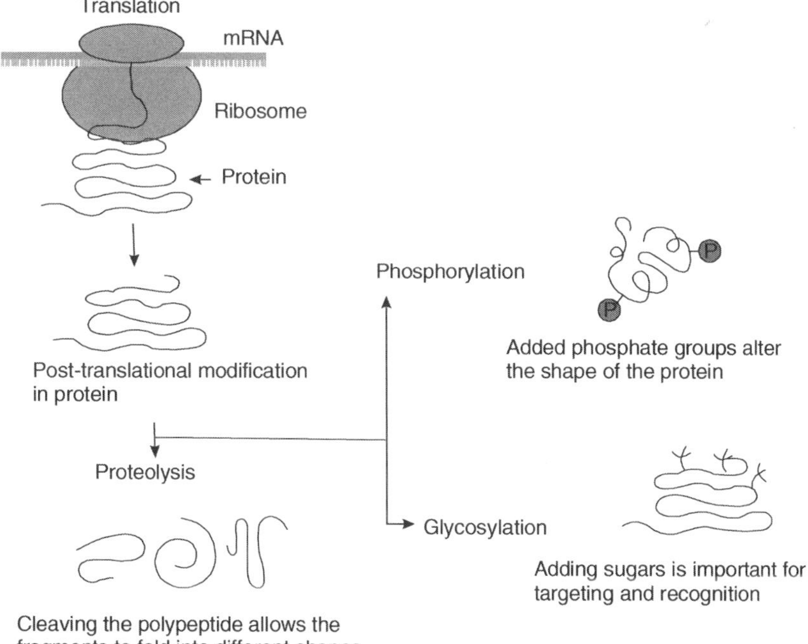

Figure 3.16   Various methods of post-translational modifications in protein

A functional protein is not necessarily the same as the polypeptide chain that is released from the ribosome. Especially in eukaryotic cells, the polypeptide may need to be moved far from the site of synthesis in the cytoplasm, destined to an organelle, or even secreted from the cells. In addition, the polypeptide is often modified by the addition of new chemical groups that have functional significance.

PROTEINS

Proteins form the part and partial of all biological systems. They are the macromolecules that carry out virtually all of a cell's activities. It is estimated that the typical mammalian cell may have as many as 10,000 different proteins having diverse array of functions. As enzymes, proteins accelerate the rate of metabolic reactions; as structural components, proteins provide mechanical support; as hormones, growth factors, and gene activators, proteins perform a wide variety of regulatory functions; as membrane receptors and transporters, proteins determine what a cell reacts to and what type of substances enter or leave the cell; as contractile elements, proteins form the machinery for biological movements; as antibodies, proteins play role in body's defense in addition to their involvement in blood clotting and in absorbance and reflection of light. Proteins could perform such versatile functions only because of their unique structures. All biologically relevant proteins are linear (except for cystines) polypeptides, which are formed by repertoire of twenty amino acids.

AMINO ACIDS

These are building blocks of proteins and enzymes. There are twenty L-amino acids and they all have (except proline, which is an imino acid) an amino group (NH_2), a carboxylic group (COOH), a hydrogen atom and a side chain (–R) all covalently bonded to the central tetrahedral α-carbon atom (Figure 3.17).

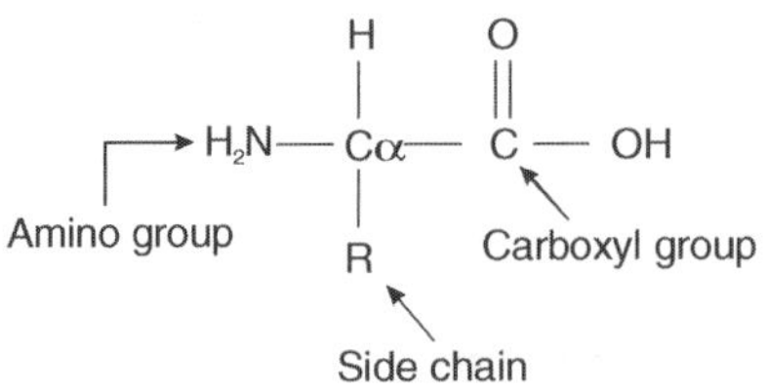

Figure 3.17   General formula of L-amino acid

Amino acids differ from each other structurally and functionally only by the side chain R group. The chemical reactivities of the R groups determine the specific properties of the amino acids. Ionization states of amino acids are pH-dependant (Figure 3.18). They exist in zwitter ionic form (amino acid with net charge zero) under the physiological pH (approximately 7) condition. Amino acids can be grouped into several categories based on the chemical and physical properties of the side chains, such as size and affinity for water. According to these properties, the side chain groups can be divided into small, large, hydrophobic, and hydrophilic categories. Within the hydrophobic set of amino acids, they can be further divided into aliphatic and aromatic. Aliphatic side chains are linear hydrocarbon chains and aromatic side chains are cyclic rings. Within the hydrophilic set, amino acids can be subdivided into polar and charged. Charged amino acids can be either positively charged (basic) or negatively charged (acidic). Each of twenty amino acids, their chemical nature and abbreviations with one letter code are given in Figure 3.19.

$$\underset{R}{\underset{|}{H_3\overset{+}{N}-\overset{\overset{\displaystyle COOH}{|}}{C}-H}} \;\underset{H^+}{\rightleftharpoons}\; \underset{R}{\underset{|}{H_3\overset{+}{N}-\overset{\overset{\displaystyle COO^-}{|}}{C}-H}} \;\underset{H^+}{\rightleftharpoons}\; \underset{R}{\underset{|}{H_2N-\overset{\overset{\displaystyle COO^-}{|}}{C}-H}}$$

	Under acidic condition	Under neutral pH (zwitter ion)	Under basic condition
Net charge:	+1	0	−1

Figure 3.18   Ionization states of amino acid at different pH

From the chemical nature of the side chains (–R) under physiological conditions, amino acids are mainly classified as:

1. Seven amino acids with nonpolar and aliphatic R group: glycine, alanine, proline, valine, leucine, isoleucine, and methionine.
2. Three amino acids with aromatic R group: phenylalanine, tyrosine, and tryptophan.
3. Five amino acids with polar and uncharged R group: serine, threonine, cysteine, asparagine, and glutamine.
4. Three amino acids with positively charged R group: lysine, arginine, and histidine.
5. Two amino acids with negatively charged R group: aspartate and glutamate.

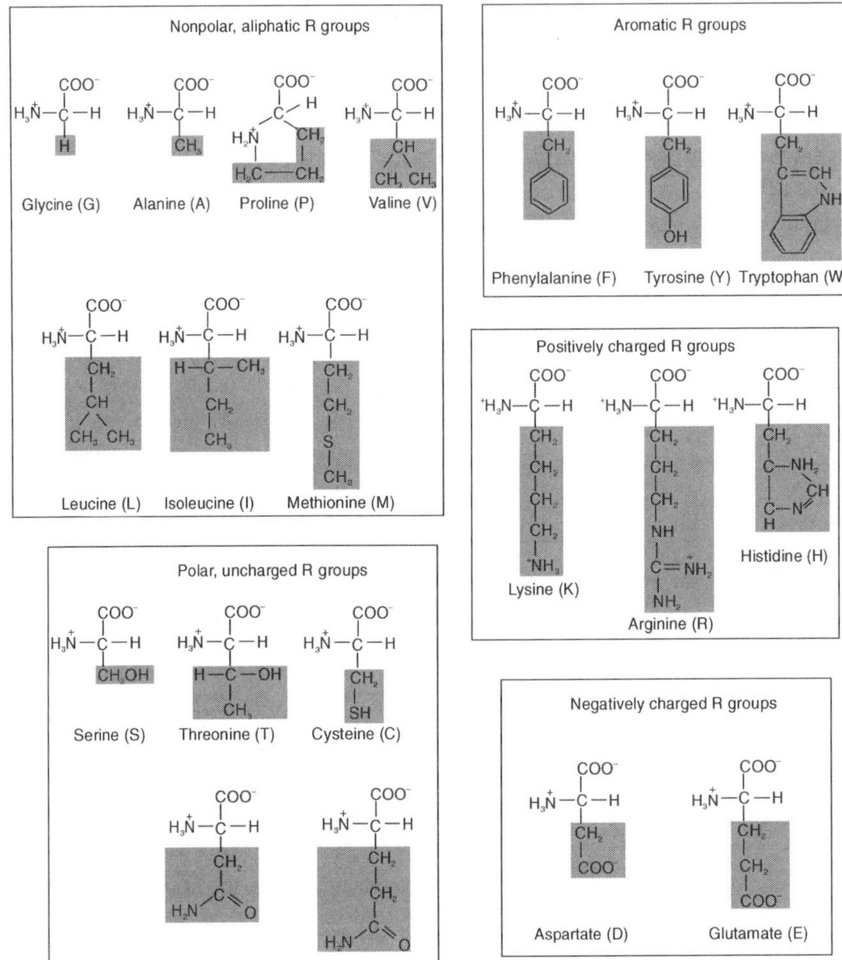

Figure 3.19   Amino acids with their side chains (–R) and their characteristics

Of particular interest within the twenty amino acids are glycine and proline. Glycine, the smallest amino acid, has a hydrogen atom as their R group. It can therefore adopt more flexible conformations that are not possible for other amino acids. Proline is on the extreme of flexibility. Its side chain forms a bond with its own backbone amino group, causing it to be cyclic. The cyclic conformation makes it very rigid, unable to occupy many of the main chain conformations adopted by other amino acids. In addition, certain amino acids are subjected to modifications after a protein is translated in a cell. This called post-translational modification as described earlier.

PEPTIDE BONDS IN POLYPEPTIDE CHAIN

A polypeptide chain is a long, continuous, and unbranched polymer of amino acids. The amino acids that comprise a polypeptide chain are joined by peptide bonds that result from covalent linkage of carboxyl group of one amino acid to the amino group of its neighbour, with the elimination of water molecule (Figure 3.20).

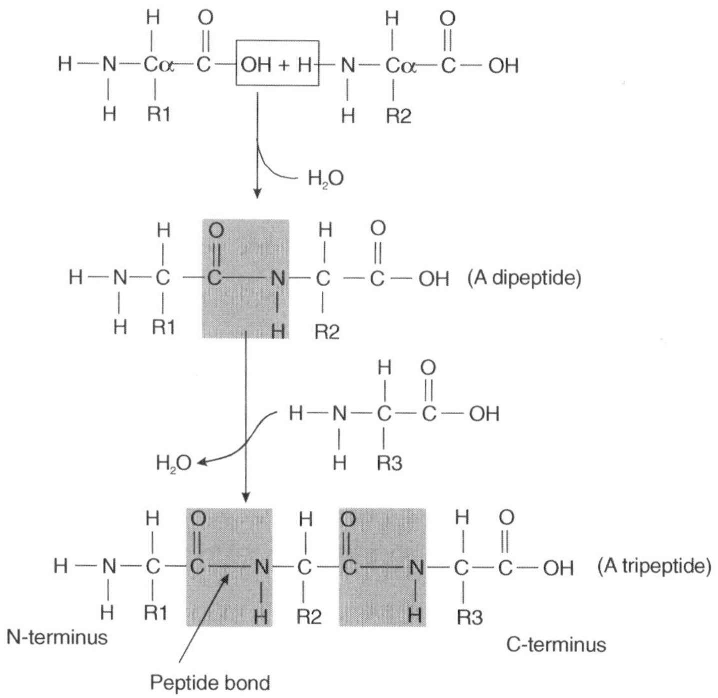

Figure 3.20 Formation of polypeptide chain from amino acids linked by peptide (CO–NH) bond

Dihedral Angles

Once incorporated into a polypeptide chain, amino acids are referred to as residues. The residue on one end of the chain, the N-terminus, contains an amino acid with a free (unbounded) α-amino group, while the residue at the opposite end, the C-terminus, has a free α-carboxyl group. The actual sequence of amino acid residues in a polypeptide determines its ultimate structure and function. Based on the series of studies, Linus Pauling and Robert Corey concluded that the peptide bonds are unable to rotate freely because of their partial double bond character. Because of the planar nature of the peptide bond and the size of the R groups, there are considerable restrictions on the rotational freedom by the two bonded pairs of atoms around the peptide bond. The angle of rotation about the bond is called **dihedral angle**. The rotation is permitted about the N–Cα and the Cα–C bonds. The backbone of a polypeptide chain can thus be pictured as a series of rigid planes with consecutive planes sharing a common point of rotation at Cα. By convention, the bond angles resulting from rotations at Cα are called φ (phi) for the N–Cα bond and ψ (psi) for the Cα–C bond. Various combinations of φ and ψ angles allow the protein to fold in many different ways.

Ramachandran Plot

Both φ and ψ are defined as 180° when the polypeptide is in its fully extended conformation and all peptide groups are in the same plane (Figure 3.21). In principle, phi and psi can have any value in between −180° to +180°, but many values are prohibited by steric interference between atoms in the polypeptide backbone and amino acid side chains. When φ and ψ angles of amino acids of a particular protein are graphically plotted against each other, the resulting diagram is called a Ramachandran plot, introduced by G.N. Ramachandran.

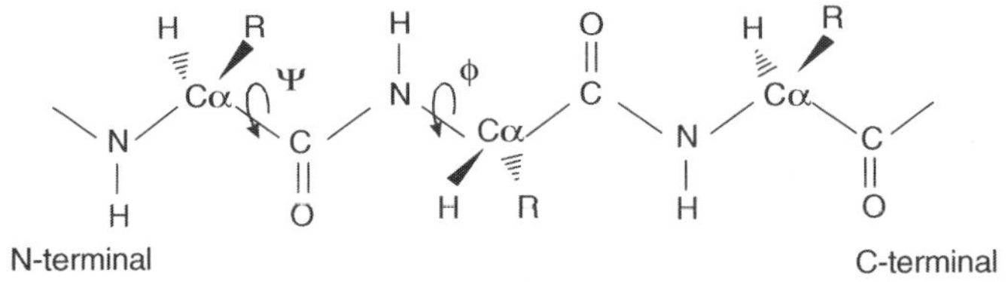

Figure 3.21　The angle of rotations in a polypeptide chain. The ψ angle for the Cα-C bond and the φ angle is the rotation about N-Cα bond.

Ramachandran plot provides the information about the entire conformational space of a peptide and shows sterically allowed and disallowed regions (Figure 3.22). It can be very useful in evaluating the quality of protein model.

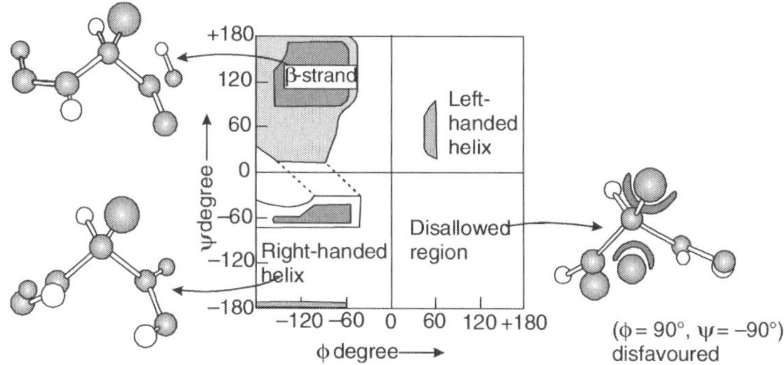

Figure 3.22 Ramachandran plot showing allowed and disallowed values of ψ and φ angles. The dark-shaded areas reflect conformations that involve no steric overlap and thus are fully allowed. The structure on the right is disallowed due to steric clashes.

THE STRUCTURE OF PROTEINS

Proteins are huge and complex molecules but their structure in any given environment is completely defined and predictable. Each amino acid residue in a polypeptide has a specific location, giving protein the structure and reactivity required to perform a specific function. Thus, every protein has three-dimensional structure that reflects its function. Protein structure can be described at several levels of organization depending on different type of interactions. The structure of protein can be organized into four levels of hierarchies with increasing complexity. These levels are primary structure, secondary structure, tertiary structure, and quaternary structure. The first, primary structure, is concerned with the amino acid sequence of a protein, while the latter three levels are concerned the organization of molecule in space (Table 3.3).

Table 3.3 Levels of protein structure

Protein structure	Description
Primary structure	The linear sequence of amino acids in a protein
Secondary structure	Local structures of a protein with regular conformation, folded either in α-helices or β-strands
Tertiary structure	The overall fold of a protein sequence, formed by the packing of its secondary structure elements
Quaternary structure	The arrangement of separate protein chains in a protein molecule with more than one subunit

Primary Structure

The primary structure of a polypeptide is the number and specific sequence of amino acids that constitute the chain. This is the simplest level with amino acid residues linked together through peptide bonds that are rigid and planar covalent bonds. Most natural polypeptide chains contain between 50 to 2000 amino acid residues and are commonly referred to as proteins. The information for the precise order of amino acids in every protein that an organism can produce is contained within the genetic inheritance of that organism. The amino acid sequence provides the information necessary to determine the molecule's three-dimensional shape and thus its function. Thus the sequence of amino acids in a polypeptide chain is very important for the molecule to perform its assigned function and any change in the sequence as a result of mutation in DNA leads to disorder. The earlier and best-studied example of this relationship is the change in the amino acid sequence of haemoglobin that causes the disease sickle-cell anaemia. This severe, inherited disease results solely from a change in single amino acid sequence within the molecule. The amino acid at 6th position in the beta chain of normal haemoglobin is glutamic acid if replaced by valine; it leads into a molecular disease in which the shape of red blood cell becomes sickle-shaped (Figure 3.23) and there is decrease in oxygen carrying capacity of RBCs in affected person.

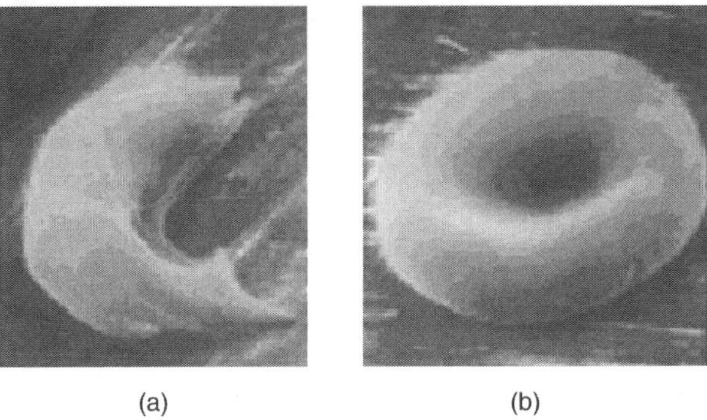

(a)   (b)

Figure 3.23   (a) Sickled red blood cells (b) Normal red blood cells

In 1953, F. Sanger determined the amino acid sequence of the protein hormone insulin that is secreted by α-cells of islets of Langerhans in pancreas (Figure 3.24). This extraordinary work showed for the first time that a protein has a precisely defined amino acid sequence. Moreover, it demonstrated that insulin consists of only L-amino acids linked by peptide bonds between α-amino and α-carboxyl groups.

In the late 1950s and early 1960s, a series of incisive studies revealed that amino acid sequences of protein are genetically determined. Knowing amino acid sequences carries significance for several reasons. First, this knowledge is essential for elucidating the mechanism of action of a particular protein. Second, amino acid sequences determine the three-dimensional structures of proteins with its specific folding pattern. Third, it helps in understanding the molecular pathology, a rapidly flourishing area of medicine. As mentioned earlier, alteration in amino acid sequences leads to fatal molecular diseases like sickle-cell anaemia and cystic fibrosis. Fourth, the amino acid sequence in a protein helps in understanding of its evolutionary history. Proteins resemble one another in amino acid sequence, only if they have a common ancestor. Ultimately it plays an indispensable role in tracing out the phylogenetic relationship among the organisms.

Secondary Structure

Every matter that exists in space has a three-dimensional arrangement or conformation, so far the spatial organization of atoms of molecule of that matter is concerned. Secondary structure of protein is an ordered segment in the form of helices or sheet of polypeptides that are hydrogen-bonded. In 1951, Linus Pauling and Robert Corey of the California Institute of Technology studied secondary structure of protein and concluded that polypeptide chains exist in preferred conformations that provide possible number of hydrogen bonds between neighbouring amino acids. Pauling and Corey postulated two ordered structures that occur in polypeptides, namely the α-helix and β-pleated sheet. Subsequently, other structures such as the β-turn and omega (Ω) loop were identified.

Alpha [α] helix It is a common secondary structure found in proteins of globular class in which the backbone of the polypeptide assumed the form of a cylindrical, twisting spiral (Figure 3.25 a & b). A tightly coiled backbone forms the inner part of the rod and the side chains extend outward in a helical array. The α-helix is stabilized by hydrogen bonds between the carbonyl oxygen (CO) and the amino hydrogen (NH) of two amino acid residues. In particular, the CO group of each amino acid (i) forms a hydrogen bond with the NH group of the amino acid that is situated four residues ($i + 4$) ahead in the sequence. Thus, except for amino acids near the ends of α-helix, all main chain CO and NH groups are hydrogen-bonded. Each residue is related to the next one by a rise of 1.5 Å along the helix axis and rotation of 100 degrees, which gives 3.6 amino acid residues per turn of helix. The pitch of the α-helix, which is equal to the product of the translation (1.5 Å) and the number of residues per turn (3.6), is 5.6 Å. The screw sense

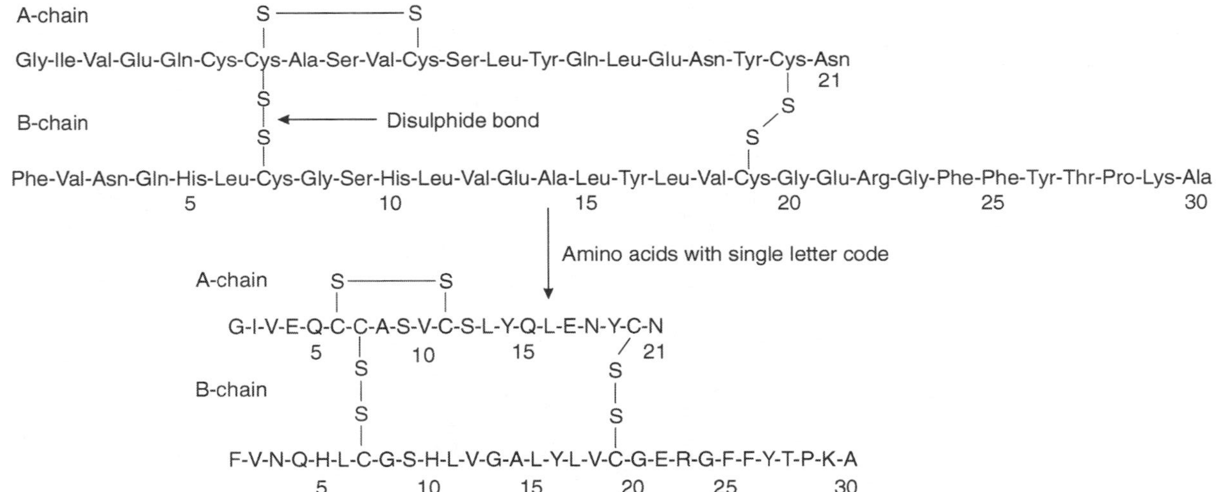

Figure 3.24   Amino acid sequence in the hormone insulin with two polypeptide chains linked by disulphide bonds

of a helix can be right-handed (clockwise) or left-handed (anticlockwise). The Ramachandran plot (Figure 3.22) reveals that both the right-handed and the left-handed helices are allowed among conformations. However, right-handed helices are energetically more favourable because there is less steric clash between the side chains and the backbone. Essentially, all α-helices found in proteins are right-handed.

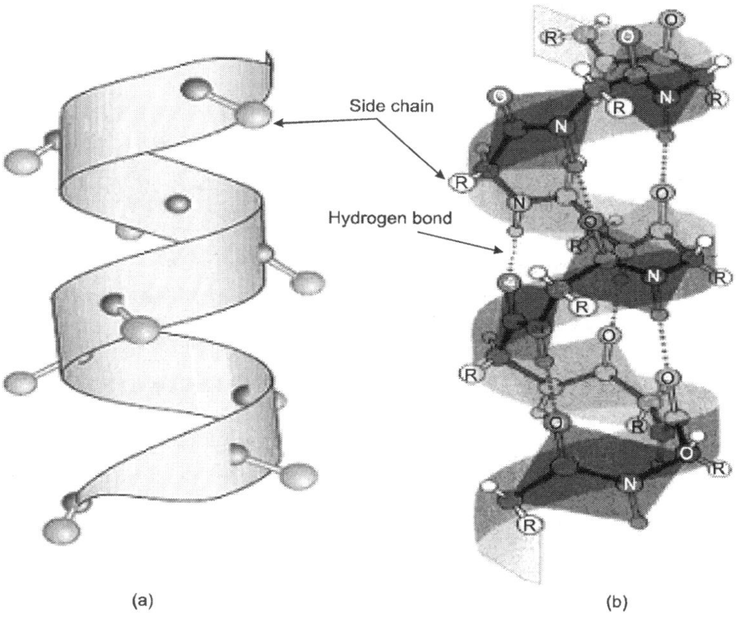

(a)   (b)

Figure 3.25 (a) The ribbon diagram of α-helix with main chain α-carbon atoms (small balls) and side chains (big balls). (b) Hydrogen bonds between the carbonyl oxygen and the amino hydrogen of two amino acid residues showed in dotted lines

The α-helical contents of proteins range widely, from nearly none to almost 100%. For example, about 75% of the residues in ferritin, a protein that helps to store iron, are in α-helices (Figure 3.26a). Single α-helices are usually less than 45 Å long. However, two or more α-helices can twist together to form a very stable structure, which can have a length of 1000 Å or more. Such α-helical-coiled coils are found in myosin and tropomyosin in muscle (Figure 3.26b), in fibrin clots, and in keratin in hair. The helical cables in these proteins serve a mechanical role in forming stiff bundles of fibres. The cytoskeleton of cell is rich in so called intermediate filaments, which also are two-stranded α-helical coiled coils.

Beta (β) pleated sheet   It was the second structural motif discovered by Pauling and Corey in which several polypeptides lying parallel to one another. Unlike the coiled, cylindrical form of α-helix, the backbone of each polypeptide (or β-strand) in a β-sheet assumes a folded or pleated conformation (Figure 3.27a). In β-sheet, the main chain is zigzag and the side chains are positioned alternately on opposite sides of the sheet. Like the α-helix, the β-sheet is also stabilized by a large number of hydrogen bonds, but these noncovalent bonds are perpendicular to the long axis of the polypeptide chain and project across from one part of the chain to another (Figure 3.27b).

The β-strands can run in the same direction to form a parallel sheet or can run every other chain in reverse orientation to form an antiparallel sheet, or a mixture of both. Like the α-helix, the β-sheet has also found in many different proteins in the form of structural elements. Silk is a protein consisting of largely of β-sheets. Fatty acid-binding proteins important for lipid metabolism are also built almost entirely from β-sheets (Figure 3.28).

Beta (β) turn and omega (Ω) loop   Most polypeptides fold to form compact, globular proteins and those portions of a polypeptide chain not organized into an α-helix or a β-sheet may consist of hinges, turns, loops, or fingerlike extensions. Often, these are the most flexible portions of a polypeptide chain and sites of greatest biological activities. The reverse turn or β-turn or hairpin bend is a common structural element formed due to reversal in the direction of their polypeptide chains (Figure 3.29). In such cases, the CO group of residue i of a polypeptide is hydrogen bonded to the NH group of residue $i + 3$. In other cases, more elaborate structures are responsible for chain reversals. These structures are called loops or sometimes omega (Ω) loops to suggest their overall shape. Unlike α-helices or β-strands, loops do not have regular, periodic structures. Nonetheless, loop structures are often rigid and well-defined.

The so-called turn structures are also classified as secondary structural elements, but unlike helices and strands, they do not have a repeating, regular geometry. Rather, they can have well-defined spatial dispositions defined by certain values of ϕ and ψ angles that often require specific residue types and/or sequences, as well as fixed hydrogen bonding patterns. Most turns are localized in the primary structure, but omega loops can have a large number of intervening residues lacking defined geometries, with the turn being defined by the conformations of residues that form the constriction that gives this turn its name (Ω).

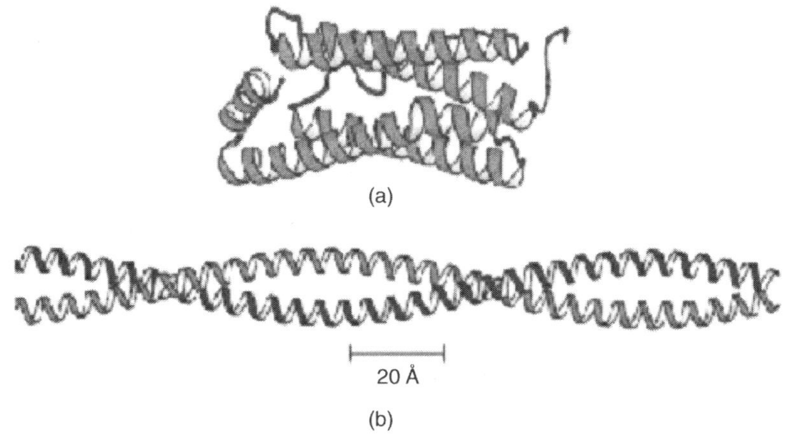

Figure 3.26 (a) Ferritin—A largely α-helical protein. (b) Tropomyosin— an α-helical coiled coil formed by two helices wrapped around one another as superhelix.

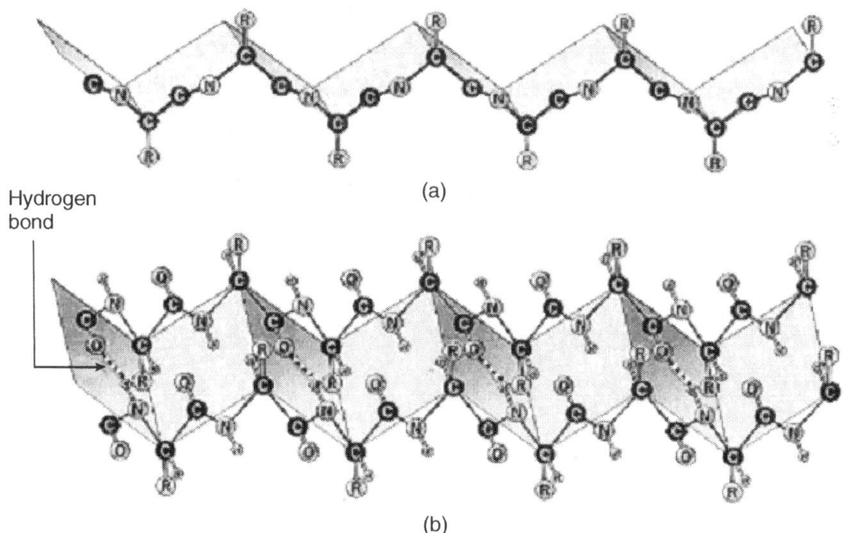

Figure 3.27 (a) Extended β-strand showing successive R-groups projecting upward and downward from the backbone of a polypeptide. (b) A β-pleated sheet consists of a number of α-strands that lie parallel to one other and are joined together by hydrogen bonds between carbonyl and imine groups of neighbouring backbones.

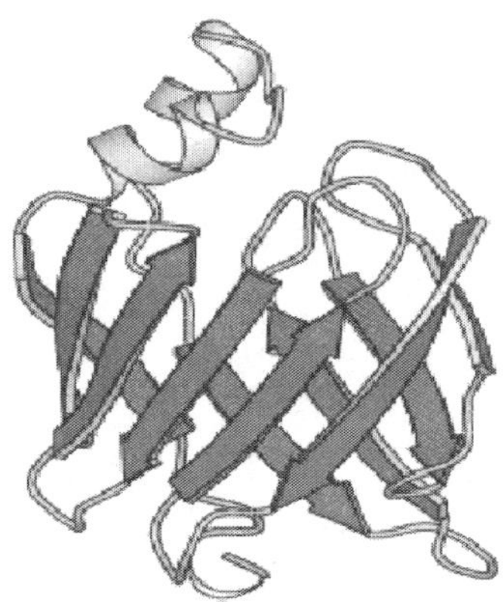

Figure 3.28   Fatty acid-binding protein rich in β-sheets

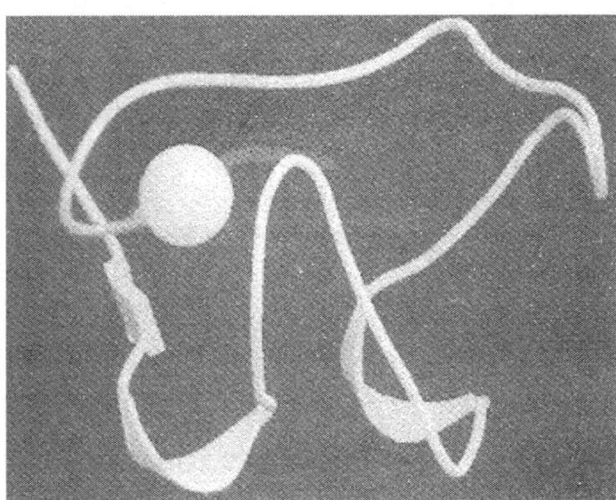

Figure 3.29   Structure of a protein (Rubredoxin Mutant G10A) rich in turns and loops

Turns are essential for allowing the polypeptide chain to fold back upon itself to form tertiary interactions. Such interactions are generally long-range, and result in compaction of the protein into a globular, often approximately spherical, form. The turn regions are thus generally located on the outside of the globular structure, with helices and/or sheets forming its core. Turns on the surfaces of proteins have a wide range of dynamics, from quite mobile in cases where they form few interactions with the underlying protein surface to quite fixed due to extensive tertiary contacts. Thus, turns are also ambiguously classified as secondary structure elements.

Supersecondary structures These are also called **motifs** or simply **folds**, are orderly organization of secondary structure. Motif is a simple combination of a few secondary structural elements with a specific geometric arrangement. Recognized motifs range from simple to complex, sometimes appearing in repeating units or combinations. A single, large motif may comprise the entire protein. Some motifs are associated with a particular biological function while others are not but rather are part of a larger structural assembly. Polypeptides with more than a few hundred amino acid residues often fold into two or more stable, globular units called **domains**. In many cases, a domain from a large protein will retain its correct three-dimensional structure even when it is separated by proteolytic cleavage from the remainder of the polypeptide chain. Folding of polypeptides is subject to an array of physical and chemical constraints. A sampling of the prominent folding rules that have emerged provides an opportunity to introduce some simple motifs as follows:

1. The β-α-β structural motif (e.g. Rossmann fold) consists of parallel β-strands at the centre pointing in different directions like arrows with α-helix wound around where hydrophobic interactions provide the stability to the protein (Figure 3.30a).

2. The α-α structural motif is the supersecondary structural motif formed by two α-helices (Figure 3.30b).

3. In β-β motifs the length of loop is generally two to five residues and connections between β-sheets cannot cross or form knots (Figure 3.30c). The connections between β-sheets are generally right-handed while left-handed are very rare (Figure 3.30d).

4. There can be a characteristic twisting of the secondary structure when many β-sheets are put together. The β-barrel and twisted β-sheet (Figure 3.30e) are two such examples that form the core of many larger structures.

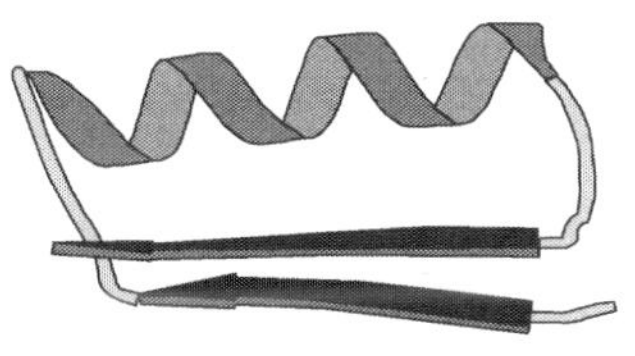

Figure 3.30 Different forms of supersecondary structures (*Continues*)

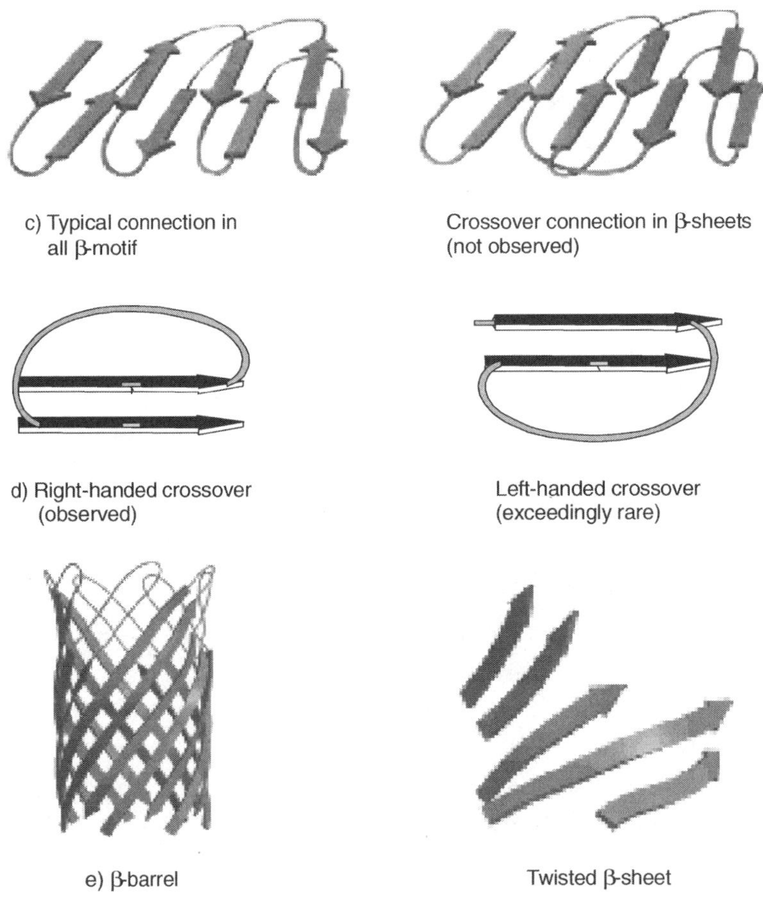

Figure 3.30   Different forms of supersecondary structures

Tertiary Structure

The overall three-dimensional arrangement of all atoms in a protein is referred to as protein's tertiary structure. In other words, it is the overall fold (spatial folding) of a protein sequence, formed by the packing of its secondary structure elements. Whereas secondary structure deals primarily with the conformation of adjacent amino acids in the polypeptide chain, tertiary structure describes the conformation of entire protein and it is unique and specific for each protein. The structural and functional features of proteins that include binding site of ligand (protein–drug interactions), the active sites of enzymes, or the binding sites for other proteins (protein–protein associations), etc. depend on their tertiary structures and hence knowledge of three-dimensional structure of a protein is very much essential to understand its structural and biochemical functions.

The first breakthrough in understanding the three-dimensional structure of a globular protein came from X-ray diffraction studies of myoglobin carried out by John Kendrew and his colleagues at Cambridge University in the 1950s. Myoglobin is a relatively small oxygen binding protein (with a molecular weight 16,700 D) present in muscle cells. It stores oxygen and facilitates oxygen diffusion in rapidly contracting muscle tissue. Myoglobin contains a single polypeptide chain of 153 amino acid residues of known sequence and a single iron protoporphyrin, or haem, as a nonpolypeptide prosthetic group. Myoglobin is particularly abundant in the muscles of diving mammals such as whales, seals, and porpoise. Storage and distribution of oxygen by muscle myoglobin permits these animals to remain submerged for long periods of time.

Myoglobin is an extremely compact molecule with the overall dimensions of $45 \times 35 \times 25$ Å, an order of magnitude less than if it were fully stretched out (Figure 3.31). Myoglobin provided early clues about the complexity of globular protein structure. The structural organization of another monomeric proteins (that are made from single polypeptide chain) such as trypsin, insulin, oxytocin, growth hormone, is complete at tertiary level.

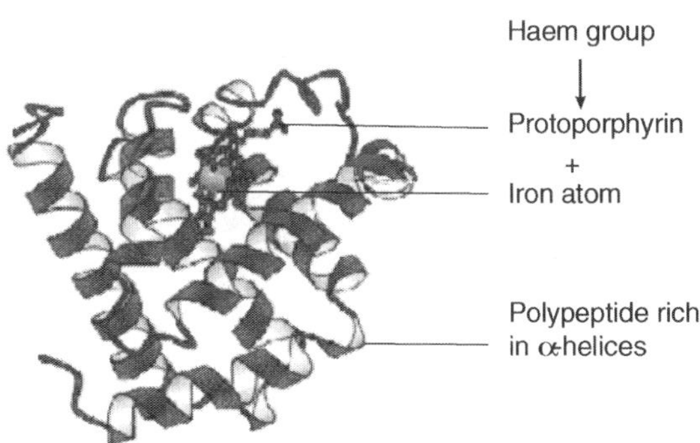

Figure 3.31   Tertiary structure of myoglobin

Quaternary Structure

The proteins that contain two or more polypeptide chains (oligomeric proteins), such as haemoglobin, cytochrome oxidase, ATPase, exhibit a fourth level of structural organization. Each polypeptide chain in such protein is a subunit. Quaternary structure refers to the spatial arrangement of subunits in a protein molecule and the nature of their interactions. Depending on the protein, the polypeptide chains may be identical or non-identical. A protein composed of two identical subunits is referred to as homodimer for example, the DNA-binding protein Cro found in bacterial virus called lambda (λ). The ribbon model of Cro protein that exhibits quaternary structure is depicted in Figure 3.32.

Figure 3.32   Cro protein present in bacteriophage λ is a dimer of two identical subunits

In contrast to homodimer, a protein composed of two non-identical subunits is a heterodimer. The protein containing more than one type of subunit shows more complicated quaternary structure. A multisubunit protein is also known as multimer. Multimeric proteins can have from two to hundreds of subunits. A multimer with just a few subunits is often called an oligomer. Human haemoglobin (Mol.wt. 64,500 D) was the first oligomeric protein for which the three-dimensional structure was determined. Haemoglobin is the oxygen-carrying protein present in RBCs, which consists of four polypeptide chains and four haem prosthetic groups containing the iron atoms are in ferrous (Fe^{2+}) state (Figure 3.33).

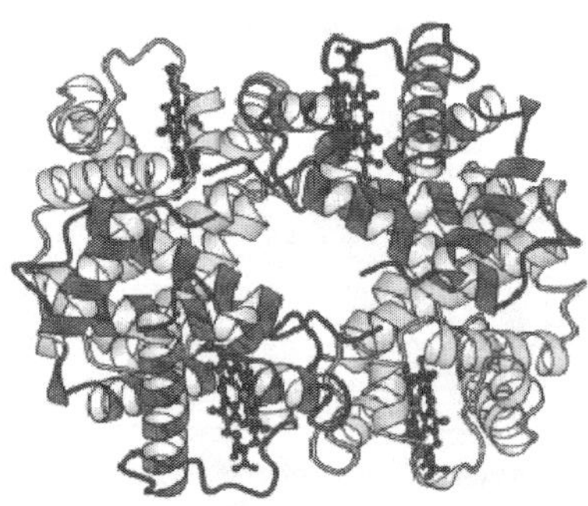

Figure 3.33   Haemoglobin—the tetramer with quaternary structure

The protein portion, called globin, consists of two subunits of one type (designated α-chain with 141 residues each) and two subunits of another type (designated β-chain with 146 residues each). Since haemoglobin is four times larger than myoglobin, to solve its three-dimensional structure by X-ray analysis took much more time and efforts. In 1959, Max Perutz, John Kendrew, and their colleagues finally achieved the task. The subunits of haemoglobin molecule are arranged in symmetric pairs, each having one α and one β subunit. Haemoglobin can therefore described as an $\alpha_2\beta_2$-tetramer. Even though haemoglobin consists of four subunits, it is still considered a protein with a single function.

REVIEW QUESTIONS

1. Explain the chemical nature of DNA.
2. What are the structural features of Watson–Crick model of DNA double helix?
3. Distinguish between nucleoside and nucleotide.
4. Give a brief account of alternative forms of DNA.
5. What is replication of DNA? Why is replication semiconservative?
6. How would you comment on semidiscontinuous replication of DNA?
7. What is replicating fork? Explain the role of various DNA polymerases in replication of DNA.
8. Explain the concept of central dogma of molecular biology. How does it modify after invention of RNA viruses?
9. Describe the chemical nature of RNA.
10. What are the different types of RNA? Explain the structural features of each type.
11. What is genetic code? Explain the role of genetic code in protein synthesis.
12. Comment on the process of transcription with the various factors involved in it.
13. What is translation? Describe its mechanism in brief.
14. How would you comment on post-translational modification of protein?
15. What are amino acids? Explain different classes of amino acids.
16. Explain the concept of dihedral angles.
17. What is Ramachandran plot? How does it help in understanding the structural features of a protein?
18. Explain the primary structure of protein with its significance.

19.   What are the various types of secondary structure of protein?

20.   Comment on the tertiary structure of protein with example.

21.   How would you comment on the quaternary structure of protein?

22.   Write short notes on:
    i.   X-ray crystallography of DNA
    ii.   Chargaff's rule for base composition in DNA
    iii.   Teminism
    iv.   RNA-dependant DNA polymerase
    v.   Transpeptidation
    vi.   Zwitter ionic form of amino acid
    vii.   Homodimer

There are various biochemical/physico-chemical and biophysical techniques used for analysis of structural and functional aspects of biomolecules. Since DNA, RNA and proteins are principal biomolecules that are involved in experimental procedures of bioinformatics—the significant techniques for physico-chemical characterization and structure determination of these biomolecules are discussed in this chapter.

CHROMATOGRAPHY

Chromatography is a versatile physical technique in which the crude mixture of dissolved components is separated as it moves through some type of porous matrix. The general principle involved in all chromatographic techniques is 1) adsorption, 2) relative affinity of substances in a mixture for partitioning, and 3) phase distribution. Chromatographic techniques are associated with two phases such as **mobile phase** (fluid or gas) that passes through a column containing a **stationary phase** (porous solid or liquid coated on a solid support). Depending on the nature of these phases, chromatographic techniques are classified into following different types.

Thin Layer Chromatography (TLC)

It is a solid–liquid technique in which the solid (stationary) phase is a thin coating of cellulose, or alumina ($Al_2O_3.7H_2O$) or silica gel ($SiO_2.7H_2O$) on a plate (adsorbent plate preferably a glass slide or plastic plate). On this sorbent plate, the spots of known and unknown samples are applied along a line at its edge and the plate is then placed in a closed chamber that contains solvent (mobile phase) in such a way that the line of loaded sample is just above the layer of solvent (Figure 4.1). As the solvent rises by capillary action up through the adsorbent on the plate, differential partitioning occurs between the components of the sample dissolved in the solvent and the stationary adsorbent phase. The distance migrated by the components is proportional to the affinity to the adsorbent and the solvent. The plate is removed just before the solvent front reaches the top edge of the plate. After drying the plate, spots are identified by photometric or other suitable methods. The relative migration, R_f , is the characteristic of analyte (or components in a mixture), which is calculated by the following formula:

$$R_f = \frac{\text{Distance travelled by analyte from the origin}}{\text{Distance travelled by solvent from the origin}}$$

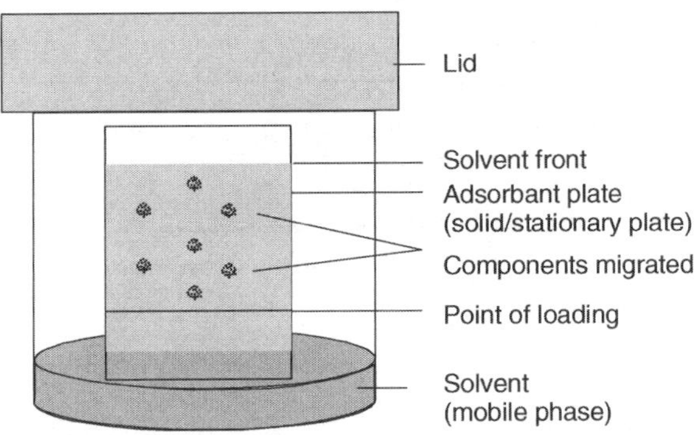

Figure 4.1   Assembly of thin layer chromatography

Thin layer chromatography (TLC) is a sensitive, fast, simple and inexpensive analytical technique that provides the analyser a quick answer as to how many components are in a mixture. TLC is also used to confirm the identity of compounds in a mixture when R_f of a unknown compound is compared with the R_f of a known compound, usually both known and unknown compounds run on the same plate.

Column Chromatography

In this chromatographic technique, the standard elements of a chromatographic column include a solid, porous material supported inside a column, which is generally made of a plastic or glass. The solid material (matrix) makes up the stationary phase through which a solution flows, the mobile phase. The solution that passes out of the column at the bottom (the effluent) is constantly replaced by solution supplied from a reservoir at the top. The mixture of proteins solution to be separated is layered on the top of the column and allowed to percolate into solid matrix. Additional solution is added on top. The protein solution forms a band within the mobile phase that is initially the depth of protein solution applied to the column. As proteins migrate through the column, they are retarded to different degree by their different interactions with the matrix material. The overall protein band thus widens as it moves through the column. Individual types of proteins (such as A, B, and C as shown in Figure 4.2) gradually separate from each other, forming bands within the broader protein band. Separation improves as the length of the column increases. However, each individual protein band also broadens with time due to diffusional spreading, a process that decreases resolution.

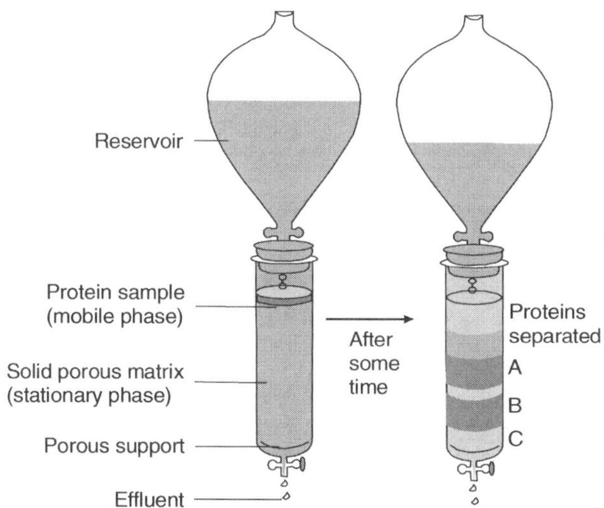

Figure 4.2 Column chromatography

Ion-exchange Chromatography

This technique exploits differences in sign and magnitude of the net electric charge on the molecule at a given pH. When the molecules are dissolved in solvent, they dissociate into charged ions and thus develop polarity that forms the basis of separation of these molecules. Differences in charges, densities and their distribution on the surface of different molecules are taken into consideration for one molecule from the other identical molecules. Ion-exchange chromatography is a powerful method used for separation of two proteins which are very similar to each other, but different from each other in having one charged amino acid. In ion-exchanger chromatography solutes are retained on an ion exchanger due to their reversible interaction with the oppositely charged groups on ion exchanger. Ion-exchanger is in the form of adsorbent column filled with insoluble, porous, and synthetic polymers (such as cellulose) containing either negatively (cation exchanger) or positively (anion exchanger) charged ions. The two most commonly employed ion-exchanger resins are diethylaminoethyl (DEAE)-cellulose and carboxymethyl (CM)-cellulose. DEAE-cellulose is positively charged and therefore acts by binding negatively charged molecules; it is an anion exchanger. CM-cellulose is negatively charged and acts as a cation exchanger. An anionic exchanger has positively charged fixed ions that are covalently attached to polymer matrix and these ions are balanced by equal and oppositely (negatively) charged mobile ions (anions) from the solution (Figure 4.3). Separation is achieved in two steps:

1. adsorption (binding) of analytes of opposite charge to column matrix as solution of analytes is passed through the column and

2. then eluting the analyte, one at a time, separated from each other, by an ionic gradient.

The affinity of each protein for the charged groups on the column is affected by pH and the concentration of free salt ions in the surrounding solution. Separation can be optimized by gradually changing the pH and/or salt concentration of the mobile phase so as to create a pH or salt gradient.

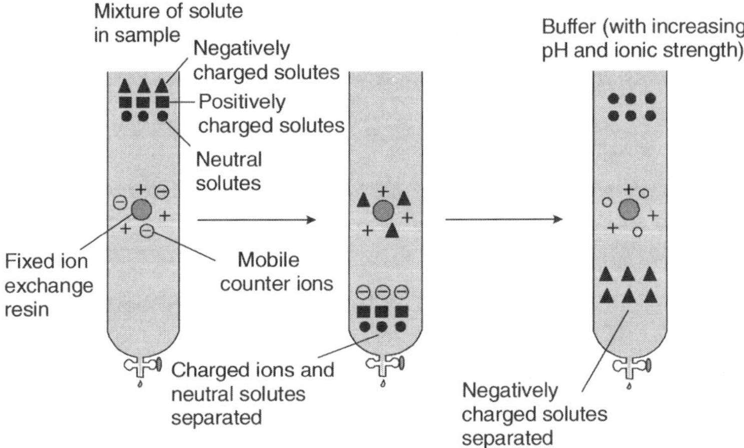

Figure 4.3   Diagrammatic representation for separation of solutes using anionic-exchanger

High-Performance Liquid Chromatography (HPLC)

It is an enhanced version of the column techniques in which the column materials themselves are much more finely divided and, as a consequence, there are more interaction sites and thus greater resolving power. Since the column is made of finer material, pressure is applied to the column to obtain adequate flow rates. As a result, there is high resolution as well as rapid separation of biomolecules (Figure 4.4).

Using HPLC technique, rapid separation of crude mixture is possible by using thin, long column through which a suitable solvent is pressurized to carry the individual component. The liquid mobile phase pressurized with the help of pump drives the sample through stationary phase packed into column. The individual components can be eluted and detected at the end of the column. HPLC is used to analyse thermally unstable compounds and non-volatile compounds. Based on the polarity between the mobile and stationary phases, HPLC is of two types:

1. Normal phase chromatography In this type, low-polarity compounds emerge out of the column initially and high-polarity compounds are eluted finally.

2. Reverse phase chromatography   Initially high-polarity compounds emerge out of the column. Solvent systems used are methanol/water, acetonitrile/water.

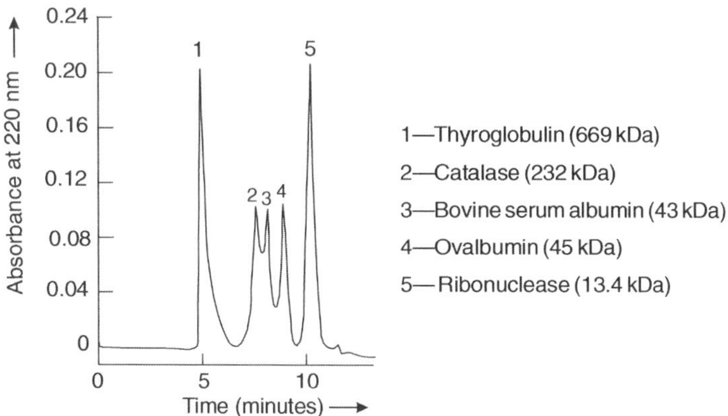

Figure 4.4   Separation of proteins by HPLC (*Source*: K.J. Wison and T.D. Sclabach. In Current Protocols in Mol. Bio. Vol.2. pp. 10–14.)

HPLC technique involves a lot of automation that eliminates or reduces human error and improves throughput, the only thing is that there should be a proper method of sample preparation. Whether working in bioanalytical setting, analysing proteins and nucleic acids, or processing environmental samples there should be careful and thorough approach for preparation of sample that is to be processed through HPLC column because by some estimates, 60 to 80% of the chromatographic difficulties are due to poor sample preparation.

Gel Filtration/Size-exclusion Chromatography

This technique is used to separate proteins (or nucleic acid) primarily by molecular weight. Like ion-exchange chromatography, the separation material consists of tiny beads that are packed into a column through which the solution of proteins slowly passes. The beads used in gel filtration are composed of cross-linked polysaccharides (dextrans or agarose) of different porosity. Molecules of size larger than the pore size cannot enter into the pores, and are excluded and eluted first. The smaller sized molecules that can enter into the pores of the beads are retarded in their rate of travel and are eluted last. Thus, gel filtration uses molecular sieves that separate the molecules in order of the molecular size.

Affinity Chromatography

This technique (Figure 4.5) uses the property, that proteins have high and specific affinity towards their substrates or cofactors or receptors or antibodies

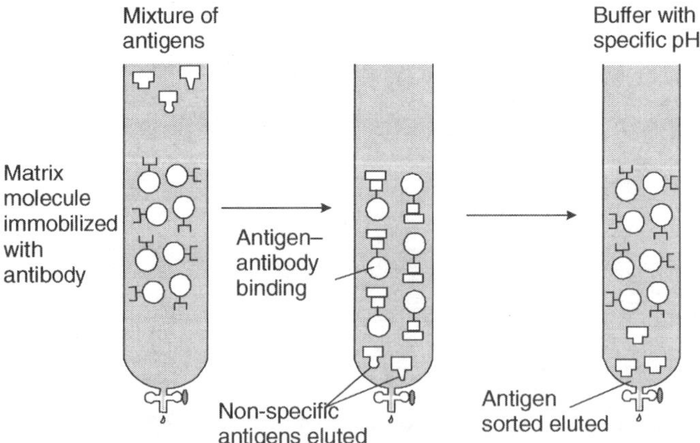

Figure 4.5   Schematic representation for steps in affinity chromatography

raised against them. Proteins interact with specific compounds—enzymes with substrates, receptors with ligands, antibodies with antigens, and so forth. The effector is immobilized on a water-insoluble carrier that is packed in a column through which the mixture is passed. The effector binds only to the molecules for which it is specific and retains it in the column; it is later recovered by elution using a buffer solution of a specific pH. For example, human leucocyte interferon is recovered in high yield and in high-purity chromatography using monoclonal antibody immobilized on a sepharose column.

ELECTROPHORESIS

It is one of the most basic but significant tools used by molecular biologists for separation of charged molecules under the influence of an electric field. The technique was pioneered by A. Tiselius in 1937. During electrophoresis, the positively charged (cationic) molecules move towards negatively charged electrode (cathode) and negatively charged (anionic) molecules travel towards positively charged electrode (anode). The speed of migration (electrophoretic mobility, μ) of a charged molecule depends on its net charge, size, shape, pore size of the separation matrix, characteristics of the buffer medium (like pH and concentration) and applied current. Electrophoresis is applied for separation of DNA, RNA and proteins. Nucleic acid molecules are negatively charged and hence migrate to anode while the charge on protein molecule depends positively and negatively charged amino acids at specific pH of the buffer.

In electrophoresis, the force moving the macromolecule is the electric potential (E). Thus electrophoretic mobility (μ) is the ratio of the velocity of particular molecule (V) to the electric potential. Electrophoretic mobility is also equal to the net charge of the molecule (Z) divided by the frictional coefficient (f), which reflects in part, a protein's shape. Thus electrophoretic mobility (μ) is defined as

$$\mu = V/E = Z/f$$

An electrophoretic set-up consists of two components—an electrophoresis unit and a power supply unit (Figure 4.6a). Depending on type of separation of biomolecules, the supporting medium can be selected in the form of a paper, cellulose acetate, agarose gel or polyacrylamide gel. A few types of electrophoresis are briefly discussed as follows:

Agarose Gel Electrophoresis

This technique is usually used to separate nucleic acids. Under the physiological conditions, the phosphate groups in phosphate–sugar backbone of nucleic acids are ionized. Polynucleotide chains of DNA and RNA are called "polyanions" and they will migrate to positive electrode (anode) when placed in electric field. Due to repetitive nature of phosphate–sugar backbone, double-stranded nucleic acids have roughly the net charge-to-mass ratio and will migrate to anode at equal velocities.

Agarose is a polysaccharide obtained from the red seaweed. When agaropectin is removed, agarose gels with melting points from 35°C to 95°C and varying degrees of electro-osmosis are obtained. Agarose dissolves in hot water. When this solution is cooled, a gel is formed with definite pore size. Gel having 0.7 to 1% (w/v) agarose will have large pore size and are widely used for the fractionation of double-stranded DNA, i.e., larger than 200 base pairs. Electrophoresis apparatus may be vertical or horizontal. Horizontal method shows less mechanical stress on gel. Better resolution is obtained using vertical apparatus but they are not easy to prepare. DNA sample treated with restriction enzymes are electrophoresed, as a result, heavier fragments remain closer to the point of loading, whereas the smaller ones migrate faster (Figure 4.6b).

Polyacrylamide Gel Electrophoresis (PAGE)

It is the most widely used electrophoresis for separation and characterization of proteins and nucleic acids. Polyacrylamide gel offers several advantages such as inertness to chemicals, superior resolution, stability over wide range of pH, temperature and ionic strength. The gel is prepared by polymerization of acrylamide ($CH_2=CH—CONH_2$) monomers with a small quantity of cross linker, bisacrylamide. Ammonium persulphate and tetra-methylethylenediamine (TEMED) are added as initiator and catalyst of polymerization. The pore size of the gel can be altered by varying the concentration of the monomer and the cross linker.

Once the gel is polymerized, the slab is suspended between two compartments containing buffers in which opposing electrodes are immersed. In a slab gel, the protein sample prepared in a solution containing sucrose or glycerol (whose density prevents the sample from mixing with buffer) is layered in slots along the top of the vertical gel electrophoresis apparatus (Figure 4.7a). A voltage is then applied between buffer compartments, and current flows across the slab, causing the proteins to move towards the oppositely

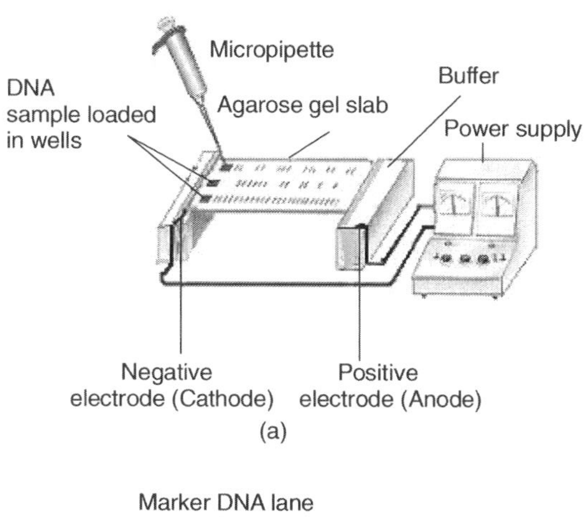

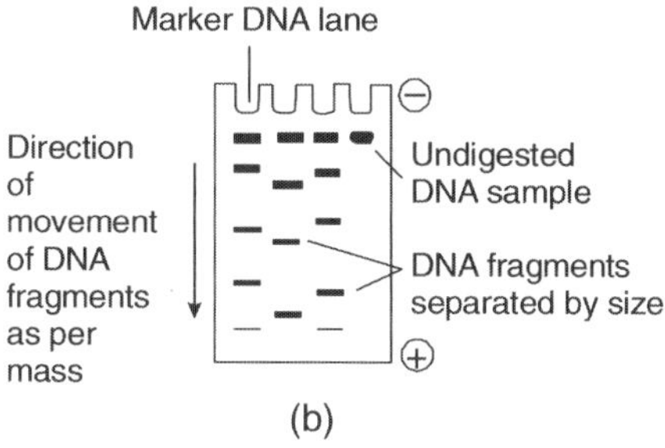

Figure 4.6   (a) Horizontal electrophoresis set-up, (b) Developed gel with DNA bands

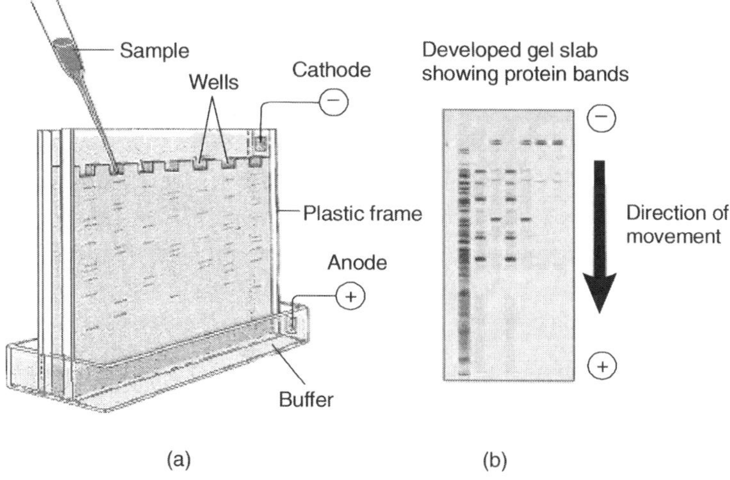

Figure 4.7   (a) Separation of proteins using vertical electrophoresis set-up (b) Polyacrylamide gel slab stained to show protein bands

charged electrode. Separations are typically carried out using alkaline buffers, which make the proteins to be negatively charged and cause them to migrate towards the positively charged anode at the opposite end of the gel (Figure 4.7b). The relative movement of proteins through polyacrylamide gel depends on the size, shape and charge density of molecules. The greater the charge density, the more forcefully the protein is driven through the gel and thus, the more rapid is its rate of migration.

Sodium Dodecylsulphate (SDS)-PAGE

Some of the proteins do not separate as their charge : mass ratio is similar, hence such proteins are treated with a negatively charged detergent called sodium dodecylsulphate (SDS) before processing to polyacrylamide gel electrophoresis. Therefore, such technique is called SDS-PAGE. The electrostatic repulsion between the bound SDS molecules causes the identical proteins to denature into their subunits. The polypeptide chains get opened and extended. The number of SDS molecules that bind to a protein is roughly proportional to the protein's molecular weight. Consequently, the proteins are separated on the basis of their mass but not the charge. In addition to separating the proteins in a mixture, SDS-PAGE can be used to determine the molecular weights of the various proteins by comparing the positions of the bands to those produced by proteins of known size.

Isoelectric Focusing

The biomolecules such as DNA, RNA and proteins have electric charges that varies from molecule to molecule and the conditions of the medium, i.e., the pH of the medium in which they are dissolved. Amino acids present in a protein molecule migrate in an electric field and this property forms the basis of their separation using isoelectric focusing. The direction and extent of migration depends to a large extent on the predominant ionic form present, and this is determined by the pH of the electrophoresis buffer. The pH at which there is zero net charge and no migration of proteins in an electric field is called isoelectric point (pI) and the isoelectric focusing is a technique that separates proteins according to their isoelectric points. A stable pH gradient is established in the gel by addition of appropriate ampholytes (low-molecular weight polymers of varying ratios of positively charged amino groups and negatively charged carboxylic groups). A protein mixture is placed in the well on the gel and a voltage is applied across the gel. As a result proteins migrate through the gel in response to electric field and are exposed to a continually changing pH that produces a continuous change in their ionic charge (I). At some point, along the gel, each protein encounters a pH that is equal to its isoelectric point at which the protein becomes a neutral molecule and subsequently causes its migration to stop. Proteins with different isoelectric points are thus distributed differently throughout the gel as shown in Figure 4.8.

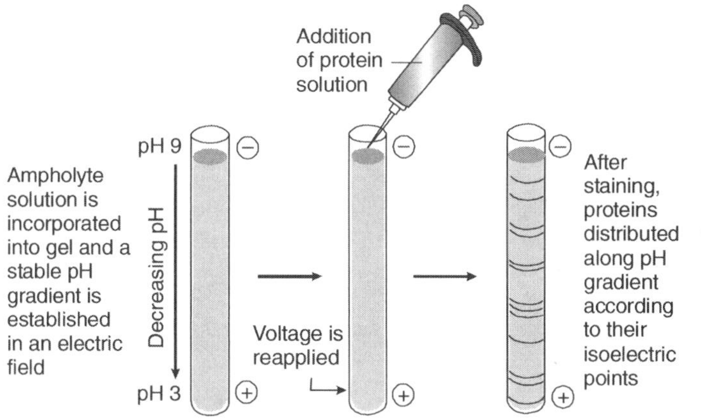

Figure 4.8 Isoelectric focusing for separation of proteins

Two-Dimensional (2D) Gel Electrophoresis

It is the technique that separates proteins in two dimensions based on different properties of the molecules. A mixture of protein is first separated according to their isoelectric points by isoelctric focusing within a tubular gel. After separation, the gel is removed and placed on to a slab of SDS-saturated polyacrylamide and subjected to SDS-PAGE. The proteins move into the slab gel and separated according to their molecular weight (Figure 4.9). The resolution of the technique is so great that virtually all the proteins present in a cell in detectable amounts can be distinguished. With the help of 2D gel electrophoresis, large number of proteins (1000–2000) in a cell extract can be separated.

Since proteins are fundamental biomolecules and carry various cellular functions, their large-scale study in relation to genome of a species is the fundamental aspect of bioinformatics specifically the proteomics. The rapidly flourishing field of proteomics relies heavily upon 2D gel electrophoresis that is followed by mass spectrometric analysis of isolated proteins which is an essential tool.

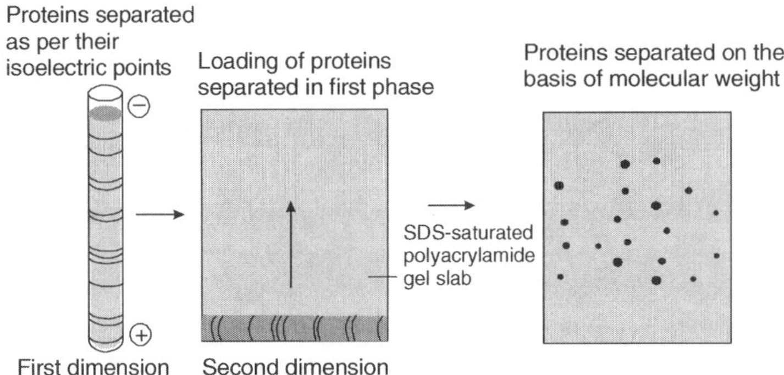

Figure 4.9 Schematic representation of 2D gel electrophoresis

BLOTTING TECHNIQUES

Different methods of gel electrophoresis help us to separate proteins in a mixture and fragments of DNA on the basis of their charges and molecular weight. We can visualize and detect these biomolecules by staining the bands in the gel at the end of the process, but these are all temporary procedures and cannot be preserved as records. Blotting techniques involve the procedures of blotting and hybridization of biomolecules separated in the gel during electrophoresis with the help of specific probes and this will help in making the record permanent. Southern blotting for DNA, northern blotting for RNA and western blotting technique for proteins are described below.

Southern Blotting Technique

E.M. Southern (1975) developed the first blotting technique that made the analysis and recording of DNA easy. Identification of sequences, correlation of DNA restriction fragments to RNA and protein mapping were facilitated by this technique. The procedure begins with the digestion of DNA sample with the help of restriction enzyme. DNA fragments of unequal lengths are eletrophoresed on the agarose gel. As a result, DNA molecules are separated based on their size. The gel containing separated DNA fragments is then denatured by alkali treatment (to make the DNA single-stranded). Gel is then placed in contact with a reservoir of buffer solution, followed by covering upper surface of the gel with nitrocellulose filter or nylon membrane overlaid with dry filter paper sheets and weight. Capillary action pulls the buffer from the reservoir and the buffer passes through the gel, carrying DNA fragments with it on to the nylon membrane. The ss-DNA fragments are trapped on the nylon membrane in the form of bands, which is like replica of original pattern on agarose gel. DNA fragments can be fixed permanently to nylon membrane by baking at 80°C or UV treatment. The membrane containing immobilized DNA molecules is then exposed to the complementary radioactive DNA probes under the appropriate hybridization conditions. Hybridized DNA molecule present on the membrane is then exposed to X-ray to get autoradiography (Figure 4.10). The autoradiographic images show the hybridized DNA molecule and thus, the sequences of DNA are identified using the probes with known complementary sequences.

Northern Blotting Technique

This method is used for detection and quantitative estimation of RNA in the sample and is an extension of Southern blotting, since RNA was not found to bind with nitrocellulose filter. Alwine *et al.* (1979) developed the technique by which RNA bands are blot-transferred from the gel on to chemically reactive paper. They used diazotized aminobenzylmethyl cellulose paper prepared from Whatman filter paper no. 540 for binding RNA that can be detected by hybridization with suitable radioactively labelled DNA probe.

Western Blotting Technique

Towbin *et al.* (1979) developed this technique to detect presence of proteins. It works on the principle of antigen–antibody reaction, hence it is also known as immunodetection technique. The procedure involves

   i. the extraction of proteins from the sample and their separation on SDS-PAGE

  ii. blotting of proteins from polyacrylamide gel to nitrocellulose filter paper

 iii. application of an electric field (15 V for 2 hours) to cause the migration of proteins from gel to nitrocellulose filter and binding on its surface

 iv. hybridization of proteins using radiolabelled specific antibodies (I^{123}-antibodies) of known structure

  v. washing of nitrocellulose filter to remove unhybridized antibodies

 vi. detection of hybridized sequences by autoradiography or by staining the membrane with specific stain. If radioactive label is not used, bound antibody may be detected by a second antibody tagged with an enzyme as in ELISA (Enzyme-Linked Immunosorbent Assay) technique.

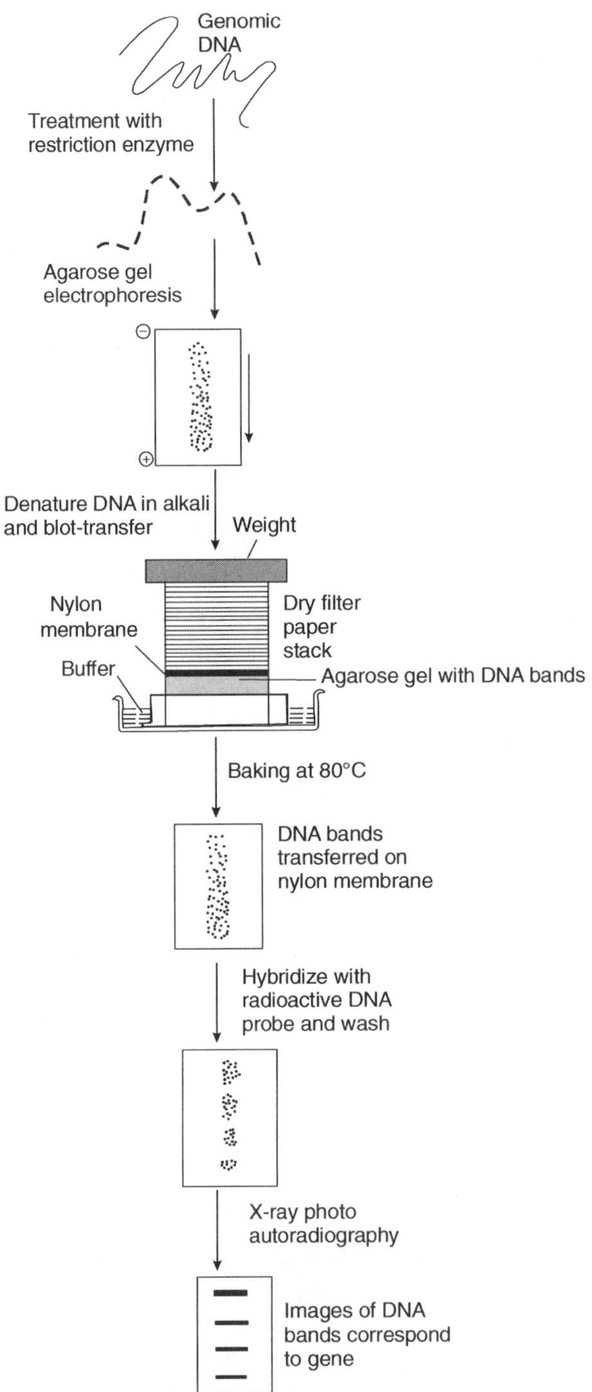

Figure 4.10   Schematic representation of Southern blotting technique

Western blots work well for studying a single protein target. However, to analyse multiple targets, the blot must be stripped and re-probed. Western blot for multiple protein targets is a time consuming process. Often some of the target protein on the blot is lost, compromising the ability to quantitate the relative amounts of different proteins of interest. The application of fluorescence-based detection of proteins on western blots provides the capacity for multiplex analyses, as well as greater signal stability relative to chemiluminescent detection. It reduces the time and labour needed for western blot processing. In addition, the size of the starting sample can be smaller.

Dot Blot Technique

It is another variety of Southern or northern blotting technique where DNA/RNA is not subjected to gel electrophoresis. Samples of DNA obtained from several tissues/individuals can be tested in a single run. The cloned or pure forms of extracted DNAs to be tested are spotted adjacent to each other on a nitrocellulose filter paper. DNA blots thus produced are immobilized and denatured so that they are bound on the filter as a single DNA blot. The DNA is first denatured and then the filter is baked at 80°C to fix the DNA on to the filter. The filter is then hybridized with single-stranded DNA (radiolabelled probe) under appropriate conditions. The dot representing sequence related to the probe will be illuminated by autoradiography (Figure 4.11). The intensity of the dot in autoradiograph corresponds to DNA or RNA sequence complementary to the probe is represented in the sample.

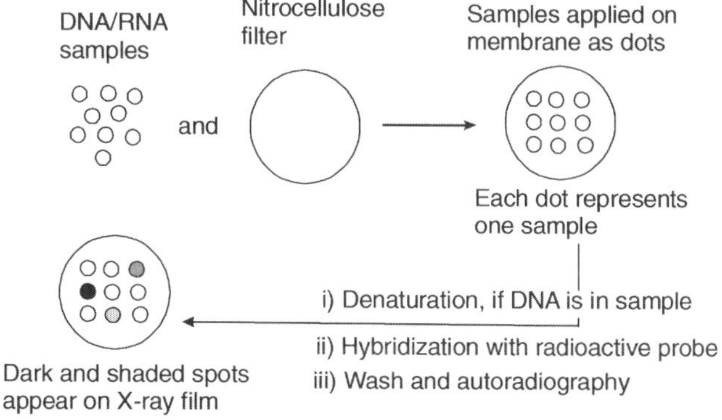

Figure 4.11 Diagrammatic representation of dot blot technique. The intensity of spots developed on the X-ray film show the differences in abundance of DNA (or RNA) sequences related to the probe.

DNA MICROARRAYS

In recent years, the competition between laboratories is largely due to speed for generating information rather than the working capabilities of their scientists. DNA microarrays/chips is one such technological advancement in the field of molecular biology that provides the genomic information with the great speed. As compared to Southern blotting or dot blots, microarrays represent high-density miniaturized arrays of molecular samples, facilitating screening of genomic DNA or cDNA samples in presence of one in 100,000 or more DNA sequences. The technology involves hybridization of an unknown

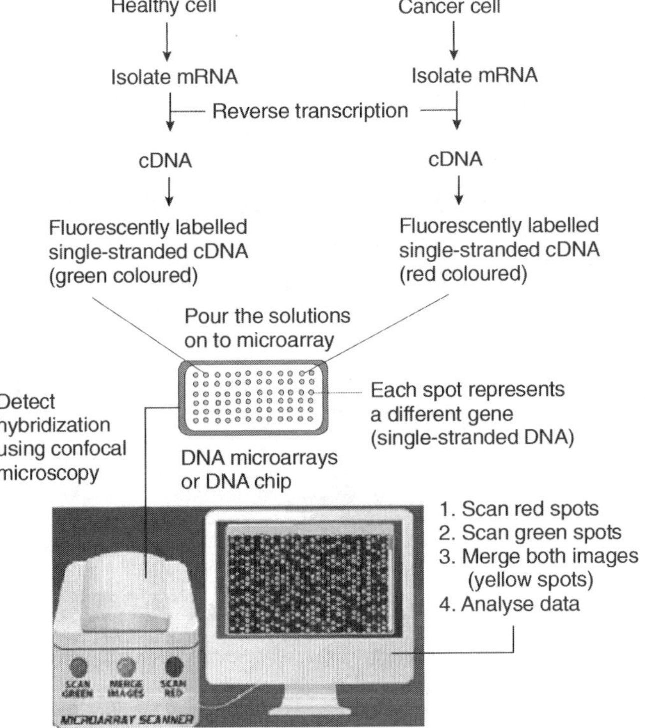

Figure 4.12   Steps involved in DNA microarray technology. The gene expression pattern in two different cells is studied with the help of DNA microarray technology. Green spots represent the genes expressed at higher levels in normal cell. Red spots represent the genes expressed at higher levels in cancer cells whereas yellow spots represent the genes that do not change their expression levels in both type of cells.

sample to an orderly array of immobilized molecules of known sequences. This produces a specific hybridization pattern that can be analysed or compared to a given standard using confocal microscopy and computer device as shown in Figure 4.12.

DNA microarrays or DNA chips allow the rapid and simultaneous screening of many thousand of genes. DNA segments from known genes are amplified by polymerase chain reaction (PCR) and placed on solid surface, using robotic devices that accurately deposit nanolitre quantity of DNA solution. Many thousands of such spots are deposited on a redesigned array on a surface area of just a few square centimetres. An alternate way to produce oligonucleotide microarrays makes the use of a photolithography with DNA-synthetic chemistry. Once DNA microarray is constructed, it can be probed with mRNA or cDNA from a particular cell type or cell culture to identify the genes being expressed in those cells. In addition, DNA chip technology helps in detection of single nucleotide polymorphisms (SNPs), in studies of the gene expression pattern of an organism as affected by the stage of development and/or environment, and in diagnosis of human genetic diseases by detecting mutant alleles such as in CFTR (cystic fibrosis), *BRCA*1 (cancer susceptible gene) and beta globin. Measurement of the level of transcripts for thousands of genes (functional genomics), and study of proteins and protein–protein interactions (proteomics) are the most wide spread uses of DNA microarrays technology.

GENE SEQUENCING TECHNIQUES

The genomic study includes isolation of a gene, amplification of gene by PCR, and the sequencing of DNA. Gene sequencing provides the knowledge of every gene, its transcript, the intron and the timing of transcript synthesis. Two common methods used for DNA sequencing are:

1. **Maxam and Gilbert's** chemical degradation process, that uses chemicals to cleave DNA at specific bases, result in formation of DNA fragments of different lengths that can be separated by gel electrophoresis and finally autoradiographed to note the sequence of nucleotides.

2. **Sanger's** dideoxyribonucleotide synthetic method, that involves use of enzyme to synthesize DNA of varying length in four different reactions, stopping the DNA replication at positions occupied by one of the four bases, and then determines the resulting fragment length by gel electrophoresis. Based on the size of fragments an order of nucleotides is determined automatically.

In addition to the above-mentioned DNA sequencing methods, high-throughput sequencing methods (required for large-scale genome projects) use robotics, automated DNA-sequencing machines and computers to achieve fast results and can handle large-scale data. Automated DNA sequencing is based on the Sanger–Coulson method, with two notable differences from the standard procedures. The first difference lies in the labelling of the products of PCR—automated sequencing methods use fluorescent labels instead of radioactive labels that are used in standard procedures. The four different flurochromes are attached to each of the four dideoxynucleotides used for chain termination and the reaction mixtures are run in a single gel lane or capillary and not in four tracks as in standard methods. The fluorescence detector identifies the signal emitted from each band produced in the gel lane or capillary. The fluorochromes are excited by a laser beam and the resulting signal is detected by a photovoltaic cell. The resulting data are fed to a computer, which in turn, converts these signals into the base sequence of the DNA molecule (Figure 4.13). This precise information of DNA sequence can be stored in the computer for future use and can be printed out; this is the second major difference from the standard Sanger–Coulson method.

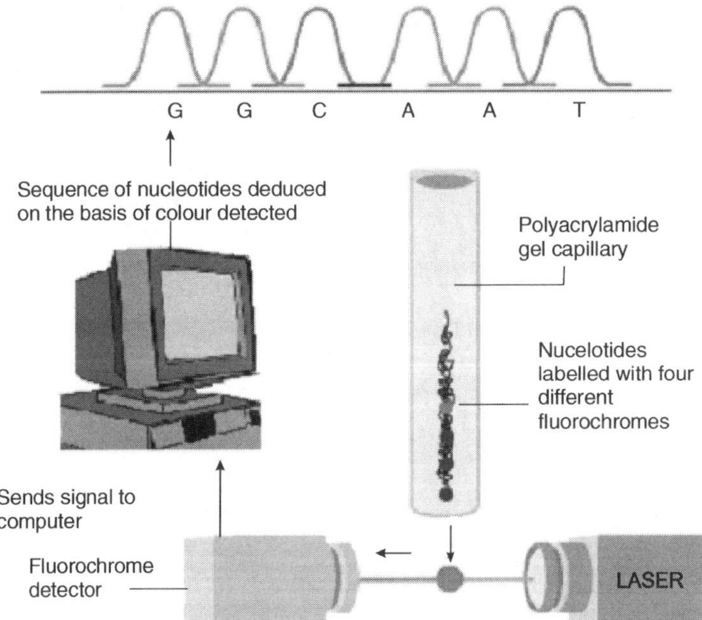

Figure 4.13   Automated DNA sequencing technique

MASS SPECTROMETRY (MS)

J. J. Thomson introduced this technique for separation of ions of different mass and energy under the fixed magnetic and/or electric field. During 1999–2000, mass spectrometry became an indispensable tool for genomic and proteomics research leading to 2002 Nobel Prize in chemistry to J.B. Fenn and K. Tanaka. Recently, 2D gel electrophoresis and mass spectrometry are coupled for the study of biomolecules like peptides, proteins and nucleic acids. The technique is used in identification of unknown compounds, quantification of known compounds and determination of structural and chemical properties of compounds when present in small amounts (10^{-6}–10^{-8} g). The identification and characterization of proteins can be done even when they are present in picomoles (10^{-12}).

The mass spectrometer basically consists of three components: the source of ion that involves the gaseous ionization of analyte to be examined, an analyser to separate ions according to their mass-to-charge ratio, and a detector to determine the molecular masses of separated ions. When the newly charged molecules are introduced into an electric and/or magnetic field, their paths in the field are the functions of their mass-to-charge ratio, *m/z*. This measured species of ionized species can be used to deduce the mass (M) of the analyte with very high precision. The MS requires that the molecules to be examined should be available in the form of charged gas. Following methods are used for ionization of biomolecules.

Matrix-Assisted Laser Desorption-Ionisation (MALDI)

This method has been successfully used to measure the mass of a wide range of macromolecules. In the year 1988, Karas and Hillenkamp developed this technique that involved co-precipitation of large excess of a matrix material (a small organic molecule) with the analyte molecule. A sub-microlitre of the mixture (matrix and analyte) is pipetted on to a metal substrate and allowed to dry. This dried mixture is then irradiated by laser pulse of the wavelength of 337 nm, which is specific for the absorbance of the selected matrix material. The irradiation causes energy transfer and desorption, resulting in the formation of gas-phase matrix ions. The charged molecular ions of analyte, produced during a gas-phase proton transfer reaction with the matrix molecule, are detected and analysed by MALDI-TOF (time-of-fight) mass spectrometer. The matrix molecules generally used in this process are α-cyano 4-hydroxycinnamic acid (CHCA), 3,5-dimethoxy 4-hydroxycinnamic acid and 2,5-dihydroxybenzoic acid.

The oligonucleotides and peptides having the mass above 500 Daltons can be successfully analysed by MALDI. The proteins are separated from crude extract by 2D-PAGE. The interested protein spots are excised and subjected to enzymatic digestion (by trypsin treatment). The peptides formed in the process are then mixed with the matrix molecule for co-precipitation on the metal substrate in the vacuum chamber of MS apparatus where the laser beam is tuned to the absorbance wavelength of the matrix. This results in the formation of positively charged peptides in the gas phase, which are attracted towards the orifice-flight tube kept at a negative bias. All peptides pass through the same electric field for the same plate-to-orifice distance (L) and then reach to the time-of-flight (TOF) tube. Once the particles are out of TOF tube, they are separated according to their mass-to-charge ratio and finally with the different time interval, particles reach the automated mass analyser where peaks in MS spectrum correspond to masses of each of the peptide analysed. MALDI-TOF-MS in combination with protein database search provides the significant information about the protein identification (Figure 4.14). It is possible to identify the protein of interest by comparing the observed mass of protein mixture in the MS spectrum with the predicted peptide mass derived from all known protein sequence databases. Proteomic study involves peptide sequencing, identification of protein, protein expression in different tissues and conditions, identification of post-translational modifications such as

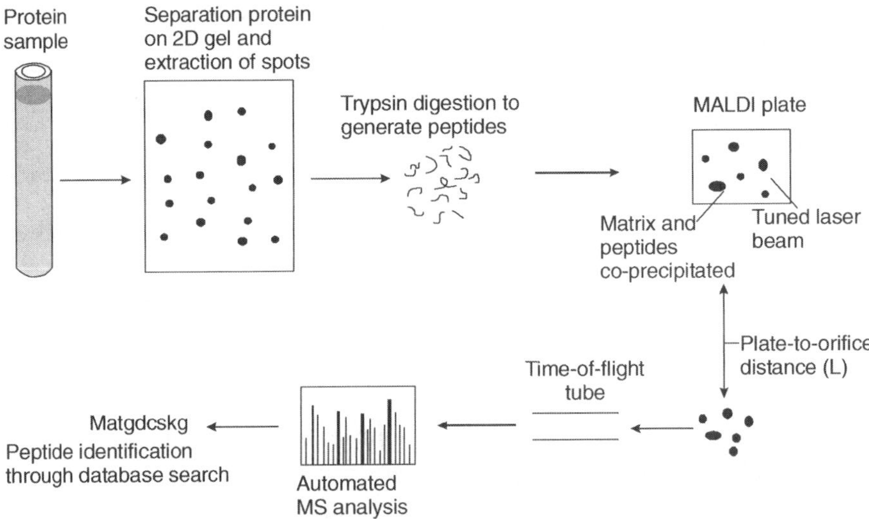

Figure 4.14   Schematic of MALDI analysis of peptide mass spectrometry

phosphorylation and glycosylation of proteins and characterization of protein interactions including protein–ligand, protein–protein and protein–DNA interaction. All these objectives are very crucial in understanding the biological processes in the post-genomic era and it is possible with the help of MALDI technology. In India, the MALDI facilities are available at Bhabha Atomic Research Center, Mumbai, Center for Cell and Molecular Biology, Hyderabad, Indian Institute of Science, Bangalore, Rajiv Gandhi Centre for Biotechnology, Thiruvananthapuram, etc.

Surface-enhanced Laser Desorption-Ionization (SELDI)

It is an ultra high-throughput screening (uHTS) system utilized for protein profiling of extracts obtained from the cells, tissue or physiological fluids. The method involves use of patented SELDI protein chip on the surface of which proteins isolated from complex biological mixture are directly applied. Identification of purified proteins is done by laser desorption/ionization time-of-flight mass spectrometric analysis. SELDI can detect and accurately calculate the masses of compounds, ranging from small molecules and peptides less than 1 kDa to proteins of 500 kDa or more, with the help of specific software package that facilitate data collection.

Electrospray Ionization Mass Spectrometry (ESI MS)

J.B. Fenn developed this unique technique for the analysis of wide range of macromolecules including proteins, oligonucleotides, sugars and lipids. In this method, the sample is dissolved in the liquid with mobile phase (water : acetonitrile or water : methanol in 1: 1). Macromolecules in solution are forced directly from liquid to the gas phase. The analyte solution is passed through the charged needle that is kept at high electric potential, dispersing the solution into a fine mist of charged microdroplets. The solvent surrounding macromolecules rapidly evaporate and impart charge on to the analyte molecules. Protons added during the passage through the needle give additional charge to analyte molecules. The *m/z* of the molecule can be analysed in the vacuum chamber (Figure 4.15). Electrospray ionization produces many more ions than MALDI and these ions are transferred into mass spectrometer with high efficiency for analysis.

Mass spectrometry provides a wealth of information for proteomic research, enzymology, and protein chemistry in general. The technique requires small amount of sample. The accurately measured molecular mass of a protein is one of the crucial parameters in characterization of protein.

Once the mass of a protein is accurately known, mass spectrometry is a convenient and accurate method for detecting changes in mass due to presence of bound cofactors, bound metal ions, covalent modifications, and so on.

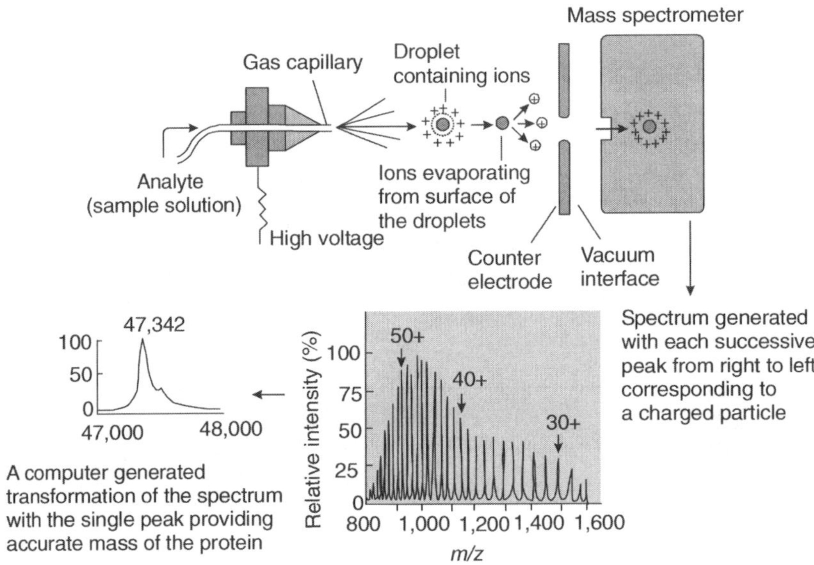

Figure 4.15　Schematic diagram of standard electrospray ionization technique

Recently, researchers tried to improve the MALDI technology since the use of matrix limits the method in several ways. Dominic S. Peterson of Los Alamos National Laboratory in New Mexico has reviewed the matrix-free method. As explained earlier matrix-assisted methods use a laser to energize the matrix, which then transfers its energy to ionize molecules in the sample. In contrast, matrix-free methods use the surface of materials to energize the molecules of interest. The first matrix-free method developed—desorption ionization on silicon—uses silicon to transfer energy to molecules. Porous silicon and silicon thin films have been employed in the experiments. Other matrix-free techniques use sol-gels and nanostructures as well as polymers and microstructures made of carbon. Researchers have used sol-gels made of silica, titanium, and zirconium. Carbon-based polymers and microstructures can have various arrangements, forms and added chemicals. For nanostructures, chemists have used carbon nanotubes, silicon nanowires, mesoporous tungsten and titanium oxides. Matrix-free laser desorption ionization technique can be applied to forensic, solid-phase synthesis, and to drug, protein and polymer analysis.

The most challenging practical objective in structure prediction of macromolecules lies in the determination of the three-dimensional (spatial) structure. X-ray diffraction (or X-ray crystallography) and nuclear magnetic resonance (NMR) spectroscopic methods are capable of providing spatial (secondary and tertiary) structural information of nucleic acids, proteins, carbohydrates, lipids and their complexes.

Nuclear Magnetic Resonance (NMR) Spectroscopy

It is the significant method for determining the three-dimensional structures of macromolecules by detecting the spinning patterns of their atomic nuclei when they are subjected to external magnetic field. Modern NMR techniques are used to determine the structures of carbohydrates, nucleic acids, and average-sized proteins. An advantage of NMR studies is that they are carried out on macromolecules in solution, whereas X-ray crystallography is limited to molecules that can be crystallized. NMR can also help in the study of protein structure, including conformational changes, protein folding, and interactions with other molecules.

NMR spectroscopy is a manifestation of nuclear spin angular momentum, a quantum mechanical property of atomic nuclei. Certain atoms of the radioisotopes like ^1H, ^{13}C, ^{15}N, ^{19}F, and ^{31}P, have a kind of nuclear spin that gives rise to an NMR signal. The strength of NMR signals not only depends on the isotope spin, but also on its abundance. The ^1H isotope with 99.98% natural abundance is the most sensitive NMR technique (^1H-NMR spectroscopy). When a strong, static magnetic field is applied to a solution containing a protein, some energy is absorbed as nuclei switch to the high-energy state, and the absorption spectrum that results, contains information about the identity of the nuclei and their immediate chemical environment. In a molecule, the nuclei of an element exist in different environment, and thus give rise to different spectral chemical shift. For example, the spatial distribution of particular nuclei (say ^1H) in different chemical environments (in CH, CH_2, CH_3 groups) in a protein molecule is determined by the "chemical shift" that forms the basis of NMR spectroscopy (Figure 4.16). In 1D NMR, the spectrum is a plot of amplitude as a function of frequency. One-dimensional (1D) NMR spectra of complex biomolecules contain a large number of peaks; some of them are close together or overlap. NMR instruments with higher field magnets or higher frequencies (e.g. 600 MHz) provide better peak resolution, helping in analysis of larger molecules.

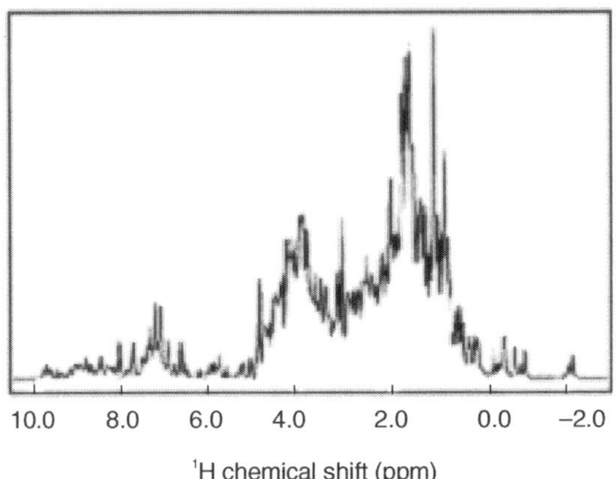

Figure 4.16 A one-dimensional NMR spectrum of a protein

Structural analysis of proteins became possible with the advent of two-dimensional (2D) NMR spectroscopy. This technique allows measurement of distance-dependent coupling of nuclear spins in nearby atoms through space the Nuclear Overhauser Effect (NOE). The NOE describes changes in the intensity of one resonance line when another line in the spectrum is perturbed by the radiation. Thus, Nuclear Overhauser effect Spectroscopy (NOESY) shows the spectrum containing peaks between hydrogen atoms (protons) that are close together in space (Figure 4.17). NOESY is a space-relaxation-mediated transformation process that provides information about the internuclear distance ($<5.0\,\text{Å}$). Total Correlation Spectroscopy (TOCSY) allows the coupling of nuclear spins in atoms connected by covalent bonds. Thus, NOESY and TOCSY are two fundamental mixing mechanisms employed for generation of 2D NMR spectra. In 2D NMR, a second time-period is added. In 2D NMR spectrum, there are diagonal peaks due to correlation of two different frequencies resulting from interaction between hydrogen atoms (protons), which are close to each other in space.

In three-dimensional NMR spectrum there are correlations of three different frequencies generated through the two different mixing times of the experiment. Hence translating a 2D NMR spectrum into a complete three-dimensional structure is a laborious process. The NOE signals provide some information about the distances between individual atoms, but for these distance constraints to be useful, the atoms giving rise to each signal must be identified. Complementary TOCSY experiments can help to identify which NOE signals reflect atoms that are linked by covalent bonds. Nowadays, modern genetic engineering can be used to prepare proteins that contain the rare isotopes ^{13}C or ^{15}N. The new NMR signals produced by these atoms,

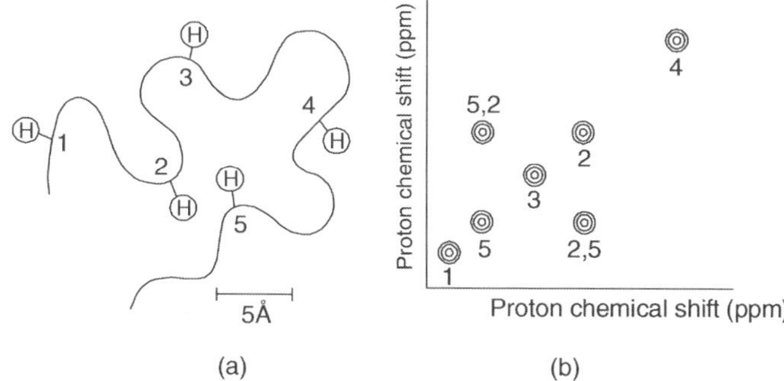

Figure 4.17 The Nuclear Overhauser Effect (NOE) identifies pairs of protons that are in close proximity. (a) Schematic representation of a polypeptide chain highlighting five particular protons. Protons 2 and 5 are in close proximity (~4 Å apart), whereas other pairs are farther apart. (b) A highly simplified NOESY spectrum. The diagonal shows five peaks corresponding to the five protons in part A. The peaks above the diagonal and the symmetrically related one below reveal that proton 2 is close to proton 5 (*Source*: C. Branden and J. Tooze, *Introduction to Protein Structure*. Garland, 1991, p. 280).

and the coupling with ^{1}H signals resulting from these substitutions, help in the assignment of individual ^{1}H NOE signals. The process is also aided by knowledge of the amino acid sequence of the polypeptide. To generate a 3D structure, researchers feed the distance constraints into a computer along with known geometric constraints such as chirality, van der Waals radii, and bond lengths and angles. The computer generates a family of closely related structures that represent the range of conformations consistent with the NOE distance constraints. The NMR techniques require the need of growing protein crystals and can help in modelling proteins and other biomolecules relatively more quickly than X-ray crystallography.

X-RAY CRYSTALLOGRAPHY

It is an indispensable tool in elucidation of three-dimensional architecture of matter in crystalline state at molecular and atomic resolution. Rosalind Franklin and Wilkins successfully studied X-ray diffraction patterns of DNA and demonstrated that DNA was a helical structure (*Refer* Figure 3.4). These findings provided significant platform for Watson and Crick to propose the model of DNA. In X-ray protein crystallography, protein needs to be grown into a crystal, which is a manifestation of regular and periodic arrangement

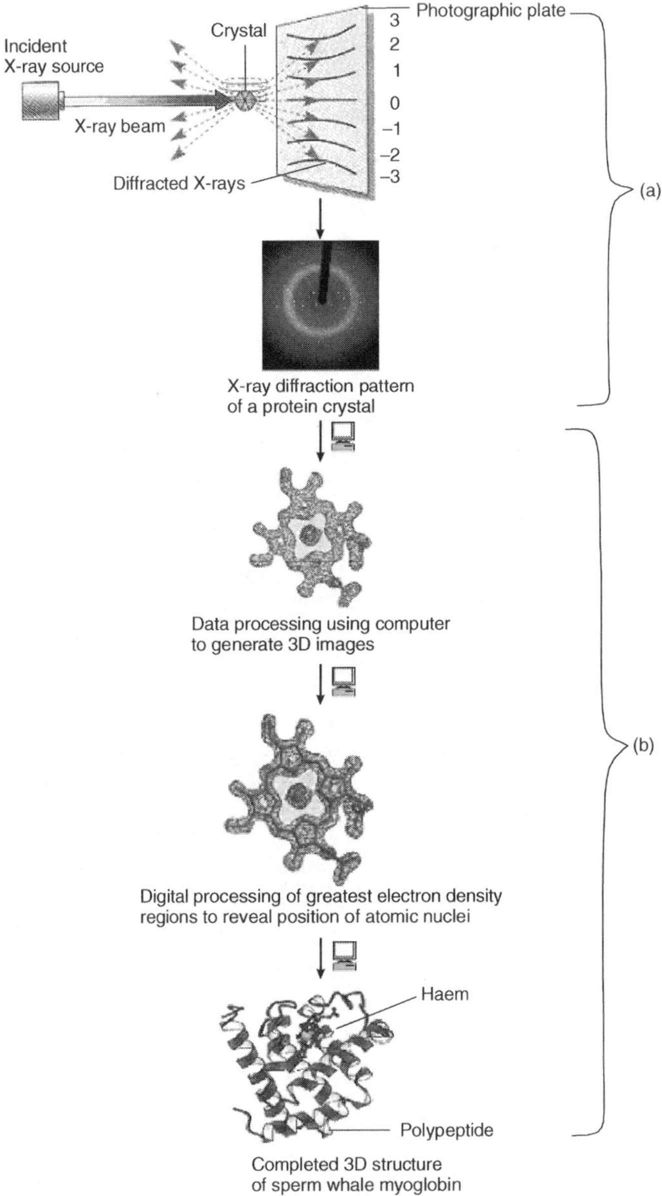

Figure 4.18 Steps in X-ray crystallography (a) X-ray diffraction pattern of a protein crystal (b) Sequential digital processing of electron density spots to construct the model for sperm whale myoglobin.

of atoms/molecules or clusters of molecules in all three-dimensions (crystal lattice). Protein crystal is then bombarded with a thin beam of X-rays of a single (monochromatic) wavelength. The radiation that is scattered (diffracted) by the electrons of the protein atoms strikes a photographic plate placed behind the crystal. The diffraction pattern is composed of thousands of tiny spots recorded on X-ray film (Figure 4.18a). The diffraction pattern can be converted into an electron density map using a mathematical procedure called Fourier transform (FT) method. These two-dimensional electron density maps can be interpreted into a three-dimensional structure by solving the phases in the diffraction data. The phases refer to the relative timing of different diffraction waves hitting the detector. The relative positions of atoms in a crystal can be determined on the basis of information of phases. Once the phases are available, protein structures can be solved by modelling with amino acid residues that best fit the electron density map.

Myoglobin was the first protein that was isolated from muscles of the sperm whale and whose structure was determined by X-ray diffraction technique by John Kendrew. The technique requires that all molecules be precisely oriented, so the first step is to obtain crystals of myoglobin. Slowly adding ammonium sulphate salt (3M) to a concentrated solution of myoglobin at pH 7 to reduce its solubility that favours the formation of highly ordered crystals in a container after several days. The X-ray diffraction pattern of crystalline myoglobin is very complex process that involved nearly 25,000 reflections. These diffraction patterns are further analysed using computer and mathematical procedures and finally the efforts lead to molecular modelling (Figure 4.18b). The computer analysis took place in stages and the resolution improved at each stage, until in 1959 the positions of all the non-hydrogen atoms in the protein had been determined virtually. The amino acid sequence of the protein, obtained by chemical analysis, was consistent with the molecular model. Later on, the structure of thousands of protein, many of them much more complex than myoglobin have been determined to a similar level of resolution. X-ray crystallography technique has a major limitation in obtaining the suitable crystals of proteins of interest.

REVIEW QUESTIONS

1. What is chromatography? Which are the chromatographic methods that employed for physico-chemical characterization of biomolecules?

2. Describe the different components of thin layer chromatography and column chromatography including their principles.

3. Explain the various types of ion exchangers with role in separation of proteins.

4. Define electrophoresis. What are the procedures followed in the electrophoretic separation of nucleic acids and proteins?

5. What are the points of differences between SDS-PAGE and 2D gel electrophoresis?

6. Explain the principle of isoelectric focusing and its advantages in separation of proteins.

7. Describe different blotting techniques. Enlist the advantages of Southern blotting method over gel electrophoresis.

8. How would you separate RNA and proteins with the help of blotting techniques?

9. Explain step-by-step procedure of DNA microarray for analysis of expression of genes in normal and diseased cells.

10. What are the advantages of DNA microarray technology in genomic analysis?

11. Explain the principle of automated DNA sequencing technique.

12. What is mass spectrometry? How can MS techniques be useful in genomic and proteomic analysis?

13. Define MALDI. Compare MALDI with other methods of ionization for mass spectrometry.

14. Explain the methodology and role of X-ray crystallography in elucidation of three-dimensional structure of biomolecules.

15. What is NMR spectroscopy? Describe the steps involved in it for 1D, 2D and 3D structure prediction of proteins.

16. Write short notes on:
    i.   The R_f in TLC
    ii. HPLC
    iii. Gel filtration chromatography
    iv. Principle of affinity chromatography
    v.   Electrophoretic mobility
    vi. Isoelectric point (pI)
    vii. DNA microarray/chip
    viii. Matrix in MALDI
    ix. Matrix-free MS
    x.   Electrospray

   xi. Fourier transform

  xii. X-ray diffraction pattern of myoglobin

 xiii. Chemical shift during NMR

 xiv. Nuclear Overhauser Effect Spectroscopy

  xv. Total correlation spectroscopy

DATABASE

It is a collection of related data, where data means recorded facts. A typical database represents some aspect of the real world and is used for specific purpose by one or more groups of users. It is a repository for a collection of computerized data files. It is a collection of persistent data that is used by the application systems of some given enterprise. Database has several advantages over traditional, paper-based methods of record keeping:

- Compactness
- Speed of operation (fast retrieval and changing data faster than human)
- Up-to-date information is available on demand at any time
- Accuracy and consistency of information

The database terms most commonly used are as follows:

- **Entity** is a person, place, thing, or condition about which data and information are collected
- **Attribute** is a category of data or information that describes an entity. Each attribute is a fact about the entity
- **Data item** is a specific detail of an individual entity that is stored in a database
- **Record** is a grouping of data items that consist of a set of data or information that describes an entity's specific occurrence
- **Relation** file is the table in a database that describes entity

DATABASE MANAGEMENT SYSTEM (DBMS)

It is a computer software designed for the purpose of managing databases based on a variety of data models. A DBMS is a complex set of software programs that controls the organization, storage, management, and retrieval of data in a database. DBMS are categorized according to their data structures or types, sometime DBMS is also known as database manager. It is a set of prewritten programs that are used to store, update and retrieve a database. A DBMS includes:

1. A modelling language to define the schema of each database hosted in the DBMS, according to the DBMS data model.

   ▣ The four most common types of organizations are the hierarchical, network, relational and object models. Inverted lists and other methods are also used. A given database management system may provide one or more of the four models. The optimal structure depends on the natural organization of the application's data, and on the application's requirements (which include transaction rate (speed), reliability, maintainability, scalability, and cost).

   ▣ The dominant model in use today is the ad hoc one embedded in SQL, despite the objections of purists who believe this model is a corruption of the relational model, since it violates several of its fundamental principles for the sake of practicality and performance. Many DBMSs also support the Open Database Connectivity that supports a standard way for programmers to access the DBMS.

2. Data structures (fields, records, files and objects) optimized to deal with very large amounts of data stored on a permanent data storage device (which implies relatively slow access compared to volatile main memory).

3. A database query language and report writer to allow users to interactively interrogate the database, analyse its data and update it according to the users privileges on data.

   ▣ It also controls the security of the database.

   ▣ Data security prevents unauthorized users from viewing or updating the database. Using passwords, users are allowed to access the entire database or subsets of it called **subschemas**. For example, an employee database can contain all the data about an individual employee, but one group of users may be authorized to view only payroll data, while others are allowed to access only work history and medical data.

▣ If the DBMS provides a way to interactively enter and update the database, as well as interrogate it, this capability allows for managing personal databases. However, it may not leave an audit trail of actions or provide the kinds of controls necessary in a multi-user organization. These controls are only available when a set of application programs are customized for each data entry and updating function.

4. A transaction mechanism, that ideally would guarantee the ACID properties, in order to ensure data integrity, despite concurrent user accesses (concurrency control), and faults (fault tolerance).

▣ It also maintains the integrity of the data in the database.

▣ The DBMS can maintain the integrity of the database by not allowing more than one user to update the same record at the same time. The DBMS can help to prevent duplicate records via unique index constraints; for example, no two customers with the same customer numbers (key fields) can be entered into the database.

The DBMS accepts request for data from the application program and instructs the operating system to transfer the appropriate data. When a DBMS is used, information systems can be changed much more easily as the organization's information requirements change. New categories of data can be added to the database without disruption to the existing system. Organizations may use one kind of DBMS for daily transaction processing and then move the detail onto another computer that uses another DBMS better suited for random inquiries and analysis. Overall systems design decisions are performed by data administrators and systems analysts. Detailed database design is performed by database administrators.

Database servers are specially designed computers that hold the actual databases and run only the DBMS and related software. Database servers are usually multiprocessor computers, with Redundant Arrays of Inexpensive Disk (RAID), used for stable storage. Connected to one or more servers via a high-speed channel, hardware database accelerators are also used in large volume transaction processing environments. DBMSs are found at the heart of most database applications. Sometimes DBMSs are built around a private multitasking kernel with built-in networking support although nowadays these functions are left to the operating system.

A database management system provides the ability for many different users to share data and process resources. But as there can be many different users, there are many different database needs. The question now is: How can a single, unified database meet the differing requirement of so many users?

A DBMS minimizes these problems by providing two views of the database data: 1) a physical view and 2) a logical view. The physical view deals with the actual, physical arrangement and location of data in the Direct Access Storage Devices (DASDs). Database specialists use the physical view to make efficient use of storage and processing resources. Users, however, may wish to see data differently; they do not want to know all the technical details of physical storage. After all, a business user is primarily interested in using the information, not in how it is stored. The logical view/user's view, of a database program represents data in a format that is meaningful to a user and to the software programs that process those data. That is, the logical view tells the user, in user terms, what is in the database. One strength of a DBMS is that while there is only one physical view of the data, there can be an endless number of different logical views. This feature allows users to see database information in a more business-related way rather than from a technical, processing viewpoint. Thus the logical view refers to the way user views data, and the physical view refers to the way the data are physically stored and processed. A DBMS also has a database query language so that report writer allows users to interactively interrogate the database, analyse its data and update it according to the users privileges on data. Thus, querying is the process of requesting attribute information from various perspectives and combinations of factors.

Advantages of DBMS

- ▣ Improved strategic use of corporate data
- ▣ Reduced complexity of the organization's information systems environment
- ▣ Reduced data redundancy and inconsistency
- ▣ Enhanced data integrity
- ▣ Application-data independence
- ▣ Improved security
- ▣ Reduced application development and maintenance costs
- ▣ Improved flexibility of information systems

◙ Increased access and availability of data and information

◙ Logical and physical data independence

BIOLOGICAL DATABASES AND INFORMATION RESOURCES

Bioinformatics deals with the management and analysis of biological information stored in databases. A database is repository of sequences of nucleotides in DNA or RNA and amino acids in protein that provide a centralized and homogeneous view of its contents. The repository is created and modified through DBMS. Each data entry in database is made according to a definite scheme, which is a set of pre-specified rules through the data definition language. Graphical User Interface (GUI) allows the user to access the contents of database and also helps in browsing through the contents of the repository. As mentioned earlier, biological databases also have a specialized query language that helps the user in querying the contents of the specific database. The data-definition language and the query language together form the data model.

A biological database is a collection of data that is organized so that its contents can easily be accessed, managed, and updated. The activity of preparing a database (Figure 5.1) can be divided into:

◙ Collection of data in a form which can be easily accessed

◙ Making it available to a multi-user system (always available for the user)

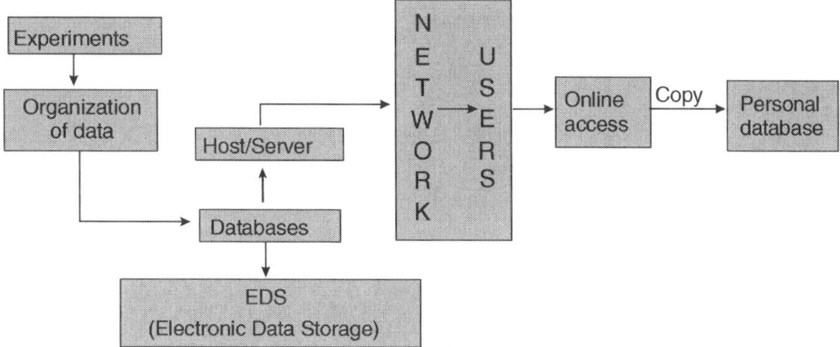

Figure 5.1　The network for production, construction and accession of a database

In an electronic environment, biological databases are at the heart of bioinformatics. In the mid-1980s, the biological databases were beginning to proliferate on Internet and the following are the public domain bioinformatics facilities:

1. In 1988, European Molecular Biology Laboratories (EMBL) used biocomputing methods in molecular biology research and established the network called EMBnet for providing up-to-date information of biological databases, services and training to users, dispersed in European laboratories.

2. On November 4, 1988 the National Centre for Biotechnology Information (NCBI) was established for the development of information system in molecular biology. It is situated in the campus of the National Institute of Health (NIH), Bethesda, USA. Since 1992, the NCBI has been maintaining GenBank, the NIH-DNA sequence database and it is the foremost repository of publicly available genomic and proteomic data.

3. In 1992, the Sanger Center was established by Wellcome Trust and Medical Research Council, UK to focus especially on mapping and sequencing the human genome. In December 2003, sequencing of human genome was completed with 99.99% accuracy in collaboration with international sequencing centres. The Sanger Center is also sequencing the DNA from various microorganisms.

4. The U.K. MRC Human Genome Mapping Project-Resource Centre receives fund from U.K. Medical Research Council. It is active in sequencing of human and mouse genome. The centre provides an online computing service and offers support and training courses.

5. In 1994, European Bioinformatics Institute (EBI) was established in Heidelberg (Germany) as an outstation of EMBL. One of the significant activities of EBI is development and distribution of EMBL Nucleotide Sequence Database, Europe's primary nucleotide data resource. This is a collaborative project with GenBank, USA and DDBJ—the DNA Data Bank of Japan (Mishima, Japan). The home page of DDBJ is shown in Figure 5.2. The EBI also has collaboration with the Swiss Institute of Bioinformatics to maintain and distribute the Swiss-Prot protein sequence database.

6. The Martinsried Institute for Protein Sequences (MIPS) (Max–Planck Institut für Biochemie, Germany) is the European partner of PIR-International Protein Sequence Database. The central activity

of MIPS is to collect, distribute and maintain up-to-date protein sequence data within Europe. The MIPS also plays a significant role in providing access to a database of aligned protein families and to dynamic database search software for retrieval of homogeneous and inspection of alignments of protein sequence.

7. The Biomolecular Structure and Modelling (BSM) unit at University College London (UCL) is a biocomputing centre with expertise in two central areas of bioinformatics, especially in protein sequence and structure analysis. UCL maintains the CATH (Class, Architecture, Topology, and Homology) database, which is a hierarchical domain classification of protein structure in addition to maintaining PDBsum (summaries and analysis of all structures in protein databanks).

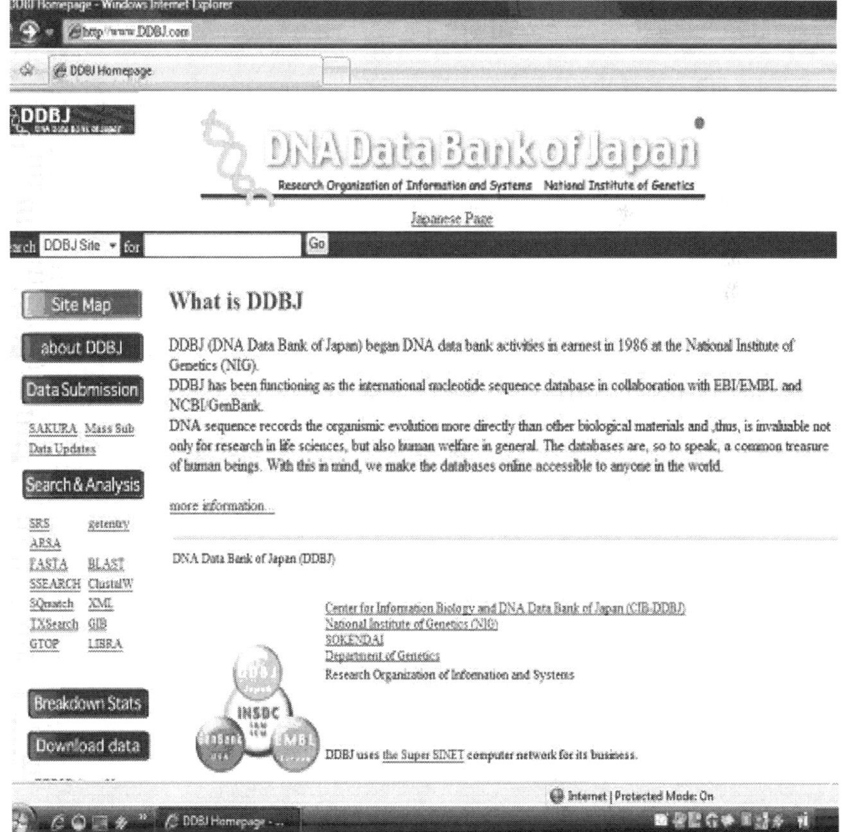

Figure 5.2   Home Page of DNA Data Bank of Japan

Other Companies, Institutes and Groups Associated with Bioinformatics

Accelrys　A company formed by MSI, Synopsys, Oxford Molecular, and GCG to provide informatic tools.

Ariadne Genomics Inc　Develops tools for systems biology—proprietary Natural Language Processing (NLP) and statistical algorithms, knowledge bases and software to analyse molecular networks.

Compugen　A pioneer group in the fields of computational genomics and proteomics. The company combines the disciplines of mathematics and computer science with molecular biology to improve the understanding of genomics and proteomics, the study of genes and proteins.

Computational and Applied Genomics Program　From Duke University.

Deep Computing Institute　IBM bioinformatic initiative.

Jena Centre for Bioinformatics (JCB)　Centre to promote inter-disciplinary research and to establish training courses in bioinformatics in the Jena region. It is to stimulate the collaboration between computer scientists and mathematicians on the one side and biologists, chemists, physicists and physicians on the other side.

Lion Bioscience　Developer for many good software like SRS (Sequence Retrieval System).

Open Bioinformatics Foundation　A non-profit, volunteer-run organization focused on supporting open source programming in bioinformatics.

Ocimum Biosolutions　Ocimum Biosolutions is a life sciences contract research and development company with competencies in bioinformatics, genomics, proteomics and custom contract research services, and operations in both the USA and India.

Physiome Sciences　Aim to help pharmaceutical companies to develop better drugs faster through the use of biological simulations.

Research Group Bioinformatics/AG Bioinformatik　Their focus is on regulatory genomic signals and regions, in particular those that govern transcriptional control. They analyse and characterize the underlying sequence elements and their context and develop database and software tools for their identification in newly unravelled genomic sequences.

Systems Biology Workbench Development Group   Their mission is to develop an integrated, easy-to-use environment, the workbench, which will enable biologists to create, manipulate, display and analyse biological models at molecular, cellular and multicellular levels.

The Bioinformatics Resource of the University of Hong Kong   It provides bioinformatics services to users in University of Hong Kong.

International Society for Computational Biology (ISCB)   A scholarly society dedicated to advancing the scientific understanding of living systems through computation. Their emphasis is on the role of computing and informatics in advancing molecular biology.

Genetic Circuits Research Group   Dr. Bernhard O. Palsson's group focuses on *in silico* modelling of genetic circuits involving metabolism and gene regulation.

Gerstein, Mark   Dr. Gerstein group is doing research in the emerging field of bioinformatics, using computation to analyse genome sequences, expression datasets, and macromolecular structures.

Classification of Biological Databases

Biological databases can be broadly classified into sequence and structure databases. Sequence databases are applicable to both nucleic acid sequences and protein sequences, whereas structure database is applicable to only proteins (Figure 5.3). The first database was created within a short period after the insulin protein sequence was made available in 1956. Incidentally, insulin is the first protein to be sequenced. The sequence of insulin consist of just 51 residues (analogous to alphabets in a sentence) which characterize the sequence. Around mid-1960s, the first nucleic acid sequence of yeast tRNA with 77 bases (individual units of nucleic acids) was found out. During this period, three-dimensional structures of proteins were studied and the well known Protein Data Bank (PDB) was developed as the first protein structure database, with only 10 entries in 1972. This has now grown into a large database with over 10,000 entries. While the initial databases of protein sequences were maintained at the individual laboratories, the development of a consolidated formal database known as Swiss-Prot protein sequence database was initiated in 1986 which now has about 70,000 protein sequences from more than 5000 model organisms, a small fraction of all known organisms. These huge varieties of divergent data resources are now available

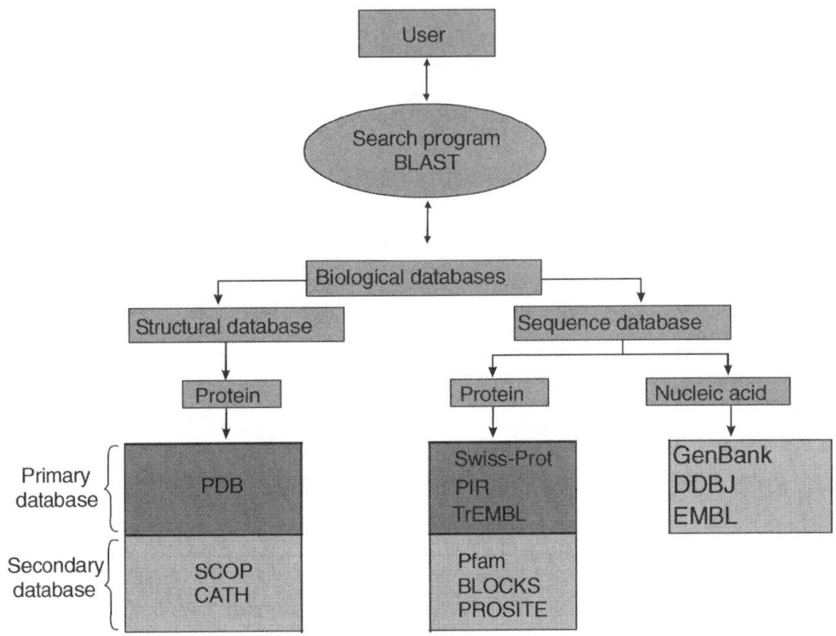

Figure 5.3   Systematic diagram of biological databases

for study and research by both academic institutions and industries. These are made available as public domain information in the larger interest of research community through Internet (www.ncbi.nlm.nih.gov) and CDROMs (on request from www.rcsb.org). These databases are constantly updated with additional entries.

Databases, in general, can be classified into primary, secondary and composite databases. A primary database contains information of the sequence or structure alone. Examples of these include Swiss-Prot and PIR for protein sequences, GenBank and DDBJ for genome sequences and the Protein Databank for protein structures.

A secondary database contains derived information from the primary database. A secondary-sequence database contains information like the conserved sequence, signature sequence and active site residues of the protein families arrived by multiple sequence alignment of a set of related proteins. A secondary-structure database contains entries of the PDB in an organized way. These contain entries that are classified according to their structure like all alpha proteins, all beta proteins, etc. These also contain information on conserved secondary-structure motifs of a particular protein. Some of the secondary databases created and hosted by various researchers at their individual laboratories include SCOP, developed at Cambridge University; CATH, developed at University College of London; PROSITE of Swiss Institute of Bioinformatics and eMOTIF at Stanford.

Composite database amalgamates a variety of different primary database sources, which obviates the need to search multiple resources. Different composite database uses different primary database and different criteria in their search algorithm. Various options for search has also been incorporated in the composite database. The National Center for Biotechnology Information which hosts these nucleotide and protein databases in their large redundant array of computer servers provides free access to the various persons involved in research. This also has link to OMIM (Online Mendelian Inheritance in Man) which contains information about the proteins that are involved in genetic diseases.

The growth of the primary databases gave rise to serious and valid questions on the format of the sequences, reliability and the comprehensiveness of the databases. To address the format issues, in-house software solutions have been developed to convert the format of one database to another. A public domain software FORCON can also be used. The newer software tools which are used for analysis accept the data in multiple formats. The problem in the reliability of the data is the possibility of misannotations. The misannotations are sometimes introduced due to the process of automation of annotation process which are carried out extensively with the help of computers. Misannotations, if introduced, multiply in subsequent additions and may accumulate to an unbelievable extent and create confusion. A possible solution to prevent this from happening is to flag the protein sequence which has been annotated by sequence comparison but whose function has not been validated by experimental methods.

Different classes of databases with their significance and URLs are given in Table 5.1.

Table 5.1 Databases with their significance and URLs

Databases	URLs and significance of databases
General databases	
DDBJ	http://www.ddbj.nig.ac.jp/ (DNA Data Bank of Japan).
EMBL	http://srs.ebi.ac.uk/ (through sequence retrieval system).
ExPASy	http://www.isb-sib.ch/ (Expert Protein Analysis System) Server maintained by Swiss Institute of Bioinformatics (SIB).
GenBank	http://www.ncbi.nlm.nih.gov/ (NCBI).
Gene Ontology Consortium	http://www.geneontology.org/ The objective of GO is to provide controlled vocabularies for the description of the molecular function, biological process and cellular component of gene products. These terms are to be used as attributes of gene products by collaborating databases, facilitating uniform queries across them.
GeneLynx	http://www.genelynx.org/ A portal to the human genome.
Genome-Phenome Superhighway	http://gps.gsc.riken.go.jp/ An integrated database designed to enable the discovery of candidate pathways related to complex traits such as susceptibility to diabetes. GPS is an integrated database that combines mapping charts of complex traits with networks of biomolecules. Biomedical researchers can analyse candidate pathways by taking into account of gene-expression profiles, the dynamics of gene interaction simulations, PPI, and the published literature.
GOLD	http://wit.integratedgenomics.com/GOLD/ (Genomes OnLine Database) A World Wide Web resource for comprehensive access to information regarding complete and ongoing genome projects around the world.
MEDLINE	http://www.ncbi.nlm.nih.gov/entrez/query.fcgi?db=PubMed
Mouse Genome Information	http://www.informatics.jax.org/mgihome/ MGI Database provides integrated access to data on the genetics, genomics and biology of the laboratory mouse. The projects contributing to this resource are: Mouse Genome Database (MGD) Project, Gene Expression Database (GXD) Project, Mouse Genome Sequence (MGS) Project and Gene Ontology (GO) Project.
OMIM	http://www3.ncbi.nlm.nih.gov/Omim/ (Online Mendelian Inheritance in Man) A catalog of human genes and genetic disorders.

(Contd.)

Table 5.1   (Continued)

Databases	URLs and significance of databases
Protein Mutant Database	http://pmd.ddbj.nig.ac.jp/
SRS	http://srs.ebi.ac.uk/ (Sequence Retrieval System).
The Genome Database	http://gdbwww.gdb.org/gdb/ Official central repository for genomic mapping data resulting from the Human Genome Initiative.

Protein families and sequence motifs databases

Databases	URLs and significance of databases
3Dee	http://jura.ebi.ac.uk:8080/3Dee/help/help_intro.html Database of Protein Domain Definitions.
Blocks	http://www.blocks.fhcrc.org/ A protein domain database.
CluSTr	http://www.ebi.ac.uk/clustr/ (Clusters of Swiss-Prot +TrEMBL proteins) Automatic classification of Swiss-Prot + TrEMBL proteins into groups of related proteins.
HPRD	http://www.hprd.org/ (Human Protein Reference Database) A centralized platform to visually depict and integrate information pertaining to domain architecture, post-translational modifications, interaction networks and disease association for each protein in the human proteome. All the information in HPRD has been manually extracted from the literature by expert biologists who read, interpret and analyse the published data. HPRD has been created using an object-oriented database in Zope, an open source web application server that provides versatility in query functions and allows data to be displayed dynamically.
InterPro	http://www.ebi.ac.uk/interpro Provides an integrated view of the commonly used signature databases.
PIR	http://www-nbrf.georgetown.edu/ PIR (Protein Information Resourse) maintained by National Biomedical Research Foundation.
Pfam	http://www.sanger.ac.uk/Pfam/ Protein families of alignments and HMMs.
PRINTS	http://www.bioinf.man.ac.uk/dbbrowser/PRINTS/ Protein Fingerprint Database.
ProDom	http://protein.toulouse.inra.fr/prodom.html The Protein domain database.

(Contd.)

Table 5.1 (Continued)

Databases	URLs and significance of databases
Protein families and sequence motifs databases	
	http://www.ebi.ac.uk/proteome/ A research-oriented initiative in order to utilise all the existing resources and provide comparative analysis of the predicted protein coding sequences of all complete genomes. The two main projects are InterPro and CluSTr.
PRF	http://www.prf.or.jp/en/ (Protein Research Foundation).
PROSITE	http://www.expasy.ch/prosite/ Database of protein families and domains.
Signal sequence/transcription elements/promoter analysis	
TFSEARCH	http://pdap1.trc.rwcp.or.jp/research/db/TFSEARCH.html Searching Transcription Factor Binding Sites.
TRANSFAC	http://transfac.gbf.de/TRANSFAC/ (The Transcription Factor Database) TRANSFAC is a database on eukaryotic *cis*-acting regulatory DNA elements and *trans*-acting factors. It covers the whole range from yeast to human.
WWW Promoter Scan	http://bimas.dcrt.nih.gov/molbio/proscan/ Predicts Promoter regions based on scoring homologies with putative eukaryotic Pol II promoter sequences.
WWW Signal Scan	http://bimas.dcrt.nih.gov/molbio/signal/ Find and list homologies of published signal sequences with the input DNA sequence.
Protein–protein interaction databases	
DIP	http://dip.doe-mbi.ucla.edu/ Database of Interacting Proteins catalogs experimentally determined interactions between proteins.
GRID	http://biodata.mshri.on.ca/grid/servlet/Index (General Repository for Interaction Data sets) A database of genetic and physical interactions. It contains interaction data from many sources, including several genome/proteome-wide studies, the MIPS database, and BIND.
Interact	http://www.bioinf.man.ac.uk/resources/interact.shtml An object-oriented database for protein–protein interactions.
Pronet Online	http://www.myriad-pronet.com/ Database that provide protein–protein interaction data to biological and medical researchers in a useful, integrated fashion. Maintained by Myriad Genetics.
Yeast protein complex database	http://yeast.cellzome.com/

(*Contd.*)

Table 5.1 (Continued)

Databases	URLs and significance of databases
Pathways databases	
BioCyc	http://biocyc.org/meta/ (**BioCyc Knowledge Library**) A collection of pathway/genome databases. Each database in the BioCyc collection describes the genome and metabolic pathways of a single organism, with the exception of the MetaCyc database, which is a reference source on metabolic pathways from many organisms. There are two main categories: (1) Literature-derived pathway/genome databases like EcoCyc and MetaCyc. (2) Computationally-derived pathway/genome databases consists of pathways from different organisms, e.g.Yeast. http://binddb.org/tCyc.
BIND	http://binddb.org/ (**Biomolecular Interaction Network Database**) A database designed to store full descriptions of interactions, molecular complexes and pathways.
CSNDB	http://geo.nihs.go.jp/csndb/ (**Cell Signaling Networks Database**) A data- and knowledge-base for signalling pathways of human cells.
Database of Quantitative Cellular Signalling	http://doqcs.ncbs.res.in/ (**DOQCS**) A repository of models of signalling pathways. It includes reaction schemes, concentrations, rate constants, as well as annotations on the models. Reference [PubMed].
EcoCyc Encyclopaedia	http://www.ecocyc.org/ Database that describes the genome and the biochemical machinery of *E. coli*, maintained by SRI International, Menlo Park, CA.
Enzymes and Metabolic Pathways (EMP) database	http://www.ecocyc.org/ Database that covers all aspects of enzymology and metabolism and represents the whole factual content of original journal publications. It contains more than 3,000 metabolic diagrams.
ExPASy Biochemical Pathways	http://www.expasy.ch/cgi-bin/search-biochem-index Digitized version of wall charts courtesy Boehringer Mannheim *et al.*, divided into metabolic pathways and cellular and molecular processes, maintained by the Swiss Institute of Bioinformatics, Geneva, Switzerland.

(Contd.)

Table 5.1   (Continued)

Databases	URLs and significance of databases
GeNet	http://www.csa.ru/Inst/gorb_dep/inbios/genet/genet.htm **(Gene Networks Database)** A database that contains the information on functional organization of regulatory genes networks acting at embryogenesis.
KEGG	http://star.scl.genome.ad.jp/kegg/ **(Kyoto Encyclopedia of Genes and Genomes)** An effort to computerize current knowledge of molecular and cellular biology in terms of the information pathways that consist of interacting molecules or genes and to provide links from the gene catalogs produced by genome sequencing projects.
Kinase Pathway Database	http://kinasedb.ontology.ims.u-tokyo.ac.jp/ An integrated database concerning completely sequenced major eukaryotes, which contains the classification of protein kinases and their functional conservation and orthologous tables among species, protein–protein interaction data, domain information, structural information, and automatic pathway graph image interface.
MetaCyc (Metabolic Encyclopedia)	http://biocyc.org/metacyc/ Description of over 450 metabolic pathways and their associated enzymes, from over 150 organisms maintained by SRI International, Menlo Park, CA.
MIPS yeast pathways	http://www.mips.biochem.mpg.de/proj/yeast/pathways/index.html
TRANSPATH	http://193.175.244.148/ An information system on gene-regulatory pathways. It focuses on pathways involved in the regulation of transcription factors. Elements of the relevant signal-transduction pathways like hormones, enzymes, complexes and transcription factors are stored together with information about their interaction. All data is extracted by experts from the scientific literature.
UM-BBD	http://umbbd.ahc.umn.edu/ **(Microbial Biocatalysis/ Biodegradatation)** Description of microbial biocatalytic reactions and biodegradation pathways primarily for xenobiotic, chemical compounds.
WIT	http://wit.mcs.anl.gov/WIT2/ A www-based system to support the curation of functional assignments made to genes and the development of metabolic models.

(Contd.)

Table 5.1   (Continued)

Databases	URLs and significance of databases
Structural databases	
Cambridge Crystallographic Data Centre	http://www.ccdc.cam.ac.uk/
CATH	http://www.biochem.ucl.ac.uk/bsm/cath/ Protein structure classification.
EBI Macromolecular Structure Database	http://www.ebi.ac.uk/msd/index.html The European project for the collection, management and distribution of data about macromolecular structures, derived in part from the Protein Data Bank (PDB).
PARTS LIST	http://bioinfo.mbb.yale.edu/partslist/ Dynamically perform comparative fold surveys; built on top of the Structural Classification of Protein (SCOP) fold classification and acts as an accompanying annotation to this system.
PDB (Protein Data Bank)	http://www.rcsb.org/pdb/ An international repository for the processing and distribution of 3D macromolecular structure data determined experimentally by X-ray crystallography and NMR.
PRESAGE	http://presage.berkeley.edu/ Database for structural genomics.
SCOP	http://scop.mrc-lmb.cam.ac.uk/scop/index.html (Structural Classification of Proteins database for 3D fold classifications.
Structural Biology Software Database	http://www.ks.uiuc.edu/Development/biosoftdb/ Software database maintained by University of Illinois.
SNPs databases	
Human Chromosome 21 cSNP Database and MAP	http://csnp.unige.ch/ A joint project between the Division of Medical Genetics of the University of Geneva Medical School and the Swiss Institute of Bioinformatics.
Histology databases	
Edinburgh Mouse ATLAS(EMAP)	http://genex.hgu.mrc.ac.uk/ Database of a series of three-dimensional models of mouse embryos at successive stages of development, linked to a standard anatomical nomenclature.

(Contd.)

Table 5.1   (Continued)

Databases	URLs and significance of databases
Standards	
OmniGene	http://omnigene.sourceforge.net/ OmniGene is an open source, open standards project aimed at helping bioinformatics professionals and students exchange biological data.
Ontology	
Gene Ontology	http://geneontology.org/ The goal of the Gene Ontology (GO) Consortium is to produce a controlled vocabulary that can be applied to all organisms even as knowledge of gene and protein roles in cells is accumulating and changing. GO provides three structured networks of defined terms to describe gene product attributes.
GOA	http://www.ebi.ac.uk/GOA/ A project run by the European Bioinformatics Institute that aims to provide assignments of gene products to the Gene Ontology (GO) resource.

The European Bioinformatics Institute (EBI) has developed and maintained a number of protein related databases. The database and URLs are given in Table 5.2.

Table 5.2   Databases with their significance and URLs (EBI)

Databases	URLs and significance of databases
CluSTr	http://www.ebi.ac.uk/clustr/ Offers an automatic classification of UniProtKB/Swiss-Prot + UniProtKB/TrEMBL.
CSA	http://www.ebi.ac.uk/thornton-srv/databases/CSA/ (Catalytic Site Atlas) A resource of catalytic sites and residues identified in enzymes using structural data.
GOA	http://www.ebi.ac.uk/GOA/ Provides assignments of proteins in UniProtKB/Swiss-Prot, UniProtKB/ TrEMBL and IPI to the Gene Ontology resource.
HPI	http://www.ebi.ac.uk/swissprot/hpi/hpi.html (Human Proteomics Initiative) An initiative, by SIB and the EBI, to annotate all known human sequences according to the quality standards of UniProtKB/Swiss-Prot.
IntEnz	http://www.ebi.ac.uk/IntEnz/index.html The Integrated Relational Enzyme Database (IntEnz) will contain enzyme data approved by the Nomenclature Committee. The goal is to create a single relational enzyme database.

(Contd.)

Table 5.2   (Continued)

Databases	URLs and significance of databases
InterPro	http://www.ebi.ac.uk/interpro/index.html An integrated documentation resource for protein families, domains and functional sites.
IPI	http://www.ebi.ac.uk/IPI/ (International Protein Index) It contains a number of non-redundant proteome sets of higher eukaryotic organisms constructed from UniProtKB/Swiss-Prot, UniProtKB/TrEMBL, Ensembl and RefSeq.
LGICdb	http://www.ebi.ac.uk/compneur-srv/LGICdb/ The Ligand Gated Ion Channel Database.
PANDIT	http://www.ebi.ac.uk/goldman-srv/pandit/ (Protein and Associated Nucleotide Domains with Inferred Trees) A collection of multiple sequence alignments and phylogenetic trees covering many common protein domains.
UniProt	http://www.ebi.ac.uk/uniprot/ (Universal Protein Resource for protein sequences) A central hub for the collection of functional information on proteins with accurate, consistent, and rich annotation, the amino acid sequence, protein name or description, taxonomic data and citation information.
UniProt Archive	http://www.ebi.ac.uk/uniparc/ A non-redundant archive of protein sequences extracted from public databases and contains only protein sequences.
UniProt/UniRef	http://www.ebi.ac.uk/uniref/ Features clustering of similar sequences to yield a representative subset of sequences.
UniProt/UniMES	mhtml:file://F:\15July\EBI Databases Protein Databases mht!ftp://ftp.ebi.ac.uk/pub/databases/uniprot/current_release /unimes/README A repository specifically developed for metagenomic and environmental data.
UniProtKB/Swiss-Prot	http://www.ebi.ac.uk/swissprot/ An annotated protein sequence database. Part of the UniProtKB.
UniProtKB/ TrEMBL	http://www.ebi.ac.uk/trembl/ A computer generated protein database enriched with automated classification and annotation. Part of the UniProtKB.

(Contd.)

Table 5.2   (Continued)

Databases	URLs and significance of databases
Some other databases EBI manage include	
EMBL Nucleotide Database	http://www.ebi.ac.uk/embl/index.html Europe's primary collection of nucleotide sequences is maintained in collaboration with Genbank (USA) and DDBJ (Japan).
UniProt Knowledgebase	http://www.ebi.ac.uk/uniprot/index.html A complete annotated protein sequence database.
Macromolecular Structure Database	http://www.ebi.ac.uk/msd/index.html European Project for the management and distribution of data on macromolecular structures.
ArrayExpress	http://www.ebi.ac.uk/arrayexpress/ For gene expression data.
Ensembl	http://www.ensembl.org/ Providing up-to-date completed metazoic genomes and the best possible automatic annotation.
IntAct	http://www.ebi.ac.uk/intact/ Provides a freely available, open source database system and analysis tools for protein interaction data.
Databases A-Z	http://www.ebi.ac.uk/Information/ databases_sitemap.html A complete listing of all the EBI databases.

PUBMED—THE CENTRAL REPOSITORY FOR BIOLOGICAL DATABASE

In 1999, Harold Varmus, the then Director of U.S. National Institutes of Health have established a centralized web-based library of scientific articles, called PubMed Central (http://www.pubmedcentral.nih.gov/). Its main objective was to provide access to peer reviewed literature relevant for biological sciences to many of the small research institutes in the developing world, which do not have access to scientific literature. MEDLINE (based at the U.S. National Library of Medicine) integrates the medical literature, including many papers dealing with subjects in molecular biology not overtly clinical in content. It is included in PubMed, a bibliographic database offering abstracts of scientific articles, integrated with other information retrieval tools of the National Center for Biotechnology Information within the

National Library of Medicine (http://www.ncbi.nlm.nih.gov/PubMed/). One of the important features of PubMed is the retrieval of information based on Medical Subject Heading (MeSH) terms, the system that includes more than 20,000 controlled and standardized vocabulary terms used for indexing articles. In addition, PubMed uses a list of tags for literature searches and some frequently used PubMed field tags are given in Table 5.3. For searching the literature, search term can be specified by the tags that are joined by Boolean operators.

Table 5.3   Selected PubMed tags with their descriptions

Tag	Name	Description
AB	Abstract	Abstract
AD	Affiliation	Institutional affiliation and address of the first author and grant numbers
AID	Article identifier	Article ID values may include the PI (publisher identifier) or DOI (Digital Object Identifier)
AU	Author	Author
DP	Publication date	The date when the article was published
JID	Journal ID	Unique journal ID in the National Library of Medicine's catalog of books, journals, and audiovisuals
LA	Language	The language in which the article was published
PL	Place of Publication	The country where the journal was published
PT	Publication type	The type of material that the article represents
RN	EC/RN number	Number assigned by the enzyme commission to designate a particular enzyme or by the Chemical Abstracts Service for Registry Numbers
SO	Source	Composite field containing bibliographic information

(Contd.)

Table 5.3   (Continued)

Tag	Name	Description
TA	Journal title abbreviation	Standard journal title abbreviation
TI	Title	The title of the article
VI	Volume	Journal volume

(Source: www.ncbi.nlm.nih.gov/entrez/query/static/help/pmhelp.html)

PubMed is one of the most valuable web resources available to a biologist to access the online biological information. Effective feature of PubMed is the option to retrieve related articles. This is a quick way to get into the literature of a topic. Combined with the use of a general search engine for websites that do not correspond to articles published in journals, fairly comprehensive information is readily available about most subjects. In collaboration with scientific journals, the NCBI is organizing the electronic distribution of the full texts of published articles. Over 4600 journals are indexed in PubMed that deal with cell and molecular biology, biotechnology, biochemistry, genetics and even clinical and experimental medicine. PubMed currently provides full-text web access to 4,058 journals. According to MedLine, the total number of scientific articles published in the peer-reviewed biomedical literature increased from 449,109 in 1998 to 491,620 in 2001. Given the global nature of the biotechnology research and development enterprise, it is unrealistic to think that biological technologies and the knowledge base, upon which they rest, can somehow be isolated within the borders of a few countries.

ENTREZ—THE DATA RETRIEVAL SYSTEM

It was the retrieval facility developed by NCBI that allows access to molecular biology databases and bibliographic citations from NCBI's integrated databases. Entrez links together a diverse set of information resources: it accesses nucleotide and protein sequences from a number of databases as well as genome data from the NCBI genomes division, three-dimensional structures from MMDB (Molecular Modelling Database of NCBI), human disease data from OMIM, genetic locus data from LocusLink, and bibliographic citations from PubMed. Entrez is available through the World Wide Web (http://www.ncbi.nlm.nih.gov/Entrez/) and in a client/server version (available by FTP at www.ncbi.nlm.nih.gov/Entrez/Network/ nentrez.overview.html or ftp://ncbi.nlm.nih.gov/entrez/README_1).

In the case of the client/server application, Network Entrez (NetEntrez), the client runs on the user's local machine and interacts with the server at NCBI. Versions of the client program are available for Macintosh, Microsoft Windows, Unix/X-windows and some other systems. Entrez for the World Wide Web (WebEntrez) works through standard www browsers. NetEntrez and WebEntrez have similar functionality, but there are differences in the way that they are used. WebEntrez has the advantage of additional links—e.g. to WebBLAST outputs and to full-text articles for some journals. The NetEntrez interface is more intuitive and is often easier to use for complex queries. Entrez can be used in many ways and some common scenarios are presented in Figure 5.4.

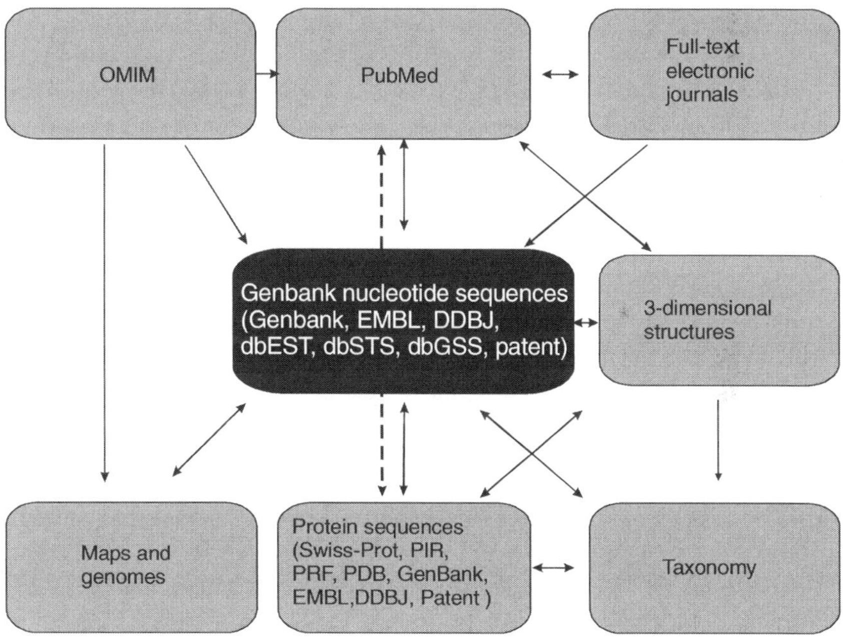

Figure 5.4   Data sources and links that constitute Entrez system

Effective use of Entrez requires an understanding of the main features of the search engine. There are several options common to all NCBI databases that help to narrow the search and some of the options are as follows:

- Limit helps to restrict the search to a subset of a particular database
- Preview/ Index connects different searches with Boolean operators and uses a string of logically connected keywords to perform new search

◙   History provides a record of the previous searches to review
Clipboard stores search results for later viewing for a limited period.

Suppose we have just done a similarity search and have observed a number of promising matches on the "hit list." The information on this list is often insufficient to allow us to reach even preliminary conclusions about the biological implications of our discovery. Thus, we want to find the publication describing this sequence, which meant, until recently, that we had to use a separate retrieval system to obtain the database record corresponding to the sequence. In this database record we would find the literature citation, and then we would have to take a trip down to the library to actually pull the printed copy of the journal. The "activation energy" for this process is high enough that one sometimes decides it is not worth it for "borderline" matches. Even if our motivation is high enough, perhaps there is no literature citation in the database record, or it involves a journal to which our library does not subscribe. In addition to finding the corresponding publication to confirm our search results, we may also want to retrieve some of the matching sequences and perform additional searches and alignments with these. All together, this is quite an involved process.

Entrez substantially lowers the activation energy by combining all of these activities in an easy-to-use system with a mouse-driven graphical user interface. Entrez establishes all of the requisite links between DNA sequences, the proteins they encode, and the literature that describes their biology. The literature component of Entrez comes from the PubMed database, which includes abstracts. While an abstract is not a substitute for the full text of a paper, one can frequently go an amazingly long way in the analysis and interpretation of sequence homology results by the information contained in an abstract. For an increasing list of journals, the web version of Entrez also has links to full-text articles. For many human disease genes, there are also links from PubMed records to the description of the disease in the OMIM database.

Entrez not only contains explicit links between different data sources, but also implicit or computed links within the data itself. Daily, the entire set of DNA and protein sequence databases are compared among one another and all significant homologies are computed and stored in the system. Thus, it is possible to instantly retrieve the homologs of any sequence in Entrez with the click of a mouse, without having to manually perform a new database search. Thus Entrez contains within it thousands of answers to questions that have not yet been asked. In one interesting case, this automated system actually "discovered" a significant homology prior to its recognition by a human biologist. In addition to precomputed sequence homologies, Entrez performs a similar operation on PubMed records. It is possible to compute a statistical relationship between two articles based on their frequency of use of significant terms. To indicate what constitutes a significant term, consider two negative examples, i.e., the terms "novel" and "important" are not significant terms because they appear in most scientific papers and thus are not of use in distinguishing one paper from another in terms of specific content. The set of "related articles" (sometimes referred to as "neighbors") is recomputed daily. This constitutes a very powerful information resource. One can enter Entrez via an author name or keyword and retrieve not only a specific paper but an entire bibliography of related material. One can then branch out (or "neighbor out") to other areas of the information space and effortlessly go back and forth between the literature and the sequences. Specific examples have appeared in reviews. Further examples can also be found on the NCBI Coffee Break website (http://www.ncbi.nlm.nih.gov/ Coffeebreak/). These pages describe recent biological discoveries and include tutorials demonstrating the linkage between various components of Entrez and other bioinformatics tools. Finding information on sequence homologs using the web version of Entrez is made even easier by the connections between Entrez and web-based BLAST output. For example, it is possible to do a BLAST search using webBLAST, and the hit list will have hypertext links to protein or nucleotide entries in Entrez. Other features of Entrez include taxonomy-based sequence retrieval; from the webpage for the NCBI Taxonomy Browser you can scan through organism names and phylogenetic trees and select the full set of nucleotide or protein sequences for any species. Now there are links in Entrez from sequences and the literature to protein three-dimensional structures in NCBI's Molecular Modelling Database. A three-dimensional structure viewer called Cn3D has also been added to the system. Cn3D can be run as a client/server program or as a "helper" application integrated into a standard web browser.

Another special viewer is available for the Entrez genomes division. The genome-level views presented by Entrez are built from sequences of complete chromosomes, from composites of sequence fragments, and from integrated genetic and physical maps (Figure 5.5). Entrez allows the user to visualize the sequence information at varying levels of detail, either graphically or as text. The chromosome views are tightly linked to the other Entrez component databases, allowing the user to jump effortlessly between maps, sequences, and bibliographic components.

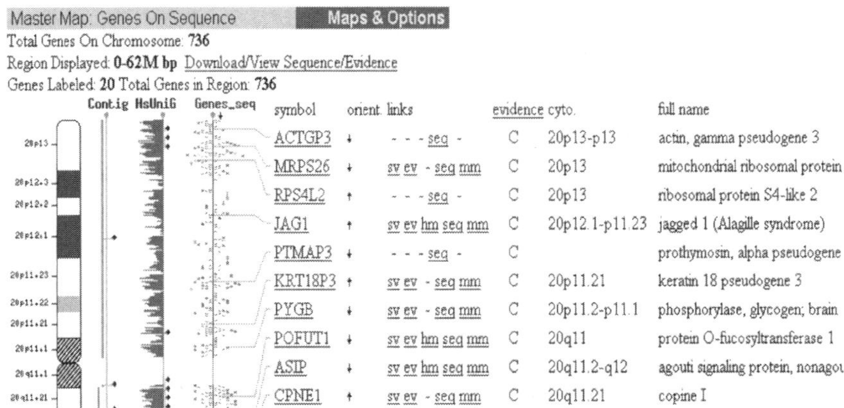

Figure 5.5   Mapviewer—the component of Entrez genome

LINKING DATABASES WITH SEQUENCE RETRIEVAL SYSTEM (SRS)

SRS is a network browser for databanks in molecular biology, integrating and linking a number of protein and nucleotide databases as well as for feeding the sequence retrieved into analytical tools such as sequence comparison and alignment programs. The sequence retrieval system originally developed by T. Etzold and Argos in 1993 that allows any flat-file database to be indexed to any other. SRS is maintained by EBI and can be compared to NCBI Entrez. It is not as integrated as Entrez, but allows the user to query the multiple databases simultaneously. SRS has the advantage that derived indices may be rapidly searched, allowing users to retrieve, link and access entries from all the interconnected resources. Using SRS one can search as many as 141 databases related to protein and nucleotide sequences, metabolic pathways, 3D structures and functions, genomes, and disease and phenotype information including small databases such as the Prosite and blocks of protein structural motifs, transcription factors, and databases specialized to certain pathogens. Typically, SRS (available at http://srs6.ebi.ac.uk) offers tight links among the resources of nucleic acid, EST, protein sequence, protein pattern, protein structure, specialist/boutique and /or bibliographic databases, enzyme, SNP, metabolic pathways (Figure 5.6).

Within the sequence category SRS has access to the following:

EMBL	Archival database of nucleotide sequences.
EMBLNEW	Updates to the latest full release of EMBL.
ENSEMBL	Annotated genomic sequences.

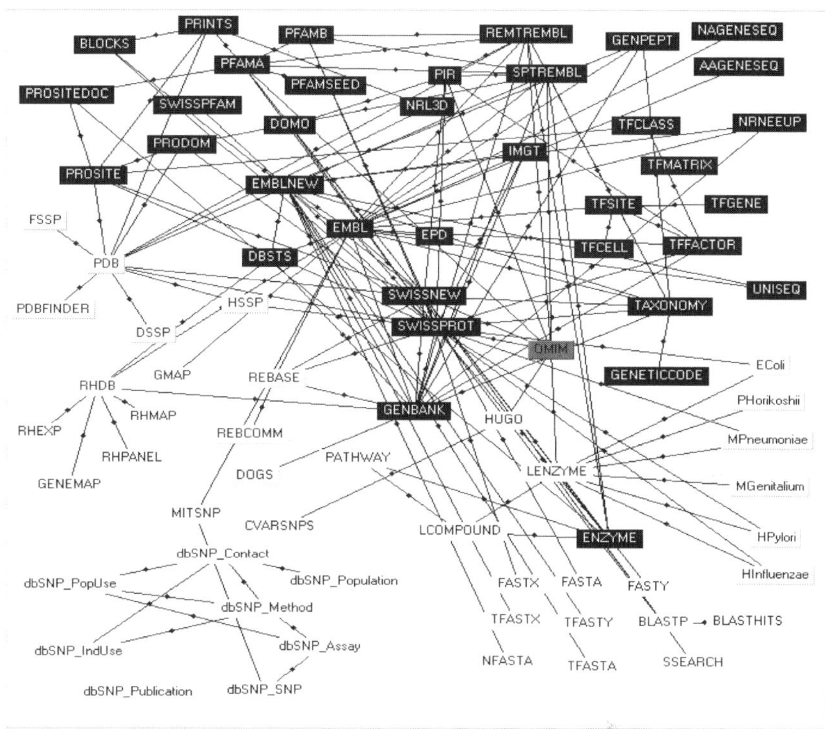

Figure 5.6   Typical network of databases linked through sequence retrieval system (SRS)

Swiss-Prot	Curated and annotated archival database of protein sequences.
SPTrEMBL	Computer-annotated protein sequence database supplementing the Swiss-Prot protein sequence database.
REMTrEMBL	Translations of coding sequences from EMBL nucleotide sequence database not destined for ultimate integration into Swiss-Prot.
TrEMBLNEW	Translation of all new and updated coding sequences in EMBL since last TrEMBL release.
SWALL	Comprehensive protein sequence database combining the full annotation in Swiss-Prot with the completeness of the weekly updated translation of all protein-coding sequences from the EMBL nucleotide sequence database.

IMGT	Integrated database specializing in immunoglobulin, T-cell receptors and Major Histocompatibility Complex (MHC) of all vertebrate species.
IMGTHLA	Sequences of human major histocompatibility complex (HLA) proteins.
InterPro	(Integrated Resource of Protein Domains and Functional Sites) A documentation resource for protein families, domains and functional sites.
OMIM	Databases of human genes and genetic disorders.

Advantages of SRS

▣ Explicit cross-references by, accession number (Swiss-Prot<->EMBL), etc. Implicit links by organism name, gene name (GenBank<->Taxonomy), small compound names, etc.

▣ Links can be performed on sets of entries or entire databanks

▣ All links become bidirectional

▣ Indirect (multistep) linking is possible

▣ Queries can be launched using "Quick Text Search"

▣ Using SRS 3D it is possible to view 3D structure of a protein as shown in Figure 5.7.

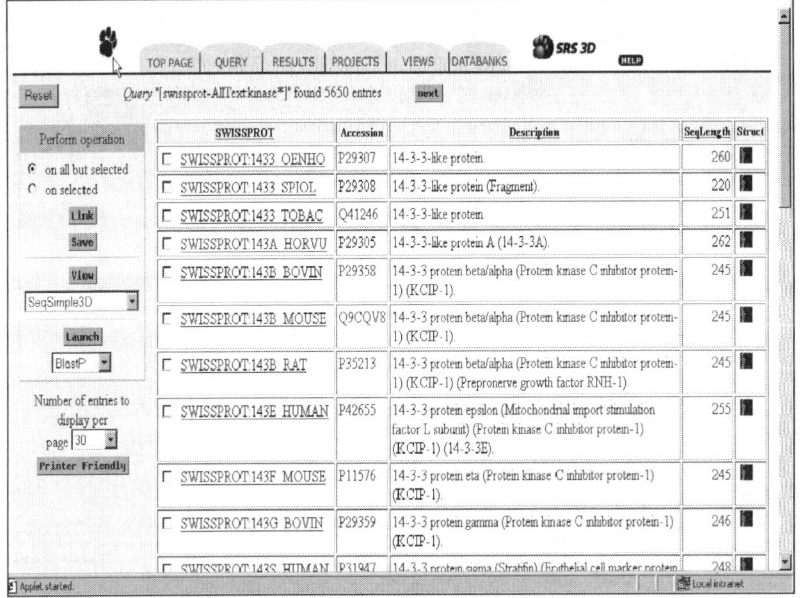

Figure 5.7   SRS 3D from sequence to structure of protein

ONLINE MENDELIAN INHERITANCE IN MAN (OMIM)

It includes databases of human diseases, textual description of human disorders and information about the genes associated with genetic disorders. The text contains numerous hyperlinks to literature citations, primary sequence records, as well as chromosome loci of the disease genes. When inherited diseases are discussed, the traditional reference is to diseases that are transmitted from one generation to the next. Many diseases show simple dominant or recessive inheritance according to Mendelian rules.

OMIM is a comprehensive, authoritative, and timely compendium of human genes and genetic phenotypes. The full-text, referenced overviews in OMIM contains information on all known Mendelian disorders and over 12,000 genes. OMIM focuses on the relationship between phenotype and genotype. It is updated daily, and the entries contain copious links to other genetics resources.

This database was initiated in the early 1960s by Dr. Victor A. McKusick as a catalog of Mendelian traits and disorders, entitled Mendelian Inheritance in Man (MIM). Twelve book editions of MIM were published between 1966 and 1998. The online version, OMIM, was created in 1985 by a collaboration between the National Library of Medicine and the William, H. Welch Medical Library at Johns Hopkins. It was made generally available on the internet starting in 1987. In 1995, OMIM was developed for the World Wide Web by NCBI. OMIM is authored and edited at the McKusick–Nathans Institute of Genetic Medicine, Johns Hopkins University School of Medicine, under the direction of Dr. Ada Hamosh.

NCBI of the U.S. National Library of Medicine has developed it into a database accessible from the web, and introduced links to other archives of related information, including sequence databanks and the medical literature. OMIM is now well integrated with the NCBI information retrieval system Entrez.

Every disease and gene is assigned a six digit number of which the first number classifies the method of inheritance (Table 5.4). If the initial digit is 1, the trait is deemed autosomal dominant; if 2, autosomal recessive; if 3, X-linked. Wherever a trait defined in this dictionary has a MIM number, the number from the 12th edition of MIM, is given in square brackets with or without an asterisk (asterisks indicate that the mode of inheritance is known; a number symbol (#) before an entry number means that the phenotype can be caused by mutation in any of two or more genes) as appropriate, e.g. Pelizaeus–Merzbacher disease [MIM*169500] is a well-established, autosomal, dominant, Mendelian disorder.

The OMIM gene map presents the cytogenetic map location of disease genes and other expressed genes described in OMIM. Whereas, the OMIM Morbid Map is a list of disease genes organized by disease. Using NCBI Entrez mapviewer and the Genome Data Base search engines, the more refined maps of genes and DNA segments can be obtained (Table 5.4).

Table 5.4   Six digit number classification method for diseases and genes

First digit	Range of MIM codes	Method of inheritance
1	100000–199999	Autosomal dominant loci or phenotypes
2	200000–299999	Autosomal recessive loci or phenotypes
3	300000–399999	X-linked loci or phenotypes
4	400000–499999	Y-linked loci or phenotypes
5	500000–599999	Mitochondrial loci or phenotypes
6	600000 onward	Autosomal loci or phenotypes

(Source: http://.ncbi.nlm.nih.gov/omim/omimfa.html)

Table 5.5   OMIM gene map of some diseases searched with genome browser database

Disorders	Symbols	OMIM	Location
Creutzfeldt-Jakob disease, 123400	PRNP	176640	20pter–p12
Crigler–Najjar syndrome, type I, 218800	UGT1A, UGT1, UGT1A1, GNT1	191740	Chr.2
Crouzon syndrome with *Acanthosis nigricans*	FGFR3, ACH	134934	4p16.3
Crouzon syndrome, 123500	FGFR2, BEK, CFD1, JWS	176943	10q26
Cryptorchidism	GTD	306190	Xp21
Currarino syndrome, 176450	HLXB9, HOXHB9, SCRA1	142994	7q36

(Contd.)

Table 5.5 (Continued)

Disorders	Symbols	OMIM	Location
Cutis laxa, 123700	ELN	130160	7q11.2
Cutis laxa, marfanoid neonatal type	LAMB1	150240	7q31.1–q31.3
Cutis laxa, neonatal	ATP7A, MNK, MK, OHS	300011	Xq12–q13
Cutis laxa, recessive, type I, 219100	LOX	153455	5q23.3–q31.2
Cyclic haematopoiesis, 162800	ELA2	130130	19p13.3
Cyclic ichthyosis with epidermolytic hyperkeratosis	KRT1	139350	12q13
Cylindromatosis, familial, 132700	CYLD1, CDMT, EAC	605018	16q12–q13
Cystic fibrosis, 219700	CFTR, ABCC7, CF, MRP7	602421	7q31.2
Cystic fibrosis, 219700	CFTR, ABCC7, CF, MRP7	602421	7q31.2
Cystinosis, nephropathic	CTNS	219800	17p13
Cystinuria, 220100	SLC3A1, ATR1, D2H, NBAT	104614	2p16.3
Cystinuria, type II	SLC7A9, CSNU3	604144	19q13.1
D-bifunctional protein deficiency	HSD17B4	601860	5q2
DECR deficiency	DECR1	222745	8q21.3

ExPASy—EXPERT PROTEIN ANALYSIS SYSTEM

ExPASy (www.expasy.org/tools/) is a comprehensive proteomic web server with a suit of programs for searching peptide information from Swiss-Prot and TrEMBL database. It is an information retrieval system and analysis system maintained by Swiss Institute of Bioinformatics in collaboration with the European Institute of Bioinformatics. The database search tools in this server dedicated to protein identification are given in Table 5.6.

Table 5.6 Protein identification and characterization

Tools	Functions
Identification and characterization of protein with peptide mass	
Aldenete	Identify proteins with peptide-mass fingerprinting data. A new, fast and powerful tool that takes advantage of hough transformation for spectra recalibration and outlier exclusion.
FindMod	Predict potential protein post-translational modifications and potential single amino acid substitutions in peptides. Experimentally measured peptide masses are compared with the theoretical peptides calculated from a specified Swiss-Prot entry or from a user-entered sequence, and mass differences are used to better characterize the protein of interest.
FindPept	Identify peptides that result from unspecific cleavage of proteins from their experimental masses, taking into account artefactual chemical modifications, Post-Translational Modifications (PTM) and protease autolytic cleavage
GlycoMod	Predict possible oligosaccharide structures that occur on proteins from their experimentally determined masses (can be used for free or derivatized oligosaccharides and for glycopeptides)
Mascot	Peptide-mass fingerprint from Matrix Science Ltd., London
PepMAPPER	Peptide-mass fingerprinting tool from UMIST, UK.
PFMUTS	Shows the possible single and double mutations of a peptide fragment from MALDI peptide-mass fingerprinting
ProFound	Search known protein sequences with peptide mass information from Rockefeller and NY Universities [or from Genomic Solutions]
Protein Prospector	UCSF tools for peptide-masses data (MS-Fit, MS-Pattern, MS-Digest, etc.)
Data identification and characterization of protein with MS/MS data	
Popitam	Identification and characterization tool for peptides with unexpected modifications (e.g. post-translational modifications or mutations) by tandem mass spectrometry (MS)
Phenyx	Protein and peptide identification/characterization from MS/MS data from GeneBio, Switzerland

(Contd.)

Table 5.6   (Continued)

Tools	Functions
Mascot	Sequence query and MS/MS ion search from Matrix Science Ltd., London
OMSSA	MS/MS peptide spectra identification by searching libraries of known protein sequences
PepFrag	Search known protein sequences with peptide fragment mass information from Rockefeller and NY Universities [or from Genomic Solutions]
ProteinProspector	UCSF tools for fragment-ion masses data (MS-Tag, MS-Seq, MS-Product, etc.)
SearchXLinks	Analysis of mass spectra of modified, cross-linked, and digested proteins whose amino acid sequence is known, from Caesar, Germany
Identification of protein with isoelectric point, molecular weight and/or amino acid composition	
AACompIdent	Identify a protein by its amino acid composition
AACompSim	Compare the amino acid composition of a UniProtKB/Swiss-Prot entry with all other entries
TagIdent	Identify proteins with isoelectric point (pI), molecular weight (MW) and sequence tag, or generate a list of proteins close to a given pI and M
MultiIdent	Identify proteins with isoelectric point (pI), molecular weight (MW), amino acid composition, sequence tag and peptide mass fingerprinting data
Other prediction or characterization tools	
ProtParam	Physico-chemical parameters of a protein sequence (amino acid and atomic compositions, isoelectric point, extinction coefficient, etc.)
Compute pI/MW	Compute the theoretical isoelectric point (pI) and molecular weight (MW) from a UniProt Knowledgebase entry or for a user sequence
GlycanMass	Calculate the mass of an oligosaccharide structure
PeptideCutter	Predicts potential protease and cleavage sites and sites cleaved by chemicals in a given protein sequence
PeptideMass	Calculate masses of peptides and their post-translational modifications for a UniProtKB/Swiss-Prot or UniProtKB/TrEMBL entry or for a user sequence

(*Contd.*)

Table 5.6   (Continued)

Tools	Functions
IsotopIdent	Predicts the theoretical isotopic distribution of a peptide, protein, polynucleotide or chemical compound
Other proteomics tools	
i. Other tools for MS data (visualization, quantitation, analysis, etc.)	
MALDIPepQuant	Quantify MALDI peptides (SILAC) from Phenyx output
MSight	Mass Spectrometry Imager
pIcarver	Visualize theoretical distributions of peptide pI (Isoelectric point) on a given pH range and generate fractions with similar peptide frequencies
ii. Other tools for 2DE data (image analysis, data publishing, etc.)	
ImageMaster/Melanie	Software for 2D PAGE analysis
Make2D-DB II	A package to build a web-based proteomics database

The servers that provide interactive access to DNA and protein sequences, and information in relation to transcription, translation, reverse transcription and mutation are listed in Table 5.7.

Table 5.7   DNA, Protein Servers

Tools	Functions
Translate	Translates a nucleotide sequence to a protein sequence
Transeq	Nucleotide to protein translation from the EMBOSS package
Graphical Codon Usage Analyser	Displays the codon bias in a graphical manner
BCM search launcher	Six frame translation of nucleotide sequence(s)
Reverse Translate	Translates a protein sequence back to a nucleotide sequence

(Contd.)

Table 5.7 (Continued)

Tools	Functions
Reverse	Transcription and Translation Tool
Genewise	Compares a protein sequence to a genomic DNA sequence, allowing for introns and frameshifting errors
LabOn Web	Elongation, expression profiles and sequence analysis of ESTs using Compugen LEADS clusters
Sequence similarity searches	
BLAST	Network Service on ExPASy
BLAST	At EMBnet-CH/SIB (Switzerland)
BLAST	At NCBI
WU-BLAST	At Bork's group in EMBL (Heidelberg)
WU-BLAST and BLAST	At the EBI (Hinxton)
BLAST	At PBIL (Lyon)
Fasta3	Fasta version 3 at the EBI
FDF	Smith/Waterman type searches on Paracel's Fast Data Finder (FDF) at EMBnet-CH
MPsrch	Smith/Waterman sequence comparison at EBI
PropSearch	Structural homolog search using a "properties" approach at Montpellier
SAMBA	Systolic Accelerator for Molecular Biological Applications
SAWTED	Structure Assignment With Text Description
Scanps	Similarity searches using Barton's algorithm
SEQUEROME	BLAST similarity search and sequence profiling at Georgetown University
SHOPS	Analysis of the genomic operon context for any group of proteins
Pattern and profile searches to predict how a sequence will fold and how a fold will match a sequence	
InterPro Scan	Integrated search in PROSITE, Pfam, PRINTS and other family and domain databases
Hits	Relationships between protein sequences and motifs
ScanProsite	Scans a sequence against PROSITE or a pattern against the UniProt Knowledge-base (Swiss-Prot and TrEMBL)
MotifScan	Scans a sequence against protein profile databases (including PROSITE)
	Pfam HMM search; scans a sequence against the Pfam protein families db [At Washington University or at Sanger Center]

(Contd.)

Table 5.7 (Continued)

Tools	Functions
Pattern and profile searches to predict how a sequence will fold and how a fold will match a sequence	
FingerPRINTScan	Scans a protein sequence against the PRINTS Protein Fingerprint Database
3of5	Complex Pattern Search
ELM	Eukaryotic Linear Motif resource for functional sites in proteins
PRATT	Interactively generates conserved patterns from a series of unaligned proteins; [at EBI/ExPASy]
PPSEARCH	Scans a sequence against PROSITE (allows a graphical output); at EBI
PROSITE scan	Scans a sequence against PROSITE (allows mismatches); at PBIL
PATTINPROT	Scans a protein sequence or a protein database for one or several pattern(s) at PBIL
SMART	Simple Modular Architecture Research Tool at EMBL
TEIRESIAS	Generate patterns from a collection of unaligned protein or DNA sequences at IBM
9aaTAD	Prediction of Nine Amino Acid Transactivation Domain
Post-translational modification prediction	
ChloroP	Prediction of chloroplast transit peptides
LipoP	Prediction of lipoproteins and signal peptides in gram-negative bacteria
MITOPROT	Prediction of mitochondrial targeting sequences
PATS	Prediction of apicoplast targeted sequences
PlasMit	Prediction of mitochondrial transit peptides in *Plasmodium falciparum*
Predotar	Prediction of mitochondrial and plastid targeting sequences
PTS1	Prediction of Peroxisomal Targeting Signal 1 containing proteins
SignalP	Prediction of signal peptide cleavage sites
DictyOGlyc	Prediction of GlcNAc O-glycosylation sites in Dictyostelium
NetCGlyc	C-mannosylation sites in mammalian proteins

(Contd.)

Table 5.7   (Continued)

Tools	Functions
NetOGlyc	Prediction of O-GalNAc (mucin-ype) glycosylation sites in mammalian proteins
NetGlycate	Glycation of epsilon amino groups of lysines in mammalian proteins
NetNGlyc	Prediction of N-glycosylation sites in human proteins
OGPET	Prediction of O-GalNAc (mucin-type) glycosylation sites in eukaryotic (non-protozoan) proteins
YinOYang	O-beta-GlcNAc attachment sites in eukaryotic protein sequences
big-PI Predictor	GPI Modification Site Prediction
DGPI	Prediction of GPI-anchor and cleavage sites (Mirror site)
GPI-SOM	Identification of GPI-anchor signals by a Kohonen Self Organizing Map
Myristoylator	Prediction of N-terminal myristoylation by neural networks
NMT	Prediction of N-terminal N-myristoylation
CSS-Palm	Palmitoylation site prediction with CSS
PrePS	Prenylation Prediction Suite
NetAcet	Prediction of N-acetyltransferase A (NatA) substrates (in yeast and mammalian proteins)
NetPhos	Prediction of Ser, Thr and Tyr phosphorylation sites in eukaryotic proteins
NetPhosK	Kinase specific phosphorylation sites in eukaryotic proteins
NetPhosYeast	Serine and threonine phosphorylation sites in yeast proteins
Sulfinator	Prediction of tyrosine sulphation sites
SulfoSite	Prediction of tyrosine sulphation sites
SUMOplot	Prediction of SUMO protein attachment sites
TermiNator	Prediction of N-terminal modification (version 3)
NetPicoRNA	Prediction of protease cleavage sites in picorna viral proteins

(Contd.)

Table 5.7　(Continued)

Tools	Functions
Post-translational modification prediction	
NetPicoRNA	Prediction of protease cleavage sites in picornaviral proteins
NetCorona	Coronavirus 3C-like proteinase cleavage sites in proteins
ProP	Arginine and lysine propeptide cleavage sites in eukaryotic protein sequences
Topology prediction	
NetNES	Leucine-rich nuclear export signals (NES) in eukaryotic proteins
PSORT	Prediction of protein subcellular localization
SecretomeP	Non-classical and leaderless secretion of proteins
TargetP	Prediction of subcellular location
TatP	Twin-arginine signal peptides
DAS	Prediction of transmembrane regions in prokaryotes using the Dense Alignment Surface method (Stockholm University)
HMMTOP	Prediction of transmembrane helices and topology of proteins (Hungarian Academy of Sciences)
PredictProtein	Prediction of transmembrane helix location and topology (Columbia University)
SOSUI	Prediction of transmembrane regions (Nagoya University, Japan)
TMAP	Transmembrane detection based on multiple sequence alignment (Karolinska Institute; Sweden)
TMHMM	Prediction of transmembrane helices in proteins (CBS; Denmark)
TMpred	Prediction of transmembrane regions and protein orientation (EMBnet-CH)
TopPred	Topology prediction of membrane proteins (France)
Primary structure analysis	
ProtParam	Physico-chemical parameters of a protein sequence (amino-acid and atomic compositions, isoelectric point, extinction coefficient, etc.)
Compute pI/MW	Compute the theoretical isoelectric point (pI) and molecular weight (MW) from a UniProt Knowledge base entry or for a user sequence

(*Contd.*)

Table 5.7 (Continued)

Tools	Functions
ScanSite pI/MW	Compute the theoretical pI and MW, and multiple phosphorylation states
MW, pI, Titration curve	Computes pI, composition and allows to see a titration curve
Radar	De novo repeat detection in protein sequences
REP	Searches a protein sequence for repeats
REPRO	De novo repeat detection in protein sequences
TRUST	De novo repeat detection in protein sequences
XSTREAM	De novo tandem repeat detection and architecture modelling in protein sequences
SAPS	Statistical analysis of protein sequences at EMBnet-CH [Also available at EBI]
Coils	Prediction of coiled coil regions in proteins (Lupas's method) at EMBnet-CH [Also available at PBIL]
Paircoil	Prediction of coiled coil regions in proteins (Berger's method)
Paircoil2	Prediction of the parallel coiled coil fold from sequence using pairwise residue probabilities with the Paircoil algorithm.
Multicoil	Prediction of two- and three-stranded coiled coils
2ZIP	Prediction of leucine zippers
PESTfind	Identification of PEST regions at EMBnet Austria
HLA_Bind	Prediction of MHC type I (HLA) peptide binding
PEPVAC	Prediction of supertypic MHC binders
RANKPEP	Prediction of peptide MHC binding
SYFPEITHI	Prediction of MHC type I and II peptide binding
ProtScale	Amino acid scale representation (Hydrophobicity, other conformational parameters, etc.)
Drawhca	Draw an HCA (Hydrophobic Cluster Analysis) plot of a protein sequence
Peptide Builder	
Protein Colourer	Tool for colouring your amino acid sequence
Three To One and One to Three	Tools to convert a three-letter coded amino acid sequence to single letter code and vice versa

(Contd.)

Table 5.7 (Continued)

Tools	Functions
Three-/one-letter amino acid converter	Tool which converts amino acid codes from three-letter to one-letter and vice versa.
Colorseq	Tool to highlight (in red) a selected set of residues in a protein sequence
HelixWheel/HelixDraw	Representations of a protein fragment as a helical wheel
RandSeq	Random protein sequence generator
Secondary structure prediction	
AGADIR	An algorithm to predict the helical content of peptides
APSSP	Advanced Protein Secondary Structure Prediction Server
GOR	Garnier *et al.* 1996
HNN	Hierarchical Neural Network method (Guermeur, 1997)
HTMSRAP	Helical TransMembrane Segment Rotational Angle Prediction
Jpred	A consensus method for protein secondary structure prediction at University of Dundee
JUFO	Protein secondary structure prediction from sequence (neural network)
nnPredict	University of California at San Francisco (UCSF)
Porter	University College Dublin
PredictProtein	PHDsec, PHDacc, PHDhtm, PHDtopology, PHDthreader, MaxHom, EvalSec from Columbia University
Prof	Cascaded Multiple Classifiers for secondary structure prediction
PSA	BioMolecular Engineering Research Center (BMERC)/Boston
PSIpred	Various protein structure prediction methods at Brunel University
SOPMA	Geourjon and Deléage, 1995
SSpro	Secondary structure prediction using bidirectional recurrent neural networks at University of California
DLP-SVM	Domain linker prediction using SVM at Tokyo University of Agriculture and Technology

(*Contd.*)

Table 5.7   (Continued)

Tools	Functions
Tertiary structure	
1) Tertiary structure analysis:	
iMolTalk	An Interactive Protein Structure Analysis Server
MolTalk	A computational environment for structural bioinformatics
qCOPS	Navigation through fold space and the instantaneous visualization of pairwise structure similarities
Seq2Struct	A web resource for the identification of sequence-structure links
STRAP	A structural alignment program for proteins
TLSMD	TLS (Translation/Libration/Screw) Motion Determination
TopMatch-web	Protein structure comparison
2) Tertiary structure prediction	
i) Comparative modeling	
SWISS-MODEL	An automated knowledge-based protein modelling server
3Djigsaw	Three-dimensional models for proteins based on homologues of known structure
CPHmodels	Automated neural-network based protein modelling server
ESyPred3D	Automated homology modelling program using neural networks
Geno3d	Automatic modelling of protein three-dimensional structure
SDSC1	Protein Structure Homology Modelling Server
ii) Threading	
3D-PSSM	Protein fold recognition using 1D and 3D sequence profiles coupled with secondary structure information (Foldfit)
Fugue	Sequence-structure homology recognition
HHpred	Protein homology detection and structure prediction by HMM-HMM comparison

(Contd.)

Table 5.7　(Continued)

Tools	Functions
Libellula	Neural network approach to evaluate fold recognition results
LOOPP	Sequence to sequence, sequence to structure, and structure to structure alignment
SAM-T02	HMM-based Protein Structure Prediction
Threader	Protein fold recognition
ProSup	Protein structure superimposition
SWEET	Constructing 3D models of saccharides from their sequences

iii) *Ab initio*

HMMSTR/Rosetta	Prediction of protein structure from sequence

Assessing tertiary structure prediction

Tools	Functions
Anolea	Atomic Non-Local Environment Assessment
	Biotech Validation Suite for Protein Structures
EVA	EValuation of Automatic protein structure prediction
LiveBench	Continuous Benchmarking of Structure Prediction Servers
NQ-Flipper	Validation and correction of asparagine and glutamine side-chain amide rotamers in protein structures solved by X-ray crystallography
PROCHECK	Verification of the stereochemical quality of a protein structure
ProSA-web	Recognition of errors in 3D structures of proteins
What If	Protein structure analysis program for mutant prediction, structure verification, molecular graphics

Quaternary structure

Tools	Functions
MakeMultimer	Reconstruction of multimeric molecules present in crystals
EBI PISA	Protein Interfaces, Surfaces and Assemblies
PQS	Protein Quaternary Structure Query form at the EBI
ProtBud	Comparison of asymmetric units and biological units from PDB and PQS

(*Contd.*)

Table 5.7 (Continued)

Tools	Functions
Molecular modelling and visualization tools	
Swiss-PdbViewer	A program to display, analyse and superimpose protein 3D structures
Ascalaph Packages	
Astex Viewer	
Jmol	
MolMol	
MovieMaker	For rapid rendering of protein motions and interactions
PyMol	
Rasmol	Free software for visualizing molecular structures
VMD	
YASARA	Molecular graphics, modelling, simulations and Learning
Prediction of disordered regions	
DisEMBL	Protein disorder prediction
GlobPlot	Protein disorder/order/globularity/domain predictor
Sequence alignment	
i) Binary	
SIM + LALNVIEW	Alignment of two protein sequences with SIM, results can be viewed with LALNVIEW
LALIGN	Finds multiple matching subsegments in two sequences
Dotlet	A Java applet for sequence comparisons using the dot matrix method
Sequence alignment	
ii) Multiple	
Decrease redundancy	Reduce a set of sequences into a non-redundant set
Nomad (Neighborhood Optimization for Multiple Alignment Discovery)	Ungapped local multiple alignment, optimized for protein sequences, even when distantly related
CLUSTALW	[At EBI, PBIL, My Hits or at EMBnet-CH]

(Contd.)

Table 5.7　(Continued)

Tools	Functions
KALIGN	An accurate and fast multiple sequence alignment algorithm [At Karolinska Institute or at EBI]
MAFFT	[At Kyushu University, EBI or at MyHits]
Muscle	[At Berkeley or at BioAssist]
T-Coffee	[At MyHits, BioAssist or at EBI]
MSA	At Genestream (IGH)
DIALIGN	Multiple sequence alignment based on segment-to-segment comparison, at University of Bielefeld, Germany
Match-Box	At University of Namur, Belgium at Washington University Multalin [At INRA or at PBIL]
MUSCA	Multiple sequence alignment using pattern discovery, at IBM
Alignment analysis	
AMAS	Analyse Multiply Aligned Sequences
Bork's alignment tools	Various tools to enhance the results of multiple alignments (including consensus building).
CINEMA	Colour Interactive Editor for Multiple Alignments
ESPript	Tool to print a multiple alignment
MaxAlign	Post-processing of alignments by removing sequences (taxa) with many gaps
PhyloGibbs	Gibbs motif sampler incorporating phylogeny and tracking statistics
SVA	Sequence Variability Analyser for multiple alignments
PVS	A protein variability server optimized for conserved epitope discovery
PVS	A protein variability server optimized for conserved epitope discovery
WebLogo	Sequence logos at Berkeley/USA
plogo	Sequence logos at CBS/Denmark
GENIO/logo	Sequence logos at Stuttgart/Germany
SeqLogo	Sequence logos at the Immunomedicine Group, Facultad de Medicina, U.C.M, Spain (The Molecular Immunology Foundation (MIF) does not exist anymore)

(*Contd.*)

Table 5.7   (Continued)

Tools	Functions
Gateways	
Biosyn Gizmo	Bundle of databases (siRNA, protein, peptide antigen) and tools
Phylogenetic analysis	
Phylogenetic programs	List of phylogenetic packages and free servers (PHYLIP pages)
PHYLIP	Server for phylogenetic analysis using the PHYLIP package
BIONJ	Server for NJ phylogenetic analysis
Evolutionary Trace Server (TraceSuite II)	Maps evolutionary traces to structures
Biological text analysis	
AcroMed	A computer generated database of biomedical acronyms and the associated long forms extracted from the recent Medline abstracts
BioMinT	Mining the biomedical literature
GPSDB	Gene and Protein Synonym Database
MedMiner	Extract and organize relevant sentences in the literature based on a gene, gene-gene or gene-drug query
XplorMed	Explore a set of abstracts derived from a bibliographic search in MEDLINE

EMBL NUCLEOTIDE SEQUENCE DATABASE

The EMBL nucleotide sequence database (also known as EMBL-Bank) from European Bioinformatics Institute constitutes Europe's primary nucleotide sequence resource. Main sources for DNA and RNA sequences are direct submissions from individual researchers, genome sequencing projects and patent applications. The database is produced in an international collaboration with GenBank (the DNA database from NCBI, USA) and DDBJ and EMBL. Each of the three groups collects a portion of the total sequence data reported worldwide, and all new and updated database entries are exchanged between the groups on a daily basis. The database is produced, maintained and distributed by DDBJ at National Institute of Genetics; sequence may be submitted to it from all corners of the world by means of a web-based data submission tool. Similarly, GenBank incorporates sequences from publicly available sources, primarily from direct submission and large-scale sequence projects. There is not only increase in size of the database but the diversity of data sources that made it convenient to split the GenBank into 17 smaller and discrete divisions (as mentioned in Table 5.8). The biggest division is the EST due to its rapid growth; it is divided into 23 pieces. This speeds up the specific searches and restricts the queries to particular database subsets.

Table 5.8 GenBank divisions

Division code	Description
PRI	Primate sequences
ROD	Rodent sequences
MAM	Other mammalian sequences
VRT	Other vertebrate sequences
INV	Invertebrate sequences
PLN	Plant, fungal, and algal sequences
BCT	Bacterial sequences
RNA	Structural RNA sequences
VRL	Viral sequences
PHG	Bacteriophage sequences
SYN	Synthetic sequences
UNA	Unannotated sequences
EST	EST sequences
PAT	Patent sequences
STS	STS sequences
GSS	GSS sequences (genome survey sequences)
HTG	HTGS sequences (high-throughput genomic sequences)

In January 1998, EMBL contained more than a million entries, representing more than 15,500 species, but with model system predominating (*Homo sapiens, Cenorhabditis elegans, Saccharomyces cerevisiae, Mus musculus* and *Arabidopsis thaliana* together constitute more than 50% of the resources. The current database releases Release 95, June 2008, release notes and user manual are available from the EBI servers.

A publication in *Nucleic Acids Research* 2008 Jan. (Database issue). Further information and details available at http://nar.oxfordjournals.org/cgi/content/full/gkm1018v1

Information can be retrieved from EMBL using the SRS; this link principal DNA and protein sequence database with motif, structure, mapping and other specialist databases, and including links to MedLine facility. Other links associated with EMBL with their explanation are as shown in Table 5.9.

The EMBL nucleotide sequence database is part of the Protein And Nucleotide Database Group (**PANDA**). This is jointly headed by Dr. Rolf Apweiler and Dr. Ewan Birney, with Dr. Birney taking responsibility for nucleotides.

Table 5.9   The link facilities available at EMBL

Link	Explanation
Access	Database queries, cmpleted genomes web server, FTP archives (EMBL release, alignments,etc.), EMBL sequence version archive (SVA),bowse by geography http://www.ebi.ac.uk/embl/Access/index.html
Submission	Primary sequence submissions, third party annotation, updates and alignment submissions http://www.ebi.ac.uk/embl/Submission/index.html
Documentation	Release notes user manual, Information for Submitters, FAQ, Release information, Forthcoming Changes , EMBL database statistics, Feature table, XML documentation, Sample entry, Accession Number Prefix Codes, Examples of annotation, EMBL Features & Qualifiers, DE line standards, Database Policies http://www.ebi.ac.uk/embl/Documentation/index.html
Publications	Group publications http://www.ebi.ac.uk/panda/EMBL.htm/
People	Group members http://www.ebi.ac.uk/panda/people.html
Contact	How to contact the EMBL nucleotide sequence database http://www.ebi.ac.uk/embl/Contact/
News	List of recent changes on this site http://www.ebi.ac.uk/embl/News/news.html

ENSEMBL

Ensembl (http://www.ensembl.org) is a joint project between EMBL, and the Wellcome Trust Sanger Institute (WTSI) to develop a software system which produces and maintains automatic annotation on selected eukaryotic genomes. The project is based at the Wellcome Trust Genome Campus, Hinxton, United Kingdom.

The Ensembl project consists of:

- a database schema and associated API to store genomic information
- extension databases to represent functional, comparative and variational genomics
- a "genebuild pipeline" which takes sequence data and builds gene models
- databases containing information for approximately 40 genomes
- a website for which users can browse this information
- an FTP site which stores the release data along with dumps of genomic sequence and it's associated annotation
- public MySQL instances containing copies of the databases behind the Ensembl website.

The Ensembl project aims to provide:

- Accurate, automatic analysis of genome data
- Analysis and annotation maintained on the current data
- Presentation of the analysis to all via the web
- Distribution of the analysis to other bioinformatics laboratories

The initial Ensembl project concentrates on vertebrate genomes. A new project Ensembl genomes based at the EBI will be extending the project into plants, bacteria, protists and metazoa. Additionally a number of other projects use some or all the parts of the Ensembl project to represent their data.

The commitments of the Ensembl project are:

- Release of data and analysis into the public domain immediately.
- Open, collaborative software development: Ensembl imposes no restrictions on access to, or use of, the data provided and the software used to analyse and present it. For more details see the code licence and disclaimer.
- Collaboration on agreed standards for distribution.
- Timely development.

Current Species in Ensembl Pipeline

Ensembl data is organized into several species-specific and multi-species MySQL databases. Each database is named using the format <species>_<database type>_<release number>_<data version>. For each supported species, a core database contains the DNA sequences, gene annotations, external references, etc. Databases of type "other features" are provided for each supported species (except for the low-coverage genomes) and include EST genes, external annotation sets and other data. Variation databases that include dbSNP and resequencing data, are provided for 10 species. This year, we introduced a functional genomics database, initially released for human and mouse, to support functional data types assayed by whole-genome tiling arrays or high-throughput sequencing (See subsequently). Comparative genomics data and the supporting data for the Ensembl BioMart data mining tool are provided in multi-species databases. The annotated genomes include most fully sequenced vertebrates and selected model organisms. All of them are eukaryotes, there are no prokaryotes. Currently this includes:

CHORDATES

Mammals

Primates Bush baby, Chimp, Human, Macaque, Mouse Lemur, Orang-utan

Rodents Guinea pig, Mouse, Pika, Rabbit, Rat, Squirrel, Tree shrew

Laurasiatheria Cat, Cow, Dog, Hedgehog, Horse, Microbat, Shrew, Pig

Afrotheria elephant Lesser hedgehog tenrec

Xenarthra

Armadillo

Marsupials and Monotremes Opossum, Platypus

Birds

Chicken

Reptiles and Amphibians

Xenopus tropicalis, Anole Lizard

Fish

Takifugu rubripes (Fugu), *Tetraodon nigroviridis* (Green spotted pufferfish), *Danio rerio* (Zebrafish), *Oryzias latipes* (Medaka), *Gasterosteus aculeatus* (Stickleback), *Petromyzon marinus* (Sea lamprey)

Ancient Relatives

Ciona intestinalis, Ciona savignyi

INVERTEBRATES

Insects

Anopheles gambiae (Mosquito), Fruitfly, *Aedes aegypti* (Mosquito)

Worm

Caenorhabditis elegans

YEAST

Saccharomyces cerevisiae (Baker's yeast)

Advantages of Ensembl

The service is used by molecular biologists and bioinformaticians around the world working with genome data of the above organisms. The predictions of coding, controlling and other elements in the genomes can be compared with primary research data and with common repositories of current genomic knowledge (biological databases). The comparison of organisms (comparative genomics or also intergenomics) with respect to their gene structures and the coded proteins is of special interest.

Ensembl is organized as an open project that encourages contributions from outside. Ensembl includes detailed and curated information about the gene functions. Ensembl also has very good links to remote databases. The Ensembl data can be queried using any sequence database accession number. The Ensembl server can also be used for converting from one numbering system to another. Data collected in Ensembl includes genes, SNPs (single nucleotide polymorphisms), repeats, and homologies. Genes may either be known experimentally, or deduced from the sequence. Because the experimental support for annotation of the human genome is so variable, Ensembl presents the supporting evidence for identification of every gene.

Very extensive link to other databases containing related information, such as OMIM, or expression databases, extend the accessible information.

Ensembl continues to make extensive use of the DAS (Distributed Annotation System) protocol. The Ensembl genome browser with DAS client functionality, which allows researchers around the world to remotely host data sources and view these on major Ensembl displays including CytoView, ContigView (representing overlapping regions of a gene), GeneView and ProtView. The client visualization support through DAS to include a colour gradient, histogram and tiling array "wiggle" format. The new visualization options are particularly applicable to dense genome data such as that produced by whole-genome tiling array experiments. The current Ensembl data is integrated into other DAS clients. Data available for integration into DAS clients includes transcripts, ditag data, markers, karyotype information, repeats and DNA and protein align features including cDNA alignments and UniProt alignments. DAS sources setup by Ensembl are also automatically registered with the DAS registry.

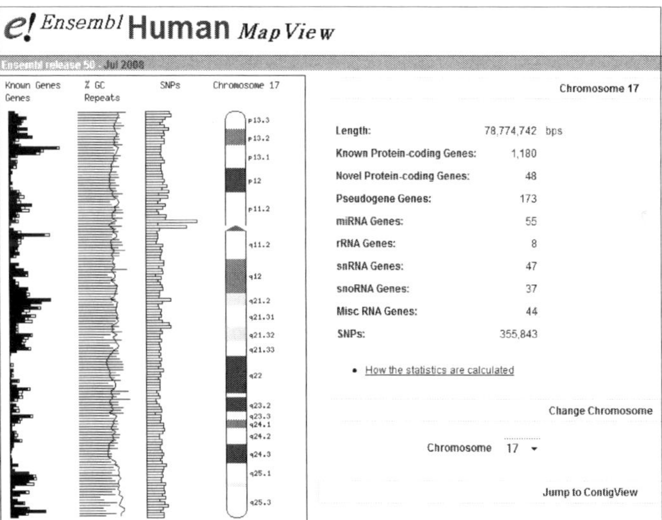

Figure 5.8   The map view of human chromosome number 17 with its known genes, %GC repeats, SNPs, and other features

Ensembl enables users to identify regions in human genome sequence via several types of lookups or searches, such as

- Browsing starting at the chromosome level then zooming in with help of control panel in the display,
- Gene name,

- Relation to diseases, via OMIM,
- Ensembl ID if the user knows it,
- Chromosomal map view,
- General text search for specific gene on the specific chromosome.

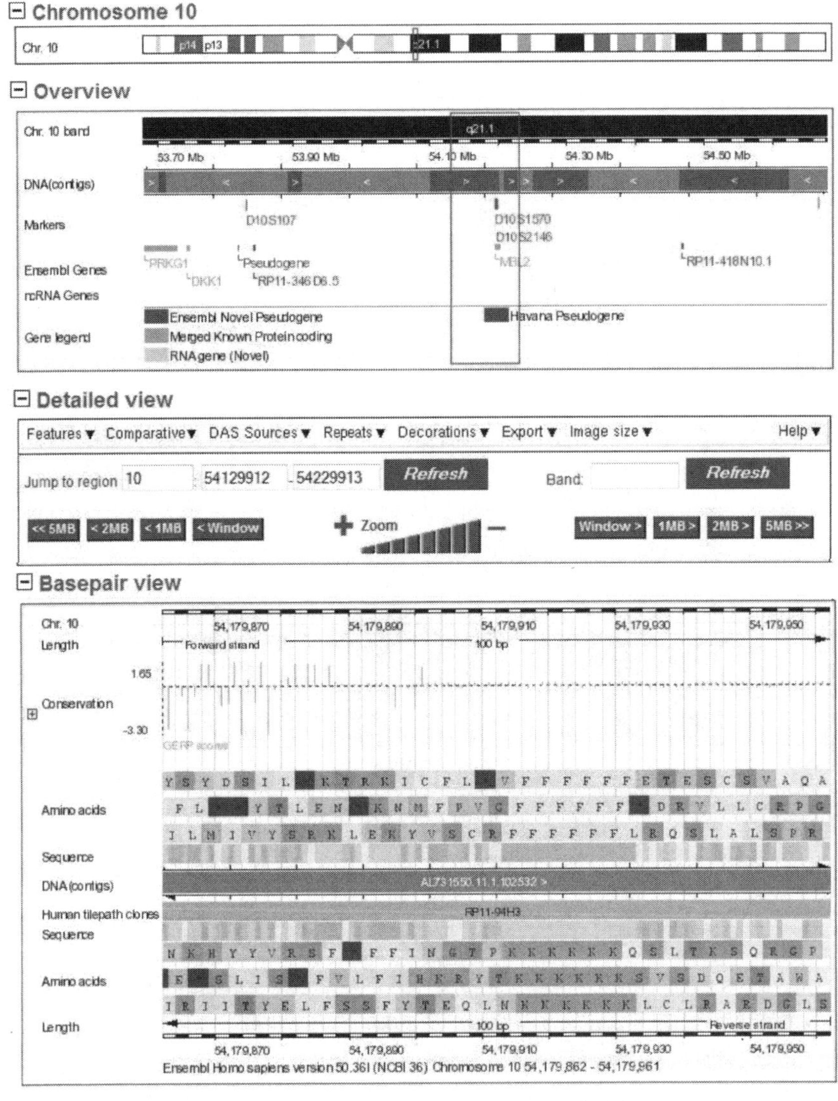

Figure 5.9 The contig view of human chromosome number 10 showing the details of the region of the chromosome between 54129912 to 54229913 loci

Ensembl Human map view provides information in relation to the known protein-coding genes, percentage of GC repeats, miRNA genes, rRNA genes, snRNA genes, SNPs and other features of a specifically numbered human chromosome (Figure 5.8). A text search in Ensembl human **contig view** for chromosome 10 produced the page as displayed in the Figure 5.9, showing the region between 54129912 to 54229913. The upper frame shows a megabase, mapped around the q21.1 band of chromosome 10. It reports markers, Ensembl novel pseudogenes, merged known protein-coding genes and RNA genes. The bottom frame shows a detailed view. Note the control panels between the two frames that permit navigation and "zooming". This view is somewhat curtailed in the display. The bottom frame shows basepair view of a 0.1 megabase region showing the length of forward and reverse strands with the corresponding sequences of amino acids.

REVIEW QUESTIONS

1. What is database? Comment on database management system with its significance.

2. Define biological database. Explain the role of the public domain bioinformatics facilities that are associated with biological databases.

3. Describe the different classes of biological databases with their importance.

4. Enlist the keywords that provide the link with protein related databases maintained by European Bioinformatics Institute.

5. Explain the role of PubMed as the Central Repository for Biological Database.

6. What is Entrez? Explain its role in linking different data sources.

7. Define Online Mendelian Inheritance in Man (OMIM). Explain six digit number classification method for diseases and genes.

8. Comment on Expert Protein Analysis System (ExPASy).

9. Enlist various database search tools in ExPASy server dedicated to protein identification with its importance.

10. What is EMBL Nucleotide Sequence Database? Describe various link facilities available with EMBL.

11. Comment on GenBank with its division codes that speed up the specific search.

12. What are the aims and commitments of Ensembl project?

13. How does a bioinformatician retrieve information of genomic database using Ensembl?

14. Write short notes on
    - i. DNA Database of Japan
    - ii. Compugen
    - iii. Ocimum Biosolutions
    - iv. PubMed tags
    - v. Human Protein Reference Database
    - vi. OMIM Morbid Map
    - vii. MIM codes
    - viii. PANDA

INTRODUCTION

Sequence alignment is a way of arranging the primary sequences of DNA, RNA, or protein to identify the regions of similarity that may be a consequence of functional, structural, or evolutionary relationships between the sequences. Aligned sequences of nucleotide or amino acid residues are typically represented as rows within a matrix. Gaps are inserted between the residues so that residues with identical or similar characters are aligned in successive columns. Sequence analysis is the problem of comparing two sequences while allowing certain mismatches between them the process which lie at the heart of bioinformatics analysis. The list of sequence alignment software is given in Table 6.1. Sequence alignment can be used as basis for prediction of structure and function of uncharacterized sequences. It provides interface for the relatedness of two sequences under study. When a sequence alignment is generated correctly, it indicates that there is evolutionary relationship between two sequences; the regions that are aligned but not identical represent that the residue might be deleted, inserted or substituted in one of the sequence during evolution.

This chapter introduces the problem of **pairwise sequence alignment** or **inexact matching**. In addition, the chapter also presents the problem and provide biological motivation with the define similarity and difference between sequences and present algorithms for computing them, with analysis of the complexity of these algorithms. The algorithms use the dynamic programming technique that determines optimal alignment by matching two sequences for all possible pairs of characters between the two sequences. For each algorithm the following information will be given:

- ▣ Intuitive explanation of the recursive process.
- ▣ Formal definition of the recursive process.
- ▣ Discussion of the complexity.

BIOLOGICALLY MOTIVATED PROBLEMS IN COMPUTER SCIENCE

A large variety of the biologically motivated problems in computer science primarily involve sequences or strings. For instance:

- ▣ Reconstructing long sequences of DNA from overlapping sequence fragments.
- ▣ Determining physical and genetic maps from probe data under various experiment protocols.
- ▣ Storing, retrieving and comparing DNA sequences in databases.
- ▣ Comparing two or more sequences for similarities.
- ▣ Searching databases for related sequences and subsequences.
- ▣ Exploring frequently occurring patterns of nucleotides.
- ▣ Finding informative elements in protein and DNA sequences.

Many of these research problems aim at learning about the functionality or structure of a protein without performing any experiments and without actually having to physically construct the protein itself. The basic idea is that similar sequences produce similar proteins. Thus, in order to predict the characteristics of a protein using only its sequence data, one can use the structure/function information on known proteins with similar sequences available in databases.

For instance, when considering protein folding, it will usually suffice for two proteins to have 25% sequence identity for their three-dimensional structures and thus more importantly their function to be almost identical. A classical example is the establishment of an association between cancer and uncontrolled cell growth by Simian sarcoma virus oncogene *v-sis*, which is derived from the gene encoding a platelet-derived growth factor. This discovery was enabled by comparing the sequence of a cancer-associated gene against the sequences of proteins which had already been known as influencing the cell growth. The correlation between these two sequences was very high, proving the connection between cancer and cellular growth.

SIMILARITY AND DIFFERENCE OF DNA

The resemblance of two DNA sequences taken from different organisms can be explained by the theory that "all contemporary genetic material has one common ancestral DNA". According to this theory, during the course of evolution mutations occurred, creating differences between families of contemporary species. Most of these changes are due to local mutations, each modifying the DNA sequence at a specific manner. These local modifications between nucleotide sequences or more generally, between strings over an arbitrary alphabet, can be either:

▣ **Insertion**   An insertion of a base (letter) or several bases to the sequence.

Example

..ATGCATGC.. ⇒ ..ATGACATGC..

is an insertion at position 4.

▣ **Deletion**   Deleting a base (or more) from the sequence.

Example

..ATGCATGC.. ⇒ ..ATCATGC..

is a deletion at position 3.

▣ **Substitution**   Replacing a base by another in the sequence.

Example

..ATGCATGC.. ⇒ ..AGGCATGC..

is a substitution at position 2.

Insertion and deletion are the reverse of one another: given two sequences, if the insertion of a character (or more) into one yields the other, then equivalently its deletion from the latter sequence transforms it to the first one. Due to this reciprocity between insertion and deletion, they are usually called **indel** (that is, insertion or deletion mutations).

The notion of distance derives its definition from the concept of mutations by assigning weights to each mutation: given two sequences, the distance between them is the minimal sum of weights for a set of mutations transforming one into the other. The notion of similarity derives its definition from the concept of one ancestral ancient DNA by assigning weights corresponding to resemblance: given two sequences, the similarity between them is the maximal sum of such weights. Very short or very similar sequences can be aligned by hand; however, most interesting problems require the alignment of lengthy, highly variable or extremely numerous sequences that cannot be aligned solely by human effort. Instead, human knowledge is primarily applied in constructing algorithms to produce high-quality sequence alignments, and occasionally in adjusting the final results to reflect patterns that are difficult to represent algorithmically (especially in the case of nucleotide sequences). Computational approaches to sequence alignment generally fall into two categories: **global alignments** and **local alignments**. Calculating a global alignment is a form of global optimization that "forces" the alignment to span the entire length of all query sequences. By contrast, local alignments identify regions of similarity within long sequences that are often widely divergent overall. Local alignments are often preferable, but can be more difficult to calculate because of the additional challenge of identifying the regions of similarity. A variety of computational algorithms have been applied to the sequence alignment problem, including slow but formally optimizing methods like dynamic programming and efficient heuristic or probabilistic methods designed for large-scale database search.

NOMENCLATURE

It is important to mention that biology and computer science use different nomenclature. Here is the table that compares the notations:

Biology	Computer science
Sequence	String, word
Subsequence	Substring (contiguous)
N/n	Subsequence
N/N	Exact matching
Alignment	Inexact matching

Subsequence (in computer science) is a non-contiguous segment of a sequence. In particular, a "subsequence" will mean a contiguous sequence of letters.

SIMPLEST MODEL—EDIT DISTANCE

The edit distance between two sequences is the minimal number of edit operations (insertions, deletions and substitutions) needed to transform one sequence into the other. Most of the changes to DNA during evolution are due to the three common local mutations: 1) insertion, 2) deletion and 3) substitution. Therefore the edit distance can be used to measure roughly the number of DNA replications that occurred between two DNA sequences. Sequence similarity is useful in hypothesizing the function of a new sequence assuming that sequence similarity implies structural and functional similarity. Sequence function analysis primarily deals with ascertaining the function of new genes or proteins based on knowledge about known genes/proteins including their sequence and function in the form of enormous sequence databases.

The general schematic could be given as:

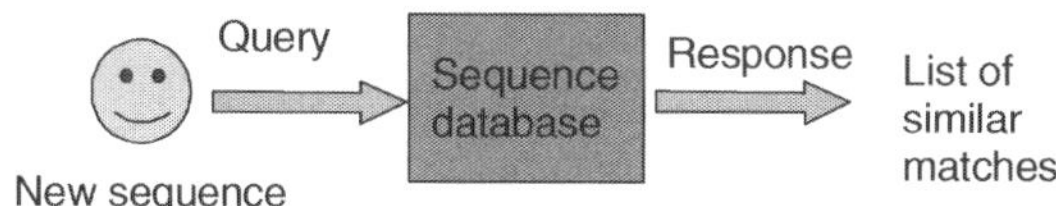

Example Given two sequences a c c t g a and a g c t a, the minimal number of edit operations required to transform one into the other is 2:

```
a c c t g a     a g c t g a

a g c t g a     a g c t - a

a g c t a
```

First, substitution of c by g is applied on the sequence a c c t g a. As a result the sequence a g c t g a is obtained. Then, an indel operation (deletion of g) is applied.

Remark The definition of edit distance implies that all operations are done on one sequence only and the representation shown above might make the false impression that the order of the changes is known.

ALIGNMENT

An alignment of two sequences *S* and *T* is obtained by first inserting chosen spaces, either into, at the ends of or before *S* and *T*, and then placing the two resulting sequences one above the other so that every character or space in either sequence opposite a unique character or a unique space in the other sequence. In the alignment model, each two character alignment and character-space alignment is given a score (weight). Usually, insert and delete (indel) operations (alignment of a character and a space) are given the same score. Using alignment algorithms, we search for the minimal scoring (or the maximum negative scoring), representing the minimal difference or maximum similarity between the two sequences. Biological models consider the significance of each mutation and score the alignment operations accordingly. Therefore, the alignment distance can be used to estimate the "biological difference" of two DNA or protein sequences (Table 6.1). The substitution matrix $S(i,j)$ represents the weight of each possible alignment.

Example   The aligned sequences:

```
SEQ 1    GTAGTACAGCT–CAGTTGGGATCACAGGCTTCT
         | | | |   | |   | | |   | | | | | |       | | | | | |     | | |
SEQ 2    GTAGAACGGCTTCAGTTG– – –TCACAGCGTTC–
```

Distance 1   Match 0, substitution 1, indel 2 $\Rightarrow$ distance = 14.

Distance 2   Match 0, $d(A,T)=d(G,C)=1$, $d(A,G)=1.5$ indel 2 $\Rightarrow$ distance = 14.5.

Similarity   Match 1, substitution 0, indel -1.5 $\Rightarrow$ similarity = 16.5.

General setup   Substitution matrix $S(i,j)$, indel $S(i,-)$ or $S(-,j)$.

Table 6.1   Sequence alignment software for database search

Name	Description	Sequence type
BLAST	k-type local search (Basic local alignment search tool)	Both
Combinatorial extension	Structural alignment search	Protein
FASTA	k-tuple local search	Both
GGSEARCH/ GLSEARCH	Global : Global (GG), Global : Local (GL) alignment with statistics	Protein
HMMER	Hidden Markov profile search	Protein/DNA
IDF	Inverse Document Frequency	Both
SAM	Hidden Markov profile search	Protein/DNA
SSEARCH	Smith–Waterman search (more sensitive than FASTA)	Both

PAIRWISE ALIGNMENT

Pairwise sequence alignment methods are used to find the best-matching piecewise (local) or global alignments of two query sequences. Pairwise alignments can only be used between two sequences at a time, but they are efficient to calculate and are often used for methods that do not require extreme precision (such as searching a database for sequences with high homology to a query). The three primary methods of producing pairwise alignments are **dot-plot method, dynamic programming**, and **word method**. Although each method has its individual strengths and weaknesses, all the three pairwise methods have difficulty with highly repetitive sequences of low information content especially where the number of repetitions differ in

Table 6.2   List of software for pairwise alignment

Name	Description	Sequence type*	Alignment t
Bioconductor Biostrings::pairwise alignment	Dynamic programming	Both	Both + Enc
BioPerl dpAlign	Dynamic programming	Both	Both + Enc
BLASTZ	Seeded pattern-matching	Nucleotide	Local
DNADot	Web-based dot-plot tool	Nucleotide	Global
DOTLET	Java-based dot-plot tool	Both	Global
GGSEARCH, GLSEARCH	Global: Global (GG), Global: Local (GL) alignment with statistics	Protein	Global in qu
JAligner	Open source Java implementation of Smith–Waterman	Both	Local
LALIGN	Multiple, non-overlapping, local similarity (same algorithm as SIM)	Both	Local non-overlap
matcher	Memory-optimized Needleman but slow dynamic programming (based on LALIGN)	Both	Local
MCALIGN2	Explicit models of indel evolution	DNA	Global
MUMmer	Suffix-tree based	Nucleotide	Global
needle	Needleman–Wunsch dynamic programming	Both	Global
Ngila	Logarithmic and affine gap costs and explicit models of indel evolution	Both	Global
PatternHunter	Seeded pattern-matching	Nucleotide	Local

(*Contd.*)

Table 6.2 (Continued)

Name	Description	Sequence type*	Alignment type**
ProbA (also propA)	Stochastic partition function sampling via dynamic programming	Both	Global
PyMOL	"Align" command aligns sequence and applies it to structure	Protein	Global (by selection)
REPuter	Suffix-tree based	Nucleotide	Local
SEQALN	Various dynamic programming	Both	Local or global
SIM, GAP, NAP, LAP	Local similarity with varying gap treatments	Both	Local or global
SIM	Local similarity	Both	Local
SLIM Search	Ultra-fast blocked alignment	Both	Both
SSEARCH	Local (Smith–Waterman) alignment with statistics	Protein	Local
stretcher	Memory-optimized but slow dynamic programming	Both	Global
tranalign	Aligns nucleic acid sequences given a protein alignment	Nucleotide	NA
water	Smith–Waterman dynamic programming	Both	Local
wordmatch	k-tuple pairwise match	Both	NA
YASS	Seeded pattern-matching	Nucleotide	Local

*Sequence type: Protein or nucleotide. **Alignment type: Local or global the two sequences to be aligned. One way of quantifying the utility of a given pairwise alignment is the "Maximum Unique Match" (MUM), or the longest subsequence that occurs in both query sequence. Longer MUM sequences typically reflect closer relatedness. The list of software for pairwise alignment is given in Table 6.2.

Dot-plot Method

It is the most basic sequence alignment method, also referred to as dot matrix method, developed by Gibbs and McIntyre (1970). It is a simple graphical picture that gives an overview of the similarities between two sequences in two dimensional matrix. In this approach, two sequences to be compared are written in the horizontal and vertical axes of the matrix. The comparison is done by scanning each residue of one sequence for similarity with all residues in the other sequence. If a residue match is found, a dot is placed within the graph. Otherwise the matrix positions are left blank. When the two sequences have substantial regions of similarity, many dots line-up to form contiguous diagonal lines, which reveal the sequence alignment (Figure 6.1a and b). If there are interruptions in the middle of a diagonal line, they indicate insertions, deletions or inverted repeats such as

ACDEFGHIIHGFEDCAACDEFGHIIHGFEDCA

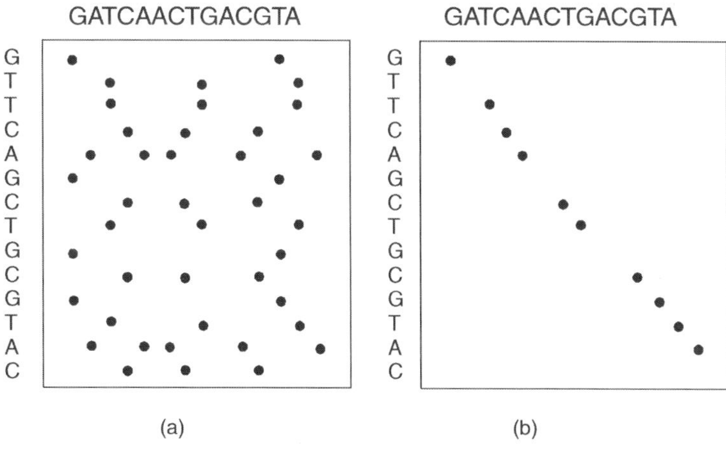

(a)   (b)

Figure 6.1   (a) Dot-plot showing the regions of identity of two sequences (b) Dot-plot with a diagonal line after eliminating random matches

Dot-plots can also be used to assess repetitiveness in a single sequence. A sequence can be plotted against itself and regions that share significant similarities will appear as lines off the main diagonal. Dot-plot showing identities between a repetitive sequence $\overrightarrow{AB}\overrightarrow{RAC}\overrightarrow{ADAB}\overrightarrow{RACAD}$ and itself. In the self comparison, the repeats appear on several subsidiary diagonals parallel to the main diagonal (Figure 6.2). The dot-plots of very closely related sequences will appear as a single line along the matrix's main diagonal.

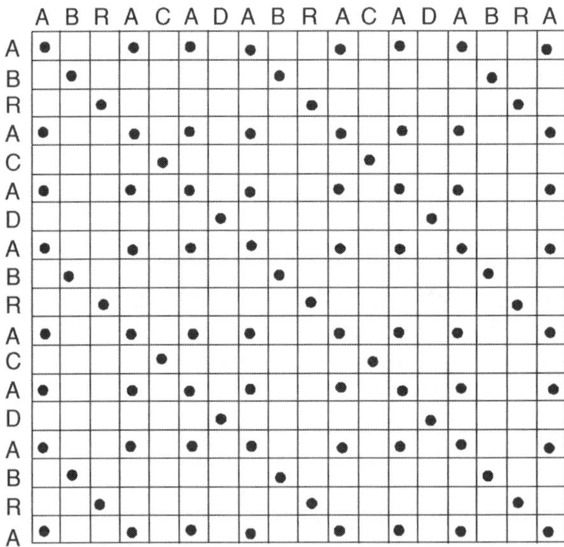

Figure 6.2 Dot-plot of self-comparison showing identities between a repetitive sequence and itself in the form of parallel diagonals to the main diagonal.

Java program to draw dot-plots The program shown below is written in the TextPad and saved as Dot-plot.java that reads:

1. A general title for the job, printed at the top of the output drawing (first line of input).
2. A dot will appear in the dot-plot if it is in the centre of a stretch of residues of length **window** such that the number of matches is ≥ **threshold**.
3. The two sequences S1 and S2 are given for sequence alignment.
4. The program draws a dot-plot similar to those shown in the text. The output is in the form of diagonal graph.

```java
//Dot-plot Technique
import java.io.*;
import javax.swing.*;
import java.awt.*;
import java.awt.event.*;
class Dot-plot
{
        public static void main(String[]args)
        {
                Dot-plotFrame ob=new Dot-plotFrame();
                ob.show();
        }
}
class Dot-plotFrame extends JFrame
{
        public Dot-plotFrame()
        {
                setSize(700,700);
                setLocation(200,10);
                setTitle("Dot-plot Technique");
          Container cp=getContentPane();
          cp.add(new Dot-plotPanel());
        }
}
class Dot-plotPanel extends JPanel implements ActionListener
{
        JTextField t1,t2;
        JButton b1,b2;
        String s1,s2;
        int l,i;
        public Dot-plotPanel()
        {
      JLabel l1=new JLabel("Enter First Sequence    :"); add(l1);
```

```
t1=new JTextField(40); add(t1);

JLabel l2=new JLabel("Enter Second Sequence :"); add(l2);

t2=new JTextField(40);add(t2);

b1=new JButton("Draw Graph");

add(b1);

b1.addActionListener(this);

b2=new JButton("Refresh");

    add(b2);

b2.addActionListener(this);
s1="KESPANKFERQHMDSGSTSSSNPTYCNQMMKRRNMTQGWCKPVNTFVHEPLADVQ
        AICLQKNITCKNGQSNCYQSSSSMHITDCRLTSGSKYPN";
s2="RESPAMKFQRQHMBSGNSPGNNPNYCNQMMMRRKMTQGRCKPVNTFVHPSLEDV
    KAVCSQKNVLCKNGRTNCYESNSTMHITBCRQTGSSKYPN";
}
public void actionPerformed(ActionEvent e)
{
  if(e.getSource()==b2)
  {
s1="KESPANKFERQHMDSGSTSSSNPTYCNQMMKRRNMTQGWCKPVNTFVHEPLADVQ
    AICLQKNITCKNGQSNCYQSSSSMHITDCRLTSGSKYPN";
s2="RESPAMKFQRQHMBSGNSPGNNPNYCNQMMMRRKMTQGRCKPVNTFVHPSLEDV
    KAVCSQKNVLCKNGRTNCYESNSTMHITBCRQTGSSKYPN";
  }
  else
  {
  s1=t1.getText();
  s2=t2.getText();
}
  repaint();
}
public void paintComponent(Graphics g)
{
    super.paintComponent(g);
```

```java
setBackground(Color.pink);
    if( s1.length()>=s2.length()   )
       l=s2.length();
else
       l=s1.length();
       int d=l*4;
g.setColor(Color.blue);
g.drawRect(50,150,d,d);
   int pointx[]=new int[l];
   int pointy[]=new int[l],j=0;
   for(i=1;i<l;i++)
{
           if(s1.charAt(i)==s2.charAt(i))
           {
             pointx[j]=50+(i*4);pointy[j]=150+(i*4);
             j++;
           }
       }
    g.setColor(Color.blue);
    for(i=0;i<j;i++)
           g.drawOval(pointx[i],pointy[i],1,1);
   }
   }
```

The above program displays an output in the form of a graph when "Run Java Application" is selected from the tool option of TextPad. The dot-plot shows many dots line-up to form contiguous diagonal line indicating that the two sequences have substantial regions of similarity (Figure 6.3a). In the same output box, when a sequence

KESWARTYSUVLQGSGRTUNWSERTVCDFERTYGTFRDNMNOPRTRRRRSSTT

is typed in the upper space as a first sequence and a sequence

KESWRTESULQGSGRETUU-------TVBDFERUGTRDNMNOPRTTRRRRSSTT

is entered in the lower space as a second sequence. After pressing the tab, draw graph, new dot-plot appears as shown in Figure 6.3b indicating the regions of similarity.

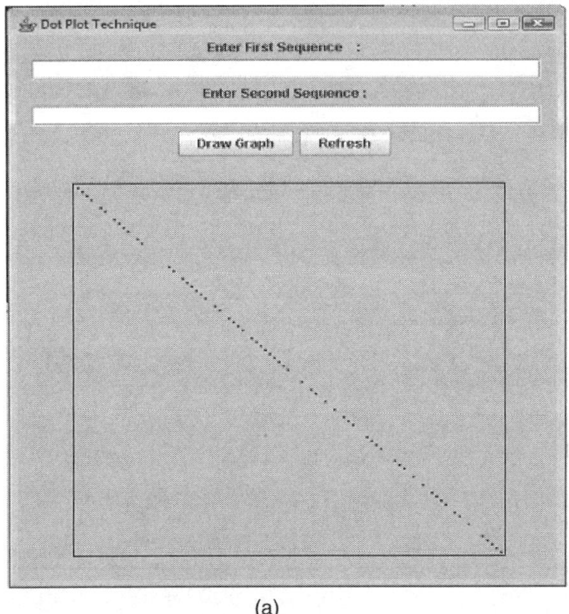

(a)

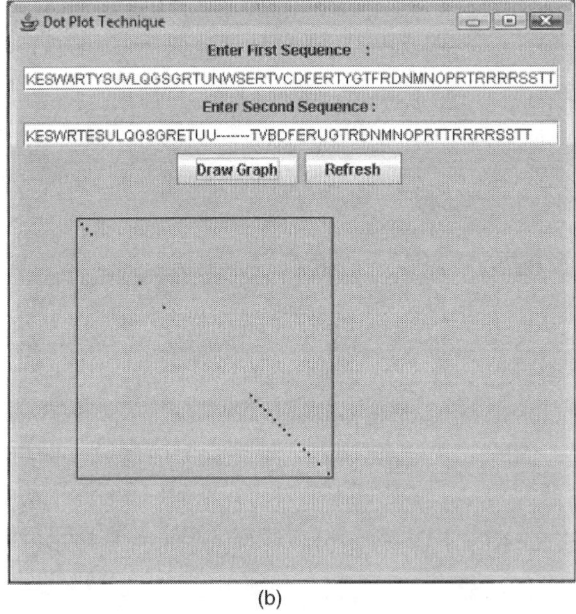

(b)

Figure 6.3   (a) Dot-plot showing diagonal line indicating that substantial regions of similarity between two sequences (b) Dot-plot resulted as the first and second sequences entered in space provided.

Dynamic Programming

In mathematics and computer science, dynamic programming is a method of solving problems exhibiting the properties of overlapping subproblems and optimal substructure. In bioinformatics, the technique of dynamic programming can be applied to produce global alignments via the Needleman–Wunsch algorithm, and local alignments via the Smith–Waterman algorithm. In typical usage, protein alignments use a substitution matrix to assign scores to amino acid matches or mismatches, and a gap penalty for matching an amino acid in one sequence to a gap in the other. DNA and RNA alignments may use a scoring matrix, but in practice, often simply assign a positive match score, a negative mismatch score, and a negative gap penalty. (In standard dynamic programming, the score of each amino acid position is independent of the identity of its neighbors, and therefore base stacking effects are not taken into account. However, it is possible to account for such effects by modifying the algorithm.)

Dynamic programming can be useful in aligning nucleotide to protein sequences, a task complicated by the need to take into account frameshift mutations (usually insertions or deletions). The frame search method produces a series of global or local pairwise alignments between a query nucleotide sequence and a search set of protein sequences, or vice versa. Although the method is very slow, its ability to evaluate frameshifts offset by an arbitrary number of nucleotides makes the method useful for sequences containing large numbers of indels, which can be very difficult to align with more efficient heuristic methods. In practice, the method requires large amounts of computing power or a system whose architecture is specialized for dynamic programming. The BLAST and EMBOSS suites provide basic tools for creating translated alignments (though some of these approaches take advantage of side-effects of sequence searching capabilities of the tools). More general methods are available from both commercial sources, such as **Frame search,** distributed as part of the accelrys GCG package available at http://www.accelrys.com/products/gcg/, and open source software such as Genewise (http://www.ebi.ac.uk/Wise2).

The dynamic programming method is guaranteed to find an optimal alignment given a particular scoring function; however, identifying a good scoring function is often an empirical rather than a theoretical matter. Although dynamic programming is extensible to more than two sequences, it is prohibitively slow for large numbers of or extremely long sequences.

Word Methods

Word methods, also known as k-tuple methods, are heuristic methods that are not guaranteed to find an optimal alignment solution, but are significantly more efficient than dynamic programming. These methods are especially useful in large-scale database searches where it is understood that a large proportion of the candidate sequences will have essentially no significant match with the query sequence. Word methods are best known for their implementation in the database search tools FASTA and the BLAST family. Word methods identify a series of short, non-overlapping subsequences ("words") in the query sequence that are then matched to candidate database sequences. The relative positions of the word in the two sequences being compared are subtracted to obtain an offset; this will indicate a region of alignment if multiple distinct words produce the same offset. Only if this region is detected do these methods apply more sensitive alignment criteria; thus, many unnecessary comparisons with sequences of no appreciable similarity are eliminated.

In the FASTA method, the user defines a value k to use as the word length with which to search the database. The method is slower but more sensitive at lower values of k, which are also preferred for searches involving a very short query sequence. The BLAST family of search methods provides a number of algorithms optimized for particular types of queries, such as searching for distantly related sequence matches. BLAST was developed to provide a faster alternative to FASTA without sacrificing much accuracy; like FASTA, BLAST uses a word search of length k, but evaluates only the most significant word matches, rather than every word match as does FASTA. Most BLAST implementations use a fixed, default word length that is optimized for the query and database type, and that is changed only under special circumstances, such as when searching with repetitive or very short query sequences. Implementations can be found via a number of web portals, such as EMBL FASTA (http://www.ebi.ac.uk/fasta33/) and NCBI BLAST (http://www.ncbi.nlm.nih.gov/BLAST/).

SCORING FUNCTION IN SEQUENCE ALIGNMENT

An intuitive idea of sequence alignment has already been given above with the formal definition of an alignment as well as its biological applications. However, it should be clear that for any two given sequences there are many possible alignments. For example, some possible alignments, which are valid under the definition of an alignment, for the sequences ATGC and ATTC could be given as:

ATGC	ATG_C	_ _ _ _ATGC	A_T_G_C_	_AT_GC
ATTC	AT_TC	ATTC_ _ _ _	_A_T_T_C	A_TT_C

All the above alignments are definitely valid. However, given the objective of an alignment as finding the possible similarities between two sequences we need to define the concept of an optimal alignment. Similar to many other problems with multiple solutions such as the travelling salesman problem we need to notion of a scoring function on two grounds:

1. To determine the best possible alignment between two sequences out of the many possible.

2. To be able to ascertain how well a sequence A aligns to a sequence B as compared to sequence A aligning to sequence C.

Design Issues for a Scoring Function

The above explanation clearly signifies the importance of a good scoring function, since without one we would not be sure of the significance of our findings. An alignment can be viewed as evidence that the two sequences in consideration have diverged from a common ancestor through the evolutionary processes of mutation and selection. The mutational operators that are considered for most alignment methods as the three sequence-edit operations given earlier as:

1. Substitutions
2. Insertions
3. Deletions

Insertions and deletions are both represented as gaps. The process of natural selection reviews the mutation of a sequence and only lets healthy mutations survive thus forming the basis of our intuition behind the alignment of two sequences.

A basic scoring function could be given as:

Score for two bases matching $= +m$

Penalty for a mismatch base at a particular position $= -s$

Penalty for introducing a gap in either sequence $= -d$

The final score of any alignment would be $= S$

$= m^*$ (No. of matches in the alignment) . s^* (No. of substitutions) . d^* (No. of gaps)

For example,

Given $m = 1$, $s = 1$, $d = 1$ and S1 = ATGC; S2 = ATTC

Some sample alignments and their scores could be:

A	T	G	C					$m = 3; s = 1; d = 0$
A	T	T	C					Score = 3 − 1 − 0 = 2
A	T	G	−	C				$m = 3; s = 0; d = 2$
A	T	−	T	C				Score = 3 − 0 − 2 = 1
−	−	−	−	A	T	G	C	$m = 0; s = 0; d = 8$
A	T	T	C	−	−	−	−	Score = 0 − 0 − 8 = −8

Additive Property of a Scoring Function

A simple description of the additive property of a scoring function could be sufficient to plug in any algorithm and proceed further. However, in order to understand the property completely, it is needed to analyse in three steps.

1. The number of total possible alignments of two sequences.
2. Dynamic programming and its role in the alignment problem.
3. Additive nature of the problem at hand.

1. *The total number of possible global alignments* The number of possible alignments of two sequences to be '0' (2^{m+n}). To be more precise, we could work out the number intuitively.

    Given: S1 of length 'm' and S2 of length 'n' (with $m = n$).

    Let k = no. of gaps in sequence S1 in an alignment.

    Then, $m + k$ = length of S1 in alignment

    Then 'n' out of these '$m + k$' positions align to characters in S2.

    And in turn k out of those n characters in S2 align with the gaps in S1 (that is the reason why those gaps are present in S1).

    Therefore, for 'k' gaps in S1, there are $(^{m+k}C_n {_*} {^n}C_k)$ possible alignments.

    However, k can vary from 0 (no gaps in S1) to 'n' (both sequences completely align to gaps).

    Therefore, the total possible alignments are $= T = \Sigma$ (for k = 0 to 'n') $(^{m+k}C_n {_*} {^n}C_k)$

2. *Dynamic programming* The above discussion was to emphasize the importance of dynamic programming. One of the major driving forces behind alignments is to align genome-length sequences. However, if all possible alignments were to be examined in order to determine the optimal alignment things would go completely out of hand.

The dynamic programming is a method to solve problems by storing partial results. A clear example was given through the case of Fibonacci series. For example, $F_3 = F_1 + F_2$ and $F_4 = F_2 + F_3$, computing each number involves computing F_2. Because both F_3 and F_4 are needed to compute F_5, a naive approach to computing F_5 may end up by computing F_2 twice or more. Wherein the normal recursive solution might have taken exponential time and the dynamic programming provided a solution linear in the input space.

3. *Additive nature of the problem at hand* The need for a dynamic programming solution was explained in order to emphasize that the score of an alignment can be obtained from the score of two sub-alignments.

For example,

Given two strings $x = x_1 \times 2.....x_M$ $y = y_1 y_2 ......y_N$

The score of aligning x_{1-M} and y_{1-N} can be

$$\text{Score}\,(x_{1-i},\, y_{1-j}) + \text{Score}\,(x_{i+1-M},\, y_{j+1-N}).$$

This additive property lends itself to dynamic programming naturally since, the partial results can be used to obtain the score for the alignment.

MODELS FOR ALIGNMENT

In this topic four alignment problems are considered, all of which are biologically motivated:

Problem 1 *[Global alignment]*

Input Two sequences S and T of roughly the same length.

Question What is the maximum similarity between them? Find a best alignment.

Problem 2 *[Local alignment]*

Input Two sequences S and T.

Question What is the maximum similarity between a subsequence of S and a subsequence of T? Find most similar subsequences.

Problem 3 *[Ends free alignment]*

Input Two sequences S and T (possibly of different length).

Question Find a best alignment between subsequences of S and T when at least one of these subsequences is a prefix of the original sequence and one (not necessarily the other) is a suffix.

Problem 4   *[Gap penalty]*

Input   Two sequences S and T (possibly of different length).

Question   Find a best alignment between the two sequences using the gap penalty function.

All the above problems are studied for various biological reasons.

GLOBAL ALIGNMENT

Definition   **Global alignment** of two sequences S and T is obtained by first inserting chosen spaces, either into or at the ends of S and T so the length of the sequences will be the same, and then placing the two resulting sequences one above the other so that every character or space in one of the sequences is matched to a unique character or a unique space in the other sequence. The term "global" emphasizes that for each sequence, the entire sequence is involved.

Example   Given a sequence "`acgctttg`" and a sequence "`catgtat`", one possible alignment would be:

```
a c - - g c t t t g

- c a t g - t a t -
```

Let us now introduce the global alignment problem:

Problem

Input   Two sequences $S = s_1...s_n$ and $T = t_1...t_m$ (n and m are approximately the same)

Question   Find an optimal alignment.

Notation   Let $\sigma\ (a, b)$ be the score (weight) of the alignment of character a with character b (including spaces).

Lemma 1   Let $V (i, j)$ be the optimal alignment score *of* $S_{1...i}$ *and* $T_{1...j}$ $(0 = i = n, 0 = j = m)$. $V(A,B)$ has the following properties:

Base conditions

$$V(i,0) = \sum_{k=0}^{i} \sigma(S_k, -)$$

$$V(0, j) = \sum_{k=0}^{j} \sigma(-, T_k)$$

Recurrence relation

$$\text{for } 1 \leq i \leq n, 1 \leq j \leq m:$$
$$V(i,j) = \max \begin{cases} V(i-1,j-1) + \sigma(S_i, T_j) \\ V(i-1,j) + \sigma(S_i, -) \\ V(i,j-1) + \sigma(-, T_j) \end{cases}$$

Proof

Base condition   The only way to align the first i elements of the sequence S with zero elements of the sequence T is to align each of the elements with a $\sigma(S_i, -)$ space in the sequence T. The score for that operation is by definition for each of the i elements and $V(i,0) = \sum_{k=0}^{i} \sigma(S_k, -)$ for the total sum.

Similarly, the expression $V(0,j) = \sum_{k=0}^{j} \sigma(-, T_k)$ follows from matching the first j elements of T with i blanks in sequence S.

Recurrence relation   Let us consider an optimal alignment of $S_{1...i}$ and $T_{1...j}$. We shall distinguish between three cases according to the three possible scoring for the three operations are:

Aligning S_i with T_j   The score in this case is the score $\sigma(S_i, T_j)$ of aligning S_i with T_j plus the score of aligning $i - 1$ elements of S with $j - 1$ elements of T, namely,

$$V(i-1, j-1) + \sigma(S_i, T_j)$$

Aligning S_i with a space character in sequence T   The score in this case is the score $\sigma(S_i, -)$ of aligning S_i with indel plus the score of aligning the previous $i - 1$ elements of S with j elements of T (Since the space is not an original character of T),

$$V(i-1, j) + \sigma(S_i, -).$$

Aligning T_j with a space character in sequence S   Similar to the previous case, the score will be $V(i, j-1) + \sigma(-, T_j)$.

Tabular Computation of Optimal Sequence Alignment   The problem can be evaluated systematically using a tabular computation (Figure 6.4). In this approach, compute $V(i, j)$ for all the possible values of i and j. Then start from smaller i, j and increase them, filling the table in a row-wise manner. Store these values in table of size $(n + 1) \times (m + 1)$. Finally, $V(n, m)$ is the required alignment score.

The following pseudocode describes the algorithm:

```
for i=0 to n do
begin
    for j=0 to m do
    begin
        Calculate V (i, j) using V (i - 1, j - 1), V (i, j - 1),
V (i - 1, j)
    end
end
```

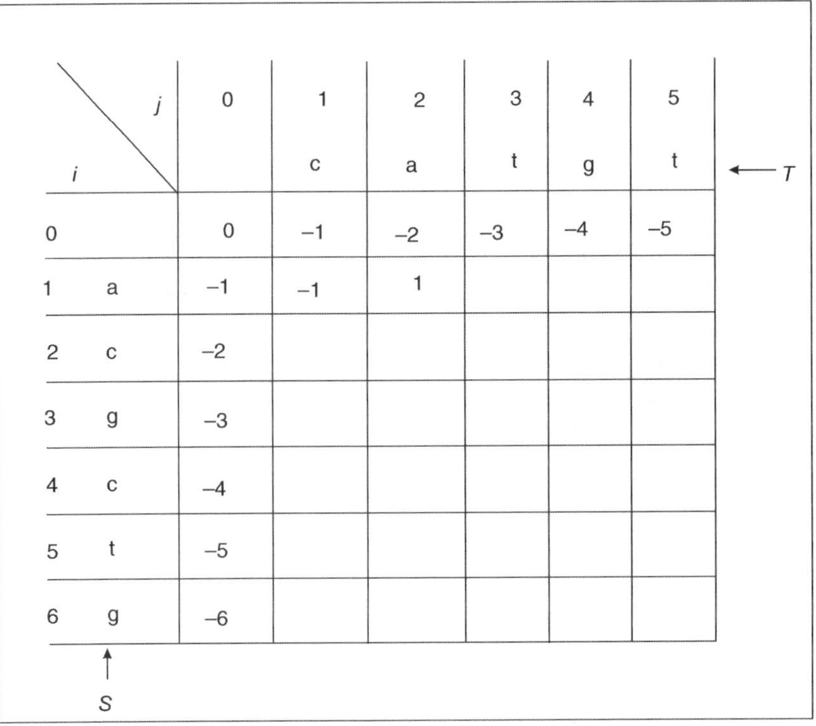

Figure 6.4   Snapshot of computing table

The Figure 6.4 demonstrates some initial stages of running an algorithm for finding global alignment. In the first column and the first row one can see results of the calculations of $V(i,-)$ and $V(-,j)$. In the second row one can see the propagation of the algorithm. Here a gap and a mismatch cost -1 and a match has value $+2$.

The Traceback

One way to trace back the alignments is to establish pointers in the cells of the table as the values are computed. The direction of the pointer in cell (i, j) indicates the value of which cell was used when $V(i, j)$ was computed (Figure 6.5).

Theorem   The time complexity of the algorithm is $O(nm)$. Space complexity is $O(n + m)$, if only $V(S, T)$ is required and $O(mn)$ for the reconstruction of the alignment.

In Figure 6.5, for each cell its outgoing arrows point to the cells from which the algorithm could arrive to the current cell. In that one can see 3 possible paths that represent alignments that give the highest score.

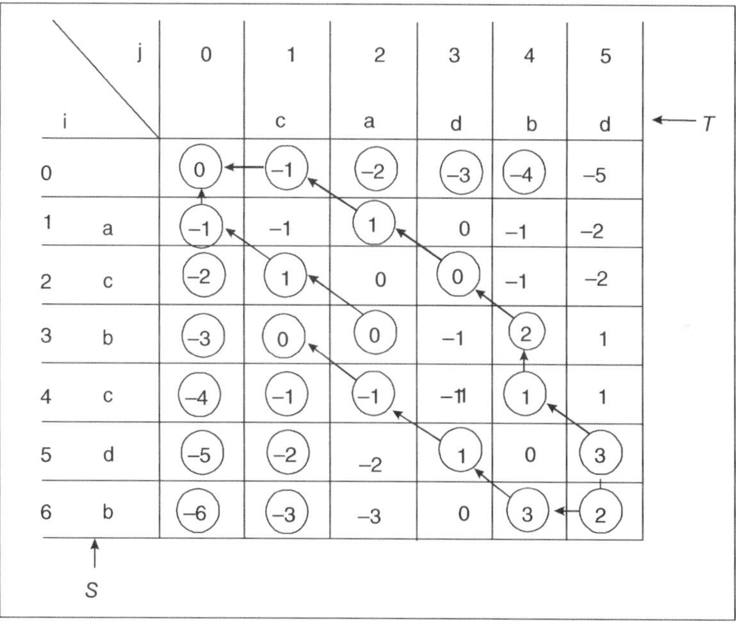

Figure 6.5   Backtracking the alignment

Proof

◙ *Time complexity*   When computing the value of a specific cell (i, j) only cells $(i - 1, j - 1)$, $(i, j - 1)$ and $(i - 1, j)$ are examined, along with two characters S_i and T_j. Hence, filling a single cell takes constant time. There are $(n + 1) \times (m + 1)$ cells in the table. So the time complexity is $O(nm)$.

◙ *Space complexity*   Using the algorithm, computing the value of cell (i, j) involves one cell $(i - 1, j)$ in row j and two cells $((i - 1, j - 1)$ and $(i, j - 1))$ in the previous row $(j - 1)$. Since the computation is performed one row at a time, when computing the values in row k only row $(k - 1)$ has to be stored, using, $O(n + m)$ space. In order to reconstruct the alignment from the recursion, pointers must be set to allow the back-tracing. Hence, the space complexity is $O(nm)$.

Sequence Alignment Graph

It is often useful to represent dynamic programming solutions of sequence problems in terms of a weighted graph.

Definition   Given two sequences S and T of lengths n and m respectively. An alignment graph is a directed graph $G = (V, E)$ on $(n+1) \times (m+1)$ nodes, each labelled with a distinct pair (i, j) $(0 = i = n, 0 = j = m)$, with the following weighted edges:

1. $((i, j), (i + 1, j))$ with weight $\sigma(S_i + 1, -)$
2. $((i, j + 1))$ with weight $\sigma(-, T_j + 1)$
3. $((i, j), (i + 1, j + 1))$ with weight $\sigma(S_i + 1, T_j + 1)$

Figure 6.6 illustrates the process of building the alignment graph. A path from node $(0, 0)$ to node (n, m) in the alignment graph corresponds to an alignment and its total weight is the alignment score. The goal is to find the heaviest path from node $(0, 0)$ to node (n, m). This alignment graph is used to map the problem of optimal alignment into the world of graphs, opening the door for new algorithms.

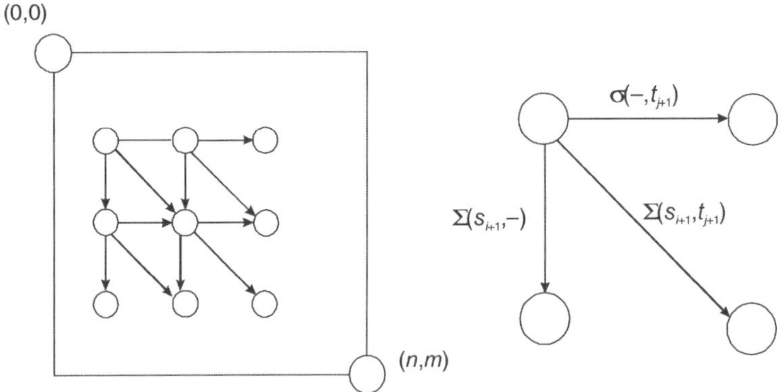

Figure 6.6   Sequence alignment graph

Figure 6.6 represents the alignment graph which is a directed acyclic graph. On the right side of this Figure one can see the segment of the alignment graph. It represents the three possibilities at each stage of the algorithm.

The global sequence alignment is a programming method which finds the best alignment of one entire sequence with another entire sequence. The first sequence alignment algorithm was developed by S.B. Needleman and C.D. Wunsch. Alignment is carried out from beginning to end of both sequences to find the best possible alignment across the entire length between the two sequences. The method is more applicable for aligning two closely related sequences of roughly the same length.

Needleman–Wunsch Algorithm for Global Alignment

Suppose that $F(i, j)$ be the score of optimally aligning two strings $x_{[1-i]}$ and $y_{[1-j]}$.

This problem can split into three cases:

i.   Match/substitution between $x[i]$ and $y[j]$:

$$X_1 \_\_\_\_ \_\_ \_ X_{i-1}\, X_i$$
$$Y_1 \_\_\_\_ \_\_ \_ y_{j-1}\, y_j$$
$$F(i, j) = F(i-1, j-1) + s(x_i, y_j)$$

Where,

$$s(x_i, y_j) = +m \text{ if } (x_i = y_j) \text{ or, } -d \text{ otherwise.}$$

ii.   Gap aligned to x_i:

$$X_1 \_\_\_\_ \_\_ \_ X_{i-1}\, X_i$$
$$Y_1 \_\_\_\_ \_\_ \_ y_j \_$$
$$F(i, j) = F(i-1, j)-d.$$

iii.   Gap aligned to y_j:

$$X_1 \_\_\_\_ \_\_ \_ X_i \_$$
$$Y_1 \_\_\_\_ \_\_ \_ y_{j-1}\, y_j$$
$$F(i, j) = F(i, j-1)-d$$

The graph/matrix shown below (Figure 6.7) clearly shows the matrix filled up with the two sequences functioning as the two axes. Each cell (i, j) contains the score $F(i, j)$ for aligning the two subsequences $x(1-i)$ and $y(1-j)$. The value in each cell depends only on 3 surrounding cells—to the left $[F(i-1, j)]$, to the top $[F(i, j-1)]$ and to the top-left $[F(i, j)]$ of the cell. Hence, any order of filling in the cells of the matrix, which ensures the availability of these 3 values for each cell that come to, is a valid way of filling up the matrix.

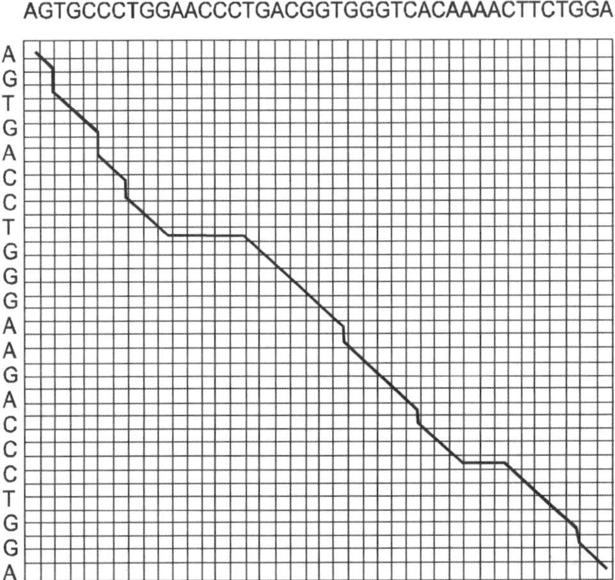

Figure 6.7 The graph of two sequences processed for global sequence alignment

In the matrix, a global sequence alignment is visualized as a non-decreasing path from the top-left cell of the matrix, to the bottom-right cell. The path in the Figure 6.7 represents the optimal alignment between the two sequences. By the term "non-decreasing", means that as we traverse the path from the top-left cell to the bottom-right cell, each elementary step that we take neither decrease the row, nor the column number. In other words, always either moving right or down or diagonally "down-right". Every time take a horizontal or vertical step along the path, introduce a gap character in the alignment, and every time take a diagonal step, line-up the two characters corresponding to that cell one below the other.

Finally, a trace back procedure from the bottom-right cell to the top-left cell makes use of additional pointers that maintains with each cell, and gives the optimal alignment.

LOCAL ALIGNMENT

In many applications two sequences may not be highly similar over their entire length, but may contain subsequences with high resemblance. Local alignment finds local regions with the highest level of similarity between the two sequences and aligns these regions without regard for the alignment of the rest of the sequence regions. The two given sequences to be aligned can be of different length. The local alignment problem is formally defined below:

Problem *[Local alignment]*

Input   Two sequences S and T (of different length)

Question   Find the local regions with highest level of similarity between two sequences (over all such pairs of subsequences.)

Reminder   A subsection is a contiguous segment of a sequence.

Motivation

Ignore stretches of non-coding DNA   In the DNA, non-coding regions (introns) are more likely to be subjected to mutations than coding regions (exons). This is since mutations in coding regions might cause a change in a protein that will have a substantial effect on the organism. A change in a non-coding region is less likely to have an effect, and therefore has a higher probability of getting "accepted". When searching for a local alignment between two stretches of DNA (from 2 different specimens), finding a best match is likely to be between 2 exons.

Protein Domains

Proteins of different kind and of different species, often exhibit local similarities called **homeoboxes**. These local similarities are most probably "functional subunits" of the protein. Finding these similarities is done by solving local alignment between different sequences of DNA.

Example   Consider the two sequences:

$S =$ g g t c t g a g

$T =$ a a a c g a

editing operations values:

match = 2   indel/substitution = -1

The best local sequence alignment is:

$\alpha =$ ctga $(\in S)$

$\beta =$ c-ga $(\in T)$

Computing Local Alignment

Given two sequences, S and T, and two indices i and j, the **local suffix alignment** problem is finding a (possibly empty) suffix α of $S_{1...i}$ and a (possibly empty) suffix β of $T_{1...j}$ such that the value of their alignment is maximal over all alignments of suffixes of $S_{1...i}$ and $T_{1...j}$.

The solution to the local sequence alignment problem is the same as the maximal solution to the local suffix alignment problem over all indices i and j of S and T.

Terminology and Restriction

▣   Use $V(i, j)$ to denote the value of the optimal local suffix alignment for a given pair i, j of indices.

▣   Limit the weights of the editing operations by the following rule:

$\sigma(x, y) = \{\geq 0$ if x, y match or

$\sigma(x, y) = \{\leq 0$ if x, y do not match or one of them is a space

The Scheme of the Algorithm

1.   Compute *local suffix alignment* (for all i and j) of $S - i = S_{1...i}$ and $T - i = T_{1...j}$.

     This is done using the global alignment algorithm with a small difference: the prefixes of S and T whose alignments are $= 0$ are discarded, thus allowing the subsequences to start from indices $= 1$.

2.   Search the results and find the indices i and j of S and T respectively, after which the similarity only decreases.

Recursive Definition

Base conditions

$V i, j.\ V(i, 0) = 0,\ V(0, j) = 0$

Recurrence relations

$$V(i, j) = \max \begin{cases} 0 \\ V(i-1, j-1) + \sigma(S_i, T_j) \\ V(i-1, j) + \sigma(S_i, -) \\ V(i, j-1) + \sigma(-, T_j) \end{cases}$$

Compute i and j

$$V(i, j) = \max 1 \leq i \leq n, 1 \leq j \leq m\, V(i, j)$$

Observe that the recurrence for computing local alignment is almost identical to the one used for computing global alignment. The only difference is the inclusion of zero in the case of local suffix alignment.

The zero in the base conditions and in the recurrence relation allows the alignment to start anywhere in the sequences, rather than forcing it to start at the beginning of each.

Searching for i and j that give the best $V(i, j)$ allows the alignment to end anywhere in the sequences, rather than at the end of each. Together, get the best alignment of two subsequences of S and T.

Example Figure 6.8 illustrates the calculation of the $n \times m$ entries table, taking σ as 2 for a match and –1 for a mismatch. As usual, pointers are created while filling in the values of the table. After cell (i, j) is found, the subsequences α and β giving the optimal local alignment of S and T are found by tracing back the pointers from cell (i, j) until reaching an entry $(i-, j-)$ that has value zero. Then the optimal local alignment subsequences are $\alpha = Si{-}...i$ and $\alpha = Tj{-}...j$.

Lemma 6 Local sequence alignment can be solved in linear space.

Proof The optimal local sequence alignment of S and T identifies subsequences α and β whose global sequence alignment has maximum value over all pairs of subsequences. Hence, if α and β can be found using only linear space, then their actual alignment can be found in linear space, using Hirschberg's method for global sequence alignment.

The score of the optimal local alignment is found in cell i, j. Those indices specify the terminating points of the sequences α and β. The values in each row can be computed in a row wise fashion and the algorithm must store values for only two rows at a time. Hence, the end positions (i, j) can be computed in linear space.

Finding the starting positions of the two subsequences can be done in linear space using reverse dynamic programming. As mentioned above, after α and β are found, calculating their alignment can be done in linear space using Hirschberg's method.

Local Sequence Alignment Algorithm Complexity

Time complexity Since it takes constant number of operation per cell to compute $V(i, j)$, it takes only $O(mn)$ time to fill in the entire table. The search for $V(i, j)$ requires only $O(nm)$ time as well. Hence the total time complexity is $O(nm)$.

Space complexity As shown in Lemma 6, the space complexity is $O(n + m)$.

j	0	1	2	3	4	5	6	
i		x	x	x	c	d	e	T
0	0	0	0	0	0	0	0	
1 a	0	0	0	0	0	0	0	
2 b	0	0	0	(0)	0	0	0	
3 c	0	0	(0)	0	(2)	1	0	
4 x	0	2	2	(2)	(1)	1	0	
5 d	0	1	1	1	1	(3)	2	
6 e	0	0	0	0	0	2	(5)	
7 x		2	2	2	1	1	1	
	S							

Figure 6.8 Finding local alignment

Smith–Waterman Algorithm for Local alignment It is first application of dynamic programming in local alignment. In this algorithm, positive scores are assigned for matching residues and zeros for mismatch. No negative scores are used.

1. *Initialization*
   $$F(0, j) = F(i, 0) = 0$$
2. *Iteration* For each i and j,
   $F(i, j) = $ Max.
   {
   0
   $$F(i-1, j-1) + s(X[i], Y[j])$$
   $$F(i-1, j) - d$$
   $$F(i, j-1) - d$$
   }

3. *Termination*

   i.   If we want the best local alignment: $F_{OPT} = \max_{i,j} F(i,j)$

   ii. If we want all local alignments scoring $>t$:

       For all i, j find $F(i,j) > t$, and trace back.

The above method is a simple tweak to the Needleman–Wunsch algorithm. The basic heuristic is to ignore badly aligned regions. The '0' in the decision pseudo-code for $F(i, j)$ ensures that whenever the alignment score drops below 0, the local alignment is not followed further. In physical terms, the previous alignment is terminated and a new alignment began from that point onwards. A local alignment can begin and end at any cell in the matrix even though it still follows the rule that any further step can only be to the "right", "down" or "diagonal lower right".

To get the alignment itself, we start our trace back from the highest $F(i, j)$ valued cell, and instead of tracing back all the way to the top-left corner of the matrix, we stop our trace back once we hit a cell with an $F(i,j)$ value of 0. Another variant of the same can be used to find the optimal alignments above a threshold score. The same method as above is followed from $F(i, j)$ cells with scores above the threshold. However, redundancy needs to be eliminated in that paths belonging to the same alignment are not retraced more than once.

END-SPACE FREE ALIGNMENT

In this variant of sequence alignment, any number of indel operations at the end or at the beginning of the alignment contributes zero weight. This drops the requirement that sequences start and end at the same place, and allows to align sequences that overlap/include one another.

Example Consider the sequences:

$S =$ c a c t g t a c

$T =$ g a c a c t t g

Assigning value of 2 for match, and −1 for indel/substitution, the best global alignment will have value 1 and will look like this:

$S =$ c a c − − t − g t a c

$T =$ g a c a c t t g − − −

While end-space free alignment will have value 9 and will look like this:

S = - - c a c - t g t a c

T = g a c a c t t g - - -

The two leading spaces at the left end of the alignment are free, as well as the three trailing spaces at the right end.

Motivation

One example where end-spaces should be free is in the "shotgun sequence assembly" procedure. In this problem, one has a large set of partially overlapping subsequences that come from many copies of an original but unknown DNA sequence. The problem is to use comparisons of pairs of subsequences to infer the original sequence (using the overlapping sections to "paste" the subsequences together).

Two subsequences that are from different parts of the original sequence will have low-global sequence alignment score, as well as low end-space free alignment score. Two overlapping subsequences from the set are unlikely to have the same starting position (nor the same end position) along the original sequence. Therefore, they will still have a low-global alignment score. On the other hand, when checking them for end-space free alignment, the two subsequences will have high score, since they posses an overlapping section. The overlap will be detected, and the subsequences will be "pasted" together using the alignment found. Similarly, the case where one subsequence contains the other can be detected by end-space free alignment Figure 6.9.

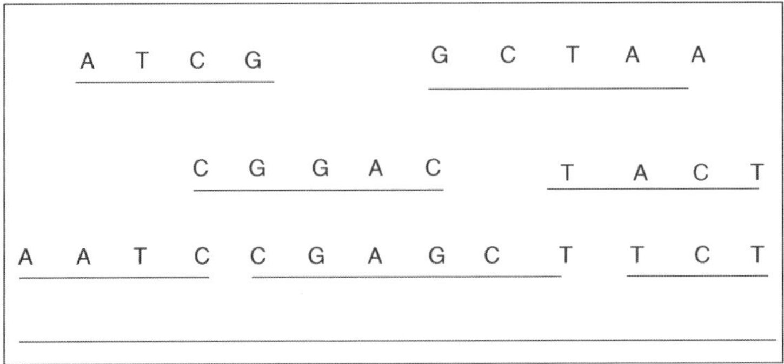

Figure 6.9   Sequence assembly

The End-space Free Alignment Algorithm

The end-space free alignment algorithm is again a variation of the global sequence alignment algorithm, they are as follows:

- We set the initial conditions to allow zero weight to leading indel operations in (at most) one of the sequences.

- After filling the table with the values of $V(i, j)$, we search for the maximal value in either of the "ending rows", thus allowing (at most) one sequence to end before the other, with zero weight for all indel operations from thereon. This value is the best value.

- The aligned sequence is tracked (using pointers created while filling the table) from cell (0, 0) in the table until the end of one sequence (bottom row/rightmost column). From thereon, all indel operations until cell (n, m) are not counted in the total value (though they are present in the table).

Recursive Definition

Base conditions $\quad V(i, j). \; V(i, 0) = 0, \; V(0, j) = 0$

Recurrence relation

$$V(i, j) = \max \begin{cases} V(i-1, j-1) + \sigma(S_i, T_j) \\ V(i-1, j) + \sigma(S_i, -) \\ V(i, j-1) + \sigma(-, T_j) \end{cases}$$

Search for i such that

$$V(i, m) = \max_{1 \le i \le n, m} V(i, j)$$

Search for i such that

$$V(n, j) = \max_{n, 1 \le j \le m} V(i, j)$$

Define alignment score

$$V(S, T) = \max \begin{cases} V(n, j) \\ V(i, m) \end{cases}$$

See Figure 6.10 for illustration of the $V(i, j)$ table over the sequences $S =$ `actgtta` and

$T =$ `gttactgt` with match = 2 and indel/substitution = −1.

Complexity

Time complexity   Computing the matrix takes $O\,(nm)$. Finding j and i takes $O\,(n + m)$. Therefore the total time complexity remains $O\,(nm)$.

Space complexity   Computing the matrix takes $O\,(n + m)$ space using Hirschberg's method. Computing the maximizing values i, j requires the last row and column to be saved, which is also $O\,(n + m)$. Therefore the total space complexity remains $O\,(n + m)$.

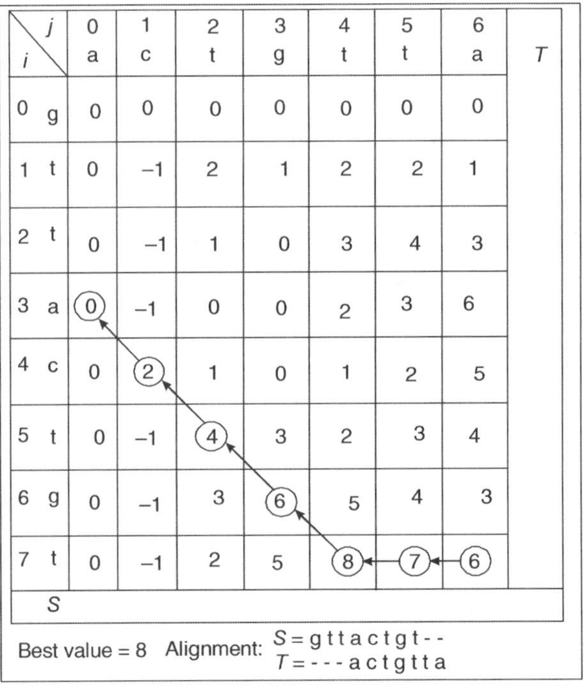

	j	0	1	2	3	4	5	6	
i		a	c	t	g	t	t	a	T
0	g	0	0	0	0	0	0	0	
1	t	0	−1	2	1	2	2	1	
2	t	0	−1	1	0	3	4	3	
3	a	⓪	−1	0	0	2	3	6	
4	c	0	②	1	0	1	2	5	
5	t	0	−1	④	3	2	3	4	
6	g	0	−1	3	⑥	5	4	3	
7	t	0	−1	2	5	⑧	⑦	⑥	
S									

Best value = 8   Alignment: $S = \text{g t t a c t g t - -}$
$T = \text{- - - a c t g t t a}$

Figure 6.10   End-space free alignment table

GAP PENALTY

Up until now the central elements used to measure the score of an alignment have been matches, mismatches and spaces. Now we introduce another important element, **gaps**. A gap is a consecutive run of spaces in an alignment. Gaps help to create alignments that better conform to underlying biological models and more closely fit patterns that one expects to find in meaningful alignments. The idea is to treat a gap as a whole, rather than give each of its spaces the same weight. There are many ways a weight/

value can be given to a gap. In this section we present several gap penalty models, and an algorithm for finding a best alignment using affine gap penalty model. The number of gaps in an alignment will be denoted by #gaps.

A gap is any maximal, consecutive run of spaces in a single sequence of a given alignment. The length of a gap is the number of indel operations in it.

A **gap penalty function** is a function that measures the cost of a gap as a (non-linear) function of its length.

Example Consider the alignment:

$S = $ a t t c - - g a - t g g a c c

$T = $ a - - c g t g a t t - - - c c

This alignment has four gaps containing a total of eight spaces. The alignment would be described as having seven matches, no mismatch, four gaps and eight spaces.

Motivation

The concept of a gap in an alignment is important in many biological applications, since the insertion or deletion of an entire subsequence often occurs as a single mutational event. Moreover, many of these single mutational events can create gaps of varying sizes. We will need to score a gap as a whole when we try to align two sequences of DNA so as to avoid assigning high cost to these mutations. At the protein level, two protein sequences might be relatively similar over several intervals but differ in intervals where one contains a protein subunit that the other does not. Again, the introduction of gaps will help us to treat these cases as "good matches", although there are long consecutive runs of indel operations in them. One concrete illustration of the use of gaps in the alignment model comes from the problem of cDNA matching.

To better understand the problem, here is a short biological background: Recall that an RNA molecule is transcribed from the DNA of a gene. The RNA transcript (pre-mRNA) is a complement of the gene's DNA, where each A in the gene is replaced by U in the RNA, each T is replaced by A, each C by G, and each G by C. Moreover, the RNA transcript spans the entire gene—introns and exons. After the pre-mRNA is created, a splicing process takes place—each intron–exon boundary is located, the RNA regions corresponding to the introns are spliced out, and the RNA regions corresponding to exons are concatenated. The mature RNA molecule is called the messenger RNA (mRNA). It includes only the regions that correspond to exons. The mRNA leaves the cell nucleus and is used to create the protein it encodes (Figure 6.11). Each cell (usually) contains a copy of all the chromosomes and hence, of all the genes of the entire individual. Yet, in each specialized cell (for example, a liver cell) only a small fraction of the genes are expressed, that is, only a small fraction of the proteins encoded in the genome are actually produced in that specialized cell.

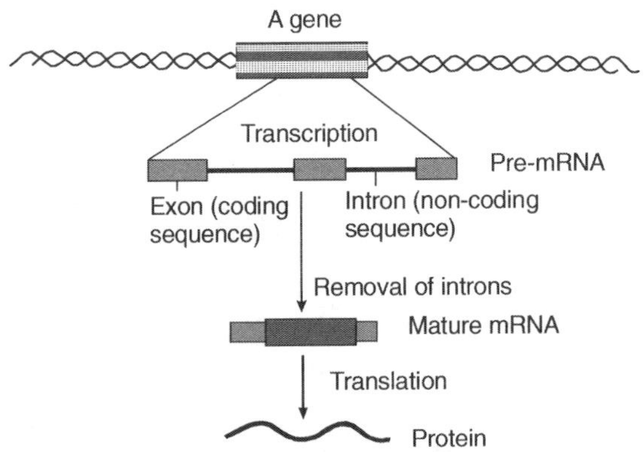

Figure 6.11 RNA splicing

The intron regions are excluded in protein synthesis. Hence, the protein sequence obtained from the protein can be viewed as the mature mRNA sequence and when aligned to the entire gene sequence should ideally have large gaps representing the intron regions that were not translated.

A standard method to determine which proteins are expressed in the specialized cell and to find the location of the encoding genes uses the mRNA. After "capturing" the mRNA, it is used to create a DNA sequence complementary to it. This sequence is called cDNA (complementary DNA). The cDNA contains the same sequence of bases as the exons on the original DNA that encode for the specific protein, with the introns missing. To determine where the gene associates with that cDNA (hence protein) resides, we now need to match the cDNA with the original DNA.

In the alignment the search is made for many long gaps that are due to introns on the DNA, which are missing on the cDNA. Using a good gap penalty model will help in preventing a very low score for these alignments (since they have many consecutive indel's in them) and allow the researcher to find the true alignment.

Constant Gap Penalty Model

The simplest choice is the constant gap penalty, where each individual space is free (has no weight), and each gap is given a weight of *Wg* independent of its length.

- ▣ Let σ denote the weights of match and mismatch only $(V \times \sigma(x,-) = \sigma(-,x) = 0)$
- ▣ Let S' and T' represent S and T after inserting spaces.
- ▣ Thus we have to find an alignment that maximizes:

$$\sum \sigma(S - i, T - i) + Wg \times \# \, gaps$$

A generalization of this model adds weight not only to an existence of a gap (*Wg*), but also another weight proportional to its length (*Ws*). This model is described below.

Affine Gap Penalty Model

In the Affine gap penalty model, a gap is given two weights. One, *Wg* is the weight to "open the gap", and the other, *Ws* is the weight to "extend the gap" with one more space.

The total penalty for a gap of length q is:

$$W_{Total} = Wg + qWs$$

The model is called "affine" after its affine formula above.

Note that the constant gap weight model is simply the affine model with $Ws = 0$, Thus the algorithm described below can be used for the constant gap penalty model as well.

Using the same symbols as above, we have to find an alignment that maximizes:

$$\sum \sigma(S - i, T - i) + Wg \times \# \, gaps + Ws \times \# \, spaces$$

Affine gap penalty algorithm To align sequences S and T, consider the prefixes $S_{1...i}$ of S and $T_{1...j}$ of T. Any alignment of these two prefixes is one of the following three types:

1. $S \rule{2cm}{0.4pt} i$

   $T \rule{2cm}{0.4pt} j$

alignment of $S_{1...i}$ and $T_{1...j}$ where characters $S(i)$ and $T(j)$ are aligned opposite to each other. This includes both the case that $S_i = T_j$ and that $S_i \neq T_j$.

2. S ——— i ················

   T ————————— j

Alignment of $S_{1..i}$ and $T_{1..j}$ where character S_i is aligned to a character strictly to the left of character T_j. Therefore, the alignment ends with a gap in S.

3. S ————————— i

   T ——— j ················

Alignment of $S_{1..i}$ and $T_{1..j}$ where character S_i is aligned to a character strictly to the right of character T_j. Therefore, the alignment ends with a gap in T.

Notation   Let us use the following notation:

$G(i, j)$ the maximum value of any alignment of type 1

$E(i, j)$ the maximum value of any alignment of type 2

$F(i, j)$ the maximum value of any alignment of type 3

$V(i, j)$ the maximum value of an alignment

Using these notations, we can recursively define the alignment table, as was done in previous algorithms. In the base conditions, we need to look at indel operations and assign the correct value: not only the weight of the spaces (qWs), but also the weight of "the opening gap" (Wg).

We will define 3 recurrence relations, one for each of $G(i, j)$, $E(i, j)$ and $F(i, j)$. Each will be calculated from previously computed values.

Take $E(i, j)$ for example. We are looking at alignments in which S ends to the left of T:

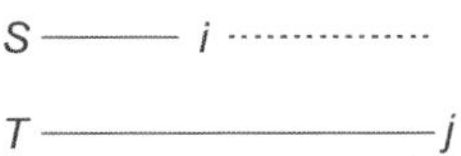

There are two possible cases for the previous alignment:

1. It looked the same, i.e., S ended to the left of T. In this case, we only need to add another "extension weight" to the value, forming the new weight $E(i, j-1) + Ws$.
2. S and T ended at the same place (type 1 alignment).
   In this case, we need to add both the gap "opening weight" and the gap "extension weight", forming the new weight $V(i, j-1) + Wg + Ws$. Taking the maximum of the two yields the value for $E(i, j)$.

Calculating $F(i, j)$ and $G(i, j)$ is done using similar arguments. $V(i, j)$ is Calculated simply by taking the maximum of the three. As in global alignment, we search for the value $V(n, m)$, and trace back the alignment by using pointers that had been created while filling the table.

Recursive Definition

Base conditions

$$V(0, 0) = 0$$

$$V(i, 0) = E(i, 0) = Wg + iWs$$

$$V(0, j) = F(0, j) = Wg + jWs$$

Recurrence relation

$$V(i, j) = \max \{E(i, j), F(i, j), G(i, j)\} \text{ where}$$

$$G(i, j) = V(i-1, j-1) + \sigma(S_i, T_j)$$

$$E(i, j) = \max \{E(i, j-1) + Ws, V(i, j-1) + Wg + Ws\}$$

$$F(i, j) = \max \{F(i-1, j) + Ws, V(i-1, j) + Wg + Ws\}$$

Complexity

Time complexity As before $O(nm)$, as we only compute four matrices instead of one.

Space complexity There's a need to save four matrices (for E, F, G, and V respectively) during the computation. Hence, $O(nm)$ space is needed for the trivial implementation.

Convex Gap Penalty Model

▣ Each additional space in a gap contributes less to the gap weight than the previous space.

▣ This model is said to better describe biological behaviour.

▣ Example: $Wg \log(q)$, where q is the length of the gap.

▣ The problem is solvable in $O(nm \log[m])$ time.

Arbitrary Gap Penalty Model

▣ Any gap weight function is acceptable (this is the most general case).

▣ Weight of a gap is an arbitrary function of its length $w(q)$.

▣ The problem is solvable in $O(nm(m + n))$ time.

Longest Common Non-Contiguous Subsequence

We finish with a classical problem in computer science that is not biologically motivated, but which we can be easily solved using the machinery developed above.

Definition   A non-contiguous subsequence of S is defined as a subset of the characters of S arranged in their original "relative" order. Formally: a non-contiguous subsequence of sequence S is specified by a list of indices $i_1 < i_2 < i_3 < \ldots < i_k$. The non-contiguous subsequence specified by this list is $S_{i1} S_{i2} \ldots S_{ik}$. Given two sequences S and T, a common subsequence is a subsequence that appears both in S and T.

Problem

Input   Two sequences S and T.

Question   What is the longest non-contiguous subsequence common to both S and T?

The longest common non-contiguous subsequence problem can be modelled and solved using the global alignment algorithm with the following scoring:

$$\sigma(x,y) = \begin{cases} 1 & x = y \\ 0 & x \neq y \end{cases}$$

$$\sigma(x,-) = \sigma(-,y) = 0$$

Or, directly compute $V(n, m)$ with:

$$V(i,0) = V(0,j) = 0$$

$$V(i,j) = \max \begin{cases} V(i-1,j-1) + \sigma(S_i, T_j) \\ V(i-1,j) \\ V(i,j-1) \end{cases}$$

Each character in the sequence S can be aligned with the same character in the sequence T or with a space in T (in this case no substitution is done). Since the goal is to find a maximum length subsequence, character matches are valued as '1', while a space match is valued '0'.

DATABASE SIMILARITY SEARCHING

Pairwise alignment is applied for retrieving biological sequences in databases based on similarity. Practically the process begins with submission of a query sequence and performing a pairwise comparison of the query sequence with all individual sequences already present in the database. Thus, database similarity searching is pairwise alignment on a large scale that helps in assigning the putative functions to newly determined sequences. However, the dynamic programming methods such as Needleman–Wunsch algorithm for global alignment and Smith–Waterman algorithm for local alignment described earlier, although accurate and reliable, are slow and impractical to use in most cases when computational resources are limited. Since the speed of searching is an important issue in the field of bioinformatics, there efforts made in the direction to invent and implement heuristic database searching methods. The heuristic algorithms perform faster searches because they examine only the fraction of possible alignments. Currently, there are two special database searches, BLAST and FASTA, which help to speed up the computational process of sequence comparison. BLAST is more time efficient than FASTA by searching only for the most significant patterns in the sequences, but with comparative sensitivity. Both methods are not guaranteed to find optimum alignment or true homology, but are 50–100 times faster than dynamic programming.

BLAST SEARCH

BLAST (Basic Local Alignment Search Tool) is a software tool that is used to determine whether a particular DNA or protein sequence matches with any other DNA or protein sequence in one of several public databases. The BLAST program was designed by Eugene Myers, Stephen Altschul, Warren Gish, David J. Lipman and Webb Miller at the NIH and was published in *J. Mol. Biol.* in 1990 and has since become one of the most popular programs for sequence analysis. It can be accessed through the Internet at the website of the National Center for Biotechnology Information at http://www.ncbi.nlm.nih.gov/BLAST. The BLAST tool allows the comparison of sequences that have no assigned function against ones that do. The websites typically contain a user interface allowing the user to input a sequence. The search algorithm then compares this sequence with all the sequences held in a database and assigns a similarity value. This allows a potential function to be assigned to the sequence in a considerably shorter time when compared to laboratory methods. The results of this type of search could potentially dictate future research avenues within the laboratory.

Depending on the type of sequences to compare, there are different BLAST programs:

1.   **BLASTP** compares an amino-acid query sequence against a protein sequence database (Protein query vs. protein database). Standard

protein–protein BLAST (BLASTP) is used for both identifying a query amino acid sequence and for finding similar sequences in protein databases. Like other BLAST programs, BLASTP is designed to find local regions of similarity. When sequence similarity spans the whole sequence, BLASTP will also report a global alignment, which is the preferred result for protein identification purposes.

2. **BLASTN** compares a nucleotide query sequence against a nucleotide sequence database (DNA query vs. DNA database).

3. **BLASTX** compares a nucleotide query sequence translated in all reading frames against a protein sequence database [DNA query (translated into protein) vs. protein database].

4. **TBLASTN** compares a protein query sequence against a nucleotide sequence database dynamically translated in all reading frames [Protein query vs. DNA database (translated into protein)].

5. **TBLASTX** compares the six-frame translations of a nucleotide query sequence against the six-frame translations of a nucleotide sequence database [DNA query (translated into protein) vs. DNA database (translated into protein)].

6. **BLASTZ** compares long stretches of nucleotides >2 kb.

7. **BLASTBGP** allows use of two new BLAST modes:

    i. **Position-Specific Iterated (PSI)-BLAST** is the most sensitive BLAST program, making it useful for finding very distantly related proteins. It uses an interactive alignment procedure to detect weak pattern matches.

    ii. **Pattern-Hit Initiated (PHI)-BLAST** is designed to search for proteins that contain a pattern specified by the user and are similar to the query sequence in the vicinity of the pattern. This dual requirement is intended to reduce the number of database hits that contain the pattern, but are likely to have no true homology to the query. This tool uses protein motifs, such as those found in prosite and other motif databases, to increase the likelihood of finding biologically significant matches.

8. **bl²seq** allows a comparison of two known sequences using BLASTP and BLASTN programmes.

9. Other new versions of **BLAST** tools include:

    i. **Reverse Position-Specific BLAST (RPS-BLAST)** is a more sensitive way of identifying conserved domains in proteins than standard BLAST searching. It compares a protein sequence against a database of Position-specific Scoring Matrices (PSSMs).

ii.   **VecScreen**, under special section, is a rapid screening tool that checks the query sequence against a non-redundant vector database, UniVec, which contains one copy of every unique sequence segment from a large number of cloning vectors. In addition, UniVec contains sequences for adapters, linkers, stuffers, and primers that are commonly used in the cloning and manipulation of cDNA or genomic DNA. Detailed information on UniVec is at: www.ncbi.nlm.nih.gov/VecScreen/ UniVec.html. This page is generally used to screen for vector contamination in sequences before their submission to GenBank. The colour-coded graphics in the result page makes the result easy to understand.

iii.   **GEOblast** allows you to blast a given set of sequences to find matches to those sequences/genes represented by entries in the GEO database. Matching hits will have links linking to corresponding entries in GEO. Different from text query on Entrez/Geo database, this page provides a way to search and retrieval of expression data through sequence similarity search by way of BLAST.

iv.   **Igblast** accesses the curated human and mouse immunoglobin gem-line sequences. The databases are updated regularly. Both protein and nucleotide sequences are available for BLASTN and BLASTP searches. In part, it functions as a replacement of the now defunct Kabat database, with extra sequence similarity search capability. Help document is linked off this page: www.ncbi.nlm.nih.gov/igblast/.

v.   **SNPblast** accesses the curated human SNPs available from NCBI's dbSNP database. The search can be against the complete set or limited to those on one or more of the selected chromosomes. For SNPblast result, displaying the result in "Query anchored with identity" format is likely to be more informative to identify the matching SNPs.

vi.   **MEGAblast** is the only BLAST web service that can accept multiple queries. There are two ways to enter batch queries in MEGABLAST. If the query sequences are not present in the NCBI Entrez system, those sequences need to be provided in FASTA format, one after another with no blank lines in between sequences.

vii. **PowerBLAST** applied for automatic analysis of genomic sequences. It combines BLAST searching with additional filtering for low complexity region and repeats. It features onto many alignment outputs showing alignment of query sequence with all matching sequences, in contrast to the earlier versions that could give only one to one alignment output.

10. **BLAST2 Sequences** is designed for direct comparison of two sequences. This program takes two input sequences and compares them directly. "Aligning Two Sequences" regards the second sequence as the database. Unlike the other BLAST programs, there is no need to format the database sequence in any special way. Since translated BLAST programs are incorporated in this program, the second sequence can be of different type so long as an appropriate BLAST program is selected. Appropriate query/program combination is listed in the Table 6.3.

Table 6.3   Appropriate query/program combinations for "BLAST2 sequences"

First query	Second query	Program to use
Nucleotide	Nucleotide	BLASTN, MEGABLAST, or TBLASTX
Nucleotide	Protein	BLASTX
Protein	Nucleotide	TBLASTN
Protein	Protein	BLASTP

BLAST, is an algorithm for comparing primary biological sequence information, such as the amino-acid sequences of different proteins or the nucleotides sequences of DNA. A BLAST search enables a researcher to compare a query sequence with a library or database of sequences, and identify library sequences that resemble the query sequence above a certain threshold. For example, following the discovery of a previously unknown gene in the mouse, a scientist will typically perform a BLAST search of the human genome to see whether humans carry a similar gene; BLAST will identify sequences in the human genome that resemble the mouse gene based on similarity of sequence.

Request ID (RID) used to format a BLAST result in different format: For each successfully submitted BLAST search request, a unique RID is issued.

This ID will be valid for 24 hours. Within this period of time, you can use the RID to retrieve the result multiple times. More importantly, the RID can be used to retrieve and display the result in different display formats to emphasize different aspect of the result and bring out the features such as identity or variation across the matches that would otherwise not stand out. Those representative display formats are described in Table 6.4.

Table 6.4   Selected display formats and their description

Format	Description
Pairwise	Query aligned to individual matched sequence, one pair at a time.
Pairwise with identity	Similar to "pairwise", with identity replaced by "."
Query anchored with identity	Pseudo multiple sequence alignment format. Identity in matched subject is replaced by "."
Flat query anchored with identity	Similar to above with whole alignment kept flat by inserting "-" into query if needed.
Hit table	Statistics for each HSP summarized in table format.
XML	Result in XML format for parsing (controlled by "format" field).

CDD and CDART searches do not have RID issued.

BLAST and Mismatches in Sequences

Degenerate bases and ambiguity codes are treated as mismatches by BLAST Uncertainties in a nucleotide sequence can be represented by a standard set of single-letter codes (Table 6.5). These codes are often used to represent degenerate bases in the third position of codons, in degenerate oligo-nucleotide primers, or sequence motifs. Even though ambiguities in the query are accepted by BLAST, BLAST webpages have a built-in functionality that screens query sequences. Too many ambiguities in a nucleotide query could make the BLAST page to mistake a nucleotide query as a protein. This will prevent the search from going through and result an error message. In alignments, BLAST treats the ambiguities in an accepted nucleotide query as mismatches. In short queries, these ambiguous bases may break the query in such a way that no valid word is available for BLAST to index the query and identify initial word hits, thus preventing BLAST from finding any matches in the database.

Table 6.5   Single-letter nucleotide code

Code	Meaning (Base)	Code	Meaning (Base)
A	Adenosine (A)	M	Amino (A or C)
C	Cytidine (C)	S	Strong (G or C)
G	Guanine (G)	W	Weak (A or T)
T	Thymidine (T)	B	Not A (G or T or C)
U	Uridine (U)	D	Not C (G or A or T)
R	Purine (G or A)	H	Not G (A or C or T)
Y	Pyrimidine (T or C)	V	Not T (G or C or A)
K	Keto (G or T)	N	Any base (A or G or C or T)
–[1]	Gap(s) of intermediate length		

[1] Dash(s) in the query will not be accepted. They will be removed before the search is submitted. To represent gaps, use a string of N's.

For those programs that use amino acid query sequences (BLASTP and TBLASTN), the IUPAC based amino acid codes are given in the Table 6.6.

Table 6.6   Single-letter code for amino acid

Code	Residue	Code	Residue
A	Alanine	P	Proline
B	Aspartate or asparagines	Q	Glutamine
C	Cysteine	R	Arginine
D	Aspartate	S	Serine
E	Glutamate	T	Threonine
F	Phenylalanine	U[1]	Selenocysteine
G	Glycine	V	Valine
H	Histidine	W	Tryptophan
I	Isoleucine	Y	Tyrosine
K	Lysine	Z	Glutamate or glutamine
L	Leucine	X	Any residue
M	Methionine	"	Translation stop
N	Asparagines	–[2]	Gap of indeterminate length

[1] BLAST cannot handle U properly in protein alignment since it was not specified in the scoring matrices used by blastp. To partially resolve thus, U in the query is replaced by an X before the search is performed.

[2] Dash(s) in the query will be removed before the search is submitted. To represent a gap, a string of X's should be used instead.

To run, BLAST requires a query sequence to search for, and a sequence to search against (or a sequence database containing multiple such sequences, also called the target sequence). BLAST will find subsequences in the database which are similar to subsequences in the query. In typical usage, the query sequence is much smaller than the database, e.g. the query may be one thousand nucleotides while the database is several billion nucleotides.

The main idea of BLAST is that there are often high-scoring segment pairs (HSP) contained in a statistically significant alignment. BLAST searches for high-scoring sequence alignments between the query sequence and sequences in the database using a heuristic approach that approximates the Smith–Waterman algorithm. The exhaustive Smith–Waterman (SW) approach is too slow for searching large genomic databases such as GenBank. Therefore, the BLAST algorithm uses a heuristic approach that is less accurate than the Smith–Waterman but over 50 times faster. The speed and relatively good accuracy of BLAST are among the key technical innovation of the BLAST programs.

Comparison of the Programs

- **Concept**

  SW tool and BLAST produce local alignments, while FASTA is a global alignment tool. BLAST can report more than one HSP per database entry, while FASTA reports only one segment (match).

- **Speed**

  BLAST > FASTA > SW

  BLAST (package) is a highly efficient search tool.

- **Sensitivity**

  SW > FASTA > BLAST (old version!)

  FASTA is more sensitive, missing less homologous sequences on the average (but the opposite can also happen, if there are no identical residues conserved, but this is infrequent). It also gives better separation between true hits and random hits.

- **Statistics**

  BLAST calculates probabilities, and it sometimes fails entirely if some of the assumptions used are invalid. FASTA calculates significance "on the fly" from the given dataset which is more relevant but can be problematic if the dataset is small.

Tips for Database Searches

▣ Use the latest database version.

▣ Run BLAST first, then depending on your results run a finer tool (FASTA, Ssearch, SW, Blocks, etc.).

▣ Whenever possible, use protein or translated nucleotide sequences.

▣ Expect value, $E < 0.05$ is statistically significant, usually biologically interesting. Check also $0.05 < E < 10$ because you might find interesting hits.

▣ Pay attention to abnormal composition of the query sequence, since it usually causes biased scoring.

▣ Split large query sequences (> 1000 for DNA, > 200 for protein).

▣ If the query has repeated segments, remove them and repeat the search.

Databases Available for BLAST Search

The BLAST pages offer several different databases for searching. Some of these, like Swiss-Prot and PDB are compiled outside of NCBI. Others like *E. coli*, dbEST and Month, are subsets of the NCBI databases.

▣ **Peptide sequence databases**

◈ **Nr**   All non-redundant GenBank CDS translations + RefSeq Proteins + PDB + Swiss-Prot + PIR + PRF

◈ **Refseq**   RefSeq protein sequences from NCBI's reference sequence project.

◈ **Swiss-Prot**   Last major release of the Swiss-Prot protein sequence database (no updates).

◈ **Pat**   Proteins from the patent division of GenPept.

◈ **PDB**   Sequences derived from the 3-dimensional structure from Brookhaven Protein Data Bank.

◈ **Month** All new or revised GenBank CDS translation+PDB+Swiss-Prot+PIR+PRF released in the last 30 days.

◈ **Env_nr**   Protein sequences from environmental samples.

▣ **Nucleotide sequence databases**

◈ **Nr**   All GenBank + Refseq nucleotides + EMBL + DDBJ + PDB sequences (excluding HTGS0,1,2, EST, GSS, STS, PAT, WGS). No longer "non-redundant".

- **Refseq_RNA** RNA entries from NCBI's reference sequence project.
- **Refseq_genomic** Genomic entries from NCBI's reference sequence project.
- **EST** Database of GenBank + EMBL + DDBJ sequences from EST divisions.
- **EST_human** Human subset of EST.
- **EST_mouse** Mouse subset.
- **EST_others** Non-mouse, non-human subset of EST.
- **GSS** Genome Survey Sequence, includes single-pass genomic data, exon-trapped sequences, and *Alu* PCR sequences.
- **HTGS** Unfinished High-Throughput Genomic Sequences: phases 0, 1 and 2 (finished, phase 3 HTG sequences are in Nr).
- **PAT** Nucleotides from the patent division of GenBank.
- **PDB** Sequences derived from the 3-dimensional structure from Brookhaven Protein Data Bank (http://www.rcsb.org/pdb/).
- **MONTH** All new or revised GenBank + EMBL + DDBJ + PDB sequences released in the last 30 days.
- **DBSTS** Database of GenBank+EMBL+DDBJ sequences from STS Divisions.
- **CHROMOSOME** A database with complete genomes and chromosomes from the NCBI Reference Sequence project.
- **WGS** A database for whole genome shotgun sequence entries.
- **ENV_NT** Nucleotide sequences from environmental samples, including those from Sargasso Sea and Mine Drainage projects.

Role of Statistical Estimators in Database Search

There are different types of statistical estimators used in the analysis of the validity of an alignment score.

Z-score The Z-score is an old, yet commonly used statistical estimator for the validity of statistical results, including alignment scores. It is defined by the number of standard deviations that separate an observed score from the average random score. In other words, it is the difference between the observed score and the average random score, normalized by the standard deviation of the distribution. A higher Z-score means that the score can be trusted with a higher confidence level.

E-value Expect value or E-value is the most frequently used statistical estimator for the validity of alignment scores. It is defined as the expected number of false positives with a score higher than the observed score. This value is dependant, obviously, on the number of random alignments, determined by the length of the query sequence and the size of the aligned sequences. A lower E-value indicates that the score has a higher confidence level.

P-value Once we have calculated the E-value, E, for a certain score, we can go one step further. The P-value is the probability of the observed score—the probability that a certain score occurred by chance. To find a formula for the P-value, let us define a random variable Y_E as the number of random records achieving an E-value of E or better. This random variable has a Poisson distribution with the parameter $\lambda = E$. The probability that no random events have a lower score then our score, i.e., $Y_E = 0$, decreases exponentially with our score 's'. Therefore, the probability that at least one random record achieved a better score then our E-value can be computed using the following simple formula:

$$P = 1 - e^{-E}$$

Like the E-value, this value is dependent on the size of the database. A lower P-value means that the score has a higher confidence level. This estimator is not widely used for determining the validity of sequence alignment scores.

Database Search Using BLASTP

Here the algorithm of BLASTP (a protein to protein search) is introduced to present the concept of BLAST.

1. Remove low-complexity region or sequence repeats in the query sequence. Low-complexity region means a region of a sequence is composed of few kinds of elements. These regions might give high scores that confuse the program to find the actual significant sequences in the database, so they should be filtered out. The regions will be marked with an X (protein sequences) or N (nucleic acid sequences) and then be ignored by the BLAST program. To filter out the low-complexity regions, the SEG program is used for protein sequences and the program DUST is used for DNA sequences. On the other hand, the program XNU is used to mask off the tandem repeats in protein sequences.

2. Make a k-letter word list of the query sequence. Take k=3 for example, we list the words of length 3 in the query protein sequence (k is usually 11 for a DNA sequence), "sequentially", until the last letter of the query sequence is included. The method can be illustrated in Figure 6.12.

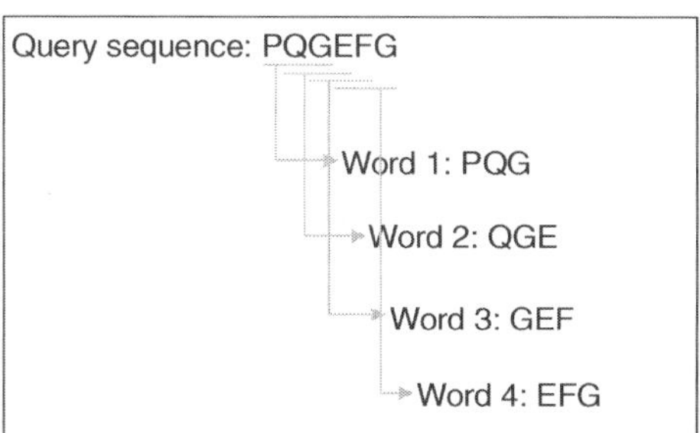

Figure 6.12   The method to establish the k-letter query word list

3. List the possible matching words This step is one of the main differences between BLAST and FASTA. FASTA cares about all of the common words in the database and query sequences that are listed in step 2; however, BLAST cares about only the high-scoring words. The scores are created by comparing the word in the list in step 2 with all the 3-letter words. By using the scoring matrix (substitution matrix) to score the comparison of each residue pair, there are 20^3 possible match scores for a 3-letter word. For example, the score obtained by comparing PQG with PEG and PQA is 15 and 12, respectively. For DNA words, a match is scored as +5 and a mismatch as –4. After that, a neighborhood word score threshold T is used to reduce the number of possible matching words. The words whose scores are greater than the threshold T will remain in the possible matching words list, while those with lower scores will be discarded. For example, PEG is kept, but PQA is abandoned when T is 13.

4. Organize the remaining high-scoring words into an efficient search tree. This is for the purpose that the program can rapidly compare the high-scoring words to the database sequences.

5. Repeat step 1 to 4 for each 3-letter word in the query sequence.

6. Scan the database sequences for exact match with the remaining high-scoring words. The BLAST program scans the database sequences for the remaining high-scoring word, such as PEG, of each position.

If an exact match is found, this match is used to seed a possible ungapped alignment between the query and database sequences.

7. Extend the exact matches to high-scoring segment pair (HSP).

   ▣ The original version of BLAST stretches a longer alignment between the query and the database sequence in left and right direction, from the position where exact match is scanned. The extension doesn't stop until the accumulated total score of the HSP begins to decrease. A simplified example is presented in Figure 6.13.

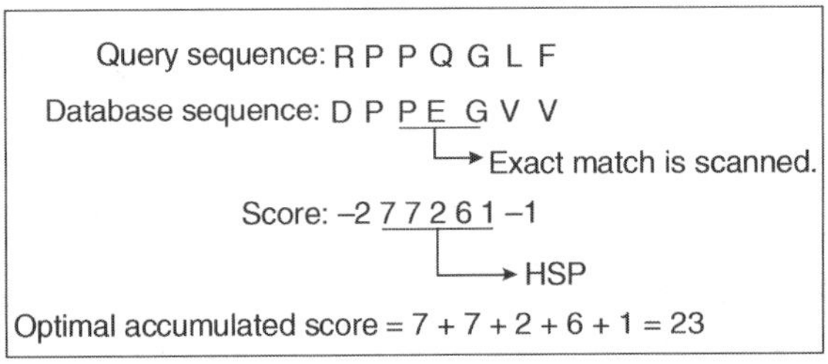

Query sequence: R P P Q G L F

Database sequence: D P P E G V V
                        └──→ Exact match is scanned.

Score: –2 7 7 2 6 1 –1
              └────→ HSP

Optimal accumulated score = 7 + 7 + 2 + 6 + 1 = 23

Figure 6.13   The process to extension the exact match

   ▣ To save more time, a newer version of BLAST, called BLAST2 or gapped BLAST, has been developed. BLAST2 adopts a lower neighborhood (closely-matching) word score threshold to maintain the same level of sensitivity for detecting sequence similarity. Therefore, the possible matching words list in step 3 becomes longer. Next, the exact matched regions, within distance A from each other on the same diagonal in Figure 6.14, will be joined as a longer new region. Finally, the new regions are then extended as the same method in the original version of BLAST, and the HSPs scores of the extended regions are then created by using a substitution matrix as before.

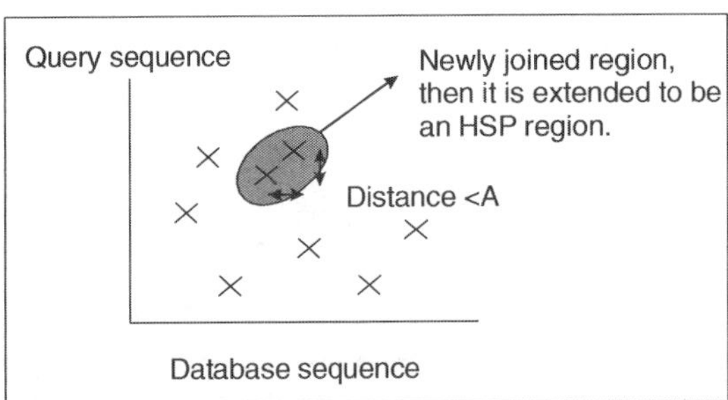

Figure 6.14   The positions of the exact matches

8. List the HSPs whose scores are greater than the empirically determined cut-off score S. By examining the distribution of the alignment scores modelled by comparing random sequences, a cut-off score S can be determined such that its value is large enough to guarantee the significance of the remained HSPs.

9. Evaluate the significance of the HSP score. Next BLAST assesses the statistical significance of each HSP score by exploiting the Gumbel Extreme Value Distribution (EVD). It is proved that the distribution of Smith–Waterman local alignment scores between two random sequences follows the Gumbel EVD, regardless of whether gaps are allowed in the alignment. In accordance with the Gumbel EVD, the probability p of observing a score S equal to or greater than x is given by the equation.

$$p(S \geq x) = 1 - \exp(-e^{-\lambda(x-\mu)}), \text{ where}$$

$$\mu = {}^{-[\log(Km'n')]} / \lambda$$

The statistical parameters λ and K are estimated by fitting the distribution of the ungapped local alignment scores, of the query sequence and a lot of shuffled versions (Global or local shuffling) of a database sequence, to the Gumbel extreme value distribution. Note that λ and K depend upon the substitution matrix, gap penalties, and sequence composition (the letter frequencies).The 'm' and 'n' is the effective length of the query and database sequence, respectively. The original sequence length is shortened to the effective length to compensate for the edge effect (an alignment start near the end of one of the query or database sequence is likely not to have enough sequence to build an optimal alignment). They can be calculated as

$$m' \approx m - {}^{(\ln Kmn)} / H$$

$$n' \approx n - {}^{(\ln Kmn)} / H,$$

where H is the average expected score per aligned pair of residues in an alignment of two random sequences. Latches and Gish gave the typical values, $\lambda = 0.318$, $K = 0.13$, and $H = 0.40$, for ungapped local alignment using BLOSUM62 as the substitution matrix. Using the typical values for assessing the significance is called the lookup table methods, and is not accurate. The expect score E of a database match is the number of times that an unrelated database sequence would obtain a score S higher than x by chance. The expectation E

obtained in a search for a database of D sequences is given by

$$E \approx 1 - e^{-p(s>x)} D$$

Furthermore, when $p < 0.1$, E could be approximated by the Poisson distribution as

$$E \approx pD$$

Note that the E value accessing the significance of the HSP score here (for ungapped local alignment) is not identical to the one in the later step to evaluate the final gapped local alignment score, due to the variation of the statistical parameters.

10. Make two or more HSP regions into a longer alignment Sometimes, two or more HSP regions in one database sequence that can be made into a longer alignment. This provides additional evidence of the relation between the query and database sequence. There are two methods, the Poisson method and the sum-of-scores method, to compare the significance of the newly combined HSP regions. Suppose that there are two combined HSP regions with the sets of score (65, 40) and (52, 45), respectively, the Poisson method gives more significance to the set with the lower score of each set is higher (45>40). However, the sum-of-scores method prefers the first set, because 65+40 (105) is greater than 52+45(97). The original BLAST uses the Poisson method; gapped BLAST and the WU-BLAST use the sum-of-scores method.

11. Show the gapped Smith–Waterman local alignments of the query and each of the matched database sequences.
    i.   The original BLAST only generates ungapped alignments including the initially found HSPs individually, even when there is more than one HSP found in one database sequence.
    ii. BLAST2 versions produce a single alignment with gaps that can include all of the initially found HSP regions. Note that the computation of the score and its corresponding E score is involved with the adequate gap penalties.

12. Report the matches whose expect score is lower than a threshold parameter E.

BLAST output format It includes a graphical overview box, a matching list and a text description of the alignment. The graphical overview box (Figure 6.15a) contains coloured horizontal bars that help in quick identification of the number of database hits and the degrees of similarity of the hits. The colour coding of the horizontal bars corresponds to the ranking of similarities of the sequence hits (red: most related, green and blue: moderately related, black: unrelated). The length of the bars represents the spans of the sequence alignments relative to the query sequence. Each bar is hyperlinked to the actual pairwise alignment in the text portion of the report. The graphical view is followed by the list of matching hits (Figure 6.15b). It is marked by hits ranked by the E-value in ascending order. Each hit includes the accession number, a partial title of the database record, bit score, and *E*-value. The list of hits are followed by the text description [Figure 6.15c] that may be divided by into three sections: the header, statistics, and alignment. The header section contains the gene index number or the reference number of the database hit plus a one-line description of the database sequence. Next to it is the summary of statistics of the search output, which include the bit score, E-value, percentages of identity, similarity ("Positives"), and gaps. In the actual alignment section, the query sequence is on the top of the pair and the database sequence is at bottom of the pair labelled as subject.

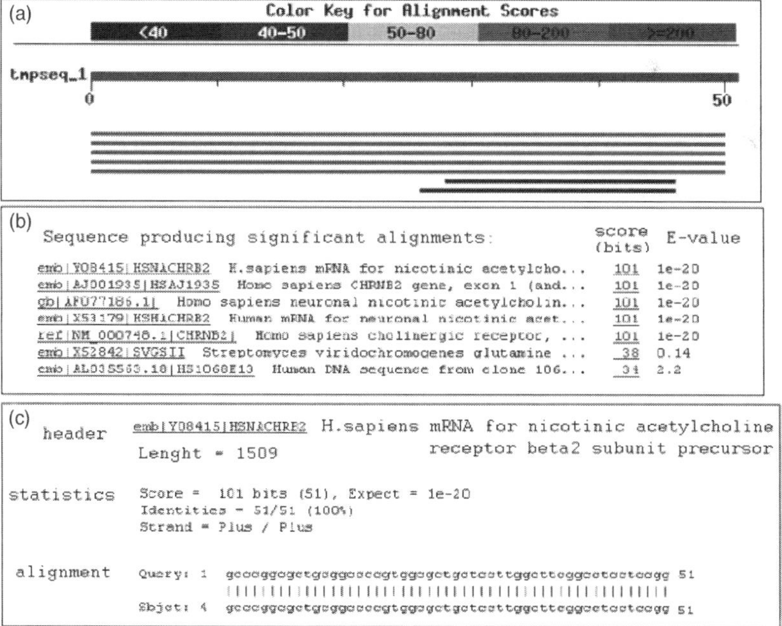

Figure 6.15 The BLAST output showing (a) Graphical overview box, (b) List of matching hits, and (c) Alignment output with text portion containing header, statistics and the actual alignment.

Improved BLAST

Altschul *et al.* suggested in 1997 an improved BLAST algorithm that allows indels in the alignment. The algorithm stages follow:

1. When considering the dynamic programming matrix to align two strings, we search along each diagonal for two w-length words such that the distance between them is $= A$ and their score is $= T$. T can be lower than in the previous algorithm. Future expansion is done only to such pairs of hits.

2. In the second stage we want to allow local alignments with indels, similarly to the FASTA algorithm. We allow two local alignments from different diagonals to merge into a new local alignment composed of the first local alignment followed by some indels and then the second local alignment. This local alignment is essentially a path in the dynamic programming matrix, composed of two diagonal sections and a path connecting them which may contain gaps. Unlike in FASTA, where we only allowed the diagonal to have local shifts, restricted to a band, here we allow local alignments from different diagonals to merge as long as the resulting alignment has a score above some threshold. This method results in an alignment structure which is much less regular. The improved version of BLAST is about 3 times faster than the original algorithm due to much less expansions made (only two-hit words are expanded).

PHI-BLAST Pattern

PHI-BLAST (Pattern-Hit Initiated BLAST) is a search program that combines matching of regular expressions with local alignments surrounding the match. Given a protein sequence S and a regular expression pattern P occurring in S, PHI-BLAST helps to answer the question.

Rules for pattern syntax for PHI-BLAST　Web PHI-BLAST search requires a pattern along with a protein sequence containing the pattern. A simple example and how to use PHI-BLAST is available on the webpage: http://www.ncbi.nlm.nih.gov/blast/producttable.shtml#phi.

The syntax for pattern specification in PHI-BLAST follows the conventions of PROSITE (Table 6.7). When using the stand-alone program, it is permissible to have multiple patterns in a file separated by a blank line between patterns. When using the webpage, only one pattern is allowed per query.

Table 6.7 Accepted PHI-BLAST pattern vocabulary

ABCDEFGHIKLMNPQRSTVWXYZU	Protein alphabet
ACGT	DNA alphabet
[ ]	Means any one of the characters enclosed in the brackets, e.g. [LFYT] means one occurrence of L or F or Y or T
-	Nothing, used as a spacer to clearly separate each position
x	Means any residue, if nothing follows
(n)	Means the preceeding residue is repeated 5 times
(m,n)	The preceeding residue is repeated between m to n times (n > m)
>	Only at the end of a pattern and means nothing it may occur before a period
.	May be used at the end, means nothing

When using the stand-alone program, the pattern should be stored in a pattern input file, with the first line starting with ID followed by 2 spaces and a text string giving the pattern a name. There should also be a line starting with PA followed by 2 spaces and then the pattern description.

All other PROSITE codes in the first two columns are allowed, but only the HI code, described below is relevant to PHI-BLAST. Here is an example from PROSITE:

```
ID CNMP_BINDING_2; PATTERN. AC PS00889;
DT OCT-1993 CREATED); OCT-1993 (DATA UPDATE); NOV-1995 (INFO
UPDATE).
DE Cyclic nucleotide-binding domain signature 2.
PA [LIVMF]-G-E-x-[GAS]-[LIVM]-x(5,11)-R-[STAQ]-A-x-[LIVMA]-x-
[STACV].
NR /RELEASE=32,49340;
NR /TOTAL=57(36); /POSITIVE=57(36); /UNKNOWN=0(0); /FALSE_POS=0(0);
NR /FALSE_NEG=1; /PARTIAL=1;
CC /TAXO-RANGE=??EP?; /MAX-REPEAT=2;
```

The line starting with ID gives the pattern a name. The lines starting with AC, DT, DE, NR, NR, and CC are relevant to PROSITE users, but irrelevant to PHI-BLAST. These lines are tolerated, but ignored by PHI-BLAST. The line starting with PA describes the pattern, which can be explained as in Table 6.8.

Table 6.8　The pattern syntax for PHI-BLAST

Pattern position	Pattern syntax	Meaning
1	[Livmf]	One of LIVMF
2	G	G
3	E	E
4	X	Any one residue
5	[Gas]	One of GAS
6	[Livm]	One of LIVM
7	X(5,11)	5 to 11 any residue
8	R	R
9	[STAQ]	One of STAQ
10	A	One A
11	X	Any one residue
12	[Livma]	One of LIVMA
13	X	Any one residue
14	[STACV]	Any one of STACV

Note: Total length of this motif/pattern is between　18 to 24 residues.

In this case the pattern ends with a period. It can end with nothing after the last specifying symbol or any number of > signs or periods or combination thereof. Given below is another example, illustrating the use of an HI line.

```
ID ER_TARGET; PATTERN.
PA [KRHQSA]-[DENQ]-E-L>.
HI (19 22)
HI (201 204)0
```

In this example, the HI lines specify that the pattern occurs twice, once from positions 19 through 22 in the sequence and once from positions 201 through 204 in the sequence. These specifications are relevant when stand-alone PHI-BLAST is used with the *seedp* option, in which the interesting occurrences of the pattern in the sequence are specified. In this case the HI lines specify which occurrence(s) of the pattern should be used to find good alignments.

In general, the *seedp* option is more useful than the standard patternp option only when the pattern occurs K > 1 times in the sequence and the user is interested in matching to J < K of those occurrences. Then using the HI lines enables the user to specify which occurrences are of interest.

FASTA

FASTA is a DNA and protein sequence alignment software package first described (as FASTP) by David J. Lipman and William R. Pearson in 1985 in the article "Rapid and sensitive protein similarity searches". The original FASTP program was designed for protein sequence similarity searching. FASTA, described in 1988 (Improved Tools for Biological Sequence Comparison) added the ability to do DNA : DNA searches, translated protein : DNA searches, and also provided a more sophisticated shuffling program for evaluating statistical significance. There are several programs in this package that allow the alignment of protein sequences and DNA sequences. FASTA is pronounced "FAST-Aye", and stands for "FAST-All", because it works with any alphabet, an extension of "FASTP" (protein) and "FASTN" (nucleotide) alignment.

The current FASTA package contains programs for protein : protein, DNA : DNA, protein:translated DNA (with frameshifts), and ordered or unordered peptide searches. Recent versions of the FASTA package include special translated search algorithms that correctly handle frameshift errors (which six-frame-translated searches do not handle very well) when comparing nucleotide to protein sequence data.

In addition to rapid heuristic search methods, the FASTA package provides search, an implementation of the optimal Smith–Waterman algorithm. A major focus of the package is the calculation of accurate similarity statistics, so that biologists can judge whether an alignment is likely to have occurred by chance, or whether it can be used to infer homology. The FASTA package is available at fasta.bioch.virginia.edu

The basic FASTA algorithm assumes a query sequence and a database over the same alphabet. Practically, FASTA is a family of programs, allowing also cross queries of DNA versus protein. The program variants are listed in Table 6.9.

Table 6.9   Variants of the FASTA algorithm

Program	Function
FASTA3	Scan a protein or DNA sequence library for similar sequences
FASTX/Y3	Compare a DNA sequence to a protein sequence database, comparing the translated DNA sequence in forward and reverse frames
TFASTX/Y3	Compares a protein a translated DNA databank
FASTS3	Compares linked peptides to a protein databank
FASTF3	Compares mixed peptides to a protein databank

Note   FASTX3 uses a simpler, faster algorithm for alignments that allows frameshifts only between codons; FASTY3 is slower but produces better alignments with poor quality sequences because frameshifts are allowed within codons.

Under different circumstances it is favourable to use different programs:

- To identify an unknown protein sequence use either FASTA3 or TFASTX3.
- To identify structural DNA sequence: (rep eated DNA, structural RNA) use FASTA3, first with ktup = 6 and then with ktup = 3.
- To identify an EST use FASTX3 (check whether the EST codes for a protein homologous to a known protein).
- Use ktup = 1 for oligonucleotides (length < 20).

FASTA3 (FASTX3, etc.) is the current version of FASTA. FASTA is available directly via the FASTA3 server, or it can be accessed through one of the retrieval systems, e.g. the GenWeb mirror site at the Weizmann Institute.

FASTA Format

- This format contains a single header line providing the sequence name, and optionally a description, followed by lines of sequence data.
- Sequences in FASTA formatted files are preceded by a line starting with a " >" symbol.
- The first word on this line is the name of the sequence. The rest of the line is a description of the sequence.

```
Term Entry Name Molecule Type Gene Name Sequence Length
e.g. FOSB_MOUSE Protein          fosB       338 bp
```

▣ The remaining lines contain the sequence itself, usually formated to 60 characters per line.

▣ Depending on the application blank lines in a FASTA file are ignored or treated as terminating the sequence.

▣ Depending on the application spaces or other non-sequence symbols (dashes, underscores, and periods) in a sequence are either ignored or treated as gaps.

▣ FASTA files containing multiple sequences are just the same, with one sequence listed right after another. This format is accepted for many multiple sequence alignment programs. A simple example of one sequence in FASTA format:

```
>FOSB_MOUSE Protein fosB. 338 bp

MFQAFPGDYDSGSRCSSSPSAESQYLSSVDSFGSPPTAAASQECAGLGEMPGSFVPTVTA

ITTSQDLQWLVQPTLISSMAQSQGQPLASQPPAVDPYDMPGTSYSTPGLSAYSTGGASGS

GGPSTSTTTSGPVSARPARARPRRPREETLTPEEEEKRRVRRERNKLAAAKCRNRRRELT

DRLQAETDQLEEEKAELESEIAELQKEKERLEFVLVAHKPGCKIPYEEGPGPGPLAEVRD

LPGSTSAKEDGFGWLLPPPPPPPLPFQSSRDAPPNLTASLFTHSEVQVLGDPFPVVSPSY

TSSFVLTCPEVSAFAGAQRTSGSEQPSDPLNSPSLLAL
```

FASTA Search Method

FASTA takes a given nucleotide or amino acid sequence and searches a corresponding sequence database by using local sequence alignment to find matches of similar database sequences.

The FASTA program follows a largely heuristic method which contributes to the high speed of its execution. It initially observes the pattern of word hits, word-to-word matches of a given length, and marks potential matches before performing a more time-consuming optimized search using a Smith–Waterman type of algorithm. The size taken for a word, given by the parameter ktup (short for k respective tuples), controls the sensitivity and speed of the program. Increasing the ktup value decreases number of background hits that are found. From the word hits that are returned the program looks for segments that contain a cluster of nearby hits. It then investigates these segments for a possible match.

There are some differences between FASTN and FASTP relating to the type of sequences used but both use four steps and calculate three scores to describe and format the sequence similarity results. The steps involved in FASTA are as follows:

1. Identify regions of highest density in each sequence comparison. Taking a ktup to equal 1 or 2 (Figure 6.16a).

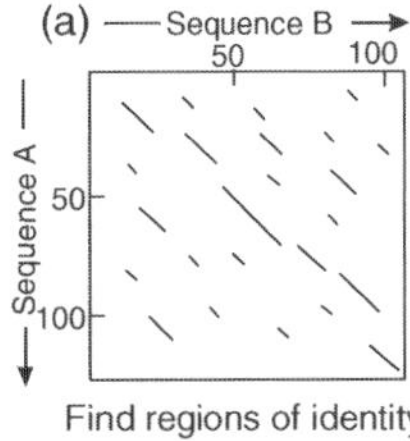

Find regions of identity

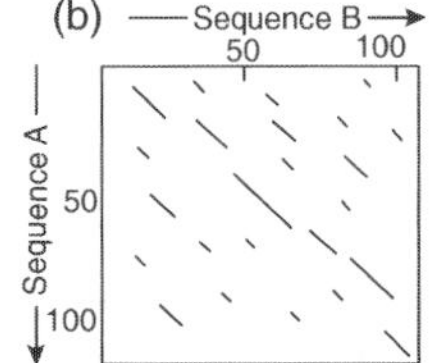

Re-scan the regions using PAM matrix and save the best initial regions. Initial regions with scores less than the joining threshold are dashed. The asteric denotes the highest scoring region reported by FASTP

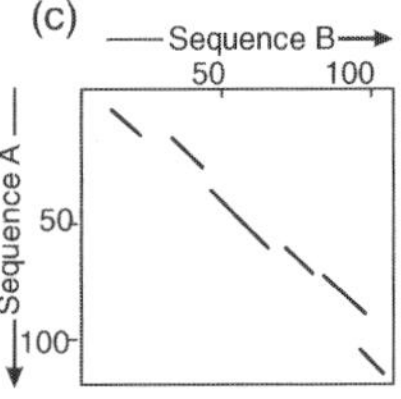

Optimally join initial regions with scores greater than a threshold. The solid lines denote regions that are joined to make up the optimized initial score.

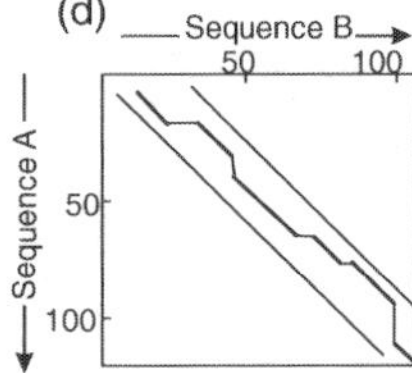

Recalculate an optimized alignment using dynamic programming in a narrow band that consists of top scoring segments

Figure 6.16   FASTA algorithm

In this step all or a group of the identities between two sequences are found using a look up table. The ktup value determines how many consecutive identities are required for a match to be declared. Thus the lesser the ktup value: the more sensitive the search. ktup =2 is frequently taken by users for protein sequences and ktup = 4 or 6 for nucleotide sequences. Short oligonucleotides are usually run with ktup = 1. The program then finds all similar local regions, represented as diagonals of a certain length in a dot-plot, between the two sequences by counting ktup matches and penalizing for intervening mismatches. This way, local regions of highest density matches in a diagonal are isolated from background hits. For protein sequences BLOSUM50 values are used for scoring ktup matches. This ensures that groups of identities with high similarity scores contribute more to the local diagonal score than to identities with low similarity scores.

Nucleotide sequences use the identity matrix for the same purpose. The best 10 local regions selected from all the diagonals put together are then saved.

2. Re-scan the regions taken using the scoring matrices. Trimming the ends of the region to include only those contributing to the highest score (Figure 6.16b). Re-scan the 10 regions taken, this time use the relevant scoring matrix while rescoring to allow runs of identities shorter than the ktup value. Also while rescoring conservative replacements that contribute to the similarity score are taken. Though protein sequences use the BLOSUM50 matrix, scoring matrices based on the minimum number of base changes required for a specific replacement, on identities alone, or on an alternative measure of similarity such as PAM, can also be used with the program. For each of the diagonal regions re-scanned in this way, a sub-region with the maximum score is identified. The initial scores found in step1 are used to rank the library sequences. The highest score is referred to as *init1* score.

3. In an alignment if several initial regions with scores greater than a CUT-OFF value are found, check whether the trimmed initial regions can be joined to form an approximate alignment with gaps. Calculate a similarity score that is the sum of the joined regions penalizing for each gap 20 points. This initial similarity score (*initn*) is used to rank the library sequences. The score of the single best initial region found in step 2 is reported (*init1*).

   Here the program calculates an optimal alignment of initial regions as a combination of compatible regions with maximal score (Figure 6.16c). This optimal alignment of initial regions can be rapidly calculated using a dynamic programming algorithm. The resulting score *initn* is used to rank the library sequences. This joining process increases sensitivity but decreases selectivity. A carefully calculated cut-off value is thus used to control where this step is implemented, a value that is approximately one standard deviation above the average score expected from unrelated sequences in the library. A 200-residue query sequence with ktup2 uses a value 28.

4. Use dynamic programming such as Smith–Waterman algorithm to calculate an optimal score for alignment (Figure 6.16d). This step uses a banded Smith–Waterman algorithm to create an optimized score (opt) for each alignment of query sequence to a database (library) sequence. It takes a band of 32 residues centred on the *init1* region of step2 for calculating the optimal alignment. After all

sequences are searched the program plots the initial scores of each database sequence in a histogram, and calculates the statistical significance of the "opt" score. For protein sequences, the final alignment is produced using a full Smith–Waterman alignment. For DNA sequences, a banded alignment is provided. Common FASTA output is a histogram indicating the number of database sequences at any given initial score (Figure 6.17).

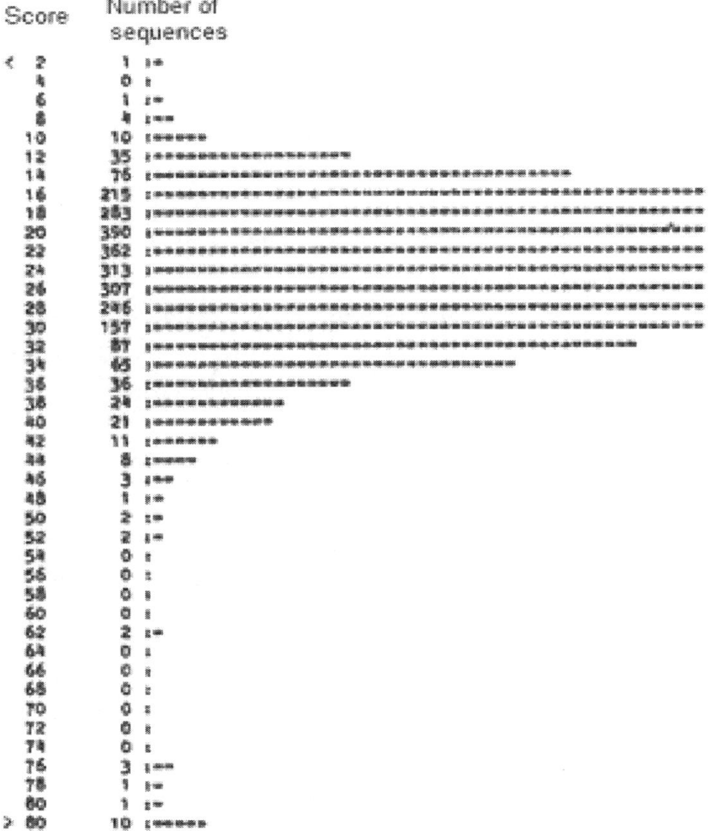

Figure 6.17   FASTA output histogram

PAM UNITS AND PAM MATRICES—

AS AMINO ACID SCORING MATRICES

The methodology of PAM units and the first specific PAM matrices were developed by Dayhoff *et al* (1978). They examined 1572 accepted mutation between 71 superfamilies of closely related sequences of proteins (no more than 15% different). During this process they noticed that the substitutions that occurred in families of closely related proteins were not random. They concluded that some amino acid substitutions occurred more readily than others, probably because they did not have a great effect on the structure and function of a protein, this meant that evolutionarily related proteins did not need to have the same amino acids at every position: They could have comparable ones. In doing alignments, this becomes very important. From such observations, the PAM matrix was born. PAM stands for "Point Accepted Mutations" or "Per cent of Accepted Mutations".

PAM Units

PAM units measure the amount of evolutionary distance between two amino acid sequences. Two sequences S1 and S2 are at evolutionary distance of one PAM, if S1 has converted to S2 with an average of one accepted point-mutation event per 100 amino acids. Here the term "accepted" means a mutation that was incorporated into the protein and passed to its progeny. Therefore, either the mutation did not change the function of the protein or the change in the protein was beneficial to the organism. Note that two strings which are one PAM unit diverged do not necessarily differ in one per cent, as often mistakenly thought, because a single position may undergo more than a single mutation. The difference between the two notions grows as the number of units does.

There are two main problems with the notion of PAM units:

1. Practically all the sequences one can obtain today are extracted from extant organisms. Researchers almost do not know any protein sequences where one is actually derived from the other. The lack of ancestral protein sequences is handled by assuming that amino acid mutations are reversible and are also likely in either direction. This assumption, together with the additivity property of the PAM units derived from its definition, imply that given two amino acid sequences S_i and S_j whose mutual ancestor is S_{ij} then

$$d(S_i, S_j) = d(S_i, S_{ij}) + d(S_{ij}, S_j)$$

   where $d(i, j)$ is the PAM distance between amino acid sequences i and j.

2. Insertions and deletions which may occur during evolution are ignored. Hence one cannot be sure of the correct correspondence between sequence positions. In order to know the exact correspondence one has to be able to identify the true historical gaps, or at least to identify

large intervals along the two sequences where the correspondence is correct. This cannot always be done with certainty, especially when the two sequences are evolutionarily distant.

PAM Matrices

PAM matrices are amino acid substitution matrices that encode the expected evolutionary change at the amino acid level. Each PAM matrix is designed to compare two sequences which are a specific number of PAM units apart. For example, the PAM120 score matrix is designed to compare between sequences that are 120-PAM units apart: The score it gives a pair of sequences is the (log of the) probabilities of such sequences evolving during 120-PAM units of evolution. For any specific pair (A_i, A_j) of amino acids the (i, j) entry in the PAM-N matrix reflects the frequency at which A_j is expected to replace A_i in two sequences that are n PAM units diverged. (One PAM unit = 1% divergence in amino acids, i.e., out of 100, one of amino acid positions have been changed). Higher PAM value indicates more divergence. These frequencies should be estimated by gathering statistics on replaced amino acids. Example of one such PAM250 scoring matrix is displayed in Figure 6.18.

A	Ala	2																		
R	Arg	-2	6																	
N	Asn	0	0	2																
D	Asp	0	-1	2	4															
C	Cys	-2	-4	-4	-5	4														
Q	Gln	0	1	1	2	-5	4													
E	Glu	0	-1	1	3	-5	2	4												
G	Gly	1	-3	0	1	-3	-1	0	5											
H	His	-1	2	2	1	-3	3	1	-2	6										
I	Ile	-1	-2	-2	-2	-2	-2	-2	-3	-2	5									
L	Leu	-2	-3	-3	-4	-6	-2	-3	-4	-2	2	6								
K	Lys	-1	3	1	0	-5	1	0	-2	0	-2	-3	5							
M	Met	-1	0	-2	-3	-5	-1	-2	-3	-2	2	4	0	6						
F	Phe	-4	-4	-4	-6	-4	-5	-5	-5	-2	1	2	-5	0	9					
P	Pro	1	0	-1	-1	-3	0	-1	-1	0	-2	-3	-1	-2	-5	6				
S	Ser	1	0	1	0	0	-1	0	1	-1	-1	-3	0	-2	-3	1	3			
T	Thr	1	-1	0	0	-2	-1	0	0	-1	0	-2	0	-1	-2	0	1	3		
W	Trp	-6	2	-4	-7	-8	-5	-7	-7	-3	-5	-2	-3	-4	0	-6	-2	-5	17	
Y	Tyr	-3	-4	-2	-4	0	-4	-4	-5	0	-1	-1	-4	-2	7	-5	-3	-3	0	10
V	Val	0	-2	-2	-2	-2	-2	-2	-1	-2	4	2	-2	2	-1	-1	-1	0	-6	-2	4
		Ala	Arg	Asn	Asp	Cys	Gln	Glu	Gly	His	Ile	Leu	Lys	Met	Phe	Pro	Ser	Thr	Trp	Tyr	Val
		A	R	N	D	C	Q	E	G	H	I	L	K	M	F	P	S	T	W	Y	V

Figure 6.18   PAM250 amino acid substitution matrix

BLOSUM (BLOcks SUbstitution Matrix)

The BLOSUM matrix is another amino-acid substitution matrix, first calculated by Henikoff and Henikoff (1992). The difference between the PAM and BLOSUM matrices is that the PAM is derived from global alignments of proteins, while BLOSUM comes from alignments of shorter sequences—

blocks of sequences that match each other at some defined level of similarity. The BLOSUM method thereby incorporates much more data into its matrices, and is therefore, presumably, more accurate. For BLOSUM matrix calculation, only blocks of similar amino acid sequences are considered. Blocks are the conserved region of a protein family with the family members aligned. These blocks represent over 500 groups of related proteins, and act as signatures of these protein families. One reason for this is that one needs to find a multiple alignment between all these sequences and it is easier to construct such an alignment with similar sequences. Another reason is that the purpose of the matrix is to measure the probability of one amino acid changing into another, and the change between distant sequences may also include insertions and deletions of amino acids. Moreover, we are more interested in conservation of regions inside protein families, where sequences are quite similar, and therefore we restrict our examination to such.

The first stage of building the BLOSUM matrix is eliminating sequences, which are identical in more than $x\%$ of their amino acid sequence. This is done to avoid bias of the result in favour of a certain protein. The elimination is done either by removing sequences from the block, or by finding a cluster of similar sequences and replacing it by a new sequence that represents the cluster. If two sequences are more than $x\%$ identical, then the contribution of these sequences is weighted to sum to one. In this way the contributions of multiple entries of closely related sequences is reduced. The matrix built from blocks with no more the $x\%$ of similarity is called BLOSUMX (e.g. the matrix built using sequences with no more then 62% similarity is called BLOSUM62 as shown in Figure 6.19).

		Ala	Arg	Asn	Asp	Cys	Gln	Glu	Gly	His	Ile	Leu	Lys	Met	Phe	Pro	Ser	Thr	Trp	Tyr	Val
A	Ala	4																			
R	Arg	-1	5																		
N	Asn	-2	0	6																	
D	Asp	-2	-2	1	6																
C	Cys	0	-3	-3	-3	9															
Q	Gln	-1	1	0	0	-3	5														
E	Glu	-1	0	0	2	-4	2	5													
G	Gly	0	-2	0	-1	-3	-2	-2	6												
H	His	-2	0	1	-1	-3	0	0	-2	8											
I	Ile	-1	-3	-3	-3	-1	-3	-3	-4	-3	4										
L	Leu	-1	-2	-3	-4	-1	-2	-3	-4	-3	2	4									
K	Lys	-1	2	0	-1	-3	1	1	-2	-1	-3	-2	5								
M	Met	-1	-1	-2	-3	-1	0	-2	-3	-2	1	2	-1	5							
F	Phe	-2	-3	-3	-3	-2	-3	-3	-3	-1	0	0	-3	0	6						
P	Pro	-1	-2	-2	-1	-3	-1	-1	-2	-2	-3	-3	-1	-2	-4	7					
S	Ser	1	-1	1	0	-1	0	0	0	-1	-2	-2	0	-1	-2	-1	4				
T	Thr	0	-1	0	-1	-1	-1	-1	-2	-2	-1	-1	-1	-1	-2	-1	1	5			
W	Trp	-3	-3	-4	-4	-2	-2	-3	-2	-2	-3	-2	-3	-1	1	-4	-3	-2	11		
Y	Tyr	-2	-2	-2	-3	-2	-1	-2	-3	2	-1	-1	-2	-1	3	-3	-2	-2	2	7	
V	Val	0	-3	-3	-3	-1	-2	-2	-3	-3	3	1	-2	1	-1	-2	-2	0	-3	-1	4
		A	R	N	D	C	Q	E	G	H	I	L	K	M	F	P	S	T	W	Y	V

Figure 6.19 Blosum62 amino acid-substitution matrix

Major Differences Between PAM and BLOSUM

Although PAM and BLOSUM are amino-acid substitution matrices, there are number of differences between them. The principal difference lies in the fact that the PAM matrices, except PAM1 are derived from an evolutionary model. In contrast, the BLOSUM matrices consist of entirely direct observations based on actual alignment. Thus, BLOSUM matrices may have less evolutionary than the PAM matrices. Hence PAM matrices are most often used to reconstructing the phylogenetic trees indicating evolutionary significance. However, in PAM matrices the mathematical extrapolation method may make the PAM values less realistic for divergent sequences. The BLOSUM matrices are entirely derived from local sequence alignments of conserved sequence blocks, whereas the PAM1 matrix is based on the global alignment of full-length sequences composed of both conserved and variable regions. This is why the BLOSUM matrices may be more advantageous in searching databases and finding conserved domains in polypeptide chain. The PAM and BLOSUM serial numbers also have opposite meanings. Matrices of high PAM numbers are used to align divergent sequences and low PAM numbers for aligning closely related sequences.

REVIEW QUESTIONS

1. Explain the concept of sequence alignment with its significance.
2. What are the different biologically motivated problems in computer science?
3. Describe the concept of the edit distance with suitable example.
4. What are the different methods for pairwise alignment of the protein or DNA sequences?
5. Is there sequence alignment software available on the web for database search?
6. Enlist the software available on the web for pairwise alignment.
7. Distinguish between local and global alignment.
8. Describe Needleman–Wunsch algorithm for global alignment.
9. Explain Smith–Waterman algorithm for local alignment.
10. What are the different gap penalty models?
11. Describe the various statistical estimators used to analyse the validity of an alignment score.
12. Define database similarity searching and add its significance.

13. What is BLAST? Enlist different BLAST programs with their uses.
14. Explain database search using the algorithm of BLASTP.
15. Describe BLAST output format with significance of E-value.
16. What is Pattern-Hit Initiated BLAST?
17. Enlist an accepted PHI-BLAST pattern vocabulary and its pattern syntax.
18. What is FASTA? Describe FASTA program variants with significance.
19. Compare BLAST and FASTA.
20. Explain FASTA format and the steps involved with FASTA search method.
21. What are amino acid scoring matrices? Explain their role in evaluation of relatedness of protein sequences.
22. Write short notes on
    i. Dot-plot method
    ii. Dynamic programming
    iii. k-tuple methods
    iv. Backtracking of the sequence alignment
    v. BLASTN
    vi. VecScreen
    vii. Points of differences between PAM and BLOSUM
    viii. Histogram as FASTA output

INTRODUCTION

Sometimes it is necessary to align a number of sequences, in order to identify regions of homology between them. From the biologist point of view, a multiple sequence alignment (MSA) is a sequence alignment of three or more biological sequences, generally protein, DNA, or RNA. MSA is a natural alignment of the pairwise alignment in which multiple related sequences are aligned to achieve optimum matching of the sequences. There is unique advantage of MSA because it reveals more biological information than many pairwise alignments can. For example, it allows the identification of conserved sequence patterns and motifs in the whole sequence family, which are not obvious to detect by comparing only two sequences. This alignment is an important tool in characterizing protein families, determining consensus sequences, finding secondary and tertiary structure of new sequences, and construction of phylogenetic trees according to the similarity level between and within the aligned sequences. There are two different approaches to multiple sequence alignment. The first—alignment of similar sequences of nucleotides or amino acids, takes into account the physio-chemical properties and mutation data. The second—alignment of sequences solely according to secondary and tertiary structure. The resulting alignment can, understandably, differ greatly between the two approaches.

In general, the input set of query sequences are assumed to have an evolutionary relationship by which they share a lineage and are descended from a common ancestor. From the resulting MSA, sequence homology can be inferred and phylogenetic analysis can be conducted to assess the sequences that shared evolutionary origins. Visual depictions of the alignment as in the image at right, illustrate mutation events such as point mutations (single amino acid or nucleotide changes) that appear as differing characters in a single alignment column, and insertion or deletion mutations (indels or gaps) that appear as hyphens in one or more of the sequences in the alignment. Multiple sequence alignment is often used to assess sequence conservation of protein domains, tertiary and secondary structures, and even individual amino acids or nucleotides.

Multiple sequence alignment also refers to the process of aligning such a sequence set. Because three or more sequences of biologically relevant length can be difficult and are almost always time-consuming to align by hand, computational algorithms are used to produce and analyse the alignments. MSAs require more sophisticated methodologies than pairwise alignment because they are more computationally complex to produce. Most multiple sequence alignment programs use heuristic methods rather than global optimization because identifying the optimal alignment between more than a few sequences of moderate length is prohibitively computationally expensive.

MULTIPLE ALIGNMENT TO A PHYLOGENETIC TREE

A tree T with a distinct string label (from a set of strings S) assigned to each leaf is called a **phylogenetic tree** on S.

Given a phylogenetic tree T on S, a *phylogenetic alignment T'* for T is an assignment of one string label to each internal node of T. Note that the strings assigned to internal nodes need not be distinct and need not be from the set S. For an example, Figure 7.1a shows a typical phylogenetic alignment and Figure 7.1b shows phylogenetic tree for mammalian evolution. Thus, when T is a star, choosing the sequence to label its central node is a phylogenetic alignment.

The phylogenetic tree T is meant to represent the established evolutionary history of a set of objects of interest, with the convention that each extant object is represented at a unique leaf of the tree. Each edge (u, v) represents some evolutionary history that transforms the string at u (assuming u is the parent of v) to the string at v.

If strings S and S' are assigned to the endpoints of an edge (i, j), then the **edge distance** of (i, j) is defined to be $D(S, S')$.

Let M_T be a phylogenetic alignment for a phylogenetic tree T. The distance of M_T is given by the sum of all the edge distances over all the edges of T.

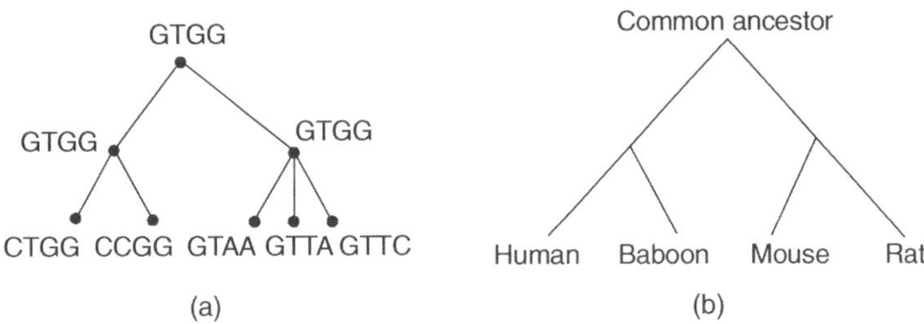

Figure 7.1   (a) A phylogenetic tree alignment (b) A phylogenetic tree showing evolutionary relationship between mammals established after progressive alignment of DNA

DYNAMIC PROGRAMMING AND COMPUTATIONAL COMPLEXITY

The most direct method for producing an MSA uses the dynamic programming technique to identify the globally optimal alignment solution. For proteins, this method usually involves two sets of parameters: a gap penalty and a substitution matrix assigning scores or probabilities to the alignment of each possible pair of amino acids based on the similarity of the amino acids' chemical properties and the evolutionary probability of the mutation. For nucleotide sequences a similar gap penalty is used, but a much simpler substitution matrix, wherein only identical matches and mismatches are considered, is typical. The scores in the substitution matrix may be either all positive or a mix of positive and negative in the case of a global alignment, but must be both positive and negative, in the case of a local alignment. In the latter case it is essential that the average score be less than 0.

For n individual sequences, the naive method requires constructing the n-dimensional equivalent of the matrix formed in standard pairwise sequence alignment. The search space thus increases exponentially with increasing n and is also strongly dependent on sequence length. In 1989, Altschul introduced a practical method that uses pairwise alignments to constrain the n-dimensional search space. In this approach pairwise dynamic programming alignments are preformed on each pair of sequences in the query set, and only the space near the n-dimensional intersection of these alignments is searched for the n-way alignment. In other words, for aligning n sequences, an n-dimensional matrix is needed to be filled with alignment scores. As the amount of computational time and memory space required increases exponentially with the number of sequences, it makes the method computationally prohibitive to use for a large data set. For this reason, fully dynamic programming is limited to small datasets of less than ten short sequences.

The MSA program optimizes the sum of all the pairs of characters at each position in the alignment (the so-called sum of pair (SP) score) and has been implemented in the MSA program. In other words, SP is the sum of scores of all possible pairs of sequences in a multiple alignment based on a particular scoring matrix. In calculating the SP scores, each column is scored by summing the scores for all possible pairwise matches, mismatches and gap costs. The score of entire alignment is the sum of all the column scores. In Figure 7.2, for the given multiple alignment of three sequences, the sum of the score is calculated as the sum of the similarity scores of every pair of sequence at each position. The scoring is based on the BLOSUM62 matrix. The total score for the alignment is $2^5 = 32$ times more likely to occur among homologous sequence than by random chance. The purpose of the most multiple sequence alignment algorithms is to achieve maximum SP scores. But MSA is still impractical for many multiple sequence alignment applications that require the simultaneous alignment of more than about 20 sequences.

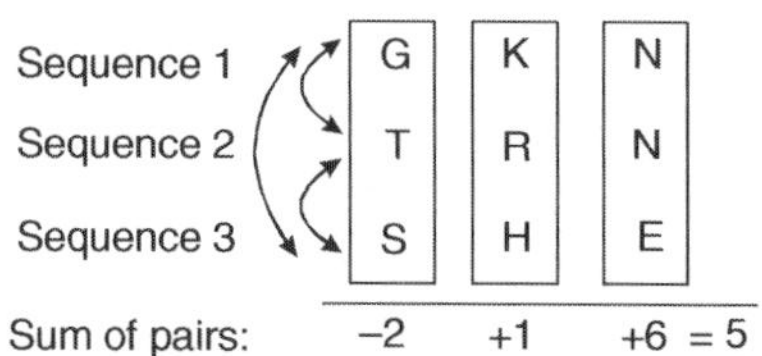

Figure 7.2   The sum of scores in a multiple alignment of three sequences

PROGRESSIVE ALIGNMENT METHOD

The most widely used method to multiple sequence alignments uses a heuristic search known as progressive technique (also known as the hierarchical or tree method), that speeds up the alignment of multiple sequences through a multistep process. It builds up a final MSA by combining pairwise alignments, beginning with the most similar pair and progressing to the most distantly related pair. All progressive alignment methods require two stages—a first stage in which the relationships between the sequences are represented as a tree, called a guide tree, and a second stage in which the MSA is built by adding the sequences sequentially to the growing MSA according to the guide tree. The tree reflects evolutionary proximity among all the sequences. The initial guide tree is determined by an efficient clustering approach such as neighbor-joining method and may use distances based on the number of identical two letter subsequences as in FASTA.

Progressive alignments cannot be globally optimal. The primary problem is that when errors are made at any stage in growing the MSA, then these errors are propagated through to the final result. Performance is also particularly bad when all of the sequences in the set are rather distantly related. Most modern progressive methods modify their scoring function with a secondary weighting function that assigns scaling factors to individual members of the query set in a non-linear fashion based on their phylogenetic distance from their nearest neighbors. This corrects for non-random selection of the sequences given to the alignment program.

Progressive alignment methods are efficient enough to implement on a large scale for many (100s to 1000s) sequences. Progressive alignment services are commonly available on publicly accessible web servers so users need not locally install the applications of interest. The most popular progressive alignment method has been the Clustal family, especially the weighted variant ClustalW to which access is provided by a large number of web portals including GenomeNet (http://align.genome.jp/), EBI, and EMBNet. Different portals or implementations can vary in user interface and make different parameters accessible to the user. ClustalW is used extensively for phylogenetic tree construction, in spite of the author's explicit warnings that unedited alignments should not be used in such studies and as input for protein structure prediction by homology modelling.

ClustalW

ClustalW is a software package for progressive multiple sequence alignments (implementing an algorithm of Thompson, Higgins, and Gibson, 1994) available at www.ebi.ac.uk/clustalw/.

The basic steps involved in ClustalW are as follows:

1. Calculate the pairwise alignment scores, and convert them to distances matrix
2. Use a neighbor-joining algorithm to build a guide tree from the distances
3. Align sequence–sequence, sequence–profile, profile–profile in decreasing similarity order

This algorithm makes use of many adhoc rules such as weighting, different matrix scores and special gap scores. Figure 7.3 shows the example of an alignment tree built by ClustalW.

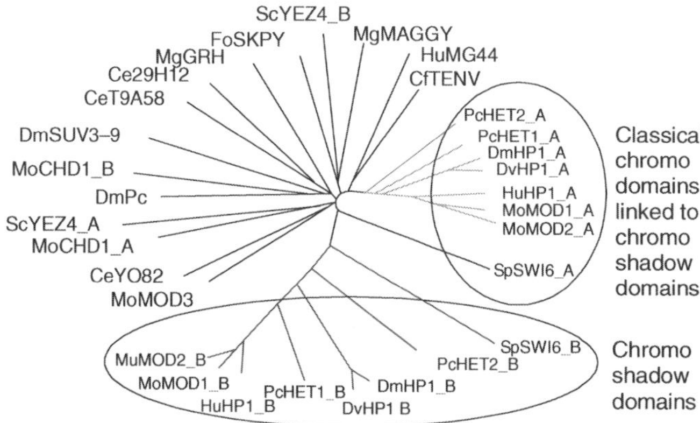

Figure 7.3   The alignment tree built by ClustalW

ALN/ClustalW 2 format   It was originated in the alignment program ClustalW 2. The file starts with word "CLUSTAL" and then some information about which Clustal program was run and the version of Clustal used, e.g. "CLUSTALW 2.1 multiple sequence alignment" The type of Clustal program is "W" and the version is 2.1. The alignment is written in blocks of 60 residues. Every block starts with the sequence names, obtained from the input sequence, and a count of the total number of residues is shown at the end of the line. The information about which residues match is shown below each block of residues they are:

"*" means that the residues or nucleotides in that column are identical in all sequences in the alignment.

":" means that conserved substitutions have been observed.

"." means that semi-conserved substitutions are observed.

Two examples of multiple sequence alignment in ClustalW 2.1 version are shown below.

```
FOSB_MOUSE ITTSQDLQWLVQPTLISSMAQSQGQPLASQPPAVDPYDMPGTSYSTPGLSAYST GGASGS 60

FOSB_HUMANITTSQDLQWLVQPTLISSMAQSQGQPLASQPPVVDPYDMPGTSYSTPGMSGYS SGGASGS 60

************************************ .***************:* .**:******

FOS_RAT PEEMSVTS-LDLTGGLPEATTPESEEAFTLPLLNDPEPK-PSLEPVKNISNMELKA EPFD 58
```

```
FOS_MOUSE PEEMSVAS-LDLTGGLPEASTPESEEAFTLPLLNDPEPK-PSLEPVKSISNVE LKAEPFD 58

FOS_CHICK SEELAAATALDLG—APSPAAAEEAFALPLMTEAPPAVPPKEPSG—SGLELKAEPFD 54

FOSB_MOUSE PGPGPLAEVRDLPG—STSAKEDGFGWLLPPPPPPP————————LPFQ 36

FOSB_HUMAN PGPGPLAEVRDLPG—SAPAKEDGFSWLLPPPPPPP————————LPFQ 36
            . . :    ** .       :.. *:.*    *   . ***:
```

Dbclustal It is a Clustal-based database search algorithm for protein sequences that combines local and global alignment features. It first performs a BLASTP search for a query sequence. The resulting sequence alignment pairs above a certain threshold are analysed to obtain anchor points that are common conserved regions, by using a program called Ballast. Subsequently a global alignment is generated by Clustal that is weighted towards the anchor points. Since the anchor points are derived from local alignments, this strategy reduces errors caused by the global alignment. The resulting multiple alignment is further evaluated by NorMD, which removes unrelated or badly aligned sequences from the multiple alignment. Thus, the final alignment should be more accurate than using Clustal alone. It also allows the incorporation of very long gaps for insertions and terminal extensions.

T-Coffee (Tree-based consistency objective function for alignment evaluation) It is another common progressive alignment method that is slower than Clustal and its derivatives but generally produces more accurate alignments for distantly related sequence sets. T-Coffee calculates pairwise alignments by combining the direct alignment of the pair with indirect alignments that aligns each sequence of the pair to a third sequence. It uses the output from ClustalW as well as another local alignment program LALIGN, which finds multiple regions of local alignment between two sequences. The resulting alignment and phylogenetic tree are used as a guide to produce new and more accurate weighting factors.

T-Coffee performs both global and local alignment for all possible pairs involved. The global pairwise alignment is performed using Clustal program. The local pairwise alignment is generated by the LALIGN program, from which the top ten scored alignments are selected. The collection of local and global sequence alignments are pooled to form a library. Later on the consistency of the alignment is evaluated. For every pair of residues in a pair of sequences, a constituency score is calculated for both global and local alignments. Each pairwise alignment is further aligned with a possible third sequence. The result is used to refine the original pairwise alignment based on a consistency criterion in a process called library extension. Based on the refined pairwise alignments, a distance matrix is built to derive a guide tree, which is then used to direct a full multiple alignment using the progressive alignment approach (Figure 7.4). Because progressive methods are heuristics that are not guaranteed to converge to a global optimum, alignment quality can be difficult to evaluate and their true biological significance can be obscure. A very recent semi-progressive method that improves alignment quality and does not use a loss-heuristic while still running in polynomial time has been implemented in the program PSAlign.

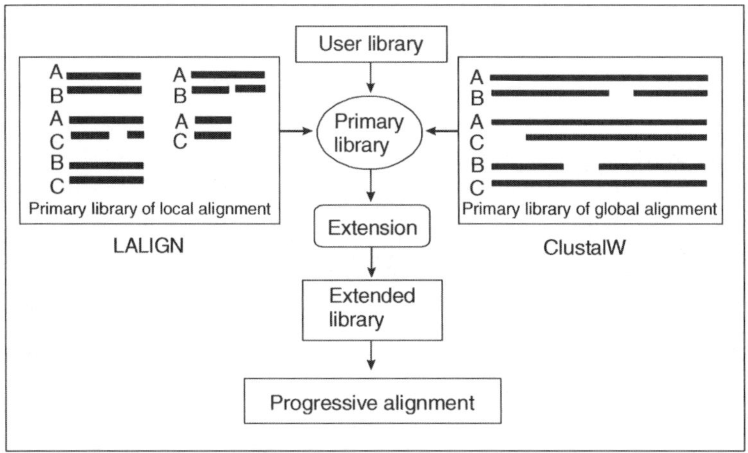

Figure 7.4   The layout of T-Coffee

In the above Figure 7.4 local and global alignments are first computed and then combined into primary library that is extended in order to be used for corrupting the multiple sequence alignment in a progressive manner.

Poa (Partial order alignments) is a progressive alignment program that does not rely on the guide tree available at http://ww.bioinformatics.ucla.edu/ poa/Poa_Tutorial.html. Instead of using regular sequence consequences, a partial order alignment uses Smith–Waterman algorithm that leads to formation of modified graph model, which is used to represent a growing multiple alignment in which identical residues in a column are condensed to a node resembling a knot on a rope and divergent residues are allowed to remain as such, allowing the rope to "bubble" (Figure 7.5). The graph profile preserves all the information from the original alignment. In addition to above described software for multiple sequence alignment, other recently developed software for the same are given in Table 7.1.

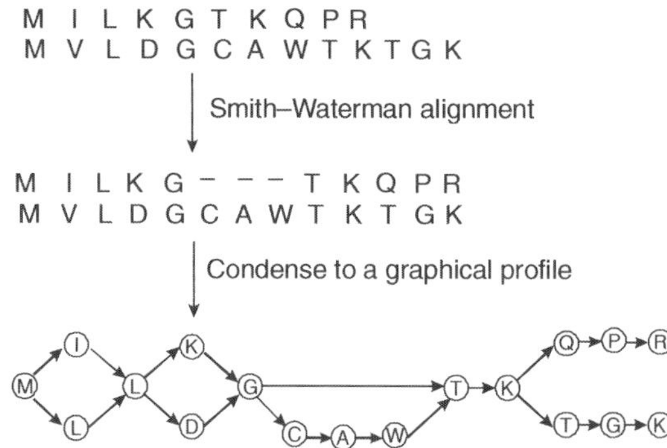

Figure 7.5   The Poa algorithm converts a sequence alignment into a graphical profile

Table 7.1   List of software for multiple sequence alignment

Name	Description	Sequence type*	Alignment type**
ABA	A-Bruijn alignment	Protein	Global
DNA Baser	Multi alignment/Batch alignment	Nucleotides	Local or global
ALE	Manual alignment; some software assistance	Nucleotides	Local
AMAP	Sequence annealing	Both	Global
BAli-Phy	Tree+Multi alignment; Probabilistic/Bayesian; Joint Estimation	Both	Global
CHAOS/DIALIGN	Iterative alignment	Both	Local (preferred)
ClustalW	Progressive alignment	Both	Local or global
CodonCode Aligner	Multi alignment; ClustalW & Phrap support	Nucleotides	Local or global
DIALIGN-TX and DIALIGN-T	Segment-based method	Both	Local (preferred) or global
DNA Alignment	Segment-based method for intraspecific alignments	Both	Local (preferred) or global

(Contd.)

Table 7.1   (Continued)

Name	Description	Sequence type*	Alignment type**
Ed'Nimbus	Seeded filtration	Nucleotides	Local
FSA	Sequence annealing	Both	Global
Geneious	Progressive/Iterative alignment; ClustalW plugin	Both	Local or global
Kalign	Progressive alignment	Both	Global
MSA	Dynamic programming	Both	Local or global
PRRN/PRRP	Iterative alignment (especially refinement)	Protein	Local or global
POA	Partial order/Hidden Markov model	Protein	Local or global
SAM	Hidden Markov model	Protein	Local or global
MAFFT	Progressive/iterative alignment	Both	Local or global
MAVID	Progressive alignment	Both	Global
MULTALIN	Dynamic programming/ clustering	Both	Local or global
Multi-LAGAN	Progressive dynamic programming alignment	Both	Global
MUSCLE	Progressive/iterative alignment	Both	Local or global
ProbCons	Probabilistic/ consistency	Protein	Local or global
PSAlign	Alignment preserving non-heuristic	Both	Local or global
SAGA	Sequence alignment by genetic algorithm	Protein	Local or global
T-Coffee	More sensitive progressive alignment	Both	Local or global
RevTrans	Combines DNA and Protein alignment, by back translating the protein alignment to DNA.	DNA/Protein (special)	Local or global

*Sequence type: Protein or nucleotide. **Alignment type: Local or global

Iterative Pairwise Alignment

This approach uses pairwise alignment scores to iteratively add one additional string to a growing multiple alignment. The process starts by aligning the two strings whose edit distance is the minimum over all pairs of strings. Then iteratively consider the string with the smallest distance to any of the strings already in the multiple alignment. A set of methods to produce MSAs while reducing the errors inherent in progressive methods are classified as "iterative" because they work similarly to progressive methods but repeatedly realign the initial sequences as well as adding new sequences to the growing MSA. One reason progressive methods are so strongly dependent on a high-quality initial alignment is the fact that these alignments are always incorporated into the final result—that is, once a sequence has been aligned into the MSA, its alignment is not considered further. This approximation improves efficiency at the cost of accuracy. By contrast, iterative methods can return to previously calculated pairwise alignments incorporating subsets of the query sequence as a means of optimizing a general objective function such as finding a high-quality alignment score. The iterative approach is based on the idea that an optimal solution can be found by repeatedly modifying existing sub-optimal solutions. Variety of iteration methods have been implemented to achieve this objective and the different approaches were available in software packages.

PRRN/PRRP (http://prrn.hgc.jp/) is the software package that uses a hill-climbing algorithm to optimize its MSA alignment score and iteratively corrects both alignment weights and locally divergent or "gappy" regions of the growing MSA. PRRP performs best when refining an alignment previously constructed by a faster method. PRRP is a web-based program that uses a double-nested iterative strategy for multiple alignment and is available at http://prrn.ims.u-tokyo.ac.jp. It performs multiple alignment through two sets of iterations: inner iteration and outer iteration. In the outer iteration, an initial random alignment is generated that is used to derive a UPGMA tree (Unweighted Pair Group Method using Arithmetic Average). Weights are subsequently applied to optimize the alignment. In the inner iteration, the sequences are randomly divided into two groups. Randomized alignment is used for each group in the initial cycle, after which the alignment positions in each group are fixed. The two groups, each treated as single sequence, are then aligned to each other using global dynamic programming. The process is repeated through many cycles until the total SP score no longer increases. At this point, the resulting alignment is used to construct a new UPGMA tree. New weights are applied to optimize alignment scores.

The newly optimized alignment is subject to further realignment in the inner iteration (Figure 7.6). The process is tested again for many cycles until there is no further improvement in the overall alignment scores.

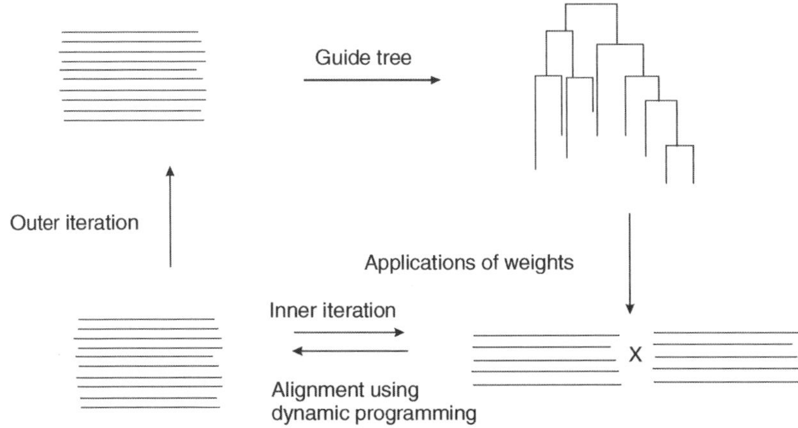

Figure 7.6   Diagrammatic representation of iterative sequence alignment approach for PRRN that involves two sets of iteration

DIALIGN is another iterative program that takes an unusual approach of focusing narrowly on local alignments between sub-segments or sequence motifs without introducing a gap penalty. The alignment of individual motifs is then achieved with a matrix representation similar to a dot-matrix plot in a pairwise alignment. An alternative method that uses fast local alignments as anchor points or "seeds" for a slower global-alignment procedure is implemented in the CHAOS/DIALIGN suite http://dialign.gobics.de/chaos-dialign-submission. DIALIGN2 is a web-based program designed to detect local similarities. It does not apply gap penalties and thus is not sensitive to long gaps. In this method, each of the sequences broken down to smaller segments and all possible pairwise alignments between segments are performed. High scoring segments, called **blocks**, among different sequences are then compiled in a progressive manner to assemble a full multiple alignment. It places emphasis on block-to-block comparison rather than residue-to-residue comparison. The sequence regions between blocks are left unaligned.

MUSCLE (multiple sequence alignment by log-expectation) (http://www.drive5.com/muscle/) is a third popular iteration-based method. It measures distances more accurately and assess the relatedness of two sequences. The distance measure is updated between iteration stages (although, in its original form, MUSCLE contained only 2–3 iterations depending on whether refinement was enabled).

MULTIPLE SEQUENCE ALIGNMENT OF RELATED SEQUENCE

Multiple sequence alignments are applied for identification of related sequences in databases by constructing Position-Specific Scoring Matrices (PSSM), profiles and Hidden Markov Models (HMMs). These are the statistical methods that reflect the frequency information of amino acid or nucleotide residues in multiple alignment.

POSITION-SPECIFIC SCORING MATRICES (PSSM)

It is in the form of a statistical table that provides probability information of amino acids or nucleotides at each position of an ungapped multiple sequence alignment. In such table, the rows represent residue positions of a particular multiple sequence alignment and the columns represent the names of the residues or vice versa. The values in the table represent log odds scores of the residue calculated from multiple alignment. The matrix is constructed on the basis of raw frequencies of each column position counted from a multiple alignment. The frequencies are then normalized by dividing positional frequencies of each residue by overall frequencies so that the scores are length and composition dependant. The values are converted to the probability values by taking to the logarithm (normally to the base of 2). Thus, the matrix values become log-odds scores of residues occurring at each alignment position. In this way the matrix is constructed (Figure 7.7) showing the positive score representing identical residue or similar residue match; a negative score representing a non-conserved sequence match. It provides the quantitative information about the degree of sequence conservation at each position of a multiple alignment.

The probabilistic model can then be used like a single sequence for database searching and alignment or can be used to test how well a particular targets sequence fits into the sequence group. Taking into consideration the above matrix derived from a DNA multiple alignment, which can be used to test the fitness of the new sequence AACTCG. The solution to this problem can be obtained by addition of the probability values of the sequence at respective positions of the matrix to produce the sum of scores as shown in the Figure 7.8. The total match score calculated for the given sequence is 6.33. Since the matrix values have been taken to logarithm to the base of 2, the score can be interpreted as the probability of the sequence fitting the matrix as $2^{6.33}$, or 80 times more likely than the random chance.

```
Position      1 2 3 4 5 6
Sequence 1    ATGTCG
Sequence 2    AAGACT
Sequence 3    TACTCA
Sequence 4    CGGAGG
Sequence 5    AACCTG
```

Convert multiple alignment to a raw frequency table →

Pos.	1	2	3	4	5	6	Overall freq.
A	0.6	0.6		0.4		0.2	0.30
T	0.2	0.2	–	0.4	0.2	0.2	0.20
G	–	0.2	0.6	–	0.2	0.6	0.27
C	0.2	–	0.4	0.2	0.6	–	0.23

Normalize the values by dividing them by overall frequency

Pos.	1	2	3	4	5	6	Overall freq.
A	2.0	2.0	–	1.33	–	0.67	0.30
T	1.0	1.0	–	2.0	1.0	1.0	0.20
G	–	0.74	2.22	–	0.74	2.22	0.27
C	0.87	–	1.74	0.87	2.61	–	0.23

Convert the values to log to base of 2

Pos.	1	2	3	4	5	6
A	1.0	1.0	–	0.41	–	–0.58
T	0.0	0.0	–	1.0	0.0	0.0
G	–	–0.43	1.15	–	–0.43	1.15
C	–0.2	–	0.8	–0.2	1.38	–

Figure 7.7   An example of position-specific scoring matrix constructed from a multiple alignment of nucleotide sequences

Match AACTCG in the following matrix
→ Find the nucleotides at respective positions of the matrix

Pos.	1	2	3	4	5	6
A	(1.0)	(1.0)	–	0.41	–	–0.58
T	0.0	0.0	–	(1.0)	0.0	0.0
G	–	–0.43	1.15	–	–0.43	(1.15)
C	–0.2	–	(0.8)	–0.2	(1.38)	–

Calculate the sum of log-odds scores

1.0 + 1.0 + 0.8 + 1.0 + 1.38 + 1.15 = 6.33

Figure 7.8 Example for testing the fitness of new sequence in the PSSM constructed in the previous figure. The matching less positions for the sequence AACTCG are circled in the matrix.

PROFILES

The profiles databases use the notion of profiles to achieve a good detection of distant sequence relationships. A profile is a scoring table with multiple alignment information for the whole sequences, not just for conserved regions. Profiles are weighted to indicate:

- ▣ What types of residues are allowed at what positions?
- ▣ Where are insertions and deletions (indels) allowed (not within core secondary structures)?
- ▣ Where are the most conserved regions located?

Profiles provide a sensitive means of detecting distant sequence relationships, where only a few residues are well conserved. The inherent complexity of profiles renders them to be highly potent discriminators. In other words, a profile is a PSSM with penalty information regarding insertions and deletions for a sequence family. To achieve an optimal alignment between a query and a profile, a series of gap parameters have to be tested.

The ISREC (Swiss Institute for Experimental Research) has created a compendium of profiles, allowing to find even distant homologues. Each of those profiles has separate data and family annotations.

PSI-BLAST (POSITION-SPECIFIC ITERATED BLAST)

Profiles can be used in database searching to find remote sequence homologue. When amino acid sequences belonging to the same protein family align together, some regions may be found very similar, with profiles showing little variance. The PSI-BLAST is another improved version of the BLAST algorithm that includes a program to establish profiles and use them to search against sequence database in an automatic way.

PSI-BLAST is the hybrid approach incorporating elements of both pairwise and multiple alignment. The innovation of the PSI-BLAST extension is that, following an initial database search, it allows automatic creation of position-specific profiles from groups of results that match the query above defined threshold. Running a program several times can further refine the profile and increase search sensitivity. Matrix is a key element in evaluating the quality of a pairwise sequence alignment which is the "substitution matrix", that assigns a score for aligning any possible pair of residues. The matrix used in a BLAST search can be changed depending on the type of sequences you are searching with. PSI-BLAST can save the PSSM constructed through iterations. The PSSM thus constructed can be used in searches against other databases with the same query by copying and pasting the encoded text into the PSSM field.

To save a PSSM file:

- ▣ Run a protein BLAST search
- ▣ Check the PSI-BLAST box on formatting page
- ▣ Click the "Format" button
- ▣ On the PSI-BLAST results page, click the "Run PSI-BLAST Iteration 2" button
- ▣ Now, on the format page, select "PSSM" from the "Show" pull down menu
- ▣ Click "Format" button
- ▣ This will display text output with the ASCII-encoded PSSM. The "Save as..." option of the browser can be used to save this to a plain text file on your hard drive

To use the PSSM in a new protein BLAST search against other databases:

- ▣ Copy the above PSSM from the browser
- ▣ Open a new protein BLAST page
- ▣ Paste the PSSM in the PSSM field page

- ▣ Provide the SAME query in the search box
- ▣ Select a different target database
- ▣ Click "BLAST" button to start the search

If the database is same as when the PSSM was stored, you'll reproduce the iteration on which you've saved the PSSM; a different database will yield a different hit list. In another approach, PSI-BLAST first uses a single query protein sequence to perform normal BLASTP search to generate initial similarity hits. The high-scoring hits are used to build a multiple sequence alignment, from which a profile is created. The profile is then used in the second round of searching to identify more members of the same family that may match with the profile. When new sequence hits are identified, they are combined with the previous multiple alignment to generate a new profile, which in turn is used in subsequent cycles of database searching. The process is repeated until no new sequence hits are found (Figure 7.9).

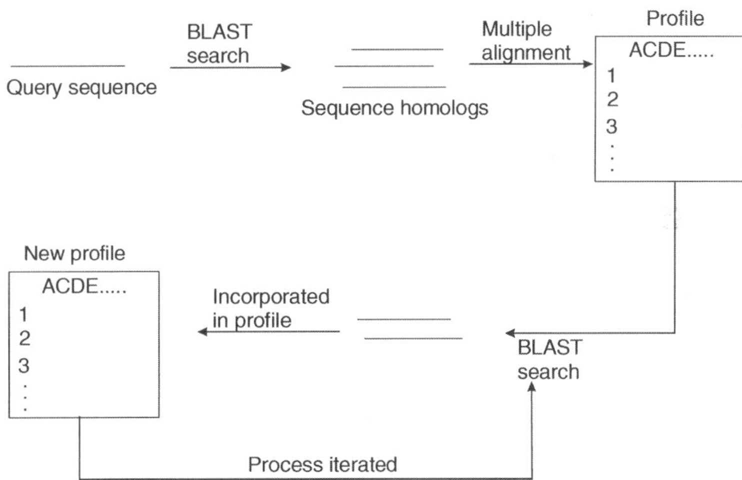

Figure 7.9 Diagrammatic representation of PSI-BLAST

MARKOV MODEL OR MARKOV CHAIN

It is a sequence of events that occur one after another in a chain. Each event determines the probability of the next event. A Markov chain can be considered as a process that moves in one direction from one state to the next with certain probability, which is known as transition probability. **Hidden Markov Models (HMMs)** are probabilistic models that can assign likelihoods to all possible combinations of gaps, matches, and mismatches to determine the most likely MSA or set of possible MSAs. HMM is a computational structure for describing the subtle patterns that define families of homologous sequences. HMMs are powerful tools for detecting distant relatives, and for prediction of protein folding patterns. They are the only methods based entirely on sequences, that is, without explicitly using structural information competitive with PSI-BLAST for identifying distant homologues. They also perform well at fold recognition, as assessed in CASP programs. HMMs are usually presented as procedures for generating sequences.

The internal structure of an HMM shows the mechanism for generating sequences (Figure 7.10). Begin at start, and follow some chain of arrows until arriving at end. Each arrow takes you to a state of the system. At each state one can emit a residue, perhaps and choose an arrow to take you to the next state. The action and the choice of successor state are governed by sets of probabilities. Associated with each state that emits a residue are one probability distribution for the twenty amino acids, and a second probability distribution for the choice of successor state. Both of these probability distributions are calibrated to encode information about a particular sequence family. In this way, the same general mathematical framework can be specialized to many different sequence families.

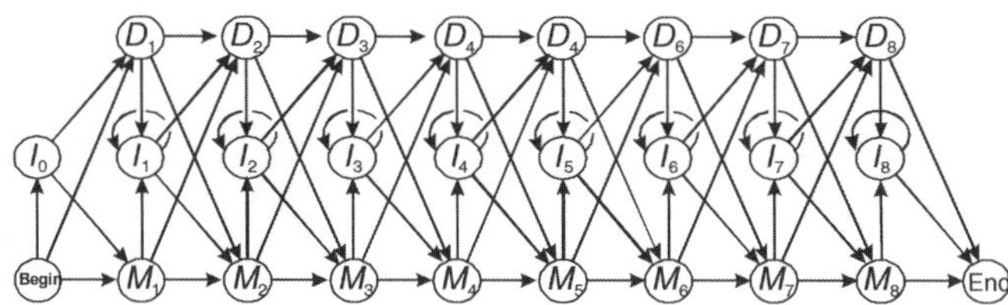

Figure 7.10   The structure of a Hidden Markov Model (HMM). Corresponding to each residue position in a multiple sequence alignment, the HMM contains a match state (M), a delete state (D) and an insert state (I) appear between residue positions, and at the beginning and end

A Profile HMM

▣ A profile HMM is a probabilistic representation of a multiple alignment.

▣ A given multiple alignment of a protein family is used to build a profile HMM.

▣ This model then may be used to find and score less obvious potential matches of new protein sequences.

▣ Multiple alignment is used to construct the HMM model.

▣ Assign each column to a match state in HMM. Add insertion and deletion state.

▣ Estimate the emission probabilities according to amino acid counts in column. Different positions in the protein will have different emission probabilities.

▣ Estimate the transition probabilities between match, deletion and insertion states.

▣ The HMM model gets trained to derive the optimal parameters.

States of Profile HMM

▣ Match states emit a residue $M_1...M_n$ (plus begin/end states)

▣ Insert states appear between two successive positions in the alignment I_0 $I_1...I_n$

▣ Delete states skip a column in the multiple sequence alignment $D_1...D_n$

Transition Probabilities in Profile HMM

▣ $\log(a_{MI}) + \log(a_{IM})$ = gap initiation penalty

▣ $\log(a_{II})$ = gap extension penalty

Emission Probabilities in Profile HMM

▣ Probability of emitting a symbol a at an insertion state I_j:
$$e_{Ij}(a) = p(a),$$
where $p(a)$ is the frequency of the occurrence of the symbol a in all the sequences.

Profile HMM Alignment

▣ Define $vMj(i)$ as the logarithmic likelihood score of the best path for matching $x1..xi$ to profile HMM ending with xi emitted by the state Mj.

▣ $vIj(i)$ and $vDj(i)$ are defined similarly.

Profile HMM Alignment—Dynamic Programming

$$vMj(i) = \log(eMj(xi)/p(xi)) + \max \begin{cases} vMj-1(i-1) + \log(aMj-1, Mj) \\ vIj-1(i-1) + \log(aIj-1, Mj) \\ vDj-1(i-1) + \log(aDj-1, Mj) \end{cases}$$

$$vIj(i) = \log(eIj(xi) / p(xi)) + \max \begin{cases} vMj(i-1) + \log(aMj, Ij) \\ vIj(i-1) + \log(aIj, Ij) \\ vDj(i-1) + \log(aDj, Ij) \end{cases}$$

Paths in Edit Graph and Profile HMM

HMMs include the possibility of introducing gaps into the generated sequence, with position-dependent gap penalties. Application of profiles requires that the multiple sequence alignment be specified up front; the pattern statistics are then derived from the alignment. HMMs carry out the alignment and the assignment of probabilities together. A conventional multiple sequence alignment table could also be used to generate sequences, by selecting amino acids at successive positions, each amino acid was chosen from a position-specific probability distribution derived from the profile. After taking the action appropriate to the type of state (M, D or I) another probability distribution governs the choice of the next state. In every possible succession of states, every column of the embedded alignment must be visited, and either matched or deleted. There is no way to traverse the network without passing through either an M state or a D state at each position.

The dynamics of the system are such that only the current state influences the choice of its successor—the system has no "memory" of its history. This is a characteristic of the processes studied by the nineteenth-century Russian mathematician A.A. Markov. Distinguish the succession of states from the succession of amino acids emitted to form the output sequence. System can generate the same sequence through several paths. Observers can see the emitted symbols of an HMM but have no ability to know which state the HMM is currently in. In other words, only the succession of characters emitted is visible; the state sequence that generated the characters remains internal to the system, which is, hidden. Hence the name, Hidden Markov Model. Thus, the goal is to infer the most likely hidden states of an HMM based on the given sequence of emitted symbols.

An HMM is a graph of connected states, each state potentially able to "emit" a series of observations. The process evolves in some dimension, often time, though not necessarily. The model is parameterized with probabilities governing the state at a time $t + 1$, given that one knows the previous states. Markov assumptions are used to truncate the dependency of having to know the entire history of states up to this point in order to assess the next state. A path through an edit graph and the corresponding path through a profile HMM is shown in Figure 7.11.

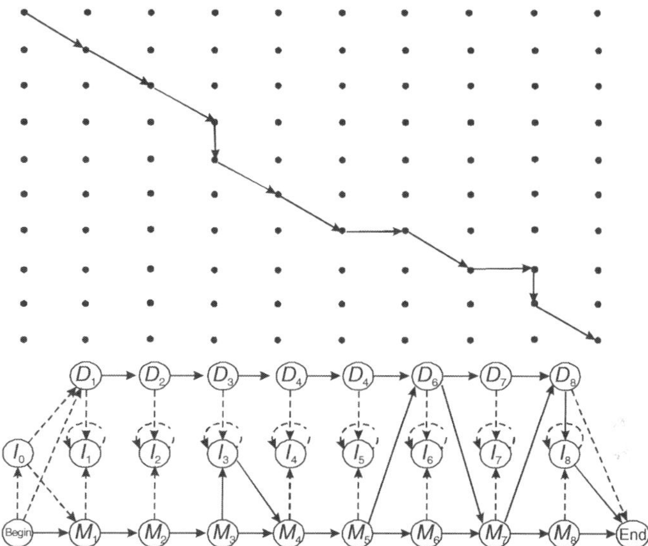

Figure 7.11   A path through an edit graph and the corresponding path through a profile HMM

Making a Collection of HMM for Protein Families

HMMs can produce a single highest-scoring output but can also generate a family of possible alignments that can then be evaluated for biological significance. A distant cousin of functionally related sequences in a protein family may have weak pairwise similarities with each member of the family and thus fail significance test. However, they may have weak similarities with many members of the family. The goal is to align a sequence to all members of the family at once. Family of related proteins can be represented by their multiple alignment and the corresponding profile. The steps for making a collection of HMM for protein family are as follows:

- use BLAST to separate a protein database into families of related proteins
- construct a multiple alignment for each protein family
- construct a profile HMM model and optimize the parameters of the model (transition and emission probabilities)
- align the target sequence against each HMM to find the best fit between a target sequence and an HMM
- application of profile HMM to modelling globin proteins
- globins represent a large collection of protein sequences

- 400 globin sequences were randomly selected from all globins and used to construct a multiple alignment
- multiple alignment was used to assign an initial HMM
- this model then get trained repeatedly with model lengths chosen randomly between 145 to 170, to get an HMM model optimized probabilities
- 625 remaining globin sequences in the database were aligned to the constructed HMM resulting in a multiple alignment. This multiple alignment agrees extremely well with the structurally derived alignment
- 25,044 proteins, were randomly chosen from the database and compared against the globin HMM
- this experiment was resulted in an excellent separation between globin and non-globin families

Although HMM-based methods have been developed relatively recently, they offer significant improvements in computational speed, especially for sequences that contain overlapping regions. Typical HMM-based methods work by representing an MSA as a form of directed acyclic graph known as a partial-order graph, which consists of a series of nodes representing possible entries in the columns of an MSA. In this representation a column that is absolutely conserved (that is, all the sequences in the MSA share a particular character at a particular position) is coded as a single node with as many outgoing connections as there are possible characters in the next column of the alignment.

In the terms of a typical Hidden Markov Model, the observed states are the individual alignment columns and the "hidden" states represent the presumed ancestral sequence from which the sequences in the query set are hypothesized to have descended. An efficient search variant of the dynamic programming method, known as the **Viterbi algorithm**, is generally used to align successively the growing MSA to the next sequence in the query set to produce a new MSA. This is distinct from progressive alignment methods because the alignment of prior sequences is updated at each new sequence addition. However, like progressive methods, this technique can be influenced by the order in which the sequences in the query set are integrated into the alignment, especially when the sequences are distantly related.

Several software programs are available in which variants of HMM-based methods have been implemented and which are noted for their scalability and efficiency, although properly using an HMM method is more complex than using more common progressive methods. The simplest is POA (Partial Order Alignment), a similar but more generalized method is implemented in the package SAM (Sequence Alignment and Modelling System). SAM has been used as a source of alignments for protein structure prediction to participate in the CASP structure prediction experiment and to develop a database of predicted proteins in the yeast species *S. cerevisiae.* HMM methods can also be used for database search with HMMER.

HMMER (hammer) is a profile-HMM package used for protein sequences, currently in version 2.2. Various programs in the HMMER suite can take in a multiple alignment as input, and produce a profile-HMM based on that alignment. Once a HMM is in place (either a user-defined HMM or one of the Pfam HMMs to be described later), sequences can be searched against the HMM to check for membership into a particular family. In addition, sequences can be emitted from a HMM based on the transition and emission probabilities. The source code for HMMER is freely available for download from the following site: http://hmmer.wustl.edu/

SAM (Sequence Alignment and Modeling System) is a profile-HMM package used for protein sequences as well. It is very similar to HMMER as far as the functionality is concerned. http://www.cse.ucsc.edu/research/compbio/sam.html

Meta-meme is a program that creates Hidden Markov Models of ungapped alignments. The benefit is that there are fewer parameters to be learned in creating the HMMs. Thus, Meta-meme is a motif-based Hidden Markov approach. http://metameme.sdsc.edu/

HMMPro is a commercial software package (free academic licence) that has been developed to add additional HMM features, including graphical interfaces, multiple topologies, and multiple training methods. http://www.netid.com/html/hmmpro.html

GENETIC ALGORITHMS AND SIMULATED ANNEALING

These are standard optimization techniques in computer science, both of which were inspired by, but do not directly reproduce, physical processes. These have been also used in an attempt to more efficiently produce quality MSAs. One such technique, genetic algorithms, has been used for MSA production in an attempt to broadly simulate the hypothesized evolutionary process that gave rise to the divergence in the query set. The method works by breaking a series of possible MSAs into fragments and repeatedly rearranging those fragments with the introduction of gaps at varying positions. A general objective function is optimized during the simulation, most generally the "sum of pairs" maximization function introduced in dynamic programming-based MSA methods. A technique for protein sequences has been implemented in the software program SAGA (Sequence Alignment by Genetic Algorithm) and its equivalent in RNA is called RAGA.

The technique of simulated annealing, by which an existing MSA produced by another method is refined by a series of rearrangements designed to find more optimal regions of alignment space than the one that the input alignment already occupies. Like the genetic algorithm method, simulated annealing maximizes an objective function like the sum-of-pairs function. Simulated annealing uses a metaphorical "temperature factor" that determines the rate at which rearrangements proceed and the likelihood of each rearrangement; typical usage alternates periods of high rearrangement rates with relatively low likelihood (to explore more distant regions of alignment space), with periods of lower rates and higher likelihoods to more thoroughly explore local minima near the newly "colonized" regions. This approach has been implemented in the program MSASA (Multiple Sequence Alignment by Simulated Annealing).

IDENTIFICATION OF MOTIFS AND DOMAINS
IN MULTIPLE SEQUENCE ALIGNMENT

A motif is a short conserved sequence pattern associated with distinct functions of a protein or DNA. It is often related with a distinct structural site performing a particular function. A typical motif, such as a zinc-finger motif, is 10 to 20 amino acids long. Zinc-finger motif is relatively small protein motifs that fold around one or more zinc ions. In addition to their role as a DNA-binding module they have recently been shown to mediate protein–protein and protein–lipid interactions. Each zinc finger binds in the major groove and interacts with about five nucleotides (Figure 7.12a), adjacent fingers interacting with contiguous stretches of DNA.

A domain is also a conserved sequence pattern, defined as an independent functional and structural unit. Domains are normally longer than motifs. Domain might be defined in other words as parts of a protein sequence with a single well-defined function (e.g. binding a particular ligand) or they might be parts of the sequence able to fold into a 3D structure, independent of the rest of the sequence. They might be defined just as parts of the protein 3-dimensional structure that appear to be geometrically distinct. One important aspect of the definition of a domain is that it must be an independent unit able to exit in many, otherwise unrelated protein sequences. Many proteins contain domains in the form of compact units within the folding pattern of a single chain, which look as if they should have independent stability. The RNA-binding protein L1 has feature typical of multidomain proteins—the binding site appears in a cleft between the two domains, and the relative geometry of the two domains are flexible, allowing for ligand-induced conformational changes (Figure 7.12b). In the hierarchy, domains fall between supersecondary structures and the tertiary structure of a protein.

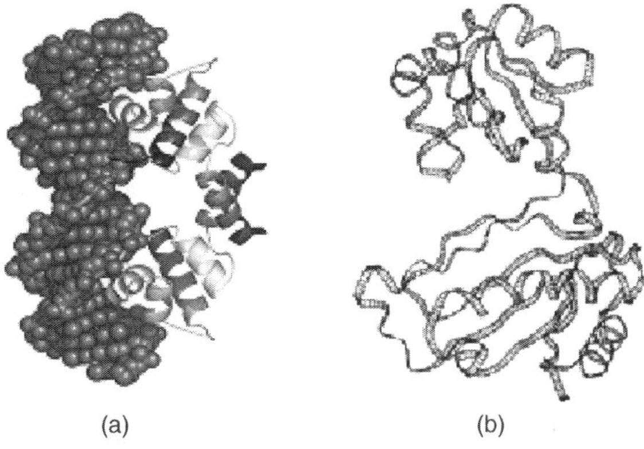

(a)   (b)

Figure 7.12   (a) Protein–DNA complex with "zinc-finger" structural motif. (b) Ribosomal protein L1 from *Methanococcus jannaschii* [ICJS] showing two domains. [ICJS] is the Protein Data Bank identification code of the entry.

The identification of motifs and domains in proteins is a significant aspect of the classification of protein sequences and functional annotation. Multiple sequence alignment as well as profiles and Hidden Markov Model construction help in identification of motifs and domains. Based on the multiple sequence alignment, commonly conserved regions can be identified, which then serve as diagnostic features for a protein family. The consensus sequence information of motifs and domains can be stored in a database for later searches of the presence of similar sequence pattern from unknown sequences. By scanning the presence of known motifs or domains in a query sequence, associated functional features in a query sequence can be revealed rapidly, which is often not possible by simply matching full-length sequence in the primary database.

Motif finding, also known as profile analysis, constructs global multiple sequence alignments that attempt to align short conserved sequence motifs among the sequences in the query set. This is usually done by first constructing a general global multiple sequence alignment, after which the highly conserved regions are isolated and used to construct a set of profile matrices. The profile matrix for each conserved region is arranged like a scoring matrix but its frequency counts for each amino acid or nucleotide at each position are derived from the conserved region's character distribution rather than from a more general empirical distribution. The profile matrices are then used to search other sequences for occurrences of the motif they characterize. In many cases when the query set contains only a small number of sequences or contains only highly related sequences, pseudocounts are added to normalize the distribution reflected in the scoring matrix. In particular, this corrects zero-probability entries in the matrix to values that are small but non-zero.

In other words, motif finding is a method of locating sequence motifs in global MSAs that is both a means of producing a better MSA and a means of producing a scoring matrix for use in searching other sequences for similar motifs. A variety of methods for isolating the motifs have been developed, but all are based on identifying short highly conserved patterns within the larger alignment and constructing a matrix similar to a substitution matrix that reflects the amino acid or nucleotide composition of each position in the putative motif. The alignment can then be refined using these matrices. In standard profile analysis, the matrix includes entries for each possible character as well as entries for gaps. Alternatively, statistical pattern-finding algorithms can identify motifs as a precursor to an MSA rather than as a derivation.

Motif databases have commonly been used to classify proteins, provide functional assignment, and identify structural and evolutionary relationships. Two databases that mainly use regular expressions for the purpose of searching sequence pattern PROSITE and eMotif.

PROSITE (www.expasy.ch/prosite)

It is a database of protein families and domains. It consists of entries describing the domains, families and functional sites as well as amino acid patterns, signatures, and profiles in them. These are manually curated by a team of the Swiss Institute of Bioinformatics and tightly integrated into Swiss-Prot protein annotation. PROSITE was created in 1988 by Amos Bairoch. It is the sequence pattern database that primarily uses a single consensus pattern or sequence signature to characterize a protein function and a sequence family. The Prosite database is based on Swiss-Prot and thus is very well annotated, but small. Characterization of protein families is done by the single most conserved motif observed in a multiple sequence alignment of known homologous. These conserved motifs are usually related to biological functions such as active sites or binding sites. The search in Prosite does not require an exact match in structure. Prosite enable searches using complex patterns. It is possible to search textually using **regular expressions** (often simply referred to as patterns) for names of known proteins. It is also possible to scan a protein sequence using Prosite for structural pattern matches. Results are checked manually to determine how well the patterns have performed; ideally, there should be only correct matches (so-called true-positives), and no incorrect matches (false-positives). Patterns whose diagnostic performance is comprised by matching too many false-positives are fine tuned, and Swiss-Prot is re-scanned. This process is repeated until an optimal pattern is created. The major problem with the Prosite patterns is that some of the sequence patterns are too short to be specific and database is relatively small and motif searches often yield no results when there are true motif matches present. Prosite is a database of protein domains that can be searched by either regular expression patterns or sequence profiles. The Prosite data can be accessed at http://www.expasy.org/prosite/.

Structure of Prosite entries   Prosite entries are deposited in two distinct files. The first file contains the pattern and lists of all matches in the pattern version of Swiss-Prot. The second is a documentation file that provides details of the characterized family and a description of the biological role of the chosen motif(s) and a supporting bibliography. In the data file, each entry contains both an identifier (ID), which is acronym for the family, and an accession number (AC). The word PATTERN tells us to expect a regular expression. A title, or description of the family, is contained in the DE line, and the pattern itself resides on PA lines. The NR line provides details about the derivation and diagnostic performance or diagnostic power of the pattern. These lines are so significant to inspect when first viewing a PROSITE entry that the large number of false-positive and false-negative results are indicative of poorly-performing patterns. In the Figure 7.13, the pattern was derived from Release 32 of Swiss-Prot, which contained 49340 sequences; it matched a total of 53 sequences, all of which are true-positive.

```
ID OPSIN; PATTERN.
AC PS00238;
DT APR-1990 (CREATED); NOV-1997 (DATA UPDATE); NOV-1997(INFO UPDATE).
DE Visual pigments (opsins) retinal binding site.
PA [LIVMW]-[PGC]-x(3)-[SAC]-K-[STALIM]-[GSACNV]-[STACP]-x(2)-[DENF]-[AP]-
PA x(2)-[IY].
NR /RELEASE=32,49340;
NR /TOTAL=53(53); /POSITIVE=53(53); /UNKNOWN=0(0); /FALSE_POS=0(0);
NR /FALSE_NEG=0; /PARTIAL=1;
CC /TAXO-RANGE=??E??; /MAX-REPEAT=1;
CC /SITE=5,retinal;
DR P06002, OPS1_DROME, T; P28678, OPS1_DROPS, T; P22269, OPS1_CALVI, T;
DR P08099, OPS2_DROME, T; P28679, OPS2_DROPS, T; P04950, OPS3_DROME, T;
DR P28680, OPS3_DROPS, T; P08255, OPS4_DROME, T; P29404, OPS4_DROPS, T;
DR P17646, OPS4_DROVI, T; P35362, OPSD_SPHSP, T; P41591, OPSD_ANOCA, T;
DR P41590, OPSD_ASTFA, T; P02699, OPSD_BOVIN, T; P32308, OPSD CANFA, T;
DR P32309, OPSD_CARAU, T; P22328, OPSD_CHICK, T; P28681, OPSD_CRIGR, T;
DR P08100, OPSD_HUMAN, T; P15409, OPSD_MOUSE, T; P35403, OPSD_POMMI, T;
DR P02700, OPSD_SHEEP, T; P29403, OPSD_XENLA, T; P22671, OPSD_LAMJA, T;
DR P31355, OPSD_RANPI, T; P24603, OPSD_LOLFO, T; P09241, OPSD_OCTDO, T;
DR P35356, OPSD_PROCL, T; P31356, OPSD_TODPA, T; P35360, OPS1_LIMPO, T;
DR P35361, OPS2_LIMPO, T; P32310, OPSB_CARAU, T; P28682, OPSB_CHICK, T;
DR P35357, OPSB_GECGE, T; P03999, OPSB_HUMAN, T; P28684, OPSV_CHICK, T;
DR P22330, OPSG_ASTFA, T; P22331, OPSH_ASTFA, T; P32311, OPSG_CARAU, T;
DR P32312, OPSH_CARAU, T; P28683, OPSG_CHICK, T; P35358, OPSG_GECGE, T;
DR P04001, OPSG_HUMAN, T; P41592, OPSR_ANOCA, T; P22332, OPSR_ASTFA, T;
DR P32313, OPSR_CARAU, T; P22329, OPSR_CHICK, T; P04000, OPSR_HUMAN, T;
DR P34989, OPSL_CALJA, T; P35359, OPSU_BRARE, T; P23820, REIS_TODPA, T;
DR P47803, RGR_BOVIN , T; P47804, RGR_HUMAN , T;
DR P17645, OPS3_DROVI, P;
DR PDOC00211 ;
//
```

Figure 7.13   Data file for visual pigment protein (opsin)—an example of regular entry from PROSITE

The comment (CC) lines provide information on the taxonomic range of family (here it means eukaryotes), the maximum number of observed repeats of the pattern (here it just one). It is followed by the list of the accession numbers and Swiss-Prot identification codes of all the true matches to the pattern (denoted by T), and any "possible" matches (denoted by P). In the above sited example, there are no false-positive or false-negative matches, when these do they occur, they are listed and are denoted by the letters F and N respectively. The final line of the file (DO) points to the associated family documentation file. Structurally, documentation file is simpler with each entry identified by its own accession number, cross reference, free-format description and appropriate bibliographic references.

eMotif (http://motif.standford.edu/emotif/emotif-search.html)

It is also known as identify, uses data from Blocks (derived ungapped alignment) and Finger prints (group of short and ungapped sequence segments related with diagnostic features of a protein family). eMotif generates consensus expressions from the conserved regions of sequence alignments. It is a motif database that uses multiple sequence alignments from both the BLOCKS and PRINTS database with an alignment collection much larger than PROSITE. eMotif adopts a "fuzzy" algorithm which allows certain amino acid alternations. Fuzzy matches, also called approximate matches, provide permissive matching by allowing more flexible matching of residues of similar biochemical properties. For example, if an original alignment contains phenylalanine only at a particular position, fuzzy matching allows other aromatic amino acid residues (either tyrosine or tryptophan) in a sequence to match the expression. This allows eMotif to find homologous sequences that other programs cannot find, but it results in a lot of noise. This is partly because the rule of matching is based on assumptions not actual observations. This trade-off shows why it is important to use multiple programs when searching for information. Some of the motif finding software are given in Table 7.2.

Table 7.2   List of software for Motif finding

Name	Description	Sequence type*
MEME/MAST	Motif discovery and search	Both
BLOCKS	Ungapped motif identification from BLOCKS database	Both
eMOTIF	Extraction and identification of shorter motifs	Both
Gibbs motif sampler	Stochastic motif extraction by statistical likelihood	Both
TEIRESIAS	Motif extraction and database search	Both
PRATT	Pattern generation for use with ScanProsite	Protein
ScanProsite	Motif database search tool	Protein
PHI-Blast	Motif search and alignment tool	Both
I-sites	Local structure motif library	Protein

*Sequence type: Protein or nucleotide

The above-described methods that use regular expressions for the purpose of searching sequence pattern do not take into account sequence probability information about multiple alignment from which they are modelled. If regular expression is derived from an incomplete sequence set, it has less predictive power because many more sequences with the same type of motifs are not represented. Unlike regular expression, the statistical models such as PSSMs, profiles, and HMMs preserve the sequence information from a multiple sequence alignment and express it with probabilistic models. Thus, these statistical models have stronger predictive power than the approach based on regular expression.

The following programs mainly use profile/HMM method extensively for construction of sequence pattern:

PRINTS

They are closely related, each representing protein families in terms of multiply aligned ungapped segments derived from the most highly conserved regions in a group of proteins or protein family. Such multiply aligned ungapped segments are termed blocks (in BLOCKS) or motifs (in PRINTS). In PRINTS, a set of such motifs represented a family is called a fingerprint. A motif represents any conserved element of a sequence alignment: it is a local alignment corresponding to a region whose function or structure is known, or its significance may be unknown. It is sufficient that it is conserved, and is hence likely to be predictive of any subsequent occurrence of such a structural/functional region in any other protein sequence.

A fingerprint is a set of motifs used to predict the occurrence of similar motifs, either in an individual sequence or in a database. Fingerprints are refined, i.e., their diagnostic potency is enhanced, by iterative scanning of the composite sequence database. Database searches with such aligned motifs are essentially frequency scans, i.e., in their most basic application, no secondary structure information, similarity data, or weighting scheme of any description is used to improve discrimination power. Unlike Prosite, fingerprints has an improved diagnostic reliability which is achieved by using more than one conserved structural motif to characterize a protein family. With fingerprints, many motifs are encoded using ungapped and unweighed local alignments. The input to fingerprints is a small multiple alignment, which has some conserved motifs. These motifs are searched for in the database, and only sequences that match all the motifs are considered for further analysis. With the new alignment, the database is searched for more sequences until no further complete fingerprint matches can be identified. These final aligned motifs constitute the refined fingerprint that is entered into the database. The development of PRINTS fingerprint database, which until 1999 was maintained in the Department of Biochemistry and Molecular Biology at University College London (UCL).

The fingerprinting method relies on the fact that, in any protein family, only parts of a sequence are held in common. These usually relate to key functional regions or to core structural elements of the fold. The starting point for fingerprint definition is thus multiple sequence alignment, which can be achieved by using the SOMAP, XALIGN or VISTAS manual alignment programs. Only small numbers need to be included in the initial alignment because the method itself is designed to add the alignment with each database scan. Once a motif, or set of motifs, has been identified, the conserved regions are excised in the form of local alignments. There are no rules regarding the juxtaposition of such motifs, other than that they should not substantially overlap. Thus motifs may occur immediately adjacent to one another, or they may be separated by any distance along the length of the alignment.

Independent database scans are made with each aligned motif, resulting in the production of a set of hit lists, one for each motif. The hit lists are then analysed, or correlated, to determine which sequences in the database have matched with all elements of the fingerprint, and which have only matched with part of it. Only those sequences that match with all elements are regarded as true matches. If the search has worked well, the true set will contain more sequences than did the original alignment. The additional sequence data from the new true set is then used to generate another set of aligned motifs, and the database is searched again, where the family being diagnosed is very large, and the redundant motifs are removed from the alignments prior to the next scan. This process is repeated until convergence, the point at which the true set remains constant between successive scans.

The final aligned motifs from this iterative procedure constitute the refined fingerprint that is entered into the PRINTS database. Good fingerprints find all true matches that exist in OWL; they exhibit clear discrimination cut-offs, and include little or no noise. Occasionally, cut-offs may be difficult to assess, this is usually the result of subfamily identification, where a sequence subset has been recognized that is characterized only by a part of a fingerprint; but it may also result from the use of only 2 or 3 motifs in the fingerprint— discrimination power improves with the number of motifs used.

The PRINTS database is of very high quality, having been created with a great deal of manual effort, and contains extensive annotation and description for the protein family and function concerned. PRINTS is a compendium of protein fingerprints. A fingerprint is a group of conserved motifs used to characterize a protein family; its diagnostic power is refined by iterative scanning of a Swiss-Prot/TrEMBL composite. Usually the motifs do not overlap, but are separated along a sequence, though they may be contiguous in 3D-space. Fingerprints can encode protein folds and functionalities more flexibly and powerfully than can single motifs.

Structure of PRINT entry   Three different aspects of a PRINTS entry are displayed in Figure 7.14. The first part ((a) in Figure 7.4) of the entry indicates the identifying code and a title that gives the family name. Associated entries are with unique accession numbers that take the form PR00000 (not shown). It is then followed by the number of motifs in the fingerprint (in the given example, it is 3). A number of database cross-links are then shown to help the user to access the related biological resources in the given family. It is followed by the date line that indicate when the entry was added to the database and when it was last updated. Finally, bibliographic information with a brief description of the characterized family is shown. In the second part ((b) in Figure 7.14)of the PRINTS entry, information in relation to the diagnostic performance both of fingerprint as a whole and of its constituent motifs is given. Initially, summary information is documented about the sequences matched and failed to match one or more motif. In the given example, there are 73 sequences that matched all three elements of the fingerprints and one motif matched only two motifs. It is followed by a table that provides additional information about the sequences matched by each individual motif. In the example displayed, the significant information gained is that the reported partial hits failed to match motif 1. In the last ((c) in Figure 7.14) part of the PRINTS entry, are listed the seed motifs used to generate the fingerprint, followed by the final motifs (not shown). Each motif is identified by its patent ID code plus a number that indicates which component of the fingerprint it is. In the example shown here, three motifs in the OPSIN fingerprints are designated OPSIN1, OPSIN2 and OPSIN3 (motif 3 is not shown). Each code is followed by the length of motif and short description of the relevant iteration number (for the initial motif it will always be 1).

```
(a)
OPSIN                  OPSIN SIGNATURE
 Type of fingerprint: COMPOUND with 3 elements
 Links:
    PRINTS; PR00237 GPCRRHODOPSN; PR00247 GPCRCAMP; PR00248 GPCRMGR
    PRINTS; PR00249 GPCRSECRETIN; PR00250 GPCRSTE2; PR00251 BACTRLOPSIN
    PROSITE; PS00238 OPSIN; PS00237 G_PROTEIN_RECEPTOR
    BLOCKS; BL00238
    SBASE; OPSD_HUMAN
    GCRDB; GCR_0085
    Creation date 20-DEC-1993; UPDATE 2-JUL-1996

    1. APPLEBURY, M.L. and HARGRAVE, P.A.
    Molecular biology of the visual pigments.
    VISION RES. 26 (12) 1881-1895 (1986).

(b)
 SUMMARY INFORMATION
   73 codes involving 3 elements
    1 codes involving 2 elements

 COMPOSITE FINGERPRINT INDEX
    3|   73    73    73
    2|    0     1     1
   --+------------------
     |    1     2     3

(c)
 INITIAL MOTIF SETS
 OPSIN1 Length of motif = 13 Motif number = 1
 Opsin motif I - 1
                        PCODE           ST      INT
 YVTVQHKKLRTPL          OPSD_BOVIN      60      60
 YVTVQHKKLRTPL          OPSD_HUMAN      60      60
 YVTVQHKKLRTPL          OPSD_SHEEP      60      60
 AATMKFKKLRHPL          OPSG_HUMAN      76      76
 AATMKFKKLRHPL          OPSR_HUMAN      76      76
 YIFATTKSLRTPA          OPS1_DROME      73      73
 VATLRYKKLRQPL          OPSB_HUMAN      57      57
 YIFGGTKSLRTPA          OPS2_DROME      80      80
 WVFSAAKSLRTPS          OPS3_DROME      81      81
 WIFSTSKSLRTPS          OPS4_DROME      77      77
 YLFSKTKSLQTPA          OPSD_OCTDO      58      58
 YLFTKTKSLQTPA          OPSD_LOLFO      57      57

 OPSIN2 Length of motif = 13 Motif number = 2
 Opsin motif II - 1
                        PCODE           ST      INT
 GWSRYIPEGMQCS          OPSD_BOVIN      174     101
 GWSRYIPEGLQCS          OPSD_HUMAN      174     101
 GWSRYIPQGMQCS          OPSD_SHEEP      174     101
 GWSRYWPHGLKTS          OPSG_HUMAN      190     101
 GWSRYWPHGLKTS          OPSR_HUMAN      190     101
 GWSRYVPEGNLTS          OPS1_DROME      187     101
 GWSRFIPEGLQCS          OPSB_HUMAN      171     101
 GWSAYVPEGNLTA          OPS2_DROME      194     101
 TWGRFVPEGYLTS          OPS3_DROME      194     100
 FWDRFVPEGYLTS          OPS4_DROME      190     100
 NWGAYVPEGILTS          OPSD_OCTDO      174     103
 GWGAYTLEGVLCN          OPSD_LOLFO      173     103
```

Figure 7.14   Three aspects of PRINTS entry showing (a) The general ID code and crosslinks, (b) A summary of the diagnostic performance of the fingerprints, and (c) Lists of seed motifs used to generate fingerprints.

BLOCKS

Fred Hutchinson Cancer Research Center (FHCRC) in Seattle provides BLOCK database, which is a multiple-motif database based on protein families contained in PROSITE. In this resource, the motifs or blocks are created by automatically detecting the most highly conserved regions of each protein family. The resulting blocks are encoded as ungapped local alignments and are calibrated against Swiss-Prot to obtain a measure of the likelihood of a chance match. Figure 7.15 displays a typical block, where each block is identified by a general ID and an accession number. The AC line shows information about the minimum and maximum distances of the block from its preceding neighbor. It is followed by DE line that indicates the title or description of the family. Next to it is BL line that provides information about the diagnostic power and some physical details of the block including the amino acid triplet (in the given example, it is RYA triplet), the width of the block, the number of sequences in it, the 99.5%-level score and finally the strength. It is then followed by block itself that indicates the Swiss-Prot IDs of the constituent sequences, the start position of the fragment, the sequence fragment itself, and a score, or weight, that provides a measure of the closeness of the relationship of that sequence to others in the block. Sequence fragments less than 80% similar are separated by blank lines.

```
ID       BACTERIAL _OPSIN_RET;      BLOCK
AC       BL00327C;  distance from previous block = (2, 4)
DE       Bacterial rhodopsin retinal binding site proteins.
BL       RYA motif; width = 29;  seqs = 6 ; 99.5% = 1296 ; strenght = 1504

BAC_ HALS1    ( 144)    LARYTWWLFSTICMIVVLYFLATSLRAAA  61

BAC_ HALS2    ( 143)    LARYTWWLFSTIAFLFVLYYLLTSLRSAA  52

BAC_ HALHA    ( 145)    SYRFVWWAISTAAMLYILYVLFFGFTSKA  73

BAC_HATPH     ( 121)    TERYALFGMGAVAFLGLVYYLVGPMTESA 100

BACH_HALSS    ( 164)    LLRWVWYAISCAFFVVVLYILLAEWAEDA  59

BACH_NATPH    ( 174)    LMRWFWYAISCACFLVVLYILLVEWAQDA  56
//
```

Figure 7.15   The structure of BLOCK entry. BL00327C is the third bacterial rhodopsin block.

BLOCK-format PRINTS

The original version of BLOCKS was created by automatic means, but now many databases (including PRINTS) are available in BLOCKS format on FHCRC web server (Henikoff *et al.*, 1998). The motifs within these databases typically cover larger regions of the sequence than PROSITE patterns, and unlike PROSITE, matching of motifs in sequences usually takes account of amino acid substitution matrices, these does not require exact matches to fixed patterns. Blocks analysis is a method of motif finding that restricts motifs to ungapped regions in the alignment. Blocks can be generated from an MSA or they can be extracted from an unaligned sequences using a pre-calculated set of common motifs previously generated from known gene families. Block scoring generally relies on the spacing of high-frequency characters rather than on the calculation of an explicit substitution matrix. The BLOCKS server provides an interactive method to locate such motifs in unaligned sequences. Figure 7.16 illustrate a typical motif in BLOCKS format. The structure is identical to that used in BLOCKS, with minor difference on the AC and BL lines. On the AC line, the PRINTS accession number is given, with an appended letter to indicate which component of the fingerprint it is (here it is PR00238A indicating that this is the first motif). On the BL line, there is the word 'adapted' indicating that motif have been taken from another database. Search engines for the databases are available from: http://bioinfo.man.ac.uk/dbbrowser/**PRINTS** and http://www.blocks.fhcrc.org/

```
ID    OPSIN; BLOCK
AC    PR00238A; distance from previous block=(31,81)
DE    OPSIN SIGNATURE
BL    adapted; width=13; seqs=73; 99.5%=858; strength=1172;

OPSD_CHICK ( 60) YVTIQHKKLRTPL 11
OPSD_RANPI ( 60) YVTIQHKKLRTPL 11
    S79840 ( 60) YVTIQHKKLRTPL 11
OPSD_BOVIN ( 60) YVTVQHKKLRTPL 11
OPSD_CANFA ( 60) YVTVQHKKLRTPL 11
OPSD_CRIGR ( 60) YVTVQHKKLRTPL 11
OPSD_HUMAN ( 60) YVTVQHKKLRTPL 11
OPSD_MOUSE ( 60) YVTVQHKKLRTPL 11
OPSD_SHEEP ( 60) YVTVQHKKLRTPL 11
     .
     .
     .

OPSD_OCTDO ( 58) YLFSKTKSLQTPA 35
OPSD_LOLFO ( 57) YLFTKTKSLQTPA 38
OPSD_TODPA ( 56) YLFTKTKSLQTPA 38
   ASOPSIN ( 50) YLFTKTKSLQTPA 38
OPSB_CHICK ( 67) FCTARFRKLRSHL 69
   HMIORH2 ( 77) YLFNKSAALRTPA 100
//
```

Figure 7.16 BLOCKS-format PRINTS entry for characterization of opsin family

Version 13.0 of the Blocks database consists of 8656 blocks representing 2101 groups documented in InterPro 3.1 keyed to SWISS-PROT 39.17 and TrEMBL obtained from the InterPro server. Release 35.0 of PRINTS contains 1750 entries, encoding 10,626 individual motifs. Overall, the database is still relatively small, largely because the detailed annotation of fingerprints is so time-consuming.

Nevertheless, the additional, unique information contained in PRINTS makes this database a useful adjunct to PROSITE, Pfam, Blocks, etc. PRINTS is released in major and minor versions. Minor releases reflect updates, bringing the contents in line with the current version of the source database(s) (formerly, PRINTS was derived from OWL, but since version 23.1, the database has changed to Swiss-Prot and TrEMBL sources). Major releases denote the addition of new material to the resource. The database growth has gathered momentum since October 1993, from which time it has shown a >10-fold increase in size. Version 14.2 of the BLOCKS database consists of 29,767 blocks representing 6,149 groups documented in InterPro 12.0 keyed to Swiss-Prot 48.3 and TrEMBL 31.3 obtained form the InterPro server.

Pfam

It is a database of multiple alignments of protein domains or conserved protein regions. The alignments represent some evolutionary conserved structure, which has implications for the protein's function. Profile Hidden Markov Models (profile HMMs) built from the Pfam alignments can be very useful for automatically recognizing that a new protein belongs to an existing protein family, even if the homology is weak. Unlike standard pairwise alignment methods (e.g. BLAST, FASTA), Pfam HMMs deal sensibly with multidomain proteins.

Pfam uses a different method for its database. High-quality seed alignments are used to create Hidden Markov Models to which sequences are aligned. Pfam has two classes of alignments, according to their credibility are given as follows:

1. Pfam-A—Non-edited seed alignments which are deemed to be accurate human crafted multiple alignments.
2. Pfam-B—Alignments derived by automatic clustering of the rest of a non-redundant protein database derived from the ProDom database. These alignments are less reliable.

Structure of Pfam-A entry   The format is compatible with PROSITE and each entry is identified by both an accession (AC) number and an ID code. DE line provides the title or description of the family, and AU line indicates the author of the entry. AL and AM lines showing information used to build seed and full automatic alignment respectively. The source database suggesting that seed members belong to one family, appropriate database cross-reference, and search program and cut-off used to build the full alignment are given in the SE, DR and GA lines as shown in the Figure 7.17.

```
AC   PF00001
ID   7tm
DE   7 transmembrane receptor (rhodopsin family)
AU   Sonnhammer ELL
AL   HMM_simulated_annealing
AM   hmma -qR
SE   Prosite
DR   Prosite; PDOC00210; [Expasy] [ SRS Japan|UK|USA]
DR   Prosite; PDOC00211; [Expasy] [ SRS Japan|UK|USA]
GA   Bic_raw 23.83 hmmfs 15
DR   http://www.gcrdb.uthscsa.edu/
```

Figure 7.17   Pfam-A entry for the 7 transmembrane receptor protein

Pfam families can be broken down into four basic types:

1. **family**—default classification, stating members are related
2. **domain**—structural unit found in multiple protein contexts
3. **repeat**—a domain that in itself is not stable, but when combined with multiple tandem repeats forms a domain or structure
4. **motif**—shorter sequence units found outside of domains

Links to the Pfam software are http://pfam.wustl.edu/ and http://www.sanger.ac.uk/Software/Pfam/index.shtml

The best way to look at the Pfam families is to jump right into the Pfam program and view some examples at http://pfam.wustl.edu/. Pfam also contains more information concerning the three-dimensional structures of proteins. The Pfam (http://www.sanger.ac.uk/Software/Pfam/) database is a collection of protein domain family multiple alignments and HMMs maintained at the Sanger Center on the Hinxton Genome Campus. Pfam is accessible for sequence searching via web server at the Sanger Center. Pfam has elucidated the full domain structure of the protein, including two PH domains and a single DEP (Disheveled, Eg1–10, Pleckstrin) domain.

ProDom (http://prodes.toulouse.inra.fr/prodom/2002.1/html/home.php)

ProDom is a database of all protein domain families automatically generated from the Swiss-Prot and TrEMBL databases. ProDom incorporates Pfam-A families as well as generating new ProDom alignments using recursive iterations of the PSI-BLAST program.

SMART (http://smartembl-heidelberg.de/)

Simple Modular Architecture Research Tool (SMART) is having HMM profiles constructed from manually refined protein domain alignments based on PSI-BLAST profiles. The human annotator checks and refines alignments before construction of HMM profile. Compared to Pfam, the SMART database is of better quality and with more extensive functional annotations in addition to presence of independent collection of HMMs.

InterPro (http://www.ebi.ac.uk/interpro/)

InterPro is a recent and very valuable development for the integration of several protein family resources into one comprehensive resource that was created in 1999. It is an integrated resource of protein families, domains, and functional sites created to handle the data from various protein family sites such as PROSITE, Pfam, PRINTS, ProDom, SMART and TIGRFAMs into a single, comprehensive resource allowing integrated search and sequence analysis. It has an intuitive interface both for text- and sequence-based searches, and since it incorporates several databases, it is much recommended. The InterPro entries use a combination of regular expressions, fingerprints, profiles and HMMs in pattern matching. Represented in InterPro are 74% of all proteins within the Swiss-Prot and TrEMBL databases. InterPro database has a graphical output to summarize motif matches and provides links to more detailed information (Figure 7.18).

Prosite Motif	Prosite: PS00018: PDOC00018: EF-hand calcium-binding domain. (2) Prosite: PS00041: PDOC00040: Bacterial regulatory proteins, araC family signature. (3) Prosite: PS00061: PDOC00060: Short-chain dehydrogenases/reductases family signature. (1) Prosite: PS00107: PDOC00100: Protein kinases ATP-binding region signature. (2) Prosite: PS00307: PDOC00278: Legume lectins beta-chain signature. (2) Prosite: PS00599: PDOC00518: Aminotransferases class-II pyridoxal-phosphate attachment site. (1) Prosite: PS00806: PDOC00523: Fructose-bisphosphate aldolase class-II signature 2.(1) Prosite: PS00681: PDOC00576: Chaperonins cpn10 signature. (1) Prosite: PS50020: PDOC50020: WW/rsp5/WWP domain profile. (2)
Pfam Domain	Pfam: PF01493: GXGXG motif(160) Pfam: PF01645: Conserved region in glutamate synthase(160) Pfam: PF00310: Glutamine amidotransferases class-II(159) Pfam: PF04898: Glutamate synthase central domain(154)
InterPro	InterPro: IPR003009: FMN/related compound-binding core InterPro: IPR001476: Chaperonin Cpn10 InterPro: IPR000005: Helix-turn-helix, AraC type InterPro: IPR002885: Pentatricopeptide repeat InterPro: IPR002198: Short-chain dehydrogenase/reductase SDR InterPro: IPR001220: Legume lectin, beta domain InterPro: IPR002110: Ankyrin InterPro: IPR002489: Glutamate synthase, alpha subunit, C-terminal InterPro: IPR002932: Ferredoxin-dependent glutamate synthase InterPro: IPR002035: von Willebrand factor, type A InterPro: IPR001202: WW/Rsp5/WWP InterPro: IPR000771: Ketose-bisphosphate aldolase, class-II InterPro: IPR000583: Glutamine amidotransferase, class-II InterPro: IPR000719: Protein kinase InterPro: IPR002048: Calcium-binding EF-hand InterPro: IPR001917: Aminotransferase, class-II InterPro: IPR006982: Glutamate synthase, central

Figure 7.18   Protein family resources

The latest release of InterPro, Release 14.0 contains 13,828 entries, 9,614 families, 3,905 domains, 232 repeats, 34 active sites, 22 binding sites and 21 post-translational modification sites.

CDART (www. ncbi.nlm.nih.gov/BLAST/)

Conserved Domain Architecture (CDART) is a domain search program that combines the results from RPS-BLAST (Reverse of PSI-BLAST that builds profiles from matched database sequences), SMART and Pfam. The resulting domain architecture of a query sequence can be graphically presented along with related sequences (Figure 7.19).

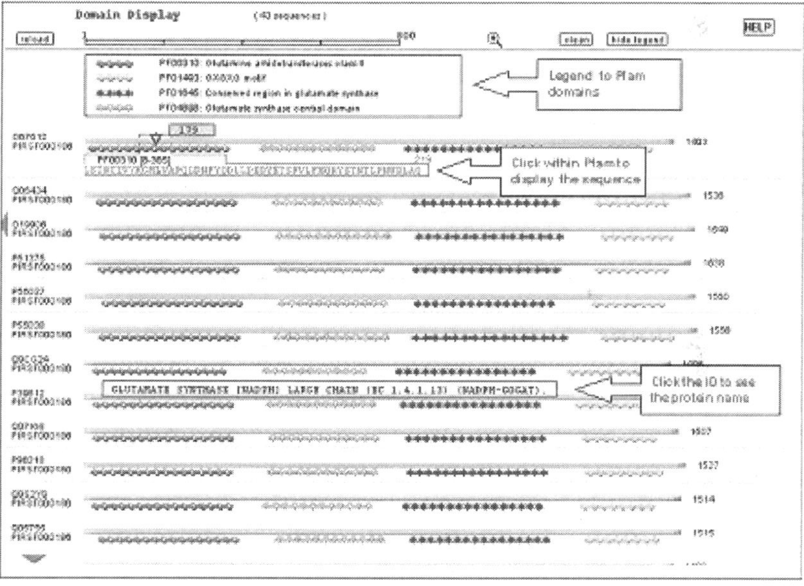

Figure 7.19 Domain architecture

PROTEIN FAMILY DATABASES

In addition to one way of classification of protein database on the basis of presence of motifs and domains there is another way of classifying proteins on the basis of full-length sequence comparison. This method requires clustering of protein based on the similarities within sequence alignment or orthologus relationship with computation of statistical scores. Cluster of Orthologus Groups (COG) and ProtoNet are two such protein family databases described in detail.

COG (www.ncbi.nlm.nih.gov/COG/)

Cluster of Orthologus Groups (COG) is a protein family database based on phylogenetic classification. COG is built by comparison of protein sequence that are mainly from prokaryotes, representing major phylogenetic lineages. Through all-against-all sequence comparison among genomes, orthologus proteins shared by three or more lineages are identified and clustered together as orthologus groups. In such group proteins considered to have descended through a vertical evolutionary scenario and if the function of one of the members is known, functionality of other member of the orthologus group can be assigned. Each group in a cluster should have at least one representative from archaea, bacteria, and eukarya. The latest release of COG database contains 4,873 clusters of proteins derived from unicellular organisms.

ProtoNet (www.protonet.cs.huji.ac.il/)

ProtoNet is a database of clusters of homologous proteins where the relatedness of protein is defined by the E-values from the BLAST alignment. In ProtoNet, there are different levels of protein similarity that yield a hierarchical organization of protein groups. Proteins having the most closely related sequences are grouped into the lowest level clusters while more distant protein groups are merged into the higher levels of clusters. This cluster merging approach leads to a tree-like structure of functional categories. ProtoNet database also provides gene ontology information for protein cluster at each level.

Expectation-maximization (EM) method and Gibbs motif sampling method are two sophisticated algorithms that are used for detecting motif in unaligned sequences. EM is used to discover hidden motif using a method different from profiles and PSSMs. In EM method, first a random or guessed alignment of the sequences is made to generate a trial PSSM, which is then used to compare with each sequence individually. The log odds scores obtained in PSSM are modified in each iteration to maximize the alignment of the matrix to each sequence. MEME (Multiple EM for motif elicitation) is web-based program available at http://meme.sdsc.edu/meme/website/meme-into.html. MEME uses expectation maximization and Hidden Markov Methods to generate motifs that are then used as search tools by its companion MAST in the combined suite MEME/MAST.

Gibbs sampling algorithm makes an initial guessed alignment of all but one sequence. Similar to EM, Gibbs sampling method built a trial PSSM to represent the alignment. Then the matrix is aligned to left-out sequence. The matrix scores are subsequently adjusted to achieve the best alignment with left-out sequence. This process is repeated for several times until there is no further improvement on the matrix scores. After number of repetitions, the most probable patterns can be incorporated into a final PSSM. Gibbs sampler (http://byesweb.wadsworth.org/gibbs/gibbs.html) is a web-based program based on Gibbs sampling technique and detects short, partially conserved gap-free segments for either DNA or protein sequences.

REVIEW QUESTIONS

1. Define multiple sequence alignments. Explain its role in making a phylogenetic tree.
2. Describe the different progressive alignment methods with significance.
3. What are the various iterative pairwise alignment approaches?
4. Explain the concept of Position-specific scoring matrices.
5. What is Position-specific Iterated BLAST?
6. Describe the Markov model and its significance in multiple sequence alignment.
7. Distinguish between motif and domain. What is the significance of identification of motifs and domains in multiple sequence alignment?
8. What is Prosite? Explain the structure of Prosite entry.
9. Define eMotif? Enlist software for motif finding.
10. What is the role of UCL in development of PRINTS fingerprint database?
11. Describe the concept of PRINTS and structure of PRINT entry.
12. What is the role of FHCRC in development of BLOCKS?
13. Explain the structure of BLOCK entry. How does it differ from that of PRINTS entry?
14. Describe the BLOCKS-format PRINTS entry for characterization of opsin family.
15. What is the role of Pfam in multiple sequence alignment?
16. Describe the structure of Pfam-A entry with a suitable example.
17. What is protein domain family? What are different related resources available on the web?

18. Explain the protein family databases with their significance.
19. Write short notes on
    i. ClustalW
    ii. T-Coffee
    iii. DIALIGN and MUSCLE
    iv. Profile HMM
    v. SMART
    vi. InterPro
    vii. COG and ProtoNet
    viii. ProDom
    ix. Conserved Domain Architecture
    x. Gibbs sampler
    xi. PRATT
    xii. Fuzzy algorithm

INTRODUCTION

A gene is a sequence of nucleotides on the DNA that encodes a polypeptide. The genome is the set of all genes plus all junk DNA that do not amount to genes. In other words, gene is a unit of heredity made up of unit length of DNA occupying specific location on a chromosome. As per the central dogma of molecular biology, DNA is transcribed to form mRNA that are further translated to synthesize a polypeptide chain. Because of redundancy, the genetic code of 64 codons is reduced to 20 distinct amino acids, and one can compare the sequence of nucleotides on DNA with that of amino acids in a protein. Twenty years ago, the low-throughput technique of chemical protein sequencing helped to determine the order of amino acids in a polypeptide chain. However, the lack of scalability of the method is a rate-determining step in generating large quantities of data. More recently, the advent of high-throughput automated fluorescent DNA sequencing technology has led to rapid accumulation of sequence information. Analysis of DNA sequences helps to detect phylogenetic relationships, genetic engineering using restriction site mapping, determination of gene structure through introns/exons prediction, inference of protein-coding sequence through Open Reading Frame (ORF) analysis. Given a piece of DNA sequence and with the knowledge of genetic code, it is possible to translate the DNA into protein by looking up successive codons in genetic code table (Chapter 3, Table 3.2); this is called the conceptual translation. It is significant to distinguish between sequences, for which the translation has biochemical support, and those that are simply derived theoretically or computationally;

the use of the term "conceptual translation" is theoretical and carries no experimental validation, whereas in reverse translation, from a given known protein one can find a gene in the genome which codes for it. Since there is rapid growth in accumulation of genomic sequence information, it is the need of time to have improved computational approaches that can accurately predict the structure of genes. It is prerequisite for detailed functional annotation of genes and genomes. The method includes detection of the location of ORFs and delineation of structures of introns as well as exons if the desired genes are of eukaryotic origin. Following is the list of ORFs detected in eukaryotic genome.

Organism name	ORF
Homo sapiens	33,133
Caenorhabditis elegans	16,639
Drosophila melanogaster	13,997
Oryza sativa	8,384
Saccharomyces cerevisiae	6,285
Neurospora crassa	1,010

A **reading frame** is a contiguous and non-overlapping set of three-nucleotide codons in DNA or RNA. There are three possible reading frames in a mRNA strand and six in a double-stranded DNA molecule due to the two strands from which transcription is possible. This leads to the possibility of overlapping genes and there may be many of these in bacteria. In rare cases a translating ribosome may shift from one frame to another, a translational frameshift. It is distinct from a frameshift mutation as the nucleotide sequence (DNA or RNA) is not altered only the frame in which it is read is altered.

Once a gene has been sequenced it is important to determine the correct ORF. Theoretically, the DNA sequence can be read in **six reading frames** in organisms with double-stranded DNA; three in the forward and three in the reverse direction.

In mRNA, if there is a sequence 5´ UCUAAAGGUGAC 3´ that has two out of three reading frames possible. This is one of the two possible mRNA sequences of the transcript that it can be read in three different ways:

- ▣ UCU AAA GGU GAC

    Its translation product will be Ser Lys Gly Asp

    In one letter code the tetrapeptide will be SKGD

▣ CUA AAG GUG, etc.

Its translation product will be Leu Lys Val

In one letter code the tripeptide will be LKV

▣ UAA AGG UGA, etc.

Since **UAA** is a stop codon, there won't be any translation and thus only two of the above three reading frames are open.

In DNA, the sequence of nucleotides is as motioned in query sequence that can undergo three forward and three reverse translations leading to six possible protein sequences, of which only one is likely to be real (Figure 8.1). The objective is to detect which of the translation product is correct.

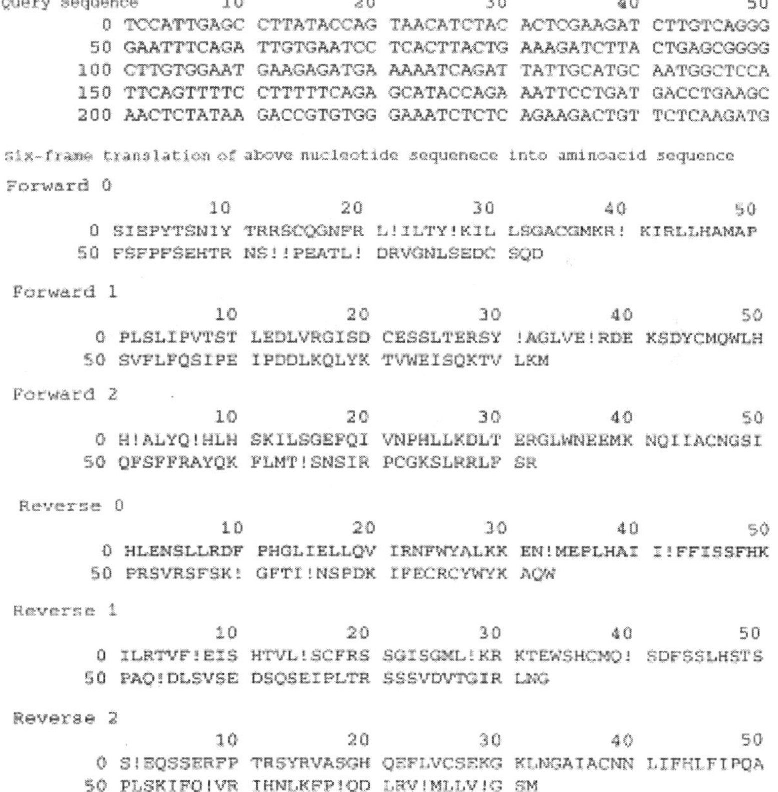

Figure 8.1 An arbitrary DNA sequence undergoes six-reading frame translations to form six possible polypeptide chains—three from forward (0,1,2) and three from reverse (0,1,2) translations. The symbol '!' indicates a stop codon.

GENE PREDICTION

Gene prediction is one of the first and most important steps in understanding the genome of a species once it has been sequenced. In its earliest days, gene prediction was based on painstaking experimentation on living cells and organisms. Statistical analysis of the rates of homologous recombination of several different genes could determine their order on a certain chromosome, and information from many such experiments could be combined to create a genetic map specifying the rough location of known genes relative to each other. Today, with comprehensive genome sequence and powerful computational resources at the disposal of the research community, gene prediction has been redefined as a computational problem that can be solved by following two approaches:

EXTRINSIC APPROACHES

In extrinsic (or evidence-based) gene finding systems, the target genome is searched for sequences that are similar to extrinsic evidence in the form of the known sequence of a messenger RNA (mRNA) or protein product. Given an mRNA sequence, it is trivial to derive a unique genomic DNA sequence from which it had to have been transcribed. Given a protein sequence, a family of possible coding DNA sequences can be derived by reverse translation of the genetic code. Once candidate DNA sequences have been determined, it is a relatively straightforward algorithmic problem to efficiently search a target genome for matches, complete or partial, and exact or inexact. BLAST is a widely used system designed for this purpose.

A high degree of similarity to a known messenger RNA or protein product is strong evidence that a region of a target genome is a protein-coding gene. However, to apply this approach systemically requires extensive sequencing of mRNA and protein products. Not only this is expensive, but in complex organisms, only a subset of all genes in the organism's genome are expressed at any given time, means that extrinsic evidence for many genes is not readily accessible in any single cell culture. Thus, in order to collect extrinsic evidence for most or all of the genes in a complex organism, many hundreds or thousands of different cell types must be studied, which itself presents further difficulties. For example, some human genes may be expressed only during development as an embryo or fetus, which might be difficult to study for ethical reasons.

Despite these difficulties, extensive transcript and protein sequence databases have been generated for human as well as other important model organisms in biology, such as mice and yeast. For example, the **Reference Sequence (Refseq)** database contains transcript and protein sequence from many different species and the Ensembl system comprehensively maps this evidence to human and several other genomes. It is, however, likely that these databases are both incomplete and contain small but significant amounts of erroneous data. The RefSeq database is an open access, annotated collection of publicly available nucleotide sequences (DNA, RNA) and their protein translations. This database is built by NCBI, and unlike GenBank, provides only one example of each natural biological molecule for major organisms ranging from viruses to bacteria to eukaryotes. For each model organism, RefSeq aims to provide separate and linked records for the genomic DNA, the gene transcripts, and the proteins arising from those transcripts. RefSeq is limited to major organisms for which sufficient data is available (almost 4,000 distinct "named" organisms as of January 2007), while GenBank includes sequences for any organism submitted (approximately 250,000 different named organisms).

AB INITIO-BASED APPROACHES

Because of the inherent expense and difficulty in obtaining extrinsic evidence for many genes, it is also necessary to resort to *ab initio* gene finding, in which genomic DNA sequence alone is systematically searched for certain tell-tale signs of protein-coding genes. These signs can be broadly categorized as either **signals**—specific sequences that indicate the presence of a gene nearby, or **content**—statistical properties of protein-coding sequence itself. *Ab initio* gene finding might be more accurately characterized as gene prediction, since extrinsic evidence is generally required to conclusively establish that a putative gene is functional.

In the genomes of prokaryotes, genes have specific and relatively well-understood promoter sequences (signals), such as the Pribnow box and transcription factor binding sites, which are easy to systematically identify (Figure 8.2). Also, the sequence coding for a protein occurs as one contiguous ORF, which is typically many hundred or thousands of base pairs long. The statistics of stop codons are such that even finding an ORF of this length is a fairly informative sign (Since 3 of the 64 possible codons in the genetic code are stop codons, one would expect a stop codon approximately every 20–25 codons, or 60–75 base pairs, in a random sequence). Furthermore, protein-coding DNA has certain periodicities and other statistical properties that are easy to detect from sequence of this length. These characteristics make prokaryotic gene-finding relatively straightforward, and well-designed systems are able to achieve high levels of accuracy.

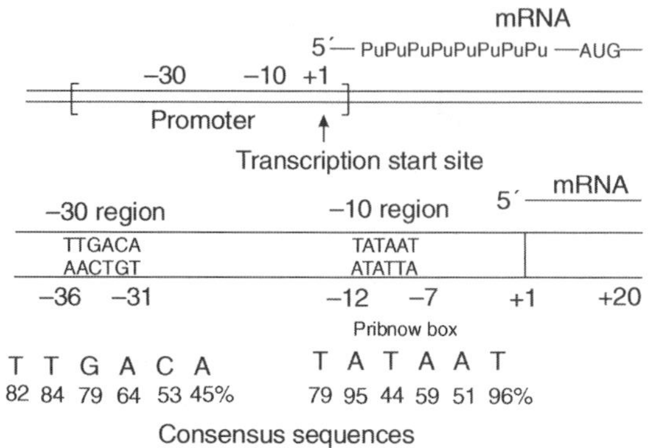

Figure 8.2   Promoter sequences in prokaryotes

Thus, the simplest method of gene predicting in prokaryotes is to search for ORFs that begins with a start (AUG) codon, and ends with one of three stop codons (UAG, UAA, and UGA). In prokaryotes, genes generally lack introns and consequently, have a somewhat simpler structure. DNA sequences coding for proteins generally are transcribed into mRNA which is translated into protein with very little modification. Therefore, in prokaryotes, locating an open reading frame from a start codon to a stop codon can give a strong suggestion into protein-coding regions. Longer ORFs are more likely to predict protein-coding regions than shorter ORFs.

Web-based Promoter Prediction Programs for Prokaryotes

BPROM (www.softberry.com/berry.phtml?topic=bprom&group=programs &subgroup=gfindb) is a web-based program for prediction of **Bacterial Promoters**. It combines the signal and content information such as consensus promoter sequence and oligonucleotide composition of the promoter sequences. BPROM first predicts a given sequence for bacterial operon structures by using an intergenic distance of 100 bp as basis for distinguishing genes to be in an operon. Once the operons are assigned, the program is able to predict putative promoter sequences. Because most bacterial promoters are located within 200 bp of the protein-coding region, the program is most effectively used when about 200 bp of upstream sequence of the first gene of an operon is supplied to increase specificity.

FindTerm (http://sun1.softberry.com/berry.phtml?topic= findterm&group= programs&subgroup= gfindb) is a program for searching bacterial ρ-independent termination signals located at the end of operons. The predictions are made based on matching of known profiles of the termination signals.

Web-based Promoter Prediction Programs for Eukaryotes

McPromoter (http://gene.mit.edu/McPromoter.html) is a web-based program that uses a neural network to make promoter predictions. This is a unique promoter model that contains six scoring segments. The program scans a window of 300 bases for the likelihoods of being in each of coding, non-coding, and promoter regions. The input for the neural network includes parameters for sequence physical properties, such as DNA bendability, plus signals such as TATA box, initiator box, and CpG islands. The hidden layer in the model combines all the features to derive an overall likelihood for a site being promoter. McPromoter is currently trained for *Drosophila* and human sequences.

TSSW (www.softberry.com/berry.phtml?topic=promoter) is another web-based program that distinguishes promoter sequences from non-promoter sequences based on combination of unique content information such as hexamer/trimer frequencies and signal information such as TATA box in the promoter region.

FirstEF (First Exon Finder) http:// rulai.cshl.org/tool/FirstEF/) is also web-based program that predicts promoters for human DNA. FirstEF integrates gene prediction with promoter prediction. A segment of DNA (15 bp) upstream of the first exon of a gene is extracted for promoter prediction on the basis of scores for CpG islands. Similarly, another web-based program **CONPRO** (http://stl.bioinformatics.med.umich.edu/conpro) uses a consensus method to identify promoter elements for human DNA.

Ab initio gene finding in eukaryotes, especially complex organisms like humans, is considerably more challenging for several reasons. First, the promoter and other regulatory signals in these genomes are more complex and less well-understood than in prokaryotes, making them more difficult to reliably recognize. Second, splicing mechanism (also called post-transcriptional modification as shown in Figure 8.3) employed by eukaryotic cells mean that a particular protein-coding sequence in the genome is divided into several parts (exons), separated by non-coding sequences (introns). In the RNA splicing mechanism, introns are removed from pre-mRNA and a typical eukaryotic monocistronic mRNA containing a single ORF is formed. When the aim is to find eukaryotic ORFs it is necessary to have a look at the spliced mRNA.

In humans, a typical protein-coding gene might be divided into a dozen exons, each less than two hundred base pairs in length, and some as short as 20 to 30. It is therefore much more difficult to detect periodicities and other known content properties of protein-coding DNA in eukaryotes. The post-transcriptional modification of pre-mRNA in eukaryotes makes it more difficult for gene prediction.

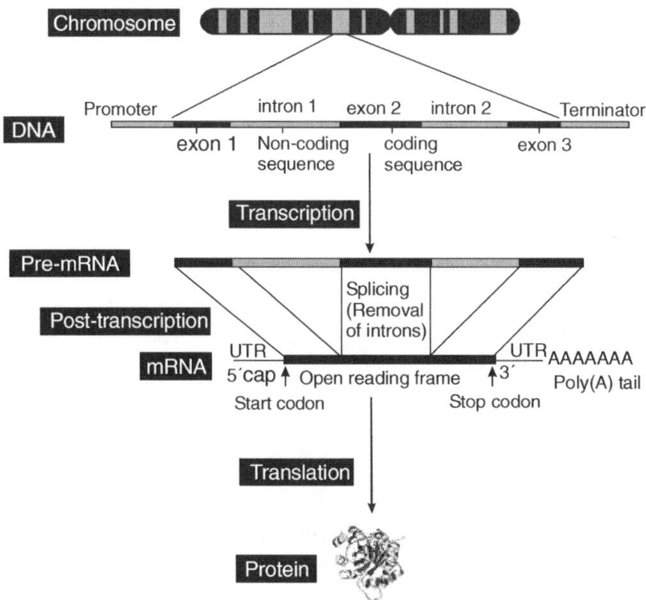

Figure 8.3   Central dogma of molecular biology with post-translational modification of pre-mRNA. (UTR — untranslated regions and poly (A)—polyadenylation)

Computational gene prediction is very complex and challenging problem since the main issue in prediction of eukaryotic gene is the identification of exons, introns, and splicing sites. The presence of split gene structure, alternative splicing, and low gene densities make the task of finding genes difficult similar to searching a needle in a haystack. There are still some conserved sequence features in eukaryotic genes that make possible the computational prediction. For example, the splice junctions of introns and exons follow the GT–AG rule in which an intron at 5′ splice junction (donor site) has a consensus motif of GTAAGT; and the 3′ splice junction (acceptor site) is a consensus motif of $(pyrimidine)_{12}NCAG$ (Figure 8.4). Given a genomic sequence and a set of candidate exons, the spliced alignment algorithm explores all possible exon assemblies and finds a chain of exons which best fits a related target protein. The set of candidate exons are constructed by considering all blocks between candidate acceptor and donor sites (i.e., between AG dinucleotide at the intron–exon boundary and GT dinucleotide at the exon–intron boundary) and further filtration of this set. To avoid losing true exons, the filtration procedure is designed to be very delicate, and the resulting set of blocks may contain a large number of false exons. Instead of trying to identify the correct exons by further pursuit of statistical methods, the algorithm considers all possible chains of candidate exons and finds a chain with the maximum global similarity to the target protein.

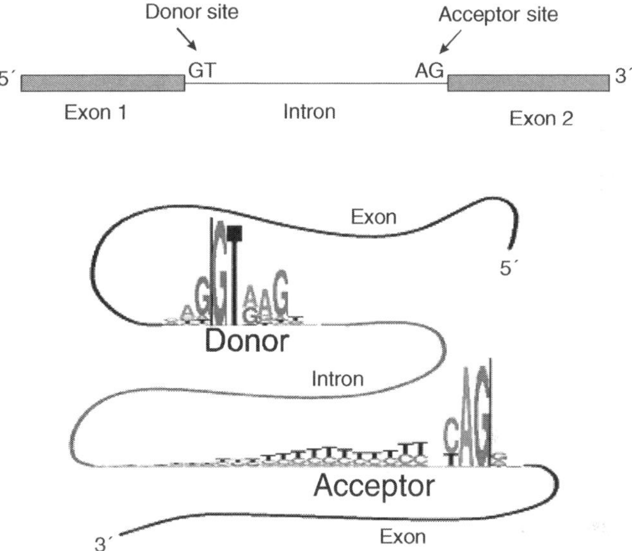

Figure 8.4   Two sequence logos representing conservation at the 5′ (donor) and 3′ (acceptor) ends of human introns. The region between black vertical bars is removed during mRNA splicing. The logos show a common pattern "CAG/GT", which suggests that, the mechanisms that recognize two ends of introns had a common ancestor. (After Stephen and Schneider, *J. Mol. Bio.* 228, 1992.)

However, there are two classic examples of signals identified by eukaryotic gene finders, CpG islands and binding sites for a poly(A) tail. Most of the vertebrate genes have a high density of CG dinucleotides near the transcriptional start site. This region is referred to as CpG island (where p stands for the phosphodiester bond that connects the two nucleotides) that helps in identification of transcriptional initiation site of a eukaryotic gene. Similarly, at the 3 ′end of mRNA where there is poly(A) signal can also assist in locating the final coding sequence of eukaryotic gene.

Advanced gene finders for both prokaryotic and eukaryotic genomes typically use complex probabilistic models, such as Hidden Markov Models, in order to combine information from a variety of different signal and content measurements. The GLIMMER system (http://www.tigr.org/software/Unveil/index.shtml) is a widely used and highly accurate gene finder for prokaryotes. GeneMark is another popular approach. Eukaryotic *ab initio* gene finders, by comparison, have achieved only limited success; notable examples are the GenScan and geneid programs. A few programs like CONTRAST also use machine learning approaches like support vector machines for successful gene prediction.

GLIMMER (GENE LOCATOR AND INTERPOLATED MARKOV MODELER) SYSTEM

GlimmerHMM is a new gene finder based on a Generalized Hidden Markov Model (GHMM). Although the gene finder conforms to the overall mathematical framework of a GHMM, additionally it incorporates splice site models adapted from the GeneSplicer program and a decision tree adapted from GlimmerM (http://www.cbcb.umd.edu/software/glimmerm/). It also utilizes Interpolated Markov Models for the coding and non-coding models. Currently, GlimmerHMM's (GHMM) structure includes introns of each phase, intergenic regions, and four types of exons (initial, internal, final, and single).

System Requirements

GlimmerHMM is released as a source code and was tested on Linux RedHat 6.x+, Sun Solaris, and Alpha OSF1, but it also works on any Unix system.

Accuracy

GlimmerHMM has been trained on species as indicated in Table 8.1 with the accuracy ranging from 86% to 99%. In the tables, "Sens" is sensitivity, which is the percentage of true exons/Nucl predicted by the system. "Prec" is precision, which is the percentage of the system's predictions that are correct.

Table 8.1 GlimmerHMM trained on the genome of various organisms

Organism	Nuc Sens	Nuc Prec	Nuc Accur	Exon Sens	Exon Prec	Exact Genes	Size of test set
Danio rerio	93%	78%	86%	77%	69%	24%	549 genes
Caenorhabditis elegans	96%	95%	96%	82%	81%	42%	1886 genes
Arabidopsis thaliana	97%	99%	98%	84%	89%	60%	809 genes
Cryptococcus neoformans	96%	99%	98%	86%	88%	53%	350 genes
Coccidioides sp.	99%	99%	99%	84%	86%	60%	503 genes
Brugia malayi	93%	98%	95%	78%	83%	25%	477 genes

GlimmerHMM has been trained on the human genome as well. The Table 8.2 presents its performance compared to GenScan on 963 human RefSeq genes selected randomly from all 24 chromosomes, non-overlapping with the training set. The test set contains 1000 bp of untranslated sequence on either side (5´ or 3´) of the coding portion of each gene.

Table 8.2 GlimmerHMM and GenScan trained on human genome

Gene finder	Nuc Sens	Nuc Prec	Nuc Acc	Exon Sens	Exon Prec	Exon Acc	Exact Genes
GlimmerHMM	86%	72%	79%	72%	62%	67%	17%
GenScan	86%	68%	77%	69%	60%	65%	13%

Glimmer3

It is version 3.02 of the Glimmer gene-finding software. This version incorporates a nearly complete rewrite of the code, resulting in improvements in both sensitivity and specificity of the predictions. In Glimmer3, a single dynamic-programming, HMM-like algorithm is used to select the highest-scoring ORFs and their start sites. This algorithm guarantees that the predictions have no overlaps longer than the specified length. Glimmer software was written for the Linux software environment. To run glimmer, a probability model of coding sequences, called an Interpolated Context Model (ICM), must be built. This is done by the program build-ICM from a set of training sequences. These sequences can be obtained in several ways:

- From known genes in the genome, e.g. genes identified by homology searches.
- From long, non-overlapping ORFS in the genome as produced by the program long-ORFs.
- From genes in a highly similar species/strain.

To obtain the best results with Glimmer, the largest possible training set of genes should be used from the same genome on which predictions are to be made. If genes are known from homology searches, they can be used. If only a few such genes are available, then they can be combined with the training set produced by the long-ORFs program (but do not include duplicate genes in the training set). Running Glimmer on small genome fragments, the genome of the nearest available evolutionary relative of the target organism can be used to provide a training set of genes.

Glimmer3 Output Formats

1. The .detail File

    The .detail file begins with the command that invoked the program and a list of the parameters used by the program. Here is a sample:

```
Command: /fs/szgenefinding/Glimmer3/bin/glimmer3 -o 50 -g 110 -
t 30 -b iterated.motif -P0.603,0.338,0.059 tpall.fna iterated.icm
iterated

Sequence file = tpall.fna

Number of sequences = 1

ICM model file = iterated.icm

Excluded regions file = none

List of orfs file = none

Input is NOT separate orfs

Independent (non-coding) scores are used

Circular genome = true

Truncated orfs = false

Minimum gene length = 110 bp

Maximum overlap bases = 50

Threshold score = 30

Use first start codon = false

Start codons = atg,gtg,ttg
```

```
Start probs = 0.603,0.338,0.059

Stop codons = taa, tag, tga

GC percentage = 52.8%

Ignore score on orfs longer than 799
```

Following that, for each sequence in the input file the FASTA-header line is echoed and followed by a list of ORFS that were long enough for Glimmer3 to score (Table 8.3). Here is a sample of the beginning of such a section:

```
>gi|15638995|ref|NC_000919.1| Treponema pallidum subsp. pallidum
str. Nichols, complete genome

Sequence length = 1138011
```

Table 8.3   List of ORFS processed through Glimmer3

		Start			Length		Scores								
ID	Frame	of Orf	of Gene	Stop	of Orf	of Gene	Raw	InFrm	F1	F2	F3	R1	R2	R3	NC
	+2	17	20	139	120	117	−4.94	0	99	0	–	0	–	–	0
	+2	140	242	361	219	117	0.99	0	87	0	–	12	–	–	0
	−1	435	417	148	285	267	5.48	2	97	–	–	2	–	–	0
	+2	668	668	790	120	120	2.89	0	99	0	–	–	–	–	0
	−3	899	839	717	180	120	−0.86	1	95	–	–	–	–	1	3
	−1	936	933	808	126	123	0.38	13	78	–	–	13	–	–	8
	−3	1124	1109	918	204	189	−1.32	0	99	–	–	–	–	0	0
0001	+1	4	4	1398	1392	1392	6.61	99	99	–	–	–	–	–	0
	−2	1750	1720	1457	291	261	−0.92	8	–	–	–	–	8	–	91
	−2	1957	1945	1751	204	192	−1.47	1	–	–	70	–	1	–	27
	−3	2078	2063	1908	168	153	−1.88	4	–	–	20	–	–	4	75
	−2	2308	2293	2174	132	117	−0.38	5	–	–	85	–	5	–	9
0002	+3	1542	1641	2756	1212	1113	3.20	99	–	–	99	–	–	–	0
	−3	2807	2774	2616	189	156	−2.08	3	0	–	–	–	–	3	96

Below is a description of the columns. All positions are counted from the beginning of the sequence with the first base being position 1.

ID   An identification number for a potential gene. Only ORFs whose in-frame (InFrm) score is above the threshold score (set by the -t option) or are longer than the ignore-score length have an entry in this column.

Frame   The reading frame of the ORF—positive for forward strand, negative for reverse strand. It is determined by the position of the leftmost base of the stop codon:

frame +1   if the stop begins in position 1, 4, 7, . . . ;

frame +2   if the stop begins in position 2, 5, 8, . . . ;

frame +3   if the stop begins in position 3, 5, 9, . . . ;

frame −1   if the stop begins in position 3, 5, 9, . . . (so the leftmost base is position 1, 4, 7, . . .);

frame −2   if the stop begins in position 4, 7, 10, . . . (left base position 2, 5, 8, . . .);

frame −3   if the stop begins in position 5, 8, 11, . . . (left base position 3, 6, 9 . . .).

Note that if the genome length is not a multiple of 3, for genes that wrap around the end of the sequence the same rules applied to the start codon position will not yield the same reading frame.

Start   The positions of the first base of the ORF and the first base of the start codon of the gene. Note that the gene start may be different for the same ORF in the .predict file.

Stop   Position of the last base of the stop codon.

Length   Number of bases in the ORF and in the gene. It does not include the bases of the stop codon.

Raw Score   This is 100 times the per-base log-odds ratio of the in-frame coding ICM score to the independent (i.e., non-coding) model score. It gives a rough quantification to how well an ORF scores that can be compared between any two ORFs.

InFrm Score   The normalized (to the range 0 . . . 99) score of the gene in its reading frame. This is just the appropriate frame value among the next six scores.

Frame Scores   The normalized (to the range 0 . . . 99) score of the gene in each reading frame. A "-" indicates the presence of a stop codon in that reading frame. The normalization compares only scores without stop codons and the independent (non-coding) NC score. If the ORF is sufficiently long, i.e., longer than the value stated in "Ignore score on ORFS longer than. . . ", the score is not used.

NC Score   The normalized independent (i.e., non-coding or intergenic) model score. This model is adjusted for the fact that the ORF, by definition, has no in-frame stop codons.

EDR Score   An additional column of scores is produced if the E-option is specified. This is the entropy-distance ratio, i.e., the ratio of the distance of the amino acid distribution from a positive model to the distance from a negative model. Scores below 1.0 are more likely to be genes; scores above 1.0 less likely to be genes. It is not currently used in the scoring process.

2. The .predict File

    This file has the final gene predictions. Its format is the FASTA-header line of the sequence followed by one line per gene. Here is a sample of the beginning of such a file:

```
>gms:3447|cmr:632 chromosome 1 {Mycobacterium smegmatis MC2}

orf00001 499 1692 +1 13.14

orf00004 1721 2614 +2 14.20

orf00006 2624 3778 +2 10.35

orf00009 3775 4359 +1 9.34
```

The columns are:

Column 1   The identifier of the predicted gene. The numeric portion matches the number in the ID column of the .detail file.

Column 2   The start position of the gene.

Column 3   The end position of the gene. This is the last base of the stop codon, i.e., it includes the stop codon.

Column 4   The reading frame.

Column 5   The per-base "raw" score of the gene. This is slightly different from the value in the .detail file, because it includes adjustments for the Position Weight Matrix (PWM) and start codon frequency.

GENSCAN (http://genes.mit.edu/GENSCANinfo.html)

GenScan algorithm is based on probabilistic model of gene structure similar to Hidden Markov Models. It uses a training set in order to estimate the HMM parameters, then the algorithm returns the exon structure using maximum likelihood approach standard to many HMM algorithms (Viterbi algorithm). GenScan was developed by Chris Burge in the research group of Samuel Karlin, Department of Mathematics, Stanford University.

Biological input for GenScan model includes (Figure 8.5): coding regions representing internal exons, *viz.*, states exon 0, exon 1 and exon 2 with different reading frames, gene structure with start and stop codons, introns such as intron 0, intron 1, and intron 2 following the exons with different reading frames, 5′end representing the region upstream of the first coding exon of the gene, 3′end representing the downstream of the last coding exon of a gene, presence of promoters, presence of genes on both strands, etc. The GenScan model is based on explicit state duration Hidden Markov Model. The Model can be divided into three components: 1) the **transition model**, which specifies the probability of moving from any one state to any other state, 2) the **duration model**, which specifies the probability of staying in a given state for a given number of nucleotides before changing to another state, and 3) the **state-specific sequence models**, which specify the probability of any given nucleotide being generated from any given state.

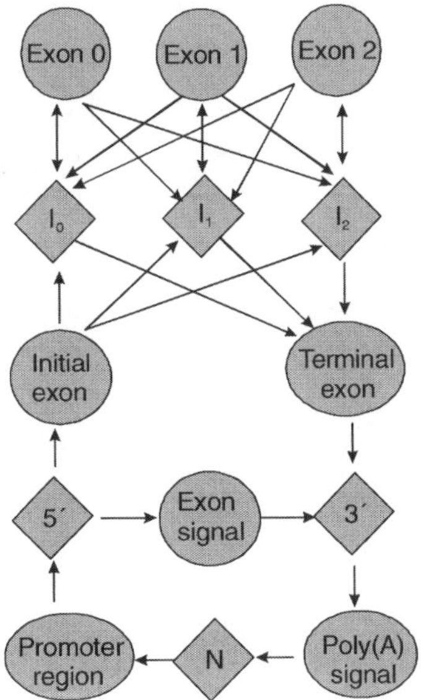

Figure 8.5   The states of GenScan model where arrows indicate transition with non-zero (N) probability for genes on forward strand. An analogous model is used for genes on opposite strand. (After Bruge and Karlin, *J. Mol. Bio.* 1997.)

Basically GenScan is an online program to identify complete gene structures in genomic DNA. It is a GHMM-based gene finder for human sequences. The web server at the MIT can be found at http://genes.mit.edu/GENSCAN.html

Important Features of GENSCAN

- Identification of complete intron/exon structures of a gene in genomic DNA.
- The ability to predict multiple genes and to deal with partial as well as complete genes.
- The ability to predict consistent sets of genes occurring on one or both the strands of the DNA.
- Predicts both optimal annotation and sub-optimal exons.

Although the results are good they are still not good enough for massive gene finding. GenScan has 80% chance for detecting an exon. If a gene has more than one exon the probability of correctly detecting all of them declines rapidly. On the positive side a more permissive figure of merit, the prediction per bp, is over 90%

GenScan Limitations

- Does not use similarity search to predict genes.
- Does not address alternative splicing.
- Could combine two exons from consecutive genes together.

TWINSCAN

It is new gene structure prediction system that directly extends the probability model of GenScan, allowing it to exploit the homology between two related genomes. Separate probability models are used for conservation in exons, introns, splice sites and UTRs reflecting the differences among their patterns of evolutionary conservation. TwinScan is specially designed for high-throughput genomic sequences containing an unknown number of genes. Thus, TwinScan is comparative gene finding method in which two sequences are aligned and each base is marked as gap (-), mismatch (:), match (|), resulting in a new alphabet of 12 letters: Ó {A-, A:, A |, C-, C:, C |, G-, G:, G |, T-, T:, T|}.

Following steps are involved to run the TwinScan gene prediction program:

- Viterbi algorithm using emissions $ek(b)$ where $b \in$ {A-, A:, A|, ..., T|}.
- Estimation of emission probabilities from human/mouse gene pairs.
- Ex. $eI(x|) < eE(x|)$ since matches are favoured in exons, and $eI(x-) > eE(x-)$ since gaps (as well as mismatches) are favoured in introns.
- Compensates for dominant occurrence of poly(A) region in introns.

The experiments conducted by Ion Korf *et al*. (2001) on high-throughput mouse genomic sequences using homologous sequences from human genome, TwinScan showed improvement over GenScan in exon sensitivity and specificity and dramatic improvement in exact gene sensitivity and specificity. This improvement in TwinScan program was attributed by researchers to modelling the patterns of evolutionary conservations in genomic sequences.

GENE PREDICTION USING NEURAL NETWORK

Neural networks are common in biological nervous system. The anatomical and physiological features of biological neural networks assist in designing of computational programs that have been applied successfully to a wide variety of pattern recognition, classification, and decision problems. In the computational model, a single neuron is a node in a directed graph, with one or more entering connections referred to as input whereas the output is in the form of a single leaving connection (Figure 8.6a and b).

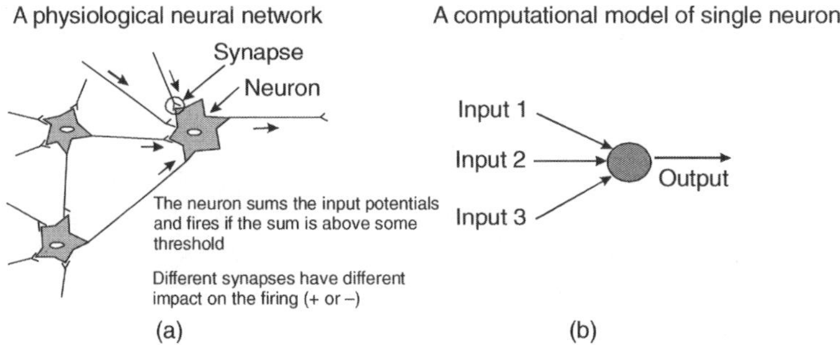

Figure 8.6   (a) A physiological neural network (b) The computational model of a single neuron where summing the input and outputting something based on that sum

Similar to biological neural network, another aspect of above computational model is its ability to "learn" and then make predictions after being trained. Such neural network has the ability to process the information and modify parameters of weight functions between variables during training period. Once it is trained, it becomes capable to predict unknown automatically. A neural network constructed for gene prediction has multiple layers, *viz.* the input, output and hidden layers (Figure 8.7). The input for neural network is the gene sequence with intron and exon sequence. The gene sequence information provided to the model is classified as hexamer frequencies, GC composition and splice sites. This input is processed by one or several hidden layers where the machine learning takes place. The machine learning process starts by feeding the model with a sequence of known gene structure. In the hidden layer the weighted functions resembling to biological synapse are adjusted to recognize the nucleotide pattern and their relationship with known structure of a gene. Once the machine acquires training, the algorithm predicts an unknown sequence and provides an output in the form of the probability of an exon structure by applying the same rules learned in training to look for the patterns associated with the gene structure.

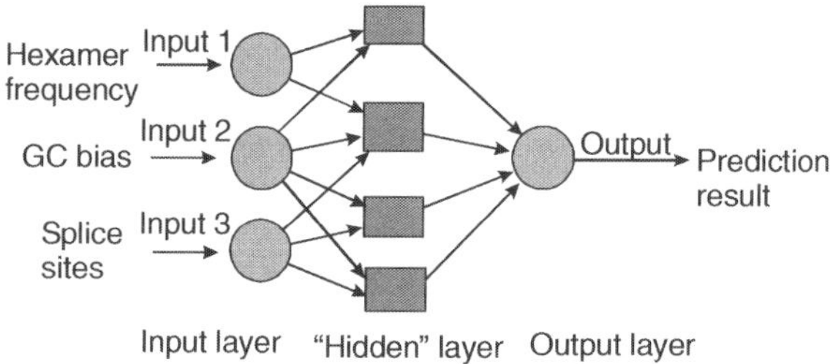

Figure 8.7   Diagrammatic representation of artificial neural network for prediction of eukaryotic gene

GRAIL   (Gene Recognition and Assembly Internet Link) (http://compbio.ornl.gov/Grail-bin/EmptyGrailForm) is the web-based program that uses neural network algorithm to predict the gene structure. The statistical features such as splice junctions, start and stop codons, poly(A) sites, promoters, and CpG islands are used to train the program. It then scans the query sequence with windows of variable length and scores for coding potentials and finally provides an output in the form of exon candidate. Currently, GRAIL program is trained for gene sequences of human, mouse, *Arabidopsis, Drosophila,* and *Escherichia coli.*

GRAIL II provides analyses of protein-coding regions, poly(A) sites, and promoters, constructs gene models, and predicts encoded protein sequences. The underlying algorithm makes a list of the most likely exon candidates and these are further evaluated using a neural network. Dynamic programming is then applied to define the most probable gene models.

Input for GRAIL II includes several indicators of sequence patterns, including a Markov model for gene recognition and inputs from two additional neural networks that evaluate the region for potential splice sites.

One important indicator in GRAIL II (and other gene prediction programs as well) is the in-frame hexamer preference score, since the occurrence of codon pairs in coding regions is not random, while in non-coding regions it is more likely to be random. A higher frequency of hexamers that are more commonly found in coding regions can be an indicator of the presence of an exon. The neural network used for GRAIL II is trained using a set of known coding sequences.

In addition to above-mentioned gene finder programs, following is the list of gene prediction projects with their specific purpose and web addresses (Table 8.4).

Table 8.4   List of gene prediction projects followed by web addresses and description

Project	URLS and description
SNAP	http://homepage.mac.com/iankorf/snap-2004-03-02.tar.gz GHMM eukaryotic gene finder
GeneZilla	(formerly TIGRscan) http://www.genezilla. org/GHMM eukaryotic gene finder
TWAIN	http://www.tigr.org/software/pirate/twain/twain.html GPHMM comparative gene finder
Shadower	http://bonaire.lbl.gov/shadower Generalized Hidden Markov Phylogeny (GHMP) gene finder
JIGSAW (formerly "Combiner")	http://www.cbcb. umd.edu/ software/ jigsaw/ Evidence combiner for eukaryotic gene prediction
Genome Threader	http://www.genomethreader.org Similarity-based gene prediction program where additional cDNA/EST and/or protein sequences are used to predict gene structures via spliced alignments
EvoGene	http://www.birc.dk/software/evogene Evolutionary HMM gene finder
ExonHunter	http://monod.uwaterloo.ca/papers/showpapers.php?topic=gene+ finding Integrative gene finding system
Unveil	http://www.tigr.org/software/Unveil/index.shtml HMM eukaryotic gene finder
CRITICA	http://www.ttaxus.com/index.php?pagename=Software Comparative prokaryotic gene finder
SGP2	http://www.ttaxus.com/index.php?pagename=Software Comparative gene finder based on geneid and TBLASTX

(Contd.)

Table 8.4 (Continued)

Project	URLs and description
Phat	http://bioinf.wehi.edu.au/Phat GHMM gene finder
geneid	http://genome.imim.es/software/geneid/index.html Hierarchically-structured gene prediction program
SLAM	http://baboon.math.berkeley.edu/~syntenic/slam.html Pair GHMM-based synthetic gene finder
EuGene	http://www.inra.fr/bia/T/eu Gene/ An open gene finder for eukaryotic organisms
AUGUSTUS	http://augustus.gobics.de/ GHMM-based gene finder for eukaryotes
SGP1	http://soft.ice.mpg.de/sgp-1/ Similarity based gene prediction program
VEIL	http://www.tigr.org/~salzberg/veil.html HMM eukaryotic gene finder (no longer supported)
MORGAN	http://www.cbcb.umd.edu/~salzberg/morgan.html A eukaryotic gene finder using OC1 decision trees
GenomeScan	http://genes.mit.edu/genomescan/ GHMM informant-based gene finder
DoubleScan	http://www.sanger.ac.uk/software/analysis Pair HMM gene finder
HMMgene	http://www.cbs.dtu.dk/services/HMMgene HMM gene finder
GRAILEXP	http://grail.lsd.ornl.gov/grailexp/ Neural-network-based gene finder (Oak Ridge Natl. Lab.)
Genie	http://www.fruitfly.org/seq_tools/genie.html GHMM-based gene finder
GeneMark™	http://opal.biology.gatech.edu/GeneMark A gene finder from Georgia Institute of Technology
FGENESH	http://www.softberry.com/ berry.phtml?topic=index & group =programs & subgroup=gfind GHMM gene finder
ORPHEUS	http://pedant.gsf.de/orpheus/ Bacterial gene finder
PROCRUSTES	http://www-hto.usc.edu/software/procrustes/ Gene finder utilizing protein evidence
Diogenes	http://analysis.ccgb.umn.edu/diogenes/ Searching for and reporting likely protein-encoding regions
WebGene	http://www.itba.mi.cnr.it/webgene/ Tools for prediction and analysis of protein-coding gene structure
GeneSplicer	http://www.cbcb.umd.edu/software/GeneSplicer A fast, flexible system for detecting splice sites in eukaryotic DNA.
ORF Finder	http://www.ncbi.nih.gov/gorf/gorf.html Program at NCBI to find ORFs in a sequence

(Contd.)

Table 8.4 (Continued)

Project	URLs and description
SplicePredictor	http://bioinformatics.iastate.edu/cgi-bin/sp.cgi A method to identify potential splice sites in (plant) pre-mRNA by sequence inspection using Bayesian statistical models
NetPlantGene	http://www.cbs.dtu.dk/services/NetPGene Neural network predictions of splice sites in *Arabidopsis thaliana* DNA
NetGene2	http://www.cbs.dtu.dk/services/NetGene2/ Neural network predictions of splice sites in human, *C. elegans* and *A. thaliana* DNA
ELPH	http://www.cbcb.umd.edu/software/ELPH Gibbs sampler for finding motifs in DNA; has been used for detecting exon splice enhancers (ESE's). Also applicable to other motif-detection tasks.
RBSfinder	http://ftp.tigr.org/pub/software/RSBfinder A program to find ribosome binding sites in prokaryotic DNA.
TransTerm	http://www.tigr.org/softwa/re/pirate/ml.shtml A program that finds rho-independent transcription terminators in bacterial genomes
OC1	http://www.cbcb.umd.edu/~salzberg/announce-oc1.html Oblique decision trees for classification
TAIL	http://www.tigr.org/software/pirate/ml.shtml Machine-learning library: neural nets, decision trees, multivariate regression, Bayesian classifiers, genetic algorithms/genetic programming, and K-nearest-neighbors with feature selection and correlation control suffix trees
tigr++	http://www.tigr.org/software/pirate/tigr++.shtml C++ class library used by several TIGR genefinders and other packages. Covers string & sequence processing, math/statistics, many efficient data structures, GFF parsing, sorting, and I/O.
PROSCAN/ PromoterScan (URL is same as above)	http://bimas.dcrt.nih.gov/molbio/proscan/index.html Predicts promoter regions based on scoring homologies with putative eukaryotic Pol II promoter sequences

GENE DISCOVERY USING EST AND cDNA

An important goal of any sequencing project is to obtain a genomic sequence and identify a complete set of genes that help in understanding of when, where and how a gene is turned on, a process commonly referred to as **gene expression**. Cells express a different range of genes at various stages in their development and functioning. This characteristic range of gene expression is the expression profile of the cell. Once it is understood where and how a gene is expressed under normal circumstances, researcher can then study what happens in an altered state, such as in disease. To accomplish the latter goal however, one must identify and study the protein, or proteins, coded for by a gene. But finding a gene that codes for a protein, or proteins, is not easy. In early days, researchers work to isolate that protein, determine its function, and locate the gene that coded for the protein. Alternatively, scientists could conduct linkage studies to determine the chromosomal location of a particular gene. Once the chromosomal location was determined, scientists would use biochemical methods to isolate the gene and its corresponding protein. Either way, these methods took a great deal of time—years in some cases—and yielded the location and description of only a small percentage of the genes.

With the advent of technology it is now possible to locate and fully describe a gene in a short span of time using Expressed Sequence Tags (ESTs). An Expressed Sequence Tag is a tiny portion of an entire gene that can be used to identify unknown genes and to map their position within a genome. ESTs provide researchers with a quick and inexpensive route for discovering new genes, for obtaining data on gene expression and regulation, and for constructing genome maps. Today, researchers using ESTs to study the human genome find themselves riding the crest of a wave of scientific discovery, the likes of which has never been seen before.

ESTs are small pieces of DNA sequence (usually 200 to 500 nucleotides long) that are generated by sequencing either one or both ends of an expressed gene. The idea is to sequence bits of DNA that represent genes expressed in certain cells, tissues, or organs from different organisms and use these "tags" to fish out a gene of a portion of chromosomal DNA by matching base pairs. The challenge associated with identifying genes from genomic sequences varies among organisms and is dependent upon genome size as well as the presence or absence of introns. DNA is transcribed to form pre-mRNA, which undergoes processing to form mRNA. Isolating mRNA is the key to finding expressed genes in the vast expanse of the genome. The mRNA is very unstable in outside of a cell, using special enzyme reverse transcriptase mRNA is converted to complementary DNA (cDNA). The cDNA is a much more stable compound and, importantly, because it was generated from the mRNA in which the introns had been removed, cDNA represents only expressed DNA sequence.

Once cDNA representing an expressed gene has been isolated, scientists can then sequence a few hundred nucleotides from either end of the molecule to create two different kinds of ESTs (Figure 8.8). Sequencing only the beginning portion of the cDNA produces what is called a 5´ EST. A 5´ EST, which is obtained from the portion of a transcript that usually codes for a protein. These regions tend to be conserved across species and do not change much within a gene family. Sequencing the ending portion of the cDNA molecule produces what is called a 3´ EST. As these ESTs are generated from the 3´ end of a transcript, they are likely to fall within non-coding or untranslated regions (UTR), and therefore tend to exhibit less cross-species conservation than do coding sequences.

If the organism, from which the genomic DNA has been extracted, has EST or cDNA sequences available, the next step is to perform a similarity search between the genomic DNA and these ESTs or cDNAs. In addition, it can be helpful to perform similarity searches with ESTs and cDNAs from other organisms as well. This step locates potential exons within genomic sequences.

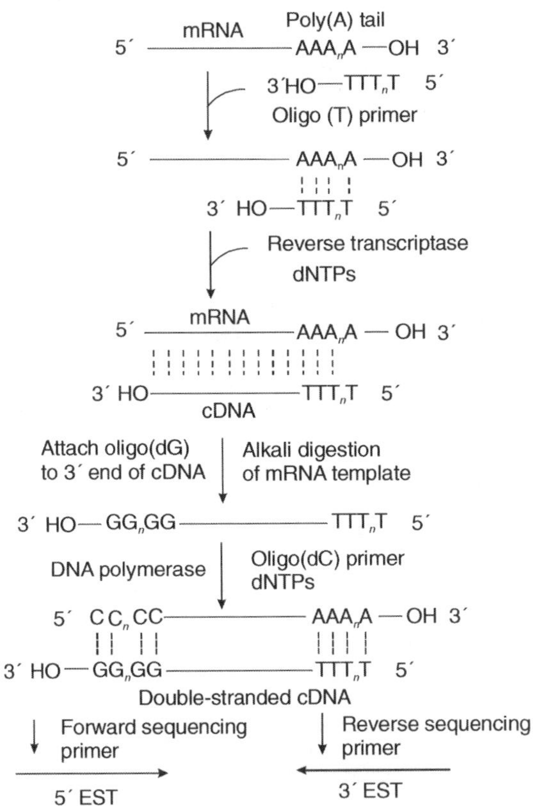

Figure 8.8   ESTs are generated by sequencing cDNA

Web-Based Programs for Searching Homology in EST or cDNA

EST2Genome (http://bioweb.pareur.fr/seqanal/interfaces/ est2genome.html) is a web-based program based on the sequence alignment approach to define intron–exon boundaries. EST or cDNA sequence with a genomic DNA sequences containing corresponding gene is compared in this program where alignment is done with the help of algorithm used in dynamic programming. This program has an ability to find very small exons and alternatively spliced exons that are very difficult to predict by any *ab initio*-type algorithm. In addition, there is no need for model training hence there is much more flexibility for gene prediction. EST2Genome program has limitation that EST and cDNA sequences often contain errors or even introns if the transcripts are not completely spliced before revere transcription.

SGP1 (http://soft.ice.mpg.de/sgp-1/) is a similarity-based web program that is used for gene prediction. It aligns two genomic DNA sequences from closely related organisms. The program translates all potential exons in each sequence and does pairwise alignment for the translated protein sequences using a dynamic programming approach. The near-perfect matches at the protein level define coding regions. Similar to EST2Genome, SGP1 model does not require training, but it has limitation that two homologous sequences should have similar genes with similar exon structure. If this condition is not met, a gene escapes detection from one sequence when there is no counterpart in another sequence.

ESTs as Genome Landmarks

Just as a person driving a car may need a map to find a destination, researchers searching for genes also need genome maps to help them to navigate through the billions of nucleotides that make up the human genome. For a map to make navigational sense, it must include reliable landmarks or markers. Currently, the most powerful mapping technique, and one that has been used to generate many genome maps, relies on STS mapping. A Sequence Tagged Site (STS) is a short DNA sequence that is easily recognizable and occurs only once in a genome (or chromosome). The NCBI's electronic PCR (e-PCR) tool, which is part of the UniSTS resource, can be used to find STS markers within a DNA fragment of interest. UniSTS (http://www.ncbi.nih.gov/genome/sts/) contains all the available data on STS markers, including primer sequences, product size, mapping information and alternative names. The 3´ ESTs serve as a common source of STSs due to their likelihood of being unique to a particular species, and provide the additional feature of pointing directly to an expressed gene.

ESTs as Gene Discovery Resources

Because ESTs represent a copy of just the interesting part of a genome—that which is expressed, they have proven themselves again and again as powerful tools in the hunt for genes involved in hereditary diseases. ESTs also have a number of practical advantages in that their sequences can be generated rapidly and inexpensively; only one sequencing experiment is needed per each cDNA generated; and they do not have to be checked for sequencing errors as mistakes do not prevent identification of the gene from which the EST was derived. The identification of ESTs has proceeded rapidly, with approximately 52 million ESTs now available in public databases (e.g. GenBank 5/2008, all species).

To find a disease gene using this approach, scientists first use observable biological clues to identify ESTs that may correspond to disease gene candidates. Scientists then examine the DNA of diseased patients for mutations in one or more of these candidate genes to confirm gene identity. Using this method, scientists have already isolated genes involved in Alzheimer's disease, colon cancer and many other diseases. So it is easy to see why ESTs will pave the way to new horizons in genetic research.

Although it is widely recognized that the generation of ESTs constitute an efficient strategy to identify genes, it is important to acknowledge that despite its advantages, there are several limitations associated with the EST approach. One is that it is very difficult to isolate mRNA from some tissues and cell types. This results in a paucity of data on certain genes that may only be found in these tissues or cell types.

Second is that important gene regulatory sequences may be found within an intron. Because ESTs are small segments of cDNA, generated from the mRNA in which the introns have been removed, much valuable information may be lost by focusing only on cDNA sequencing. Despite these limitations, ESTs continue to be invaluable in characterizing the human genome, as well as the genomes of other organisms. They have enabled the mapping of many genes to chromosomal sites and have also assisted in the discovery of many new genes.

ESTs and NCBI

Due to their utility, speed with which they may be generated, and the low cost associated with this technology, many individual scientists as well as large genome sequencing centres have been generating hundreds of thousands of ESTs for public use. Once an EST was generated, scientists were submitting their tags to GenBank, the NIH sequence database operated by the NCBI. With the rapid submission of so many ESTs, it became difficult to identify a sequence that had already been deposited in the database. It was becoming increasingly apparent to NCBI investigators that if ESTs were to be easily accessed and useful as gene discovery tools, they needed to be organized in a searchable database that also provided access to other genome data. Therefore, in 1992, scientists at the NCBI developed a new database designed to serve as a collection point for ESTs. Once an EST that was submitted to GenBank had been screened and annotated, it was then deposited in the new database, called **dbEST** (http://www.ncbi.nlm.nih.gov/dbEST/index.html).

Submission Format for EST Sequence Database

The following is a specification for flat file formats for delivering EST and related data to the NCBI EST database:

- The format consists of colon delineated capitalized field tags, followed by data.
- The data fields should appear on the same line as the tag, with no line wrapping. Exceptions to this are the TITLE and AUTHORS fields of the Publication file; Description (DESCR) field of the Library file; and the CITATION, COMMENT, and SEQUENCE fields of the EST file. In these fields, the data begins on the line following the field tag and the lines can be wrapped.
- Note that some fields are obligatory.
- The TYPE field is obligatory at the beginning of each entry, even if there are multiple entries of a given type in a file.
- If data is not available for a non-obligatory field, the field can either be omitted entirely, or the tag may be included with an empty data field. Please do not put "*", "-", etc. to indicate missing data.
- Each record (including the last record in the file) should end with a double-bar tag (||) to indicate the end of the record.

Each EST file needs to reference the Publication, Library, and Contact data. Therefore the Publication, Library, and Contact files must be in the database when the EST file is entered. Once these files have been submitted and entered, they do not need to be re-submitted for additional EST files that have the same Publication, Library, or Contact. Following are the Publication and EST files format.

Publication Files

These are the valid tags and a short description used in Publication file:

```
TYPE     :  Entry type - must be "Pub" for publication entries.
            **Obligatory field**
MEDUID   :  Medline unique identifier.
            Not obligatory, include if you know it.
TITLE    :  Title of article. (Begin on line below tag,
            use multiple lines if nec.)
           **Obligatory field**
AUTHORS :  Author name, format: Name,I.I.; Name2,I.I.; Name3,I.I.
            (Begin on line below tag, use multiple lines if nec.)
  **Obligatory field**
JOURNAL :  Journal name
VOLUME   :  Volume number
SUPPL    :  Supplement number
ISSUE    :  Issue number
I_SUPPL :  Issue supplement number
PAGES    :  Page, format: 123-9
YEAR     :  Year of publication.
            **Obligatory field**
STATUS   :  Publication status.
  1=unpublished, 2=submitted, 3=in press, 4=published
  **Obligatory field**
||
```

Examples of Publication file to be submitted to EST database

```
TYPE     :  Pub
MEDUID   :  92347897
```

```
TITLE     : Expressed sequence tags and chromosomal localization
            of cDNA clones from a subtracted retinal pigment
            epithelium library
AUTHORS : Gieser, L.; Swaroop, A.
JOURNAL : Genomics
VOLUME    : 13
ISSUE     : 2
PAGES     : 873-6
YEAR      : 1992
STATUS  : 4
||
```

EST Files

These are the valid tags used in EST files with their short description:

```
TYPE : Entry type - must be "EST" for EST entries.
 **Obligatory field**
STATUS : Status of EST entry - "New" or "Update".
 **Obligatory field**
CONT_NAME : Name of contact
          (must be identical string to the contact entry)
             **Obligatory field**
CITATION : Journal citation.
             (Must be identical string to the publication title)
             Begins on line below tag - use continuation lines
             if necessary.
        **Obligatory field**
LIBRARY : Library name.
  (Must be identical string to library name entry.)
  **Obligatory field**
EST# : EST id assigned by contact lab.
   For EST updates, this is the string we match on.
   **Obligatory field**
GB#: GenBank accession number
GB_SEC: Secondary GenBank accessions
```

GDB#: Genome database accession number

GDB_DSEG: Genome database Dsegment number

CLONE: Clone id.

SOURCE: Source providing clone e.g. ATCC

SOURCE_DNA:Source id number for the clone as pure DNA

SOURCE_INHOST: Source id number for the clone stored in the host.

OTHER_EST: Other ESTs on this clone.

DBNAME: Database name for cross-reference to another database

DBXREF: Database cross-reference accession

PCR_F: Forward PCR primer sequence

PCR_B: Backward PCR primer sequence

INSERT: Insert length (in bases)

ERROR: Estimated error in insert length (bases)

PLATE: Plate number or code

ROW: Row number or letter

COLUMN: Column number or letter

SEQ_PRIMER:Sequencing primer description or sequence.

P_END: Which end sequenced e.g. 5'

HIQUAL_START:Base position of start of highest quality sequence
   (default=1)

HIQUAL_STOP:Base position of last base of highest quality
   sequence.

DNA_TYPE: cDNA (default), Genomic, Viral, Synthetic, Other

PUBLIC: Date of public release.
   Leave blank for immediate release.
   Format: 9/11/1994 (MM/DD/YYYY)
   Obligatory field

PUT_ID: Putative identification of sequence by submitter.

TAG_LIB: Name of library whose tag is found in this sequence.

TAG_TISSUE:Tissue that was source for the tagged library, if
   a library tag was found.

TAG_SEQ: The actual sequence of the library tag found in the EST
   read. If the tag was searched for and not found, put

```
  'Not found' in this field.
POLYA: Y or N to indicate if a polyA tail was or was not found
  in the EST sequence.
COMMENT: Comments about EST.
  Starts on line below COMMENT: tag.
SEQUENCE: Sequence string.
  Starts on line below SEQUENCE: tag.
  **Obligatory field**
||
```

Examples of EST file to be submitted to EST database:

```
TYPE            : EST
STATUS          : New
CONT_NAME       : Kerlavage AR
EST#            : HHC189f
CLONE           : HHC189
SOURCE          : ATCC
SOURCE_INHOST   : 65128
OTHER_EST       : HHC189r
CITATION        : Complementary DNA sequencing: expressed sequence
    tags and human genome project
SEQ_PRIMER      : M13 Forward
P_END           : 5'
HIQUAL_START    : 1
HIQUAL_STOP     : 285
DNA_TYPE        : cDNA
LIBRARY         : Hippocampus, Stratagene (cat. #936205)
PUBLIC          :
PUT_ID          : Actin, gamma, skeletal
COMMENT         : This is a comment about the sequence.
                  It may span several lines.
SEQUENCE:
AATCAGCCTGCAAGCAAAAGATAGGAATATTCACCTACAGTGGGCACCTCCTTAAGAAGCTG
ATAGCTTGTTACACAGTAATTAGATTGAAGATAATGGACACGAAACATATTCCGGGATTAAA
CATTCTTGTCAAGAAAGGGGGAGAGAAGTCTGTTGTGCAAGTTTCAAAGAAAAAGGGTACCA
```

GCAAAAGTGATAATGATTTGAGGATTTCTGTCTCTAATTGGAGGATGATTCTCATGTAAGGT

TGTTAGGAAATGGCAAAGTATTGATGATTGTGTGCTATGTGATTGGTGCTAGATACTTTAAC

TGAGTATACGAGTGAAATACTTGAGACTCGTGTCACTT

||

dbEST—A Descriptive Catalog of ESTs

dbEST is a division of GenBank established in 1992. As for GenBank, data in dbEST is directly submitted by laboratories world-wide and is not curated. Scientists at NCBI created dbEST to organize, store, and provide access to the great mass of public EST data that has already accumulated, and that continues to grow daily. Using dbEST, a scientist can access not only data on human ESTs, but information on ESTs from over 300 other organisms as well. Whenever possible, NCBI scientists annotate the EST record with any known information. For example, if an EST matches a DNA sequence that codes for a known gene with a known function, that gene's name and function is placed on the EST record. Annotating EST records allows public scientists to use dbEST as an avenue for gene discovery. By employing a database search tool, such as NCBI's BLAST, any interested party can conduct sequence similarity searches against dbEST.

dbEST as on 11 Dec 1999: 3,292,593 entries

        Human 1,642,565

        Mouse 860,918

        Rat 128,950

        Nematode 101,232

        Fruit fly 86,121

        Zebra fish 46,428

dbEST as on 17 Oct 2008: 57,425,912 entries

Homo sapiens (human)	8,138,094
Mus musculus + domesticus (mouse)	4,850,602
Danio rerio (zebrafish)	1,379,829
Arabidopsis thaliana (thale cress)	1,526,133
Glycine max (soybean)	1,351,356
Oryza sativa (rice)	1,220,876
Triticum aestivum (wheat)	1,051,300
Rattus norvegicus + sp. (rat)	1,009,934

```
Xenopus laevis (African clawed frog)              677,784

Gallus gallus (chicken)                           599,610

Drosophila melanogaster (fruit fly)               573,749

Canis lupus familiaris (dog)                       365,909

Caenorhabditis elegans (nematode)                 346,109

Saccharum officinarum (sugarcane)                 246,373

Nicotiana tabacum (tobacco)                       240,440

Bombyx mori (domestic silkworm)                   237,962

Solanum tuberosum (potato)                        231,116

Ovis aries (sheep)                                209,808

Sorghum bicolor (sorghum)                         209,814

Vigna unguiculata (cowpea)                        183,658

Macaca mulatta (rhesus monkey)                     58,410

Saccharomyces cerevisiae (baker's yeast)           34,915

Cyprinus carpio (common carp)                      32,046

Poecilia reticulata (guppy)                        16,206
```

NEW GENE PREDICTION METHOD
CAPITALIZES ON MULTIPLE GENOMES

Researchers at Stanford University report a new approach to computationally predicting the locations and structures of protein-coding genes in a genome. Gene finding remains as a important problem in biology as scientists are still far from mapping fully the set of human genes.

Furthermore, gene maps for other vertebrates, including important model organisms such as mouse, are much more incomplete than the human annotation. The new technique, known as CONTRAST (CONditionally TRAined Search for Transcripts), works by comparing a genome of interest to the genomes of several related species.

CONTRAST exploits the fact that the functional role of protein-coding genes play a specific part within a cell and are therefore subjected to characteristic evolutionary pressures. For example, mutations that alter an important part of a protein's structure are likely to be deleterious and thus selected against. On the other hand, mutations that preserve a protein's amino acid sequence are normally well tolerated. Thus, protein-coding genes can be identified by searching a genome for regions that show evidence of such patterns of selection. However, learning to recognize such patterns when more than two species are compared has proved difficult.

Previous systems for gene prediction were able to effectively make use of one additional "informant" genome. For example, when searching for human genes, taking into account information from the mouse genome led to a substantial increase in accuracy. But, no system was able to leverage additional informant genomes to improve upon state-of-the-art performance using mouse alone, although it was expected that adding informants would make patterns of selection clearer.

CONTRAST solves this problem by learning to recognize the signature of protein-coding gene selection in a fundamentally different way from previous approaches. Instead of constructing a model of sequence evolution, CONTRAST directly learns which features of a genomic alignment are most useful for recognizing genes. This approach leads to overall higher levels of accuracy and is able to extract useful information from several informant sequences.

In a test on the human genome, CONTRAST exactly predicted the full structure of 59% of the genes in the test set, compared with the previous best result of 36%. Its exact exon sensitivity of 93%, compared with a previous best of 84%, translates into many thousands of exons correctly predicted by CONTRAST but missed by previous methods. Importantly, CONTRAST's accuracy using a combination of eleven informant genomes was significantly higher than its accuracy using any single informant. The substantial advance in predictive accuracy represented by CONTRAST will further efforts to complete protein-coding gene maps for human and other organisms.

REVIEW QUESTIONS

1. What is open reading frame? Explain the relationship between open reading frame and coding sequence of a gene.
2. How would you explain the concept of six-reading frame?
3. Describe the various approaches for prediction of gene.
4. What are the different promoter prediction programs for prokaryotes?
5. Explain the role of McPromoter web-based program.
6. Describe the post-translational modifications of pre-mRNA and splice junctions.

7. What is GLIMMER system? Explain its role as gene finder for prokaryotes.

8. Describe the version 3.02 of the Glimmer gene-finding software with its detail file and .predict file format.

9. Define GenScan. Explain the components of GenScan model that identify complete gene structures in genomic DNA.

10. What is biological neural network? Explain similarities and differences between biological and artificial neural network.

11. Define Express Sequence Tags. Describe the method for generation of ESTs.

12. What is cDNA? Explain its role in synthesis of ESTs.

13. Describe the web-based programs for searching homology in EST or cDNA.

14. Explain submission format for EST Sequence Databases.

15. What is dbEST? Enlist the latest dbEST entries made for various organisms.

16. Write short notes on
    i. ORF
    ii. Six reading frames
    iii. *Ab initio* gene prediction
    iv. RNA splicing
    v. GLIMMER
    vi. GenScan
    vii. EST and STS
    viii. Biological neural network
    ix. dbEST
    x. cDNA
    xi. EST2Genome
    xii. CONTRAST
    xiii. UTR
    ix. Sequence logos
    xx. BPROM

INTRODUCTION

As mentioned earlier in Chapter 3, proteins are the macromolecules that carry out virtually all of a cell's activities. Every cell has as many as 10,000 different proteins having diverse array of functions including enzymes, as structural and mechanical components, as hormones, growth factors, and gene activators, as membrane receptors and transporters, as contractile elements, as antibodies, as blood-clotting factors and as proteins involved in absorbance and reflection of light. Proteins could perform such versatile functions only because of their unique structure, and their large-scale study is proteomics. The entire complement of proteins, including the modifications made to a particular set of proteins, produced by an organism or system constitutes the proteome. Proteomes vary with time and distinct requirements, or stresses, that a cell or organism undergoes. Proteomics include functional descriptions for the gene products in a cell or organism using the technique of two-dimensional polyacrylamide gel electrophoresis by which thousands of proteins can be distinguished in a single experiment. The analytical techniques make use of antibody affinities or physical and chemical properties of protein sequences. A technique called mass spectrometry analyses a trait of proteins known as the mass-to-charge ratio, which essentially enables the sequence of amino acids comprising the protein to be determined. Techniques exist that detect modifications after protein manufacture, such as the addition of phosphate groups. Analogous to DNA chips, so-called protein microarrays have been developed. In these, a solid support holds various molecules (antibodies and receptors, as two examples) that will specifically bind protein. The binding pattern of proteins to the support can help to determine what proteins are being made and when they are synthesized.

Proteomics can be widely applied to research of diverse microbes. For example, the yeast *Saccharomyces cerevisiae* is being studied to reveal the proteins produced and their functional associations with one another. The task of sorting out all the proteins that can be produced by a bacterium or yeast cell is formidable. Targeting of the research effort is essential. For example, the comparison of the protein profile of a bacterium obtained directly from an infection (*in vivo*) with populations of the same microbe grown under defined conditions in the lab (*in vitro*) could identify proteins that are unique to the infection. Some of these could become targets for diagnosis, therapy, or for prevention of the infection. Proteomics involves the systematic study of proteins in order to provide a comprehensive view of the structure, function and regulation of biological systems. Advances in instrumentation and methodologies have fuelled an expansion of the scope of biological studies from simple biochemical analysis of single protein to measurements of complex protein mixtures. Proteomics is rapidly becoming an essential component of biological research. Coupled with advances in bioinformatics, this approach to comprehensively describing biological systems has a major impact on understanding of the phenotypes of both normal and diseased cells.

Initially, proteomics focused on the generation of protein maps using two-dimensional polyacrylamide gel electrophoresis. The field has since expanded to include not only protein expression profiling, but the analysis of post-translational modifications and protein–protein interactions. Protein expression, or the quantitative measurement of the global levels of proteins, may still be done with two-dimensional gels, however, mass spectrometry has been incorporated to increase sensitivity, specificity and to provide results in a high-throughput format. There are varieties of platforms available to conduct protein expression studies. The study of protein–protein interactions has been revolutionized by the development of protein microarrays. Analogous to DNA microarrays, these biochips are printed with antibodies or proteins and probed with a complex protein mixture. The intensity or identity of the resulting protein–protein interactions may be detected by fluorescence imaging or mass spectrometry. Other protein capture methods may be used in place of arrays, including the yeast two-hybrid system or the isolation of proteins/protein complexes by affinity chromatography or other separation techniques.

Experimental methods used in protein study are as follows:

- 2D electrophoresis—separating expressed proteins
- Mass spectroscopy—identifying expressed proteins

- ▣ Protein arrays/microarrays—identifying expressed proteins. Analogous to DNA microarrays, protein biochips are printed with antibodies or proteins and probed with a complex protein mixture.

- ▣ X-ray crystallography

- ▣ NMR spectroscopy

- ▣ Edman degradation

Proteomics also includes:

- ▣ Quantization of gel images
- ▣ Determination of protein identity
- ▣ Data storage and sharing
- ▣ Integration with genomic data

METHODS OF STUDYING PROTEINS

Determining the Post-translationally Modified Proteins

One way in which a particular protein can be studied is to develop an antibody which is specific to that modification. For example, there are antibodies which recognize only certain proteins when they are tyrosine-phosphorylated; also, there are antibodies specific to other modifications. These can be used to determine the set of proteins that have undergone the modification of interest such as glycosylation of proteins, certain lectins have been discovered which bind sugars.

A more common way to determine post-translational modification of interest is to subject a complex mixture of proteins to electrophoresis in "two dimensions", which simply means that the proteins are electropheresed first in one direction, and then in another, this allows small differences in a protein to be visualized by separating a modified protein from its unmodified form.

Determining the Existence of Proteins in Complex Mixtures

Classically, antibodies to particular proteins or to their modified forms have been used in biochemistry and cell biology studies. These are among the most common tools used by practising biologists today. For more quantitative determination of protein amounts, techniques such as ELISAs (Enzyme-Linked ImmunoSorbant Assays) can be used. For proteomic study, techniques such as Matrix-Assisted Laser Desorption/Ionization (MALDI) have been employed for rapid determination of proteins in particular mixtures.

Establishing Protein–Protein Interactions

Most proteins function in collaboration with other proteins, and one goal of proteomics is to identify which proteins interact. This is especially useful in determining potential partners in cell signalling cascades. Several methods are available to probe protein–protein interactions. The traditional method is yeast two-hybrid analysis. New methods include protein microarrays, immunoaffinity chromatography followed by mass spectrometry, and combinations of experimental methods such as phage display and computational methods.

PROTEOMICS DATABASES

Protein databases are at the heart of proteomics. These databases are usually freely available to everyone and are maintained by NCBI. The Molecular Information Agent and DBGET are the graphical portals that show interconnected protein and gene databases.

VARIETIES OF PROTEIN DATABASES

General information about proteins—the most famous ones are Swiss-Prot, PIR, and NCBI. Amino acid sequences in the format (FASTA) were used with mass spectrometry measurements to identify proteins and constitute protein sequence databases. These are combined in UniProt—the world's most comprehensive catalogue of information about proteins. A search on the Bioinformatic Harvester returns all that is known about a protein and its related gene in fifteen different databases.

- **Proteomics databases**—data collected in proteomics experiments such as the protein identifications in PeptideAtlas (Figure 9.1), the open proteomics database, and the global proteome machine. It also includes databases of experimental 2D gels.

- **Three-dimensional structures of proteins**—the most well-known is PDB.

- **Protein–protein interactions**—which proteins interact and with whom.

- Gene Ontology—a database of terms that classify protein functions, processes and subcellular locations. GoBrowser, FatiGo and Gene Finder are websites that link proteins to their Gene Ontology functions.

- Databases that relate proteins and genes to diseases—the most well-known is OMIM.
- Pubmed—the database of abstracts of all journal articles relevant to biology, medicine, and proteomics.

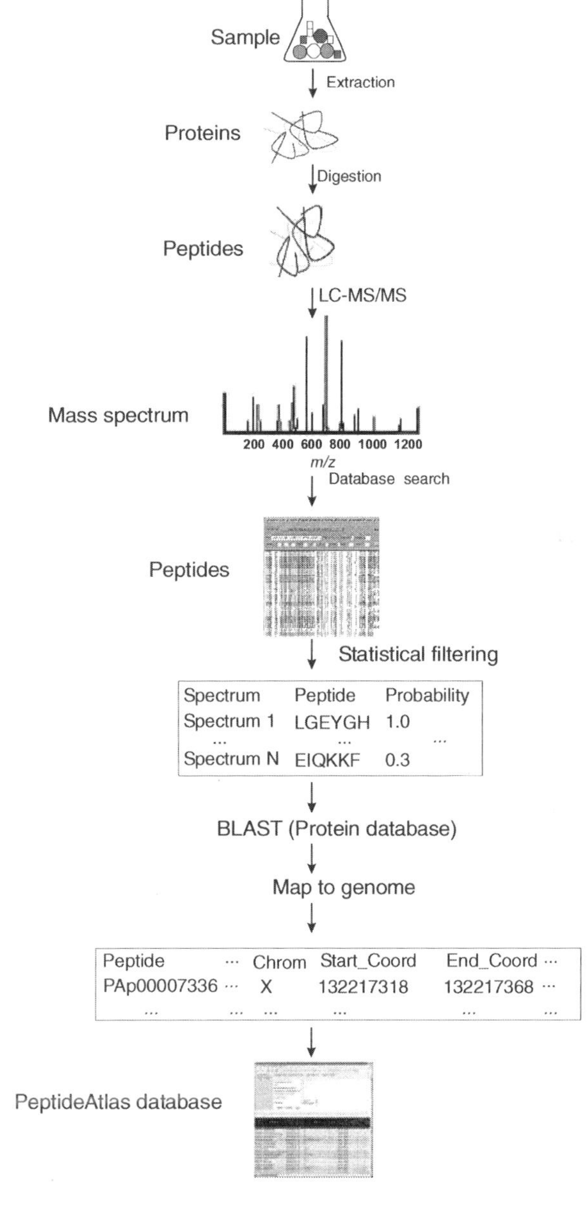

Figure 9.1   Steps involved in construction of PeptideAtlas database

PROTEIN SEQUENCE DATABASES (PSD)

Determination of protein/peptide sequences is a basic requirement for biomedical research, including cancer research. It is absolutely essential for characterizing and identifying proteins or peptides. Imagine a biologist, who has discovered an unknown peptide, perhaps theoretically translated from a nucleotide sequence, or isolated from a gel, which biologist have had sequenced and wants to try and find out as much as he can about it. The first step in this process is to look for similarities with already discovered peptide sequences/proteins. This is accomplished by comparing the novel sequence with those contained in protein databases, the most important of these being UniProt (http://www.uniprot.org/).

The UniProt Knowledgebase (accessible as UniProt on the EBI SRS server http://srs.ebi.ac.uk/) is a central database of protein sequence and function created by joining the information contained in:

- UniProtKB/Swiss-Prot (http://www.ebi.ac.uk/2can/databases/protein2.html),
- UniProtKB/TrEMBL (http://www.ebi.ac.uk/2can/databases/protein3.html),
- PIR-PSD (http://pir.georgetown.edu/).

Until recently, the EBI/SIB UniProtKB/Swiss-Prot + UniProtKB/TrEMBL databases and the PIR-PSD co-existed as protein databases with differing protein sequence coverage and annotation priorities. In 2002, the EBI, SIB, and PIR (at the Georgetown University Medical Center and National Biomedical Research Foundation) joined forces as the UniProt consortium.

Consisting of richly annotated entries, the UniProtKB is the centrepiece of the consortium activities. Initially, the Knowledgebase derived from the merge of UniProtKB/Swiss-Prot, UniProtKB/TrEMBL and PIR-PSD protein sequences with annotations of sequence and functional information. Future Knowledgebase entries will be derived from the UniProt archive sequences we see as essential for the UniProt Knowledgebase. For example, sequences for which novel functional, structural, and biochemical data has been published have high annotation priority. The UniProt Knowledgebase consists of two parts, a section containing fully manually annotated records resulting from information extracted from literature- and curator-evaluated computational analyses, and a section with computationally analysed records awaiting full manual annotation. For the sake of continuity and name recognition, the two sections are referred to as "UniProtKB/Swiss-Prot" and "UniProtKB/TrEMBL" respectively.

UniProtKB/Swiss-Prot Format

UniProtKB/Swiss-Prot is an annotated protein sequence database. The UniProtKB/Swiss-Prot protein Knowledgebase consists of sequence entries. Sequence entries are composed of different line types, each with their own format. For standardization purposes, the format of UniProtKB/Swiss-Prot follows as closely as possible that of the EMBL nucleotide sequence database. The UniProtKB/Swiss-Prot user manual is available at http://ca.expasy.org/sprot/userman.html (Figure 9.2). The entries in the UniProtKB/Swiss-Prot database are structured so as to be usable by human readers as well as by computer programs. The explanations, descriptions, classifications and other comments are in ordinary English. Wherever possible, symbols familiar to biochemists, protein chemists and molecular biologists are used. Each sequence entry is composed of lines. Different types of lines, each with their own format, are used to record the various data that make up the entry.

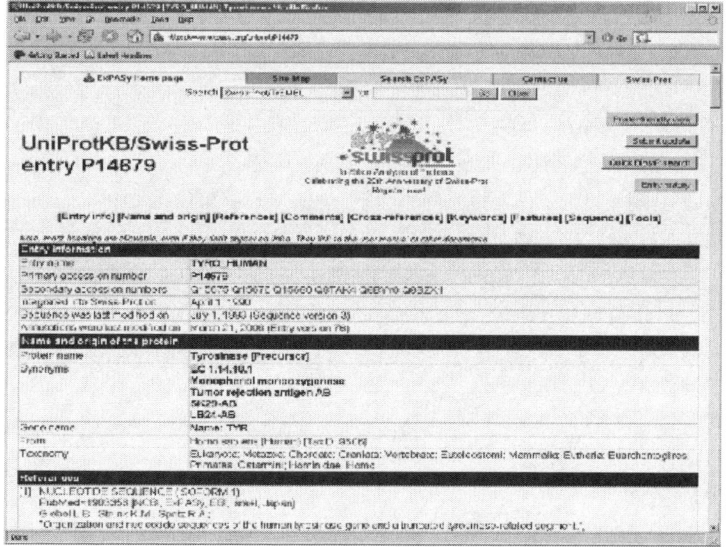

Figure 9.2 ExPASy Home Page with UniProtKB/Swiss-Prot user manual

◨ The ID (IDentification) line is always the first line of an entry. The general form of the ID line is:

Term	ID	ENTRY_NAME	STATUS	SEQUENCE_LENGTH.
e.g.	ID	FOSB_MOUSE	Reviewed	338 AA

◙ **Entry name**   The first item on the ID line is the entry name of the sequence. This name is a useful means of identifying a sequence. The entry name consists of up to ten uppercase alphanumeric characters.

◙ **Status**   To distinguish the fully annotated entries in the Swiss-Prot section of the UniProt Knowledgebase from the computer-annotated entries in the TrEMBL section, the "status" of each entry is indicated in the first line (ID) of the entry. The two defined classes of entries are:

- ◈ Reviewed entries that have been manually reviewed and annotated by UniProtKB curators (Swiss-Prot section of the UniProt Knowledgebase).

- ◈ Unreviewed computer-annotated entries that have not been reviewed by UniProtKB curators (TrEMBL section of the UniProt Knowledgebase).

◙ **Length of the molecule**—The sequence length in amino acids.

- ◈ The AC (**AC**cession number) line lists the accession number(s) associated with an entry.

- ◈ The DT (**DaTe**) lines show the date of creation and last modification of the database entry.

- ◈ The DE (**DE**scription) lines contain general descriptive information about the sequence stored.

- ◈ The GN (**G**ene **N**ame) line contains the name(s) of the gene(s) that code for the stored protein sequence.

- ◈ The OS (**O**rganism **S**pecies) line specifies the organism(s) which was (were) the source of the stored sequence.

- ◈ The OG (**O**r**G**anelle) line indicates if the gene coding for a protein originates from the mitochondria, the chloroplast, a cyanelle, or a plasmid.

- ◈ The PR (**PR**oject) line shows the International Nucleotide Sequence Database Collaboration (INSDC) project identifier that has been assigned to the entry.

- ◈ The OC (**O**rganism **C**lassification) lines contain the taxonomic classification of the source organism.

❖ The OX (Organism taxonomy **Cross**-Reference) line is used to indicate the identifier to a specific organism in a taxonomic database.

❖ The RN (Reference Number) line gives a sequential number to each reference citation in an entry.

❖ The RP (Reference Position) line describes the extent of the work carried out by the authors of the reference cited.

❖ The RC (Reference Comment) lines are optional lines which are used to store comments relevant to the reference cited.

❖ The RX (Reference **Cross**-Reference) line is an optional line which is used to indicate the identifier assigned to a specific reference in a bibliographic database.

❖ The RA (Reference Author) lines list the authors of the paper (or other work) cited.

❖ The RT (Reference Title) lines give the title of the paper (or other work) cited.

❖ The RL (Reference Location) lines contain the conventional citation information for the reference.

❖ The CC (Comment) lines are free text comments on the entry, and are used to convey any useful information.

❖ The DR (Database cross-Reference) lines are used as pointers to information related to UniProtKB/Swiss-Prot entries and found in other data collections.

❖ The KW (KeyWord) lines provide information that can be used to generate indexes of the sequence entries based on functional, structural, or other categories.

❖ The FT (Feature Table) lines provide a precise but simple means for the annotation of the sequence data. The table describes regions or sites of interest in the sequence. In general, the feature table lists post-translational modifications, binding sites, enzyme active sites, local secondary structure or other characteristics are reported in the cited references.

❖ The SQ (SeQuence header) line marks the beginning of the sequence data and gives a quick summary of its content.

❖ The sequence data line has a line code consisting of two blanks rather than the two-letter codes used until now. The sequence counts 60 amino acids per line, in groups of 10 amino acids, beginning in position 6 of the line.

◈ The // (terminator) line contains no data or comments and designates the end of an entry.

The entries in the UniProtKB/Swiss-Prot database are structured as given below.

```
ID    FOSB_MOUSE Reviewed; 338 AA.

AC    P13346;

DT    01-JAN-1990, integrated into UniProtKB/Swiss-Prot.

DT    01-JAN-1990, sequence version 1.

DT    20-FEB-2007, entry version 54.

DE    Protein fosB.

GN    Name=Fosb;

OS    Mus musculus (Mouse).

OC    Eukaryota; Metazoa; Chordata; Craniata; Vertebrata;
      Euteleostomi;

OC    Mammalia; Eutheria; Euarchontoglires; Glires; Rodentia;
      Sciurognathi;

OC    Muroidea; Muridae; Murinae; Mus.

OX    NCBI_TaxID=10090;

RN    [1]

RP    NUCLEOTIDE SEQUENCE [MRNA].

RX    MEDLINE=89251612; PubMed=2498083;

RA    Zerial M., Toschi L., Ryseck R.-P., Schuermann M., Mueller
      R.,

RA    Bravo R.;

RT    "The product of a novel growth factor activated gene, fos
      B, interacts

RT    with JUN proteins enhancing their DNA binding activity.";

RL    EMBO J. 8:805-813(1989).

RN    [2]

RP    NUCLEOTIDE SEQUENCE [GENOMIC DNA].

RX    MEDLINE=92158623; PubMed=1741260; DOI=10.1093/nar/20.2.343;

RA    Lazo P.S., Dorfman K., Noguchi T., Mattei M.-G., Bravo R.;

RT    "Structure and mapping of the fosB gene. FosB downregulates
      the
```

```
RT   activity of the fosB promoter.";
RL   Nucleic Acids Res. 20:343-350(1992).
CC   -!- FUNCTION: FosB interacts with Jun proteins enhancing
CC       their DNA binding activity.
CC   -!- SUBUNIT: Heterodimer (By similarity).
CC   -!- SUBCELLULAR LOCATION: Nucleus.
CC   -!- INDUCTION: By growth factors.
CC   -!- SIMILARITY: Belongs to the bZIP family. Fos subfamily.
CC   -!- SIMILARITY: Contains 1 bZIP domain.
DR   EMBL; X14897; CAA33026.1; -; mRNA.
DR   EMBL; AF093624; AAD13196.1; -; Genomic_DNA.
DR   PIR; S35477; TVMSFB.
DR   UniGene; Mm.248335; -.
DR   HSSP; P01100; 1FOS.
DR   SMR; P13346; 157-215.
DR   DIP; DIP:1067N; -.
DR   TRANSFAC; T00291; -.
DR   Ensembl; ENSMUSG00000003545; Mus musculus.
DR   KEGG; mmu:14282; -.
DR   MGI; MGI:95575; Fosb.
DR   ArrayExpress; P13346; -.
DR   GermOnline; ENSMUSG00000003545; Mus musculus.
DR   InterPro; IPR011700; bZIP_2.
DR   InterPro; IPR008917; Euk_TF_DNA_bd.
DR   InterPro; IPR000837; Leuzip_Fos.
DR   InterPro; IPR004827; TF_bZIP.
DR   Pfam; PF07716; bZIP_2; 1.
DR   PRINTS; PR00042; LEUZIPPRFOS.
DR   SMART; SM00338; BRLZ; 1.
DR   PROSITE; PS50217; BZIP; 1.
DR   PROSITE; PS00036; BZIP_BASIC; 1.
KW   DNA-binding; Nuclear protein.
FT   CHAIN 1 338 Protein fosB.
```

```
FT    /FTId=PRO_0000076477.

FT    DOMAIN 183 211 Leucine-zipper.

FT    DNA_BIND 161 179 Basic motif.

SQ    SEQUENCE 338 AA; 35977 MW; E9D031A4BEAE48EC CRC64;

MFQAFPGDYD SGSRCSSSPS AESQYLSSVD SFGSPPTAAA SQECAGLGEM PGSFVPTVTA
ITTSQDLQWL VQPTLISSMA QSQGQPLASQ PPAVDPYDMP GTSYSTPGLS AYSTGGASGS
GGPSTSTTTS GPVSARPARA RPRRPREETL TPEEEEKRRV RRERNKLAAA KCRNRRRELT
DRLQAETDQL EEEKAELESE IAELQKEKER LEFVLVAHKP GCKIPYEEGP GPGPLAEVRD
LPGSTSAKED GFGWLLPPPP PPPLPFQSSR DAPPNLTASL FTHSEVQVLG DPFPVVSPSY
TSSFVLTCPE VSAFAGAQRT SGSEQPSDPL NSPSLLAL
//
```

Swiss-Prot is a manually curated biological database of protein sequences. Swiss-Prot was created in 1986 by Amos Bairoch during his Ph.D. and developed by the Swiss Institute of Bioinformatics and the European Bioinformatics Institute. Swiss-Prot strives to provide reliable protein sequences associated with a high level of annotation (such as the description of the function of a protein, its domain structure, post-translational modifications, variants, etc.), a minimal level of redundancy and high level of integration with other databases.

In 2002, the UniProt consortium was created. It is the collaboration between the Swiss Institute of Bioinformatics, the European Bioinformatics Institute and the Protein Information Resource, funded by the National Institutes of Health. Swiss-Prot and its automatically curated supplement TrEMBL, have joined with the Protein Information Resource protein database to produce the UniProt Knowledgebase, the world's most comprehensive catalogue of information on proteins. As of 3rd April 2007, UniProtKB/Swiss-Prot release 52.2 contains 263,525 entries and the UniProtKB/TrEMBL release 35.2 contains 4,232,122 entries.

The UniProt consortium produced 3 database components, each optimized for different uses:

The UniProt Knowledgebase (UniProtKB (Swiss-Prot + TrEMBL)).

The UniProt non-redundant Reference (UniRef) databases, which combine closely related sequences into a single record to speed similarity searches.

The UniProt Archive (UniParc), which is a comprehensive repository of protein sequences, reflecting the history of all protein sequences.

Swiss-Prot and structure of Swiss-Prot entries Swiss-Prot, as a curated-protein sequence database, offers a wide range of annotations, covering areas such as function, domain parsing, post-translational modifications, and variants. Swiss-Prot can be accessed at http:// www.expasy.ch/sprot/sprot-top.html (home server, Switzerland), http:// expasy.nhri.org.tw/sprot/sprot-top.html (Taiwan), and http:// expasy.proteome.org.au/sprot/sprot-top.html (Australia). Swiss-Prot is integrated into many other databases, such as SRS.

Human vitronectin is used as an example in the following text for searching protein sequence databases. To locate the Swiss-Prot entry for this protein, one can search either the entry name (VTNC_HUMAN) or the accession number (P04004) obtained from a BLAST search. Alternatively, one can use the full-text search at the Swiss-Prot webpage to search by protein name (human vitronectin) or keywords (e.g. serum spreading, as vitronectin is also called serum spreading factor). A combination of several entries can be used in a search.

The entry name in Swiss-Prot has the general format X_Y, where X is a mnemonic code of up to four characters indicating the protein name (in this case, VTNC), and Y is a mnemonic species identification code of up to five characters for the biological source of the protein. Some codes used for Y are full English names, e.g. HORSE, HUMAN, MAIZE, MOUSE, PIG, RAT, SHEEP, and WHEAT. Some are abbreviations, including BOVIN (bovine), CHICK (chicken), ECOLI (*Escherichia coli*), PEA (garden pea, *Pisum sativum*), RABIT (rabbit), SOYBN (soybean, *Glycine max*), TOBAC (common tobacco, *Nicotina tabacum*), and YEAST (baker's yeast, *Saccharomyces cerevisiae*). An entry name may have several accession numbers if they have been merged. An accession number is always conserved from release to release, and therefore allows unambiguous citation.

Each entry contains the following items shown in table format in the NiceProt View layout: 1) general information about the entry, 2) name and origin of the protein, 3) references providing the protein sequence, 4) comments (e.g. annotated functions, subunits, similar proteins), 5) cross-references (links to other databases), 6) keywords, 7) features, and 8) sequence information. The text in the comments entry provides a function annotation for the protein (e.g. "vitronectin is a cell adhesion and spreading factor found in serum and tissues. Vitronectins interact with glycosaminoglycans and proteoglycans."). Cross-references list the annotations of the protein by other databases, such as GeneCards and ProDom. GeneCards, a database of

human genes, shows chromosomal location and the involvement of the
protein in certain diseases (if applicable). ProDom contains protein domain
families by automated sequence comparisons. Clicking the link to ProDom
from Swiss-Prot leads to a nice graphic view for domain parsing for
vitronectin. Various research results are given under features. Some of the
features items for VTNC_HUMAN are as follows:

```
SIGNAL         1 19

CHAIN          20 398 V65 SUBUNIT.

CHAIN          399 478 V10 SUBUNIT.

PEPTIDE        20 63 SOMATOMEDIN B.

DOMAIN         150 287 HEMOPEXIN-LIKE 1.

DOMAIN         288 478 HEMOPEXIN-LIKE 2.

SITE           64 66 CELL ATTACHMENT SITE.

SITE           398 399 CLEAVAGE.

MOD_RES        75 75 SULFATATION.

CARBOHYD       86 86

DISULFID       293 430

BINDING        362 395 HEPARIN.

CONFLICT       50 50 C -> N (IN REF. 5).
```

where,

> SIGNAL   represents the extent of a signal sequence (pre-peptide),
>
> MOD_RES   indicates a post-translationally modified residue
> (sulphatation, in this case),
>
> CARBOHYD   shows the glycosylation site,
>
> DISULFID   means that a disulphide bond exists between the two
> indicated residues (293 and 430), and
>
> CONFLICT   shows that different papers report differing sequences.

Swiss-Prot + TrEMBL

▣   Swiss-Prot is a protein sequence database with high level of annotations,
a minimal level of redundancy and high level of integration with other
databases (124464 entries).

▣ Translated **EMBL** (TrEMBL) is a computer-annotated supplement of Swiss-Prot that contains all sequence entries not yet integrated in Swiss-Prot (828210 entries). (http://us.expasy.org/sprot/)

TrEMBL is a very large protein database in Swiss-Prot format generated by computer translation of the genetic information from the EMBL nucleotide sequence database. Computer translation is not entirely perfect, so proteins predicted by the TrEMBL database can be hypothetical, and many TrEMBL entries are poorly annotated. In contrast to Swiss-Prot which contains only proteins actually found in the wild, and PIR which is entirely unchecked. TrEMBL is currently being combined with the above two databases in the UniProt project.

UniProtKB/TrEMBL

UniProtKB/TrEMBL (http://www.ebi.ac.uk/trembl/), consists of computer-annotated entries derived from the translation of all coding sequences (CDS) in the EMBL (http://www.ebi.ac.uk/embl/index.html) nucleotide sequence database, except for those already included in UniProtKB/Swiss-Prot (http://www.ebi.ac.uk/uniprot/index.html).

UniProtKB/TrEMBL is split in two main sections, SP-TrEMBL and REM-TrEMBL. SP-TrEMBL contains the entries, which should be eventually incorporated into UniProtKB/Swiss-Prot. REM-TrEMBL (REMaining TrEMBL) contains the entries that will not get included in UniProtKB/Swiss-Prot. REM-TrEMBL contains sequences that are not destined to be included in Swiss-Prot. They include synthetic sequences, truncated, pseudogenes, patented sequences, small fragments containing less than eight amino acids, immunoglobulins, T-cell receptors, and translation of codon that do not encode real proteins. TrEMBL was designed to address the need for a well-structured Swiss-Prot-like resource.

PIR-PS Database

The final release for the PIR-PSD made through the Release 80.00 on 31st December 2004 that was the world's first database of classified and functionally annotated protein sequences that grew out of the Atlas of Protein Sequence and Structure (1965–1978) edited by Margaret Dayhoff. This database was produced and distributed by the Protein Information Resource in collaboration with MIPS (Martinsried Institute for Protein Sequence) and JIPID (Japan International Protein Information Database), PIR-PSD has been the most comprehensive and expertly curated protein sequence database in the public domain for over 20 years. In 2002, PIR joined EBI (European Bioinformatics Institute) and SIB (Swiss Institute of Bioinformatics) to form the UniProt consortium. PIR-PSD sequences and annotations have been integrated into UniProt Knowledgebase. Bi-directional cross-references between UniProt (UniProt Knowledgebase and/or UniParc) and PIR-PSD (http://pir.georgetown.edu/) are established to allow easy tracking of former PIR-PSD entries. PIR-PSD unique sequences, reference citations, and experimentally verified data can now be found in the relevant UniProt records. There are 283308 comprehensive and annotated protein sequence database entries in the public domain.

Protein information resource   Protein Information Resource (PIR) was established in 1984 by the National Biomedical Research Foundation (NBRF) located at Georgetown University Medical Center (GUMC-http://www-nbrf.georgetown.edu/pirwww/search/textpsd.html) as a resource to assist researchers in the identification and interpretation of protein sequence information. Prior to that, the NBRF compiled the first comprehensive collection of macromolecular sequences in the Atlas of Protein Sequence and Structure, published from 1965–1978 under the editorship of Margaret Dayhoff. Dayhoff *et al.* pioneered in the development of computer methods for the comparison of protein sequences, for the detection of distantly related sequences and duplications within sequences, and for the inference of evolutionary histories from alignments of protein sequences.

Since 1988, the Protein Sequence Database has been maintained collaboratively by PIR-International, an association of macromolecular sequence data collection centres—the consortium included the PIR, the International Protein Information Database of Japan (JIPID), and the Martinsried Institute for Protein Sequence (MIPS: http://www.mips.gsf.de). For four decades, PIR has provided many protein databases and analysis tools freely accessible to the scientific community, including the Protein Sequence Database (PSD), the first international database, which grew out of Atlas of Protein Sequence and Structure. PIR is an integrated public bioinformatics resource to support genomic and proteomic research, and scientific studies.

The PIR is an effective combination of a carefully curated database, information retrieval access software, and a workbench for investigations of sequences. The PIR also produces the Integrated Environment for Sequence Analysis (IESA). Think of this as an analysis package sitting on top of a retrieval system. Its functionality includes browsing, searching and similarity analysis, and links to other databases. Users may:

- ▣ browse by annotations
- ▣ search selected text fields for different annotations, such as superfamily, family, title, species, taxonomy group, keywords and domains
- ▣ analyse sequences using BLAST or FASTA searches, pattern match, multiple alignment
- ▣ use global and domain search, and annotation-sorted search
- ▣ view statistics for superfamily, family, title, species, taxonomy group, keywords, domains, features
- ▣ view links to other databases, including PDB, COG, KEGG, WIT, and BRENDA
- ▣ select specialized sequence groups such as human, mouse, yeast and *E. coli* genomes

In 2002, PIR along with its international partners, EBI (European Bioinformatics Institute) and SIB (Swiss Institute of Bioinformatics), were awarded a grant from NIH to create UniProt, a single worldwide database of protein sequence and function, by unifying the PIR-PSD, Swiss-Prot, and TrEMBL databases.

PROTEIN FAMILY DATABASES

Proteins are significant biomolecules present in the cell or an organism performing vital functions. Protein can be categorized on the basis of their evolutionary, structural, or functional relationships. A protein family is much more informative than the single protein itself. For example, residues conserved across the family often indicate special functional roles. Two proteins classified in the same functional family may suggest that they share similar structures, even when their sequences do not have significant similarity. There is no unique way to classify proteins into families. Boundaries between different families may be subjective.

Overall, proteins are classified into three categories based upon sequence similarity, structure, or function-based methods as follows:

- i. *Sequence-based methods* are applicable to any proteins whose sequences are known.
- ii. *Structure-based methods* are limited to the proteins of known structures.
- iii. *Function-based methods* depend on the functions of proteins being annotated.

Sequence- and structure-based classifications can be automated and are scalable to high-throughput data, whereas function-based classification is typically carried out manually. Structure- and function-based methods are more reliable, while sequence-based methods may result in a false positive result when sequence similarity is weak (i.e., two proteins are classified into one family by chance rather than by any biological significance). In addition, since protein structure and function are better conserved than sequence, two proteins having similar structures or similar functions may not be identified through sequence-based methods.

Databases for Sequence-based Protein Families

Sequence-based protein families are classified according to a profile derived from a multiple sequence alignment. The profile can be shown across a long domain (typically 100 residues or more) or can be revealed in short sequence motifs. Classification methods based on profiles across long domains tend to be more reliable but less sensitive than those based on short sequence motifs. Several sequence-based methods focus more on profiles across long domains, including Pfam, ProDom, and Clusters of Orthologous Group (COG). These methods differ in the techniques used to construct families. Pfam builds multiple sequence alignments of many common protein domains using Hidden Markov models. The ProDom protein domain database consists of homologous domains based on recursive PSI-BLAST searches. COG aims towards finding ancient conserved domains by delineating families of orthologs across a wide phylogenetic range. An example of Pfam for the GRIP domain whose accession number is PF01465 with some useful information for the Pfam entry is as follows:

The GRIP (golgin-97, RanBP2alpha, Imh1p and p230/golgin-245) domain is found in many large coiled-coil proteins. It has been shown to be sufficient for targeting the Golgi. The GRIP domain contains a completely conserved tyrosine residue. In addition, Pfam gives the alignment between the family members. They are:

```
O15045/1511-1558 SAANLEYLKNVLLQFIFLKPG—SERERLLPVINTMLQLSPEEKGKLAAV
O15045

YNP9_CAEEL/633-681          NEKNMEYLKNVFVQFLKPESVP-
AERDQLVIVLQRVLHLSPKEVEILKAA P34562

Q06704/864-909  KNEKIAYIKNVLLGFLEHKE—QRNQLLPVISMLLQLDSTDEKRLVMS
Q06704
```

```
Q92805/691-737 REINFEYLKHVVLKFMSCRES--EAFHLIKAVSVLLNFSQEEENMLKET
Q92805

O42657/703-748  MLIDKEYTRNILFQFLEQRD--RRPEIVNLLSILLDLSEEQKQKLLSV
O42657

O70365/1161-1205 EPTEFEYLRKVMFEYMMGR---ETKTMAKVITTVLKFPDDQAQKILER
O70365

Q21071/692-741D PAEAEYLRNVLYRYMTNRESLGKESVTLARVIGTVARFDESQMKNVISS
Q21071

Q18013/574-623 STSEIDYLRNIFTQFLHSMGSPNAASKAILKAMGSVLKVPMAEMKIIDKK
Q18013
```

The alignment of Pfam shows accession numbers and the range of each sequence in the family. One can identify some features of the family through this pattern (i.e., from particularly conserved residues at specific alignment positions). Some methods are based on "fingerprints" of small conserved motifs in sequences, as with PROSITE, PRINTS, and BLOCKS. In protein sequence families, some regions have been better conserved than others during evolution. These regions are generally important for the function of a protein or for the maintenance of its three-dimensional structure, and hence are suitable for fingerprinting. The fingerprints can be used to assign a newly sequenced protein to a specific family. Fingerprints are derived from gapped alignments in PROSITE and PRINTS, but are derived from ungapped alignments (corresponding to the highly conserved regions in proteins) in BLOCKS. A fingerprint in PRINTS may contain several motifs from PROSITE, and thus may be more flexible and powerful than a single PROSITE motif. Therefore, PRINTS can provide a useful adjunct to PROSITE. It should be noted that some functionally unrelated proteins may be classified together due to chance matches in short motifs. Other sequence-based protein family databases consist of multiple sources. The ProClass database is a non-redundant protein database organized according to family relationships as defined collectively by PROSITE patterns and PIR superfamilies.

The MEGACLASS server provides classifications by different methods, including Pfam, BLOCKS, PRINTS, and ProDom. The MOTIF search engine at http://www.motif.genome.ad.jp/ includes PROSITE, BLOCKS, ProDom, and PRINTS.

Databases for Structure-based Protein Families

The hierarchical relationship among proteins can be clearly revealed in structures through structure–structure comparison. Structure families often provide more information on the relationship between proteins than what sequence families can offer, particularly when two proteins share a similar structure but no significant sequence identity. Sometimes, sequence similarity between two proteins exists but is not strong enough to produce an unambiguous alignment. The alignment between two structures of given proteins can generate better alignment in terms of biological significance, and thus may pinpoint the active sites more accurately. Different structure–structure comparison methods yield different structure families.

Databases for Function-based Protein Families

There are various protein functional families classified from different perspectives. The ENZYME databank contains the following data for each enzyme: EC number, recommended name, alternative names, catalytic activity, cofactors, pointers to the Swiss-Prot entry, and pointers to any disease associated with a deficiency of the enzyme. PROCAT is a database of three-dimensional enzyme active site templates. Protein Disease Database (PDD) correlates diseases with proteins observable in serum, urine, and other common human body fluids based on biomedical literature. There are also a growing number of databases dedicated to special types of proteins, such as antibodies, G-protein-coupled receptors, HIV proteases, glycoproteins, and RNases.

Other Databases Related to Proteins

Protein Binding Databases include protein–substrate docking and protein–protein association. Re-LiBase is a database system for analysing receptor–ligand complexes in the PDB. Database of Interacting Proteins (DIP) records protein pairs that are known to bind with each other in addition to providing information related to signalling pathways, multiple interactions, and complex systems.

Protein Energetics Databases can be found in ProTherm (thermodynamic database for proteins and mutants). It contains thermodynamic data on mutations, including Gibbs free energy, enthalpy, heat capacity, and transition temperature. These data are important for understanding the structure and stability of proteins.

Bibliographic Databases search for protein information through traditional bibliographic databases, such as MedLINE or Grateful Med, can be rewarding. In addition, some bibliographic reference databases dedicated to proteins may provide certain information more directly. For example, SeqAnalRef stores papers dealing with sequence analysis.

Combined Databases, by integrating different types of protein databases together, a database of databases (or a data warehouse) can be built. Such combined databases not only serve as "one-stop shopping," but also provide cross-references between entries in different databases. Two combined databases, Entrez and SRS, have been very successful.

PROTEIN DATA BANK (PDB)

Protein Data Bank (http://home.rcsb.org/pdb) is the only central repository for 3D structural data of proteins and nucleic acids. These data, typically obtained by X-ray crystallography or NMR spectroscopy and submitted by biologists and biochemists from around the world, are released into the public domain, and can be accessed for free. PDB was founded in 1971 by Edgar Meyer and Walter Hamilton Brookhaven National Laboratory (BNL). PDB was transferred to members of the Research Collaboratory for Structural Bioinformatics (RCSB; http://home.rcsb.org/). In addition to providing structural information of biomolecules, the PDB website gives a number of services for structural submission and data searching and retrieval. The three-dimensional structures of proteins not only define their biological functions, but also hold a key in rational drug design. Traditionally, protein structures were solved at a low-throughput mode. However, recent advances in new technologies, such as synchrotron radiation sources and high-resolution nuclear magnetic resonance (NMR), accelerate the rate of protein structure determination substantially. There is an overwhelming consensus in the structural biology community that protein structures can be solved en masse (an effort called structural genomics), in a similar fashion as for determining DNA sequences, and that impact of this approach can be compared with that of the Human Genome Project. When the PDB was originally founded it contained just 7 protein structures. Since then it has undergone an approximate exponential growth in the number of structures.

Most of the structures in the PDB were determined experimentally by X-ray crystallography ($>82\%$) and NMR ($>16\%$). A small number of structures ($>2\%$) were derived from theoretical models, which may provide approximate structures but may not be accurate. The PDB also contains some structures of chemical ligands and nucleotides. Each PDB entry is represented by a four-character identifier (PDB ID), where the first character is always a number from 0 to 9 (e.g. 1cau, 256b). The PDB can be accessed through the home server (Figure 9.3; http://www.rcsb.org/pdb/ in the USA) or through one of its mirror sites from around the world, such as:

http://molmod.angis.org.au/pdb/ (Australia)

http://www.pdb.ufmg.br (Brazil)

http://www.ipc.pku.edu.cn/npdb/ (China)

http://pdb.weizmann.ac.il/ (Israel)

http://pdb.ccdc.cam.ac.uk/ (United Kingdom).

The PDB offers three search methods: search by 1) PDB ID, 2) SearchLite, and 3) SearchFields. SearchLite is a simple keyword search that uses, for instance, protein name or author's name. A search using SearchFields, as an advanced search engine, allows a user to specify features of the protein, such as Enzyme Commission (EC) number, name of binding ligand, range of protein size, range of resolution in the X-ray structure, and secondary structure content. The PDB is a database consisting of a set of ASCII files, each containing the Cartesian atomic coordinates describing the three-dimensional structure of a protein, nucleic acid or other bio-macromolecule. The principal method to access and search the holdings of the PDB is via its web interface, using your favourite Internet browser. Direct your browser either to the main RCSB PDB location at URL: http://www.rcsb.org or the local mirror copy, which uses the old Brookhaven interface and browser, at URL: http://pdb.weizmann.ac.il

In either place, the next step is to select "search the PDB" to bring up the browser. The Brookhaven interface supports two search tools, the full PDB Browser, and the simplified PDB-lite browser for casual users and lab biologists. Whichever browser is in use, one has to fill in the blanks with appropriate search questions, based on the header lines in the PDB files. Options exist to serch by PDB ID-code (the unique 4-character ID the PDB assigns to each entry), or by sequence in a FASTA-like interface (which of course returns only sequences with structures in the PDB).

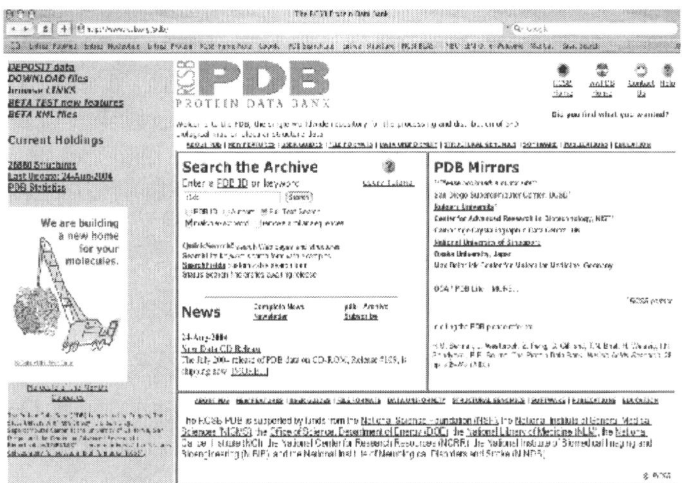

Figure 9.3 PDB Home Page with URL

Worldwide Protein Data Bank (wwPDB)

wwPDB (http://www.wwpdb.org/index.html/) consists of organizations that act as deposition, data processing and distribution centers for PDB data. The founding members are RCSB PDB (USA), the Macromolecular Structure Database at the EMBL's European Bioinformatics Institute (MSD-EBI) (http://www.ebi.ac.uk/msd) and PDBj (Japan; http://www.pdbj.org/). The BMRB (USA) group joined the wwPDB in 2006. The PDB is a canonical research resource that transcends both scientific and political boundaries.

The mission of the wwPDB is to maintain a single Protein Data Bank archive of macromolecular structural data that is freely and publicly available to the global community. The PDB is a key resource in structural biology and is critical to more recent work in structural genomics. Countless derived databases and projects have been developed to integrate and classify the PDB in terms of protein structure, protein function and protein evolution. The collaboration in wwPDB reflects the growing international and interdisciplinary nature of scientific research, and formalizes the global character of the PDB, which has been used as an international resource for the collection and sharing of three-dimensional information on proteins and other large molecules since its inception 32 years ago. The formation of the wwPDB will be transparent to users and will ensure the overall quality and consistency of data directly available through the PDB. The wwPDB members

RCSB PDB, MSD-EBI, and PDBJ are working together to make the data uniform across the archive. Some believe this to be desirable; others argue that, without a universal repository of information (i.e., a common dictionary), it is not possible to draw comparisons.

Each structure published in PDB receives a four-character alphanumeric identifier, its PDB ID. This should not be used as an identifier for biomolecules, since often several structures for the same molecule (in different environments or conformations) are contained in PDB with different PDB IDs. If a biologist submits structure data for a protein or nucleic acid, wwPDB staff reviews and annotates the entry. The data are then automatically checked for plausibility. The source code for this validation software has been released for free. The main database accepts only experimentally derived structures, and not theoretically predicted ones.

Entries in PDB

On 22nd April 2003, PDB has 20747 structures. As of 24 June 2008, the database contained 51,491 released atomic coordinate entries (or "structures"), 47,526 of that proteins, the rest being nucleic acids, nucleic acid–protein complexes, and a few other molecules (Table 9.1). About 5,000 new structures are released each year. Data are stored in the mmCIF format specifically developed for the purpose.

Note that the database stores information about the exact location of all atoms in a large biomolecule (although, usually without the hydrogen atoms, as their positions are more of a statistical estimate); if one is only interested in sequence data, i.e., the list of amino acids making up a particular protein or the list of nucleotides making up a particular nucleic acid, the much larger databases from Swiss-Prot and the International Nucleotide Sequence Database Collaboration should be used.

PDB holdings list at RCSB (http://www.rcsb.org/pdb/statistics/holdings) as on September, 2005 was as follows:

Proteins, peptides	29,876
Nucleic acids (NA)	1,500
Protein/NA complexes	1,338
Carbohydrates	13
Total	32,727

Table 9.1   PDB entries for different biomolecules as on 24 June 2008.

Technique used	Proteins	Nucleic acids (NA)	Protein/NA complexes	Others	Total
X-ray diffraction	40912	1046	1873	24	43855
NMR	6401	09	138	7	7355
Electron microscopy	124	11	47	0	182
Others	89	4	4	2	99
Total	47526	1870	2062	33	51491

It is estimated that the size of the PDB archive will triple to 150,000 structures by the year 2014.

PDB File Format

The PDB stores structural information in two formats: the PDB file format and the macromolecular crystallographic information file (mmCIF) format in addition to the Molecular Modeling DataBase (MMDB) format (http://www.ncbi.nih.gov/Structure/MMDB/mmdb.shtml).

The PDB file format is the dominant format used in the protein community. It contains three parts: annotations, coordinates, and connectivities. The connectivity part, which shows chemical connectivities between atoms, is optional. It is listed at the end of the PDB file, beginning the line with the keyword CONECT. The coordinate part uses each line for a three-dimensional coordinate of an atom, starting from ATOM (for standard amino acids) parts that refer to protein atom information or HETATM (for heteroatom groups) part that refer to atoms of cofactor or substrate molecules. The annotation part of the PDB file format contains dozens of possible record types, including: HEADER (name of protein and release date), COMPND (molecular contents of the entry), SOURCE (biological source), AUTHOR (list of contributors), SSBOND (disulphide bonds), SLTBRG (salt bridges), SITE (groups comprising important sites), HET (nonstandard groups or residues heterogens, MODRES (modifications to standard residues), SEQRES (primary sequence of backbone residues),

HELIX (helical substructures), SHEET (sheet substructures), and REMARK (other information and comments). The PDB file format is fairly easy to read and simple to use.

The following diagrams (Figure 9.4a to j) show different entries in the PDB file format

```
HEADER   OXIDOREDUCTASE                03 OCT 02  1MXT
TITLE    ATOMIC RESOLUTION STRUCTURE OF CHOLESTEROL OXIDASE
TITLE   2 (STREPTOMYCES SP. SA-COO)
COMPND   MOL_ID: 1;
COMPND  2 MOLECULE: CHOLESTEROL OXIDASE;
COMPND  3 CHAIN: A;
COMPND  4 SYNONYM: CHOD;
COMPND  5 EC: 1.1.3.6;
COMPND  6 ENGINEERED: YES;
COMPND  7 OTHER_DETAILS. FAD COFACTOR NON-COVALENTLY BOUND TO THE
COMPND  8 ENZYME
```

Figure 9.4a   Portion of PDB file of bacterial enzyme cholesterol oxidase showing header section

Each line shows the atom serial number, atom type, residue type, chain identifier (in case of multichain structure), residue serial number, orthogonal coordinates (three values), occupancy, temperature factor, and segment identifier. Header section provides information about protein and structure date of the entry, references, crystallographic data, contents and positions of secondary structure elements, altogether about the quality of structure of a protein.

```
SOURCE   MOL_ID: 1;
SOURCE  2 ORGANISM_SCIENTIFIC: STREPTOMYCES SP.;
SOURCE  3 ORGANISM_COMMON: BACTERIA;
SOURCE  4 GENE: CHOA;
SOURCE  5 EXPRESSION_SYSTEM: ESCHERICHIA COLI;
SOURCE  6 EXPRESSION_SYSTEM_COMMON: BACTERIA;
SOURCE  7 EXPRESSION_SYSTEM_STRAIN: BL21(DE3)PLYSS;
SOURCE  8 EXPRESSION_SYSTEM_VECTOR_TYPE: PLASMID;
SOURCE  9 EXPRESSION_SYSTEM_PLASMID: PCO202
```

Figure 9.4b   Portion of PDB file of bacterial enzyme cholesterol oxidase showing source

```
AUTHOR    A.VRIELINK,P.I.LARIO
REVDAT   1   25-FEB-03 1MXT    0
JRNL        AUTH   P.I.LARIO,N.SAMPSON,A.VRIELINK
JRNL        TITL   SUB-ATOMIC RESOLUTION CRYSTAL STRUCTURE OF
JRNL        TITL 2 CHOLESTEROL OXIDASE: WHAT ATOMIC RESOLUTION
JRNL        TITL 3 CRYSTALLOGRAPHY REVEALS ABOUT ENZYME MECHANISM AND
JRNL        TITL 4 THE ROLE OF FAD COFACTOR IN REDOX ACTIVITY
JRNL        REF    J.MOL.BIOL.              V. 326 1635 2003
JRNL        REFN   ASTM JMOBAK  UK ISSN 0022-2836
```

Figure 9.4c   Portion of PDB file of bacterial enzyme cholesterol oxidase showing author, review date, journal name with title of the entry, its volume, pages and date of publishing

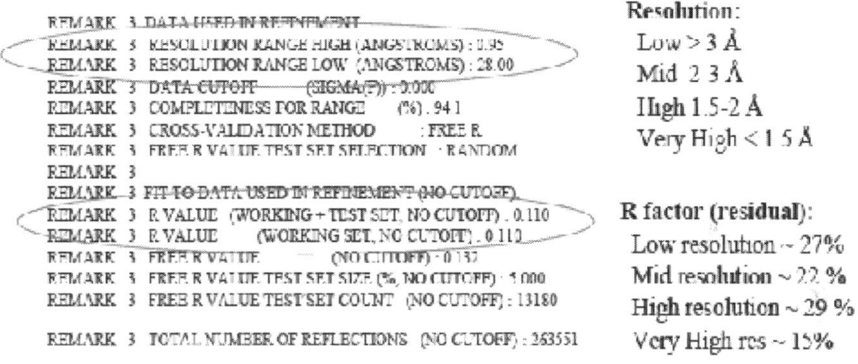

Figure 9.4d   Portion of PDB file of bacterial enzyme cholesterol oxidase showing remarks

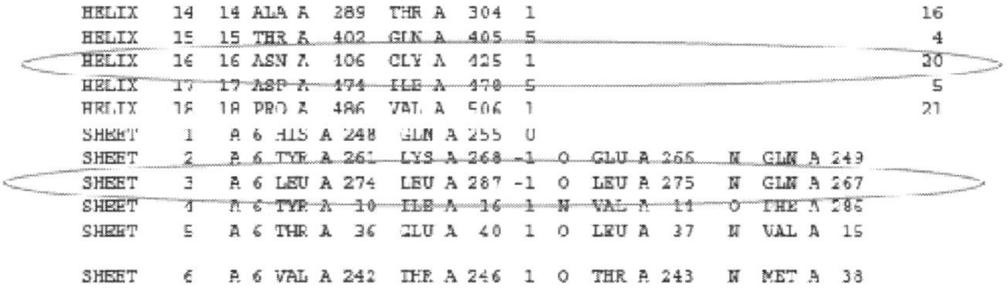

Figure 9.4e   Portion of PDB file of bacterial enzyme cholesterol oxidase showing helix and sheet structures present in the concerned protein

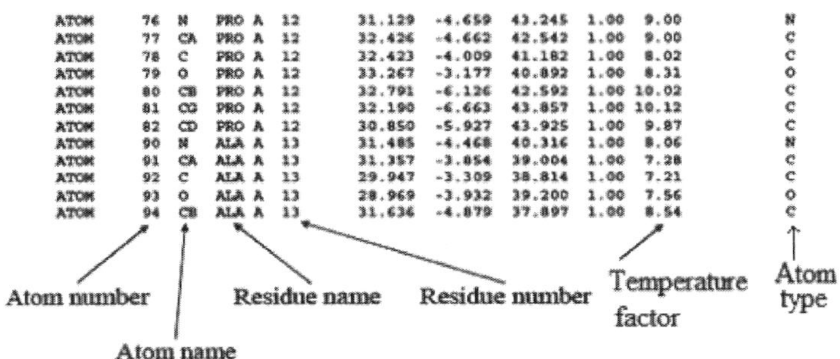

Figure 9.4f Portion of PDB file of bacterial enzyme cholesterol oxidase showing information about the atoms in the structure

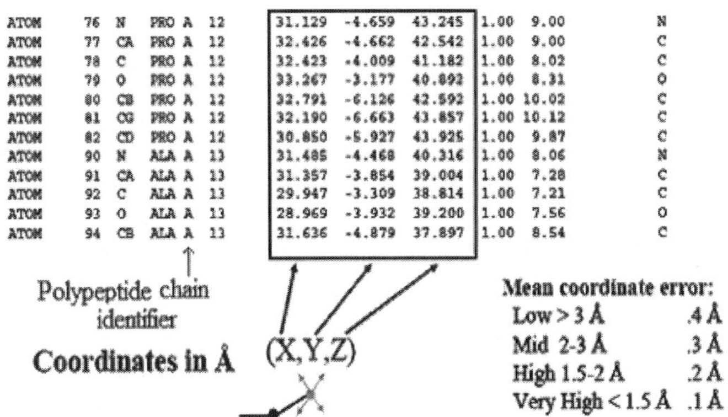

Figure 9.4g Portion of PDB file of bacterial enzyme cholesterol oxidase showing information about the X, Y, and Z coordinates representing the spatial position of each atom of the structure

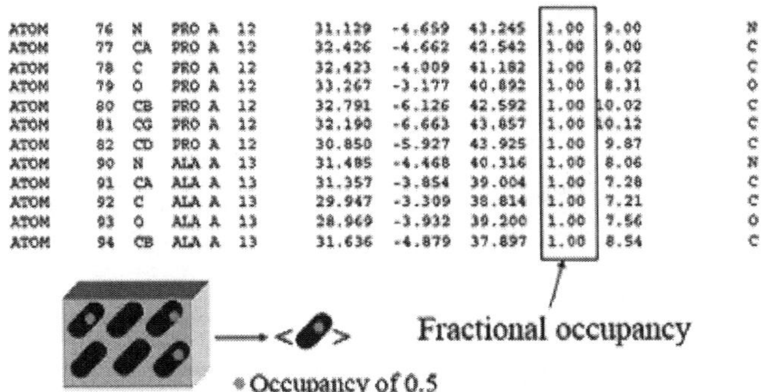

Figure 9.4h Portion of PDB file of bacterial enzyme cholesterol oxidase showing information about the fractional occupancy

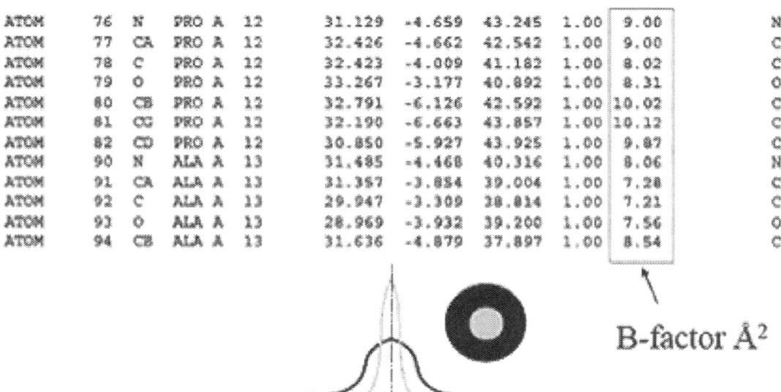

Figure 9.4i   Portion of PDB file of bacterial enzyme cholesterol oxidase showing information about the B-factor

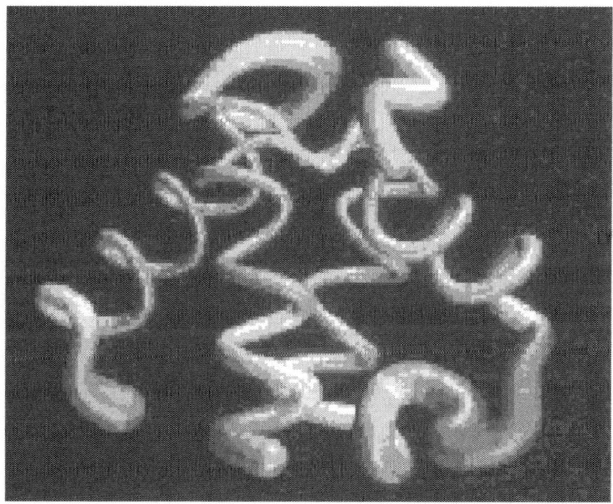

Figure 9.4j   Bacterial enzyme cholesterol oxidase visualized in PDB viewer

It is easy to retrieve a structure if its identifier is known. From the RCSB Home Page, entering a PDB ID and selecting Explore gives a 1-page summary of the entry. Figure 9.5 shows summary page for another protein, myosin isolated from *Dictyostelium discoideum,* identifier 1MND.

The PDB allows a user to view a molecule structure interactively through a Virtual Reality Modeling Language (VRML) viewer, RasMol, Chime, or QuickPDB (a Java applet for viewing sequence and structure) when the browser is configured to support these free rendering tools. The PDB provides related information about the protein, such as secondary structure assignment and geometry.

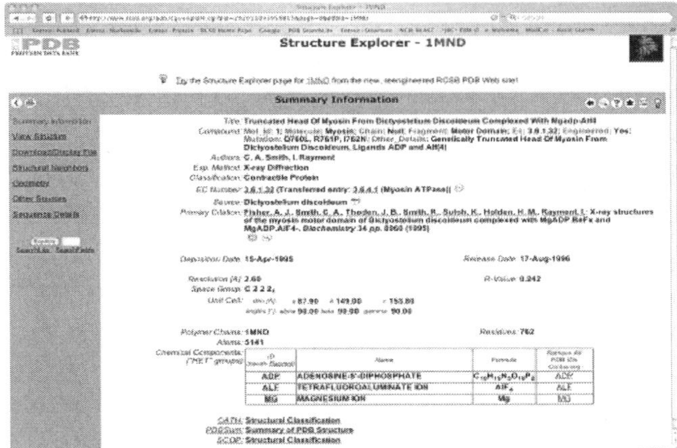

Figure 9.5   Summary Page for the PDB entry 1MND, myosin from *Dictyostelium discoideum*

MACROMOLECULAR CRYSTALLOGRAPHIC INFORMATION FILE (mmCIF) FORMAT

The mmCIF format for a protein is similar to the format for relational database in which a set of tables are used to record databases . Each table or field of information is explicitly assigned by a tag and linked to other fields through a special syntax. The example of mmCIF file format is shown in Figure 9.6, where a single line of description in the header section of PDB is divided into many lines or fields with each having clear and obvious assignment of item names and item values.

```
PDB          HEADER PLANT SEED PROTEIN 11-OCT-91 1CBN
```

```
mmCIF    _struct.entry_id '1CBN'
         _struct.title 'PLNAT SEED PROTEIN'
         _struct_keywords.entry_id '1CBM'
         _struct_keywords.text 'plant seed protein'
         _database_2.database_id 'PDB'
         _database_2.database_code '1CBN'
         _database_PDB_rev.rev_num 1
         -database_PDB_rev.date_original '1991-10-11
```

Figure 9.6.   PDB and mmCIF file formats in two boxes showing the differences in header section. *(Courtesy:* Jin Ziong, *Essential Bioinformatics;* 2006.)

Each line in the mmCIF file starts with an underscore character followed by category name and keyword description separated by a period. The annotation in the same figure indicates that the data items belong to the category of "struct" or "database". Following the keyword tag, a short text string enclosed by quotation marks is used to assign values for the keyword. Thus, there is one-to-one relationship between item names and item values that are entered in mmCIF file format that provides much more flexibility for information storage and retrieval.

MMDB FILE FORMAT

Molecular Modeling DataBase (MMDB) (http://www.ncbi.nih.gov/structure/mmdb/mdb.shtml) is the project to treat experimentally determined macromolecular structures. MMDB format is developed by NCBI to compute and sort pieces of information in PDB (Figure 9.7).

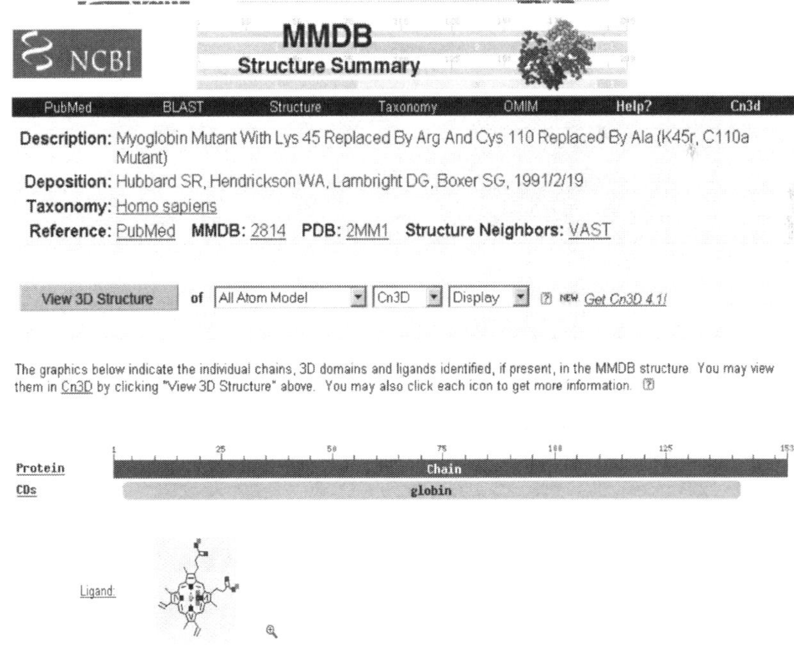

Figure 9.7   Three-dimensional structure of protein viewed through MMDB

The goal of the project is to allow the information to be more easily integrated with GenBank and MedLine through Entrez. The MMDB file uses ASN.1 (and an XML conversion of this format) and it has the information in a record structured as a nested hierarchy. In addition, to the faster retrieval rate than mmCIF and PDB, the MMDB format also contains information about the bond connectivity for each molecule, called a "chemical graph" that is recorded in the ASN.1 file and it helps to draw the 3D structure of a protein molecule easily.

The Home Page of the PDB contains links to data files themselves, to expository and tutorial material including short news items and PDB newsletter that facilitates deposition of new entries and specialized search software for retrieving structure. Each PDB entry also links to a wide range of annotations from secondary databases, including (1) summary and display databases such as Graphical Representation and Analysis of Structure Server (GRASS), Image Library, MMDB in Entrez, PDBsum, and Sequence to and within Graphics (STING); (2) domain partition information from 3Dee; (3) the MEDLINE bibliography; (4) structure quality assessment in PDB REPORT from WHAT IF; (5) protein movements recorded in Database of Macromolecular Movement (MolMovDB); and (6) structure families (CATH, SCOP, DALI, and FSSP). Swiss-Prot, NCBI, EMBL and others provide gateways to access PDB file, while the databases in CATH, SCOP, DALI, FSSP and others interpret PDB files (Figure 9.8).

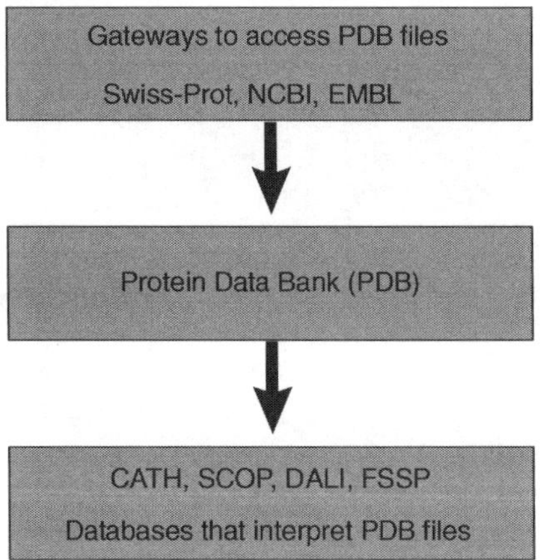

Figure 9.8   Data-processing centres and interpreting tools for PDB file

Several structure databases that are not linked by the PDB can also provide useful information. WPDB can be used to visualize and analyse a PDB entry from Microsoft Windows. **BioMagResBank** (University of Wisconsin, 1999) is a repository for NMR spectroscopy data on proteins, peptides, and nucleic acids. Particularly, it provides partial NMR data (e.g. chemical shifts) before the full structure can be solved.

Visualization of Protein Structure

Once a protein structure has been solved, the next task is to visualize the 3D structure of a protein based on the Cartesian coordinates (X, Y, and Z coordinates of atoms as indicated in Figure 9.4g for a given protein). In the absence of visualization software, molecular structures were represented by physical models of metal wires, rods and spheres. But, nowadays there is tremendous growth and development of computer hardware and software technology that has lead to sophisticated computer graphic programs that help to visualize and manipulate 3D structures of biomolecules.

The following graphics (Figure 9.9) of the small protein (hen egg-white lysozyme) illustrate the different display options available for viewing molecular structures in Protein Explorer. The cartoons, ribbons, and strands display options are useful for viewing protein secondary structure (α-helices and β-pleated sheets).

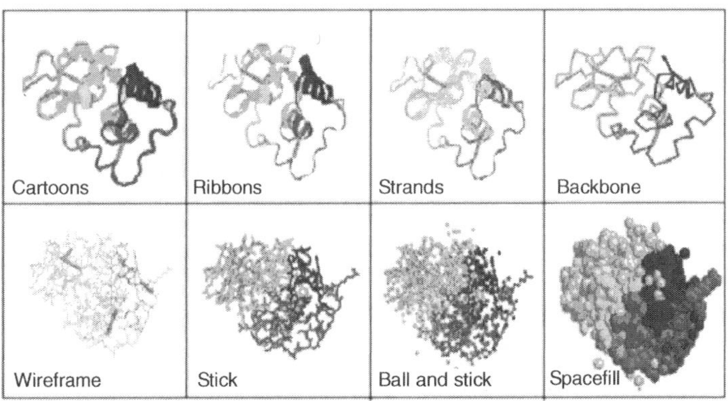

Figure 9.9 Different displays for visualization of protein structures

Software Tools for 3D Molecular Graphic Visualization

Structures and domains are classified according to the type and arrangement of secondary structural elements. Different forms of graphical representation are: RasMol, Swiss-PDBViewer, Ribbons, Grasp, Chime, PyMOL, Sirius, Garlic (http://garlic.mefos.hr/garlic/) STING, JmolViewer (Java-based interactive molecular viewer), QuteMol (high-quality interactive molecular viewer), and StarBiochem (Java-based interactive molecular viewer with integrated search of protein databank).

Today, computer graphics hardware and software are sufficiently fast to allow interactive representation of even the largest biomolecules. Excellent commercial and free software is available for visualizing molecular structures. Some of the most popular packages include:

RasMol   The name **RasMol** (http://www.umass.edu/microbio/rasmol/) comes from **Ras**ter display of **Mol**ecules. A flexible scripting language allows choice of representation styles and colouration and selective display of portions of a molecule. RasMol is a compact, self-contained program for the display of molecular structures. In Figure 9.10, lysozyme is displayed in RasMol. Bond diagram (Figure 9.10a), shows every covalent bond linking atoms in the structure. Space-filling diagram (Figure 9.10b) represents each atom as a sphere. Ribbon diagram (Figure 9.10c) is a schematic representation of the topology of the protein chain. Note the advantages and disadvantages of each representation. Bond diagrams allow exploration of the detailed geometry of the molecule but are often too complicated for easy comprehension. Space-filling diagrams give a good feeling for the shape and size of the protein. Ribbon diagrams are excellent for understanding biomolecular folding and topology.

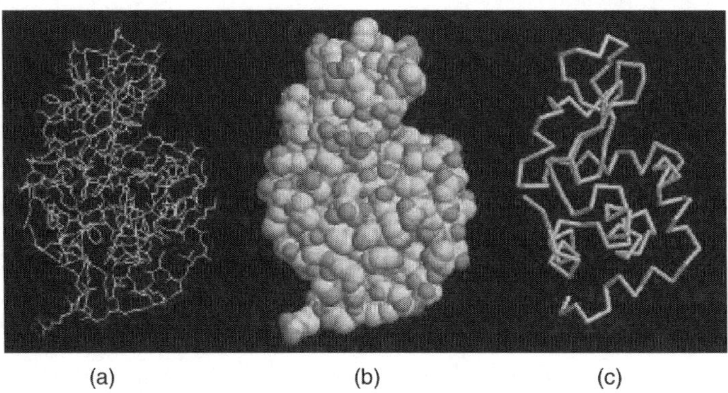

Figure 9.10   (a) Bond diagram (b) Space-filling diagram (c) Ribbon diagram of lysozyme, as drawn in the viewing program RasMol

RasMol is a free software for visualizing molecular structures that can be run on Windows-based, Macintosh, or Unix PCs. With it one can view three-dimensional structures of many types of molecules, including proteins and nucleic acids. Two versions of the RasMol manual are available: 1) a frames version, which is easy to navigate and use online, and 2) a single-page version, which you can download and use when you are not connected to the internet. If the computer has a Windows or Mac OS environment and enough memory, it is convenient to run RasMol and Netscape at the same time and thus have the manual available when needed. RasMol reads in molecular coordinate files in a number of formats and interactively displays the molecule on the screen in a variety of colour schemes and representations. Currently supported input file formats include Brookhaven Protein Databank, Tripos' Alchemy and Sybyl Mol2 formats, Molecular Design Limited (MDL) Mol file format, Minnesota Supercomputer Center's (MSC) XMol XYZ format, CHARMm format, MOPAC format, CIF format and mmCIF format files. If bond connectivity information and/or secondary structure information about the molecule to be displayed is not contained in the file this is calculated automatically by the supportive visualization program. The loaded molecule may be shown as wireframe, cylinder, stick bonds, alpha-carbon trace, space-filling spheres, macromolecular ribbons (either smooth shaded, solid ribbons or parallel strands), hydrogen bonding and dot surface. Atoms may also be labelled with arbitrary text strings (Figure 9.11).

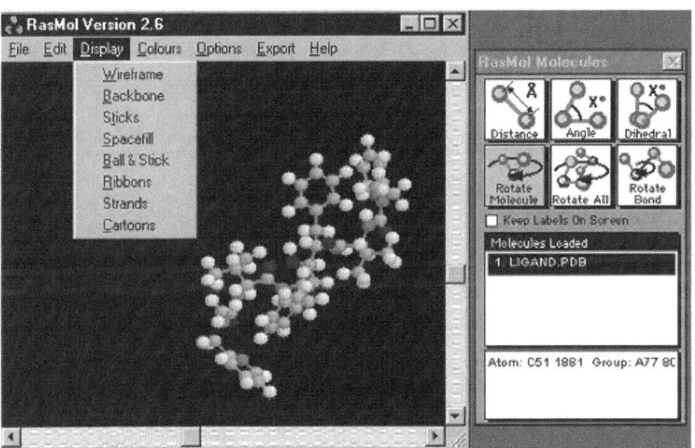

Figure 9.11   RasMol version 2.6 displaying various options

Alternate conformers and multiple NMR models may be specially coloured and identified in atom labels. Different parts of the molecule may be displayed and coloured independently of the rest of the molecule or shown in different representations simultaneously. The space-filling spheres can even be shadowed. The displayed molecule may be rotated, translated, zoomed; z-clipped (slabbed) interactively using the mouse, the scroll bars, the command line or an attached dials box. RasMol can read a prepared list of commands from a script file (or via interprocess communication) to allow a given image or viewpoint to be restored quickly. RasMol can also create a script file that contains the commands required to regenerate the current image. Finally the rendered image may be written out in a variety of formats including both raster and vector PostScript, GIF, PPM, BMP, PICT, Sun rasterfile or as a MolScript, input script or Kinemage.

Versions of RasMol have run on a wide range of architectures and systems including SGI, sun4, sun3, sun386i, SGI, DEC, HP and E and S workstations, IBM RS/6000, Cray, Sequent, DEC Alpha (OSF/1, OpenVMS and Windows NT), IBM PC (under Microsoft Windows, Windows NT, OS/2, Linux, BSD386 and BSD), Apple Macintosh (System 7.0 or later), PowerMac and VAX VMS (under DEC Windows). UNIX and VMS versions require an 8 bit, 24 bit or 32 bit X Windows frame buffer (X11R4 or later). The X Windows version of RasMol provides optional support for a hardware dials box and accelerated shared memory rendering (via the XInput and MIT-SHM extensions) if available.

Basic operations and steps in RasMol to visualize the molecule

- RasMol converts X, Y, and Z coordinates of a molecule into 3D graphics in several display modes.
- Molecule can be moved and rotated to view it in different angle.
- Individual bond can be measured and rotated.
- Input file can be in different format (the most popular are MDL Mol format: file.mol, and Brookhaven databank format: file.pdb).
- Output graphics in several standard formats (such as BMP and GIF).
- Open a XYZ coordinate file: File=>Open=>List Files of Type=>MDL Mol File Format=>(Select File)
- Move and rotate the molecule: Use the mouse.
- Change the display mode: Display=>(Make selection).
- Export image: Export=>(Select the desired format such as GIF)
- Measure and rotate chemical bond:

    Step 1: Select item on the side window.
    Step 2: Click on the relevant atoms.
    Step 3: To rotate a bond, hold down the left button of the mouse and move.

Advantages of RasMol

- Many instructive RasMol files, tutorials, views, animations, and other resources for RasMol users already available on the web.
- Available files include small molecules and VSEPR examples.

- There are many ways to render models for active exploration such as wireframe, backbone, spacefill, ribbons, strands, cartoons, dotted surface.
- Command line window allows logical selection and rendering of parts of molecule.
- User can write "scripts" to reproduce complex images as a canned views for teaching.
- Advanced script writers can prepare animations.
- Available in a web browser plug-in form, called Chime. If you can use RasMol, you can use Chime.

Disadvantages of Rasmol

- Awkward centering and zooming.
- User must learn command-line, UNIX-like, Boolean language for advanced use.
- Complex process for making scripts portable—scripts must be opened and edited in a word processor, then file type and creator must be restored, requiring a utility program like FileTyper (http://www.ugcs.caltech.edu/~dazuma/filetyper/index.html).
- Finds hydrogen bonds in main chain only.
- Displays only one molecule at a time.
- Structural comparison is impossible.
- No links to other molecular biology tools.
- No built-in viewing of the PDB structure file.

Swiss-PdbViewer   Swiss-PdbViewer (SPdbV; www.expasy.ch/spdbv/) is a structure visualization, analysis and modelling software tool (Figure 9.12). The excellent and comprehensive Swiss Institute of Bioinformatics website contains extensive information on protein structure. From it one can obtain protein sequences and three-dimensional structures of proteins, as well as the versatile Swiss-PdbViewer software, which has several advanced capabilities not found in RasMol.

Advantages of SPdbV

- It is faster software tool for visualization of 3D structure of a protein.
- More graceful handling of basic viewing features; for example, the "help" key automatically resizes and recentres display, eliminating much fumbling.
- Program under active development, with new features appearing frequently.

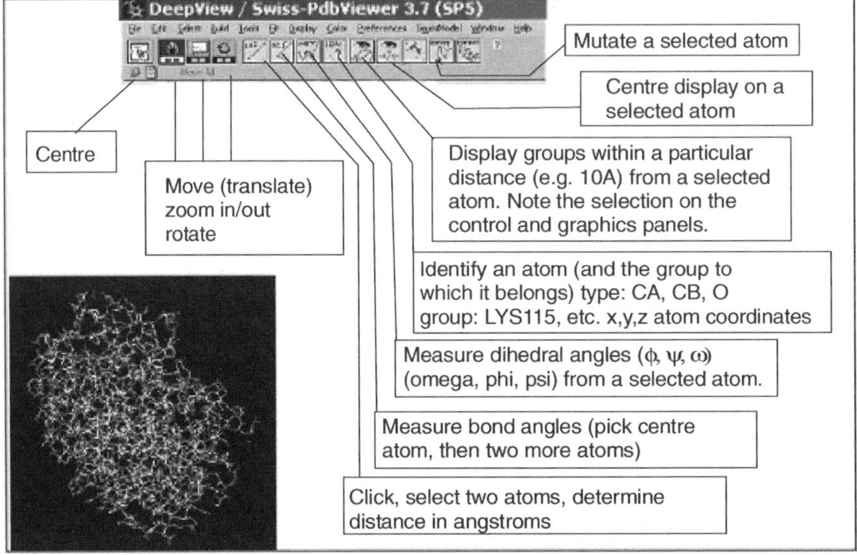

Figure 9.12   Swiss-PdbViewer 3.7 software tool for visualization of 3D structure of a protein

▣ Few bugs and active bug correction.

▣ Developer responsive to e-mail for help and suggestions.

▣ More intuitive visual means for selecting and displaying parts of molecules.

▣ Entirely menu-driven, no command line language to learn.

▣ Control Panel Window shows selectable list of PDB file contents at all times.

▣ Full contents of PDB file available at a click.

▣ Finds all hydrogen bonds, including those to non-protein groups, with user selectable criteria.

▣ Display, superimpose, and compare up to 10 molecules at a time.

▣ Full homology modelling from raw sequence, build unknown structure onto homologous proteins as templates, and submit model automatically to Swiss-Model for energy minimization.

▣ Automatic link to Swiss-Prot for structural and sequence comparisons and for finding homology templates.

▣ View electron-density maps for teaching and crystallographic refinement.

▣ Allows building full biological form of oligomeric proteins from coordinates of unique subunits.

▣ Allows display of symmetry-related molecules, and building of full contents of crystallographic unit cell.

▣ Allows loading of multiple NMR models in a single PDB file. You can decide how many models to load.

▣ Direct links to ExPASy molecular biology server for structural and sequence comparisons.

▣ Render beautiful and exportable QuickDraw 3D images or ray-traced scenes for publication or display in other applications.

▣ Many file modifications possible, including combining files and renaming (and subdividing) chains.

▣ Files can be saved with all settings retained, equivalent to a RasMol script, but including coordinates of all or part of model.

▣ Saved files fully portable and self-contained—just double click them to open.

Disadvantages of SPdbV

▣ Fewer resources already on the web.

▣ Fewer rendering styles for interactive model (wireframe, surface dots, ribbons), but others available for rendering still views.

▣ Not available in Netscape plug-in form.

▣ No animation of the molecule.

Chime To fully appreciate how proteins function in a cell, it is helpful to have a three-dimensional view of how proteins interact with other cellular components. In addition to RasMol, this is possible using online protein databases and the three-dimensional molecular viewing utilities Chime and Protein Explorer.

Chime (www.mdlchime.ucsf.com/chime/) is a Java plug-in for display of molecular structure within HTML pages. It is a plug-in for web browser that allows interactive display of graphics of protein structure. Chime is a web browser-based molecular visualization application that was written by researchers at MDL Information Systems. The images it produces look very similar to those of RasMol, many of the menu-based commands that are available in RasMol are also available in Chime through pop-up menus. What Chime lacks, however, is RasMol's command line, which can be used to type in commands that are not found in the menus. Chime does, however, understand RasMol commands, which can be sent to it by using embedded buttons. This makes Chime a very valuable tool for developing web-based tutorials and figures.

PROTEIN STRUCTURE CLASSIFICATION

There are two major classification schemes, SCOP and CATH. These approaches to classifying proteins have been incorporated into structural databases. There is considerable overlap between the schemes, but important differences. As indicated in the introduction to the RCSB, it is easy to gain access to the databases; however it is important to understand how these schemes evolved.

CATH

CATH is a database derived with a semi-automatic procedure for classifying proteins. There are five levels in the CATH hierarchy.

- **Protein class (C)**   Refers four classes of domain mainly α , mainly β, α-β , few SS- low-secondary structure content.

- **Architecture (A)**   Describes gross arrangement of secondary structures and ignore their connectivity.

- **Topology (T)**   Refers overall shape and connectivity of secondary structural elements based on structure comparison algorithm.

- **Homologous superfamily (H)**   Groups all the domains that have high structural and functional similarity and considers that they are evolved from a common ancestor.

- **Sequence family**   Provides final level within hierarchy where domains have sequence identities >35% that indicate identical structure and function, but exceptions exist. Proteins with similar sequence and structure are expected to share function.

CATH is a hierarchical classification of protein-domain structures (Figure 9.13). CE (Combinatorial Extension of the optimal path) provides structural neighbors of the PDB entries with structure–structure alignments and three-dimensional superpositions.

As mentioned earlier, the protein classes are mainly α, mainly β, α-β, and few SS. CATH numbering scheme for representative structures from globin-like fold family is mainly α class. Four of the seven levels within the CATH database are shown in Figure 9.14. Each level is associated with a unique number. The A, T, and H levels are numbered in bins of ten to allow expansion of the database.

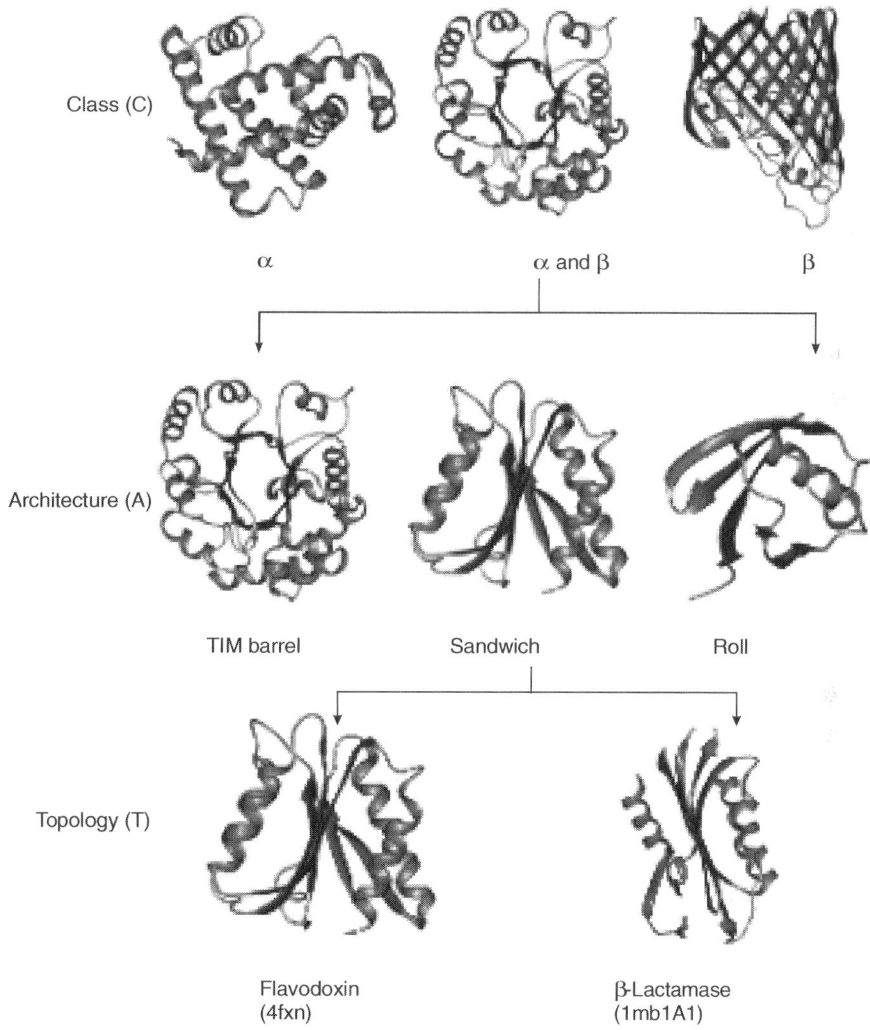

Figure 9.13 Diagrammatic representation of class (C), architecture (A), and topology (T) level in the CATH database. α-helices and β-sheets are shown in the diagram. The barrel, three-layer sandwich and roll architectures (A-level) are for the α-β class. Two representatives, falvodoxin and β-lactamase, from fold families in the three-layer sandwich architecture are shown.

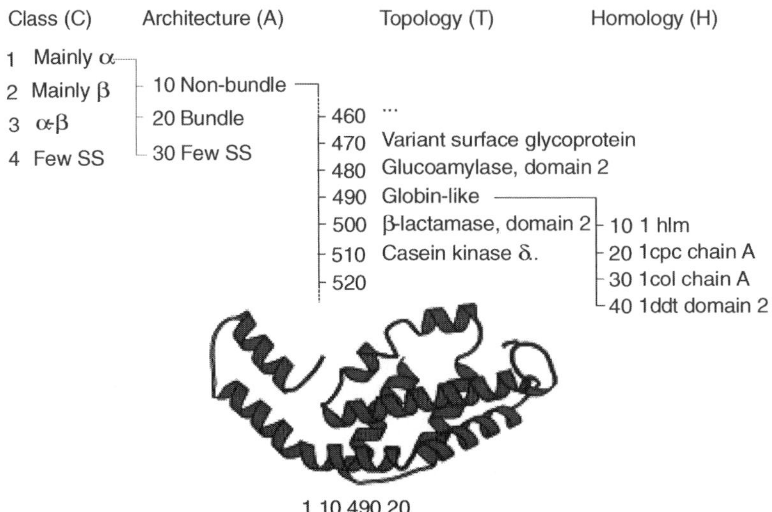

Figure 9.14   CATH numbering scheme globin-like fold family in mainly α class and whose CATH code is 1.10.490.20

For a given protein, CATH PDB code search can be performed online that results into CATH classification entries with domain, CATH code, the length of the protein chain and its 3-dimensional image. The CATH PDB code search for the protein domain 1MND01 to 1MND05 with one preliminary data entry for a given protein is displayed in Figure 9.15.

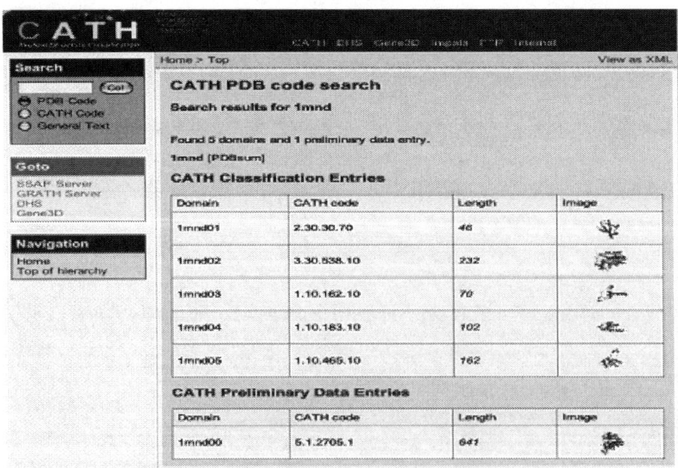

Domain	CATH code	Length	Image
1mnd01	2.30.30.70	46	
1mnd02	3.30.538.10	232	
1mnd03	1.10.162.10	70	
1mnd04	1.10.183.10	102	
1mnd05	1.10.465.10	162	

Domain	CATH code	Length	Image
1mnd00	5.1.2705.1	641	

Figure 9.15   CATH PDB code search with CATH code, length and image of particular domain

SCOP

Structural Classification of Proteins (SCOP) is a database of structural motifs that was generated manually. It has a similar hierarchy to that of CATH, but the site has a different "feel". The domains are defined manually in SCOP, whereas CATH utilizes a semi-automatic algorithm that gives smaller subdomains.

- Class
- All alpha, all beta, alpha/beta, alpha+beta, multidomain
- Fold
- Superfamily
- Family
- Protein
- Species

SCOP (http://scop.berkeley.edu/) uses augmented manual classification, class, fold, superfamily, and family classification (Figure 9.16). Table 9.2 shows the most frequent SCOP superfamilies detected in *E. coli*. Vector Alignment Search Tool (VAST) contains representative structure alignments and three-dimensional superpositions. Among these five databases, SCOP provides more function related information. However, due to the manual work involved, SCOP is not updated as frequently as the others, whereas FSSP and CATH follow the PDB updates closely. SCOP is used here as an example to show the features of structure-based families.

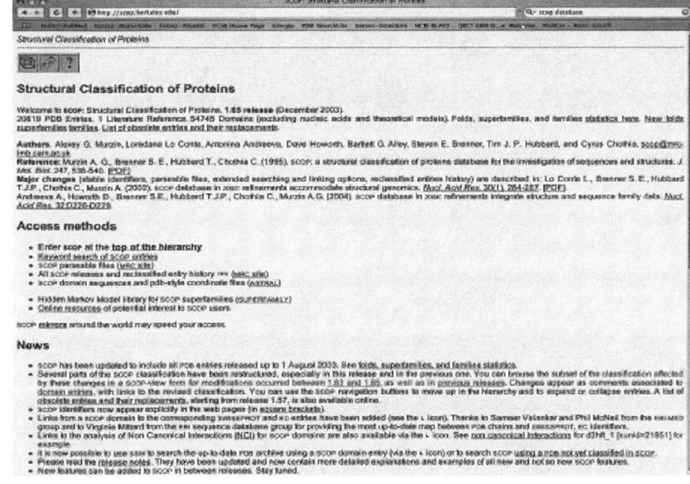

Figure 9.16   SCOP Page showing links, access method and news

Table 9.2 Most frequent SCOP superfamilies detected in *E. coli*

Superfam	Counts	Superfam description
52540	277	P-loop containing nucleoside triphosphate hydrolases
51735	123	NAD(P)-binding Rossmann-fold domains
46785	120	"Winged helix" DNA-binding domain
53850	89	Periplasmic-binding protein-like II
46689	67	Homeodomain-like
53067	52	Actin-like ATPase domain
53335	46	S-adenosyl L-methionine-dependent methyltransferases
50249	45	Nucleic acid-binding proteins
52172	41	CheY-like
53383	41	PLP-dependent transferases
51905	40	FAD/NAD(P)-binding domain
46894	38	C-terminal effector domain of the bipartite response regulators
53098	35	Ribonuclease H-like
52833	35	Thioredoxin-like
49401	34	Bacterial adhesins
54862	34	4Fe-4S ferredoxins
55874	33	ATPase domain of HSP90 chaperone/DNA topoisomerase II/histidine kinase
56784	28	HAD-like
53822	27	Periplasmic-binding protein-like I
47413	27	lambda repressor-like DNA-binding domains
52402	26	Adenine nucleotide alpha hydrolases-like
56935	26	Porins
55729	25	Acyl-CoA N-acyltransferases (Nat)
53474	25	alpha/beta-Hydrolases
53448	24	Nucleotide-diphospho-sugar transferases
52518	24	Thiamin diphosphate-binding fold (THDP-binding)
51445	23	(*Trans*)glycosidases

(Contd.)

Table 9.2 (Continued)

Superfam	Counts	Superfam description
51569	23	Aldolase
47384	21	Homodimeric domain of signal transducing histidine kinase
48179	21	6-phosphogluconate dehydrogenase C-terminal domain-like
55781	21	GAF domain-like
50129	20	GroES-like

SCOP can be accessed through its home server in the UK (http://scop.mrc-lmb.cam.ac.uk/scop/). It is also widely mirrored around the world. However, only the home server has the BLAST sequence-similarity search capacity. SCOP describes the hierarchical relationship among proteins through the major levels of (homologous) family, superfamily, and fold (Figure 9.17).

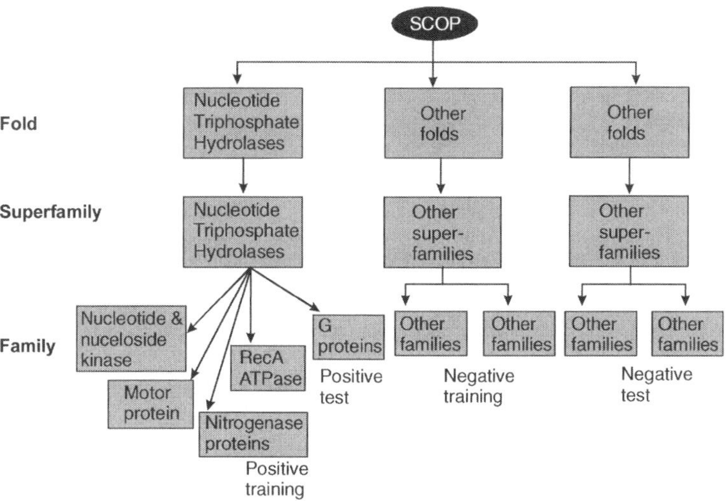

Figure 9.17 SCOP describing the hierarchical relationship among proteins through the major levels of fold, superfamily, and family

Proteins are clustered together into a (homologous) family if they have significant sequence similarity. Different families that have low sequence similarity but whose structural and functional features suggest a common evolutionary origin are placed together in a superfamily. Different superfamilies are categorized into a fold if they have the same major secondary structures in the same arrangement and with the same topological connections (the peripheral elements of secondary structure and turn regions may differ in size and conformation). Two superfamilies in the same fold may not have a common evolutionary origin. Their structural similarities may arise from the physics and chemistry of proteins favouring certain packing arrangements and chain topologies.

The SCOP database is developed as an evolutionary classification, in which the main focus is to place the proteins in a coherent evolutionary framework, based on their conserved structural features. The database aims to provide a comprehensive and detailed description of the relationships between all proteins whose 3D structures have been determined. A fundamental unit of classification in the SCOP database is the protein domain. A domain is defined as an evolutionary unit observed in nature either in isolation or in more than one context in multidomain proteins. This implies that a domain is defined as an entity that could form a stable structural entity in its own right. In SCOP, a domain is a folding unit that can exist independent of other secondary structural components. An advantage of the SCOP database is that it embeds a theory of protein evolution as defined by human experts rather than by empirical rules implemented in a variety of bioinformatics algorithms and tools. It actually groups the proteins by structural similarity.

DALI

The DALI method, or Distance matrix ALIgnment, is a fragment-based method for constructing structural alignments based on contact similarity patterns between successive hexapeptides in the query sequences. It can generate pairwise or multiple alignments and identify a query sequence's structural neighbours in the PDB. It has been used to construct the FSSP structural alignment database (Fold classification based on Structure–Structure alignment of Proteins, or Families of Structurally Similar Proteins). A DALI web server can be accessed at EBI DALI (http://www.ebi.ac.uk/dali/ and the FSSP is located at the DALI database (http://ekhidna.biocenter.helsinki.fi/dali/start).

The DALI server is a network service for comparing protein structures in 3D. When we submit the coordinates of a query protein structure, DALI compares them against those in the PDB. A multiple alignment of structural neighbors is e-mailed back to us. In favourable cases, comparing 3D structures may reveal biologically interesting similarities that are not detectable by comparing sequences. If we want to know the structural neighbors of a protein already in the PDB, we can find them in the DALI database. The DALI Domain Dictionary is a numerical taxonomy of all known structures in the PDB, we can view the entire DALI classification by following the link to the DALI Domain Dictionary.

FSSP

Fold classification based on Structure–Structure alignment of Proteins (http://www.ebi.ac.uk/dali/fssp/) and Families of Structurally Similar Proteins (FSSP). It features a protein family tree and a domain dictionary, in addition to whole-chain-based classification, sequence neighbours, and multiple structure alignments. The FSSP classification scheme is based on an exhaustive 3D structure comparison of protein structures in the Protein Data Bank. Each FSSP entry includes the structure alignments of the search structure with its sequence homologs. Classification and structure alignments are continuously updated when new structures are available at the Protein Data Bank. Compared to other classification databases, FSSP is unique in that it employs a purely automated method of classification.

SSAP

SSAP (Sequential Structure Alignment Program) is a dynamic programming-based method of structural alignment that uses atom-to-atom vectors in structure space as comparison points. It has been extended since its original description to include multiple as well as pairwise alignments, and has been used in the construction of the CATH hierarchical database classification of protein folds. The CATH database can be accessed at CATH protein structure classification (http://www.cathdb.info/).

Protein Structure Bioinformatics Resources

- 123D+ threads a sequence through a set of 3D structures. It combines sequence profiles, secondary-structure prediction, and contact capacity potentials to find the most compatible fold among the 3D structures, and the best alignment of the sequence with that fold. (http://123d.ncifcrf.gov/123D+html)

- 3D-PSSM attempts to predict 3D structure and probable function from a protein sequence in part by threading with known structures and scoring for compatibility. (http://123d.ncifcrf.gov/123D+.html)

- **Columba-DB** (Protein Structure Annotation), is a meta-server that provides an integrated summary with links to details for information from the PDB, KEGG, ENZYME, ExPASy, DSSP, CATH, SCOP, Swiss-Prot, NCBI taxonomy, GO, and PISCES. (http://www.columba-db.de/)

- **CAPRI** (Critical Assessment of PRedicted Interaction) assesses the capacity of protein–protein docking methods to predict protein–protein interactions. The CAPRI is an ongoing series of events in which researchers throughout the community attempt to dock the same proteins, as provided by the assessors. Rounds take place approximately every 6 months. Each round contains between one and six target protein–protein complexes whose structures have been recently determined experimentally. The coordinates are held privately by the assessors, with the cooperation of the structural biologists who determined them. (http://capri.ebi.ac.uk/)

- **CASP** (Critical Assessment of techniques for protein Structure Prediction) is an annual competition in which information is solicited from X-ray crystallographers and NMR spectroscopists on structures about to be solved. Before the empirical structures are known, modellers submit the structure predictions in three categories: 1) comparative modelling, 2) fold recognition or threading, and 3) *ab initio* folding. CASP includes ten most wanted proteins (http://tmw.llnl.gov/) that to provide structural and functional insights into biologically important proteins, particularly those that are intractable to experimental structural determination. CASP is a community-wide experiment for protein structure prediction taking place every two years since 1994. It provides research groups with an opportunity to assess the quality of their methods for protein structure prediction from the primary structure of the protein. As a consequence, CASP provides the research community with an assessment of the state of the art in this field. Protein structures that are either expected to be solved shortly or that have been recently solved, but not yet discussed in public, are used as targets for the prediction. If the given sequence is found (for example, using sequence alignment methods such as BLAST or FASTA) to be similar to a protein sequence of known structure, comparative protein modelling may be used to predict the tertiary structure. Otherwise, other methods such as protein threading or de novo protein structure prediction are applied. (http://predictioncenter.llnl.gov/)

- **CASTp** is a server that identifies pockets and cavities in proteins, and quantifies their volumes. Atoms lining each pocket or cavity can be displayed in Chime, RasMol, or MAGE. (http://cast.engr.uic.edu/)

▣ **Combinatorial Extension** finds 3D similarities in protein structures without reference to sequence. Automated intelligent alignment of protein chains and ligands are discarded. (http://cl.sdsc.edu/)

▣ **Comparative Modelling** (Homology Modelling) servers are continuously and automatically evaluated by EVA. (http://molvis.sdsc.edu/protexpl/psbiores.htm#eva)

▣ **ConSurf** is an automated identification of functional regions in proteins of known 3D structure based upon evolutionary conservation. (http://consurf.tau.il/)

▣ **DisEMBL** is an intrinsic protein disorder prediction server. It is a computational tool for prediction of disordered/unstructured regions within a protein sequence. Avoiding potentially disordered segments in protein expression constructs can increase expression, foldability and stability of the expressed protein. (http://dis.embl.de/)

▣ **DisProt.org** (Database of Protein Disorder) is a curated database that provides information about proteins that lack fixed 3D structure in their putatively native states, either in their entirety or in part. (http://www.disprot.org/)

▣ **EVA** continuously and automatically analyses protein structure prediction servers in "real time". Evaluates servers for secondary structure prediction, comparative modelling, inter-residue distances and contacts, and threading. (http://cubic.bioc.columbia.edu/eva/)

▣ **GRASS** (Graphical Representation and Analysis of Structure Server), using the widely acclaimed **GRASP** software, renders a PDB file in a manner selected by the user from menus, and offers visualization in Chime or VRML as well as GRASP. Notably, molecular surfaces can be coloured by electrostatic potential, hydrophobicity, distances from a ligand, surface curvature. (http://honiglab.cpmc.columbia.edu/GRASS/main.html)

▣ **Helmholtz Network for Bioinformatics** is an integrative web portal for bioinformatics resources, including protein structure prediction, alignment generation, threading, and structure-similarity searching (all under protein structure). (http://www.hnbioinfo.de/)

▣ **MaxSprout** generates protein backbone and side chain coordinates from a C (alpha) trace. (http://ebi.ac.uk/maxsprout/)

▣ **MolProbity** conducts a new and powerful structure validation based on all-atom contacts analysis, producing automatic corrections (rotamer flips). The flips can be inspected and accepted or rejected in a java-based Kinemage-style viewer. The corrected PDB file, containing added hydrogen atoms, is provided. For secure use, free software that performs the all-atom contacts analysis can be downloaded and run locally. The server also offers very useful Ramachandran plots (identifying residues outside the favourable regions), C-beta deviations, and rotamer analysis. (http://kinemage.biochem.duke.edu/molprobity/index-king.html)

▣ **OCA** is a powerful alternative mechanism for searching the world structure database in the Protein Data Bank. (http://bioportal.weizmann.ac.il/oca-docs/oca-home.html).

▣ **PDB_Select** generates sequence non-redundant subsets of chains in the Protein Data Bank used by OCA for non-redundant subsets. (http://homepages.fh-giessen.de/~hg12640/pdbselect/)

▣ **PDBSProtEC** is a resource to link PDB chains with Swiss-Prot codes and EC numbers. Enter one of a PDB identification code, an EC number, or a Swiss-Prot identification code(s), and get the other two. Updated daily. (http://www.rubic.rdg.ac.uk/~andrew/bioinf.org/pdbsprotec/)

▣ **PDBsum** summarizes and analyses the PDB structures. It is a pictorial database providing an at-a-glance overview of every macromolecular structure deposited in the Protein Data Bank. It includes Jmol view and rotatable (RasMol or Chime) images of surfaces and clefts in surfaces, and a Raster3D image generating form. Highlights and lists the oldest/newest PDB entries, and those with the largest/smallest molecules, number of protein chains, longest chains, metal ions, most/largest ligands and their interactions, fold cartoons, R-factor, highest and lowest resolution, etc. PDBsum is accessible for keyword interrogation via UCL's Bimolecular Structure and Modelling Unit Web server. PDBsum is updated whenever any new structures are released by the PDB and is freely accessible via http://www.biochem.ucl.ac.uk/bsm/pdbsum. (http://www.ebi.ac.uk/thornton-srv/databases/pdbsum/)

▣ **PISCES** is a Guoli Wang and Roland Dunbrack's site that culls entries from the PDB based on resolution and sequence identity, among other things. (http://www.fccc.edu/research/labs/dunbrack/pisces/)

▣ **Probable Quaternary Structure (PQS)** uses an empirically derived weighted score to estimate whether protein–protein contacts that occur in crystals are specific oligomer interfaces or crystal artifacts. For those deemed specific oligomers, the oligomer is provided. When multiple copies of the molecule occur in the asymmetric unit, single copies are provided. PQS also provides complete virus capsids, using symmetry operations typically provided by the authors in the PDB file (virus capsid examples). (http://pqs.ebi.ac.uk/)

- ▣ **PROCAT** is a database where you can search for 3D enzyme active site template motifs in a protein structure. (http://www.biochem.ucl.ac.uk/bsm/PROCAT/PROCAT.html)

- ▣ **ProMode** calculates vibrational dynamics of protein molecules and among other things, displays the results as an animation in MDL Chime. (http://prodome.socs.waseda.ac.jp/)

- ▣ **ProSAT** maps ProSite patterns and Swiss-Prot features onto 3D protein structures. (http://projects.villa-bosch.de/dbase/pdba/)

- ▣ **PubMed** is a searchable journal article abstracts from the US National Library of Medicine. (http://www.pubmed.gov/)

- ▣ **RDP** threads by Recursive Dynamic Programming, available through the Helmholtz Network under Protein Structure. (http://molvis.sdsc.edu/protexpl/psbiores.htm#helmholtz)

- ▣ **SPAM** (Systematic Protein Annotation and Modeling) is a multi-institutional initiative to provide tools to assist in selecting and registering targets for structural genomics and for associating structure with function. (http://spam.sdsc.edu/)

- ▣ **SSAP** is a web server that uses an intermolecular distance-based method in which matrices are built based on the C-β distances of all residue pairs. (www.biochem.ucl.ac.uk/cgi-bin/cath/GetSsapRasMol.pl)

- ▣ **STAN** (STructure ANalysis server at University of Uppsala, Sweden) includes Ramachandran analysis and routines to detect water molecules that should be metal cations, and *trans*-peptides that should be *cis* structure searching and comparison—finding structure neighbours without reference to sequence. (http://xray.bmc.uu.se/cgi-bin/gerard/rama_server.pl)

- ▣ **STAMP** is a UNIX program that uses the intermolecular approach to generate protein structure alignment. It involves use of the iterative alignment based on dynamic programming to obtain the best superposition of two or more structures. (www.compbio.dundee.ac.ul/Software/Stamp/stamp.html)

- ▣ **SuMo** (Surfing the Molecules) is a automated search for similar sites in proteins, "to screen the Protein Data Bank (PDB) for finding ligand binding sites matching your protein structure or inversely, for finding protein structures matching a given site in your protein. This method is neither based on amino acid sequence nor on fold comparisons. Priority is given to biological relevance." SuMo result for 1LPM (entire protein).(http://sumo-pbil.ibcp.fr/)

▣ **SWISS-MODEL** is an automated comparative protein modelling server. (http://www.expasy.ch/swissmod/SWISS-MODEL.html)

▣ **Swiss-Prot** is a curated protein **sequence** database which strives to provide a high level of annotations (such as the description of the function of a protein, its domains structure, post-translational modifications, variants, etc.), a minimal level of redundancy and high level of integration with other databases. To get Swiss-Prot codes from a PDB code (or EC number), use PDBSProtEC. (http://us.expasy.org/sprot/)

▣ **Threading servers** include 123D and RDP available through the Helmholtz Network (under Protein Structure), and 3D-PSSM. Threading servers are continuously and automatically evaluated by EVA. (http://molvis.sdsc.edu/protexpl/psbiores.htm#helmholtz)

▣ **TOPS** (Services provided Glasgow University that includes protein topology diagrams, topology matching between domains, structure comparisons, motif matching. (http://balabio.dcs.gla.ac.uk/tops/index.html)

▣ **Transmembrane Protein PDB** identifies the transmembrane proteins in the PDB. Searchable, browsable, and downloadable.(http://www.enzim.hu/PDB_TM/)

▣ **VAST** (Vector Alignment Search Tool) is the web server that performs alignment using both the inter-and intramolecular approaches. Optimum alignment between two structures is defined by the highest degree of vector matches. (http://www.ncbi.nih.gov/Structure/VAST/vast.shtml)

▣ **WhatIF WWW Interface** by Gert Vriend offers some powerful structure validation and modelling routines, including addition of hydrogen, and crystal contacts (http://molvis.sdsc.edu/protexpl/xtlcon.htm). (http://swift.cmbi.kun.nl/WIWWWI)

▣ **World Index of Molecular Visualization Resources, in which** visitors contribute links to online resources in categories including ready to use tutorials on specific molecules (including proteins, organic chemicals, inorganic crystals), K12 resources, resources in German, French, Italian, Japanese, Spanish, etc., technical resources on how to create Chime websites, DeepView (SwissPDB-Viewer) resources, sources of molecules (PDB files; http://molvis.sdsc.edu/protexpl/igloss.htm#pdb), galleries, email lists, free and commercial software. Over 200 entries are described in detail and cross-indexed by subjects and authors. (http://molvisindex.org/)

PROTEIN STRUCTURE PREDICTION

Protein structure prediction is one of the most important unsolved problems in biology today. With the wealth of genome data now available, it is of great interest to determine the structures of the proteins that genes encode. Proteins are composed of long chains of amino acid residues, of which there are twenty natural varieties. A gene encodes a specific amino acid sequence, which, when translated, folds into a unique three-dimensional conformation. The protein structure prediction problem is to predict this conformation (the protein's tertiary structure) from the amino acid sequence (the protein's primary structure). The biological function of a protein is dependent on its structure, so structure prediction is an important step towards function prediction. Potential applications of structure prediction range from elucidation of cellular processes to vaccine design.

Experimental methods for protein structure determination are costly and time-intensive, and the number of known protein sequences now far outstrips the capacity of experimentalists to determine their structures. Computational methods have been improving steadily and are approaching the level of resolution attainable in experiments. Structure prediction methods fall into two broad camps: comparative modelling, in which solved protein structures are known for one or more proteins with sequences similar to the target sequence ("homologs"), and *ab initio* modelling, in which no homologs are known. Comparative modelling, also known as homology modelling, is a class of methods for constructing an atomic-resolution model of a protein from its amino acid sequence (the "query sequence" or "target"). Almost all homology modelling techniques rely on the identification of one or more known protein structures (known as "templates" or "parent structures") likely to resemble the structure of the query sequence, and on the production of an alignment that maps residues in the query sequence to residues in the template sequence. The sequence alignment and template structure are then used to produce a structural model of the target (Figure 9.18). If necessary, alignment and model building are repeated until a satisfactory result is obtained. Finally, the statistical evaluation of the model can be achieved by web-based approaches, ANOLEA (Atomic Non-Local Environment Assessment; http://protein.bio.puc.cl/cardex/server/anolea/index.html) or Verify3D (www.doe-mbi.ucla.edu/Services/Verify3D). They perform energy calculations for atomic interactions in a protein chain and compare these interaction energy values with those compiled from a database of protein X-ray structure.

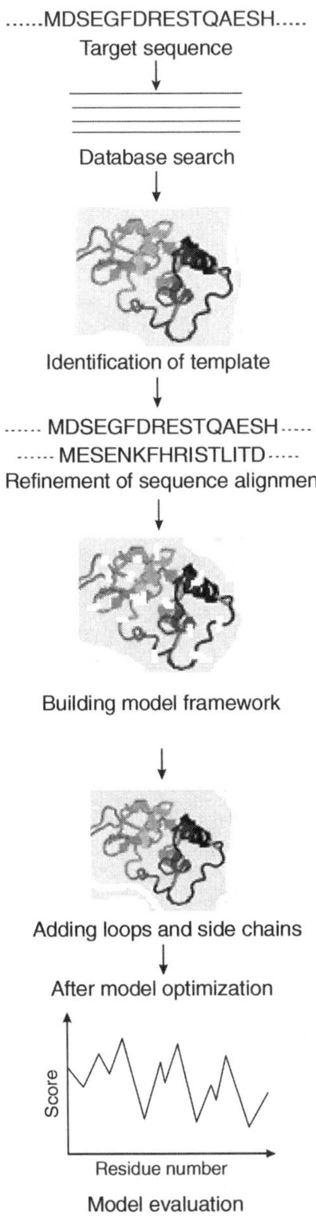

Figure 9.18   Steps involved in homology modelling

Ab initio- or de novo-protein modelling methods are based on physical principles rather than (directly) on previously solved structures. There are many possible procedures that either attempt to mimic protein folding or apply some stochastic method to search possible solutions (i.e., global optimization of a suitable energy function). These procedures tend to require vast computational resources, and have thus only been carried out for tiny proteins. To predict protein structure de novo for larger proteins will require better algorithms and larger computational resources like those afforded by either powerful supercomputers (such as Blue Gene or MDGRAPE-3).

Following Table 9.3 is the list of protein structure prediction software that summarizes commonly used software tools in protein structure prediction, including homology modelling, protein threading, *ab initio* methods, secondary structure prediction, and transmembrane helix and signal peptide prediction.

Table 9.3 Protein structure prediction software

Software	Method
Homology modelling	
3D-JIGSAW	Fragment assembly
CABS	Reduced modelling tool
CPHModel	Fragment assembly
ESyPred3D	Template detection, alignment, 3D modelling
GeneSilico	Consensus template search/fragment assembly
Geno3D	Satisfaction of spatial restraints
HHpred	Template detection, alignment, 3D modelling
LIBRA I	LIght Balance for Remote Analogous proteins, ver. I
MODELLER	Satisfaction of spatial restraints
ROSETTA	Rosetta homology modelling and *ab initio* fragment assembly with Ginzu domain prediction
SWISS-MODEL	Local similarity/fragment assembly
TIP-STRUCTFAST	Automated Comparative Modelling
WHAT IF	Position specific rotamers
Threading/fold recognition	
PSI-BLAST	Iterative sequence alignment for fold identification
HHpred	Template detection, alignment, 3D modelling
3D-PSSM	3D-1D sequence profiling

(Contd.)

Table 9.3   (Continued)

Software	Method
SUPERFAMILY	Hidden Markov modelling
Bioingbu	Evolutionary information recognition
mGenTHREADER/ GenTHREADER	Sequence profile and predicted secondary structure
LOOPP	Multiple methods
***Ab initio* structure prediction**	
ROSETTA	Rosetta homology modelling and *ab initio* fragment assembly with Ginzu domain prediction
Rosetta@Home	Distributed-computing implementation of Rosetta algorithm
CABS	Reduced modelling tool
Bhageerath	A computational protocol for modelling and predicting protein structures at the atomic level.
Secondary structure prediction	
GOR	Information theory/Bayesian inference
Jpred	Neural network assignment
Meta-PP	Consensus prediction of other servers
PREDATOR	Knowledge-based database comparison
PredictProtein	Profile-based neural network
PSIPRED	Two feed-forward neural networks which perform an analysis on output obtained from PSI-BLAST
YASSPP	Cascaded SVM-based predictor using PSI-BLAST profiles
Transmembrane helix and signal peptide predicction	
HMMTOP	Hidden Markov Model
PHDhtm in PredictProtein	Multiple alignment-based neural network system
Phobius	Homology supported predictions

Source: "http://en.wikipedia.org/wiki/List_of_protein_ structure_ prediction _software"

ROSETTA

One of the key challenges in computational biology is prediction of three-dimensional protein structures from amino acid sequences. For most proteins, the "native state" lies at the bottom of a free energy landscape. Protein structure prediction involves varying the degrees of freedom of the protein in a constrained manner until it approaches its native state. In addition to CASP as a protein structure prediction program there is a Rosetta protein structure prediction protocol in which a large number of independent folding trajectories are simulated, and several lowest-energy results are likely to be close to the native state. Least Angle Regression (LAR) is used to identify structural features that best account for energy variation in the initial set of models. Resampling improves on unbiased Rosetta sampling if the number of viable runs (runs in which no non-native features are enforced) is sufficiently high that the benefits from the enforcement of native features for a given protein are visible (Figure 9.19). Rosetta is the state of the art in protein structure prediction, and it has undergone years of incremental advances and optimizations.

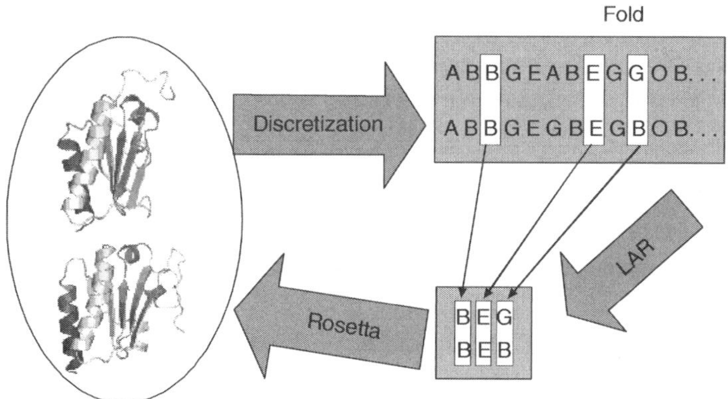

Figure 9.19   Flowchart of resampling method

Rosetta is developed in David Baker's research group at the University of Washington as a software package and a protein structure prediction algorithm. Rosetta is a web server that predicts protein three-dimensional conformations using *ab initio* method. It uses a Monte Carlo search procedure to minimize an energy function that is sufficiently accurate that the conformation found in nature (the "native" conformation) is generally the conformation with lowest Rosetta energy. Finding the global minimum of the energy function is very difficult because of the high dimensionality of the search space and the very large number of local minima. Rosetta employs a number of strategies to combat these issues, but the primary one is to perform a large number of random restarts. Thanks to a very large-scale distributed computing platform called Rosetta@home, composed of more than three hundred thousand volunteer computers around the world, up to several million local minima of the energy function ("decoys," in Rosetta parlance) can be computed for each target sequence. Rosetta@home is a distributed computing project for protein structure prediction on the Berkeley Open Infrastructure for Network Computing (BOINC) platform. It also aims to predict protein–protein docking and design new proteins with the help of over 86,000 volunteered computers processing over 68 teraFLOPS on average as of September 7, 2008. Though much of the project is oriented towards basic research on improving the accuracy and robustness of the proteomics methods, Rosetta@home also does applied research on malaria, Alzheimer's disease and other pathologies. It has been used in designing multiple possible vaccines for human immunodeficiency virus (HIV). RosettaDock, a Rosetta-based application, which was used to model docking between an antibody (Ig G) and a surface protein expressed by herpes simplex virus 1 (HSV-1) which serves to degrade the antiviral antibody and it was also used in conjunction with experimental methods to model interactions between three proteins—lethal factor (LF), oedema factor (EF) and protective antigen (PA) that makes up anthrax toxin.

Protein Folding

Proteins show a great variety of three-dimensional conformations that are necessary to support their very diverse structural and functional roles. The amino acid sequence of a protein dictates its three-dimensional structure. For each natural amino acid sequence, there is a unique stable native state that under proper conditions is adopted spontaneously and it is essential for its normal biological functioning. If a purified protein is heated, or otherwise brought to conditions far from the normal physiological environment, it will unfold to a disordered and biologically-inactive structure. When normal conditions are restored, protein molecules will generally readopt the native structure.

The spontaneous folding of proteins to form their native states is the point at which nature makes the giant leap from the one-dimensional world of genetic and protein sequences to the three-dimensional world we inhabit. There is a paradox: "The translation of DNA sequences to amino acid sequences is very simple to describe logically; it is specified by the genetic code." The folding of the polypeptide chain into a precise three-dimensional structure is very difficult to describe logically. However, translation requires the immensely complicated machinery of the ribosome, tRNAs and associated molecules; but protein folding occurs spontaneously.

There are two principles that might control the folding of a protein into the correct higher order structure.

- Folding is an intrinsic feature of the primary sequence. In this case, the final structure must always be the most stable thermodynamic and can be generated at any time after synthesis of the polypeptide chain is complete.

- The correct structure can be generated only during the synthesis of the polypeptide. Then it becomes possible that an intrinsically less stable structure could prevail because the protein becomes "trapped" in it during synthesis.

The relationship between higher order structures and the primary structure may be revealed when a protein is denatured by heating or by chemical treatments that disrupt the protein conformation. Most denaturing events involve the breakage of hydrogen and other noncovalent bonds. An exception is the disruption of S–S bridges that result from treatment with reducing agents. However, all of these changes affect the conformation; the primary sequence of amino acids in the polypeptide chain remains unaltered.

In some cases, the higher order structure follows ineluctably from the primary sequence, e.g. enzyme ribonuclease. After the protein has been denatured, its active conformation can be regained by reversing the denaturing procedure. All the information necessary to form the secondary structure resides in the primary sequence. Thus the production of active ribonuclease is an inevitable event whenever the intact primary chain is placed in the appropriate conditions.

In other cases, proteins can be irreversibly denatured. Thus under certain (non-physiological) conditions, a protein may have alternative stable conformations. In some cases, the correct conformation probably can be attained only during synthesis of the protein. The conformation could depend on specific interactions between regions of the protein that can occur only in the absence of other regions (that is, those that have not yet been synthesized). This is probably the more common situation.

In some instances, a cofactor that is part of the active protein (such as the iron-binding heme group of the cytochromes) must be present in order for the polypeptide chain to take up its proper conformation. In the case of multimeric proteins, it may be necessary for one subunit to be present in order for another to acquire the proper conformation. Protein folding is usually rapid *in vivo*, occurring within seconds or less. It begins even before a protein has been completely synthesized. Probably it involves a sequential folding mechanism, in which the reaction passes through discrete (although highly transient) intermediates. The process is initiated by the collapse of hydrophobic side chains into the "core" of the protein; this occurs within milliseconds. Units of secondary structure, largely α-helices and β-sheets, form on the same time scale. The transition from this structure to the final tertiary structure is slower. The process appears to be cooperative, so that formation of one region of secondary structure enhances formation of the next region, and so on.

Necessarily, the acquisition of structure when a protein is synthesized is not a spontaneous process, but may require assistance of enzymes that catalyse specific isomerization steps and factors that act stoichiometrically to influence folding directly. The formation of disulphide bonds has a major effect on the conformation of the protein. It is influenced by both environment and specific accessory proteins. The process is catalysed by an enzyme, protein disulphide isomerase (PDI), which is a curious protein, participating in a variety of functions concerned with protein modification. Proline has a major effect upon protein structure because of the restrictions imposed by its ring structure. Proteins-containing proline fold slowly because the peptidyl-proline link does not necessarily form in correct stereochemical conformation. The enzyme peptidyl-prolyl isomerase (PPI) catalyses the *cis–trans* conversion, and by this means significantly accelerates the folding reaction.

Proteins that act stoichiometrically on the folding of other proteins are called molecular chaperones. A chaperone forms a complex with a protein during folding, but is required only during assembly, and is not part of the mature structure. The major role of a chaperone is to prevent the formation of incorrectly folded structures, in which the substrate protein might otherwise become trapped during folding. For example, a protein from *E. coli*, called Do or DegP or HtrA, acts as a chaperone (catalysing protein folding) at low temperatures, but at 42° turns into a proteinase. The rationale seems to be— under normal conditions or moderate heat stress the goal is to rescue proteins that are having difficult folding; under more severe heat stress, when salvage is impossible, to recycle them.

Protein-folding Disorders

Defects in protein folding may be the molecular basis for a wide range of human genetic disorders. For example, cystic fibrosis is caused by defects in a membrane-bound protein called Cystic Fibrosis Transmembrane conductance Regulator (CFTR), which acts as a channel for chloride ions. The most common cystic fibrosis-causing mutation is the deletion of a Phe residue at position 508 in CFTR, which causes improper protein folding. Many of the disease related mutations in collagen also cause defective folding. An improved understanding of protein folding may lead to new therapies for these and many other diseases. A misfolded protein appears to be the causative agent of a number of rare degenerative brain diseases in mammals. Perhaps the best known of these is mad-cow disease (Bovine Spongiform Encephalopathy, BSE). Related diseases include kuru and Creutzfeldt–Jacob disease in humans, scrapie in sheep, and chronic wasting disease in deer and elk. These diseases are also referred to as spongiform encephalopathies, because the diseased brain frequently becomes riddled with holes. Typical symptoms include dementia and loss of coordination. The diseases are fatal. Alzheimers, Parkinson's disease, one of the inherited forms of emphysema, type 2 Diabetes, and even some cancers are all related to the abnormal folding of proteins within the body. These devastating diseases cause untold misery, extracting both monetary and human tolls.

PROTEIN FUNCTIONS

According to a biochemist, protein function means the biochemical role of an individual protein: if it is an enzyme, function refers to the reaction catalysed; if it is a signalling protein, function refers to the interactions that the protein makes. In the view of the geneticist or cell biologist, protein function includes these roles but will also encompass the cellular roles of the protein, such as the phenotype of its deletion, the pathway in which it operates, among others. A physiologist or developmental biologist may have an even broader view of function, including tissue specificity and expression during the life cycle of the organism. In general, proteins perform several functions that are classified in major categories involving 1) energy that is produced or consumed in metabolic and biosynthetic pathways, 2) information that is executed due to the proteins participating in replication, transcription and translation, and 3) communication and regulation that is possible due to proteins present in cell membrane.

In case of the yeast genome, which contains 5885 predicted protein-coding genes, $\sim$140 genes for rRNAs, 40 genes for small nuclear RNAs, and

275 tRNA genes including 3408 genes for known proteins performing specific functions. About 1000 more contain some similarity to known proteins in other species. Another ~800 are similar to ORFs in other genomes that correspond to unknown proteins. Many of these homologues appear in prokaryotes. Only approximately one-third of yeast proteins have identifiable homologues in the human genome. In taking censuses of genes, it has been useful to classify their functions into broad categories. The following classification of yeast protein functions is taken from http://www.mips.biochem.mpg.de/proj/yeast/catalogues/funcat/:

- Metabolism
- Energy
- Cell growth, cell division and DNA synthesis
- Transcription
- Protein synthesis
- Protein destination
- Transport facilitation
- Cellular transport and transport mechanisms
- Cellular biogenesis
- Cellular communication/signal transduction
- Cell rescue, defence, cell death and ageing
- Ionic homeostasis
- Cellular organization
- Transposable elements, viral and plasmid proteins
- Unclassified

Prediction of Protein Function

One of the most challenging problems of the post-genomic era is to determine the function of the protein. The availability of entire genome sequences and of high-throughput capabilities to determine gene co-expression patterns has shifted the research focus from the study of single proteins or small complexes to that of the entire proteome. In this context, the search for reliable methods for assigning protein function is of primary importance. There are various approaches available for deducing the function of proteins of unknown function using information derived from sequence similarity or clustering patterns of co-regulated genes, phylogenetic profiles, protein–protein interactions, and protein complexes.

In every protein, the flow of information is from the sequence $\rightarrow$ structure $\rightarrow$ function. However, although one can be confident that similar amino acid sequences in the protein will produce similar protein structures, the relation between structure and function is more complex. Proteins of similar structure and even of similar sequence can be recruited for very different functions. Very widely diverged proteins may retain similar functions. Moreover, just as many different sequences are compatible with the same structure; unrelated proteins with different folds can carry out the same function.

As proteins evolve they may

- retain function and specificity,
- retain function but alter specificity,
- change to a related function, or a similar function in a different metabolic context,
- change to a completely unrelated function.

It is difficult to define the idea of difference in function quantitatively. When are two different functions more similar to each other than two other different functions? In some cases, altered function may conceal similarity of mechanism. For example, the enolase superfamily contains several homologous enzymes that catalyse different reactions with shared mechanistic features. This group includes enolase itself, mandelate racemase, muconate lactonizing enzyme I, and D-glutarate dehydratase. Each acts by abstracting an alpha proton from a carboxylic acid to form an enolate intermediate. The subsequent reaction pathway, and the nature of the product, varies from enzyme to enzyme. These proteins have very similar overall structures but have different TIM-barrel folds (a type of supersecondary structure). Different residues in the active site produce enzymes that catalyse different reactions.

Automated Protein Function Prediction (PFP)

It is a web-based program (http://dragon.bio.purdue.edu/) that is operated and maintained by Kihara Laboratory of Bioinformatics at Purdue University, USA. PFP is designed to predict Gene Ontology (GO) annotations for a query protein sequence beyond what can be found by searching conventional databases. The PFP algorithm has been shown to increase coverage of sequence-based function annotation more than five-fold by extending a PSI-BLAST search to extract and score GO terms individually and include information from distantly related sequences. It applies a novel data mining tool, the Function Association Matrix (FAM), to score significantly associating pairs of annotations. Prediction scores for top 5 predictions in each GO category processed through automated PFP version 2.0 Beta Release are given in Table 9.4.

Table 9.4   Prediction scores for top 5 predictions in each GO category

GO term	Short definition	Raw score
GO0008152	metabolism	1931.09
GO0030435	sporulation	1885.44
GO0006807	nitrogen compound metabolism	912.88
GO0007275	development	810.57
GO0000074	regulation of cell cycle	606.73
GO0004356	glutamate-ammonia ligase activity	1360.54
GO0016772	transferase activity, transferring phosphorus-containing groups	1105.99
GO0003674	molecular function	751.65
GO0016301	kinase activity	456.44
GO0001760	aminocarboxymuconate-semialdehyde decarboxylase activity	64.95
GO0005838	proteasome regulatory particle (sensu Eukaryota)	1249.37
GO0005575	cellular component	358.31
GO0005829	cytosol	23.10

Each GO term can be accessed with its URL, the first GO term has http://amigo.geneontology.org/cgi-bin/amigo/go.cgi?action=query&view=query&query = 0008152&search_constraint=terms

Diversity in Protein Functions

Trypsin is a proteolytic enzyme present in digestive tract of mammals where it catalyses the hydrolysis of peptide bonds adjacent to a positively charged residue, Arg or Lys. Enzymes with similar sequence, structure, function, and specificity exist in many species, including human, cow, Atlantic salmon, and even *Streptomyces griseus* (Figure 9.20). The similarity of the *S. griseus* enzyme to vertebrate trypsins suggests a lateral gene transfer. For the three vertebrate enzymes, each pair of sequences has $\geq$ 64% identical residues in the alignment, and the bacterial homologue has $\geq$30% identical residues with the others; all have very similar structures. These enzymes are called orthologues that is homologous proteins present in different species. (Other bacterial homologues are very different in sequence.)

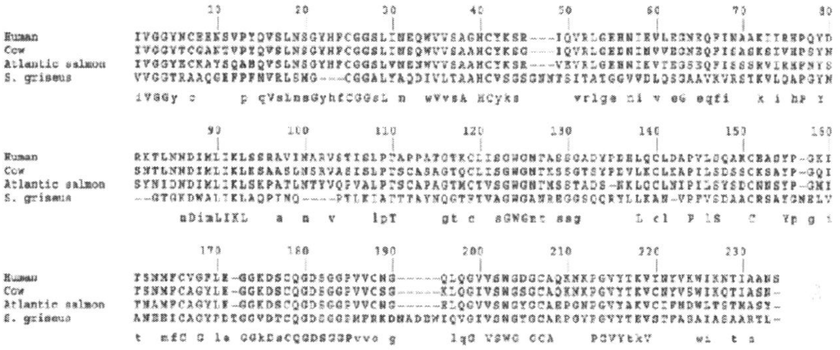

```
                    10        20        30        40        50        60        70        80
                    |         |         |         |         |         |         |         |
Human           IVGGYNCEEKSVPYQVSLNSGYHFCGGSLINEQWVVSAGHCYKSR----IQVRLGEHNIEVLEGNEQFINAAKIIRHPQYD
Cow             IVGGYTCGAKTVPYQVSLNSGYHFCGGSLINSQWVVSAAHCYKSG----IQVRLGEDNINVVEGNEQFISAGKSIVHPSYN
Atlantic salmon IVGGYECKAYSQAHQVSLNSGYHFCGGSLVNENWVVSAAHCYKSR----VEVRLGEHNIEVTEGSSQFISSSRVIKHPNYS
S. griseus      VVGGTRAAQGEFPFNVRLSNG----CGGALYAQDIVLTAAHCVSGSGNNTSITATGGVVDLQSGAAVKVRSTKVLQAPGYN

                iVGGy  c     p  qVslnsGyhfCGGsL n  wVvsA HCyks      vrlge ni v eG eqfi    x i hP Y

                    90       100       110       120       130       140       150       160
                    |         |         |         |         |         |         |         |
Human           RKTLNNDIMLIKLSSRAVINARVSTISLPTAPPATGTKCLISGWGNTASSGADYPEHLQCLDAPVLSQAKCEASYP-GKI
Cow             SNTLNNDIMLIKLKESAASLNSRVASISLPTSCASAGTQCLISGWGNTKSSGTSYPEVLKCLKAPILSDSSCKSAYP-GQI
Atlantic salmon SYNIDNDIMLIKLSKPATLNTYVQFVALPTSCAPAGTMCTVSGWGNTMSSTADS-NKLQCLNIPILSYSDCNNSYP-GNI
S. griseus      --GTGKDWALTKLAQPTNQ-----PTLKIATTAYNQGTFTVAGWGANREGGSQQRYLLKAH-VPFVSDAACRSAYGNELV

                nDimLIKL    a   n   v     lpT      gt c  sGWGnt ssg       L cl  P lS   C   Yp g i

                   170       180       190       200       210       220       230
                    |         |         |         |         |         |         |
Human           YSNMFCVGFLE-GGKDSCQGDSGGPVVCNG-------QLQGVVSNGDGCAQKNKPGVYTKVYNYVKWIKNTIAANS
Cow             YSNMPCAGYLE-GGKDSCQGDSGGPVVCSG-------KLQGIVSNGSGCAQKNKPGVYTKVCNYVSWIKQTIASE-
Atlantic salmon TNAMPCAGYLE-GGKDSCQGDSGGPVVCNG-------RLQGVVSNGYGCAEPGNPGVYAKVCIFHDWLTSTMASY-
S. griseus      ANEEICAGYPDTGGVDTCQGDSGGPMFRKDNADEWIQVGIVSNGTGCARPGYPGVYTEVSTFASAIASAARTL-

                t  mfC G le GGkDsCQGDSGGPvvc g         lqG VSWG GCA       PGVYtkV      wi  t s
```

Figure 9.20 Amino acid sequences of trypsins from human, cow, Atlantic salmon and *Streptomyces griseus*

In Figure 9.20 lines under the blocks of alignment sequences, uppercase letters indicate absolutely conserved residues and lowercase letters indicate residues conserved in three of the four sequences (in most but not all cases *S. griseus* is the exception).

It is also an evolution that has created related enzymes in the same species with different specificities. Similar to trypsin, chymotrypsin and pancreatic elastase are other digestive enzymes that cleave peptide bonds, but next to different residues: Chymotrypsin cleaves adjacent to large flat hydrophobic residues (Phe, Trp) and elastase cleaves adjacent to small residues (Ala). The change in specificity of the enzyme is due to the mutations of residues at the specific locations in the polypetide chain.

In addition to above-mentioned homologue, leukocyte elastase is essentially an enzyme needed for phagocytosis and defence against infection. Under certain conditions it is responsible for lung damage leading to emphysema. Homologous proteins in the same species are called paralogues. Trypsin, chymotrypsin and pancreatic elastase are proteolytic enzymes of the digestive tract. Another set of paralogues facilitates the cascade of events that occur during the process of blood coagulation. Although all are proteinases, the requirements for activation and control are very different for digestion and blood coagulation, and the families have diverged and become specialized for these respective functions.

It is important to note that some homologues of trypsin have developed entirely new functions. For example, haptoglobin is a chymotrypsin homologue that has lost its proteolytic activity. It acts as a chaperone, preventing unwanted aggregation of proteins. Haptoglobin forms a tight complex with haemoglobin fragments released from erythrocytes, with several useful effects including preventing the loss of iron. Similarly, the insects immune protein, scolexin is a distant homolog of serine proteinases that induce coagulation of haemolymph in response to infection.

In case of the chymotrypsin family there is the retention of structure with similar functions in closely-related proteins, and progressive divergence of function in some but not in all distantly-related ones. It indicates that the overall folding pattern of a protein is an unreliable guide to predicting function, especially for very distant homologues. For correct prediction of function in distantly-related proteins it is necessary to focus on the active site present in the protein under investigation. For example, the viral 3C proteinases were recognized as distant chymotrypsin homologues, despite the fact that the serine of the catalytic triad is changed to cysteine.

PROTEIN-PROTEIN INTERACTIONS

The cell or an organism has thousands of proteins and the association of protein molecules within the possesser constitutes protein–protein interactions. The study of these associations from the point of biochemical pathways, their role in signal transduction and other networks is of prime importance in proteomics. The subject of protein–protein interactions represents a vast ensemble of results from biological, biochemical and biophysical studies carried out to date and cannot be treated in its entirety in any reasonable fashion. Imagine a cell in which, suddenly, the specific interactions between proteins would disappear. This unfortunate cell would become deaf and blind, paralytic and finally would disintegrate, because specific interactions are involved in almost any physiological process. Detection of extracellular signals is a matter of receptor to adaptor interactions and the shape of the cell is maintained by an intricate network of structural protein interactions. Finding interprotein interactions involved in common cellular functions is a way to get a broader view of how proteins work cooperatively in a cell.

The interactions among proteins are operative at almost every level of cell function, in the structure of sub-cellular organelles, the transport machinery across the various biological membranes, packaging of chromatin, the network of sub-membrane filaments, muscle contraction, and signal transduction, regulation of gene expression, to name a few. Aberrant protein–protein interactions have implicated in a number of neurological disorders such as Creutzfeld–Jacob and Alzheimer's disease. This will be concerned mainly with the biophysical aspects of the finely tuned specific interactions between proteins involved in the regulation of cell function, namely those proteins implicated in signal transduction and transcriptional regulation. Signal transduction plays a fundamental role in many biological processes and in many diseases (e.g. cancer). Because of their importance in development and disease, these systems have been the object of intense research for many years. It has emerged from these studies that nature has employed in many instances a strategy of mixing and matching of domains that specify particular classes of protein–protein interactions, modifying the amino acid sequence in order to confer specificity for particular target proteins. The regulation of cell function, brought about by the interactions of these proteins, is delicately balanced by the relative affinities of the various protein partners and the modulation of these affinities by the binding of ligands, other proteins, nucleic acids, ions such as Ca^{2+}, and covalent modification, such as specific phosphorylation or acetylation reactions. Specificity and the strength of signal transduction is encoded by the exact amino acid sequence of the domain, and it is this relationship between sequence, structure, dynamics, energetics and function that constitutes the fundamental issue for the biophysics of protein–protein interactions. Thus, protein–protein interactions are so significant in virtually every process in a living cell that can assist in the interpretation of diseases and can provide altrenate way to treat the disorder.

Predication Methods for Protein–Protein Interactions

Considering the significance of interprotein interactions in proteomics, it is very much essential to predict structural and functional aspects of the components in the interactions using various methods. There are several approaches to map out the protein–protein interactions and each of the approaches has its own strength and weakness. X-ray and NMR techniques assist in identification of components of the interactions. Following are the alternative approaches that help in the better understanding of the functional aspects of the components in protein–protein interactions:

- Yeast two-hybrid (Y2H) screen system detects the interaction between artificial fusion proteins inside the nucleus of yeast. In 1991, the Y2H method was developed by Stan Fields at University of

Washington, before the term proteomics was coined. Y2H method is designed to use a protein of interest as bait in order to discover proteins that physically interact with the bait protein; these proteins are called prey. Its principle is displayed in Figure 9.21. In this approach, every protein in a proteome can be tested individually for its potential to interact with the bait. Hence Y2H method is systematically applied to study interactions at the whole proteome level. In addition, the greatest benefit of Y2H is that yeast cells can express genes from almost any species, which means this is a powerful proteomics method for *Drosophila, C.elegans, Arabidopsis,* zebra fish, mice, humans, and of course yeast too.

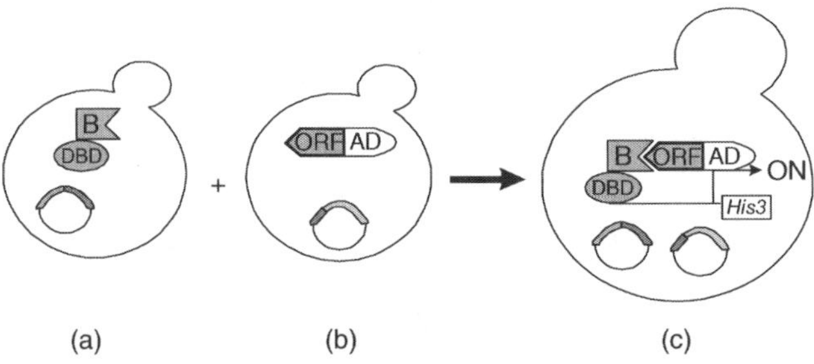

Figure 9.21   Yeast two-hybrid method. a) The DNA-binding domain (DBD) is fused onto the protein of interest "B". This construct, the "bait", is encoded by the plasmid shown in the same yeast cell. b) The "prey" is encoded by its own plasmid and is composed of an ORF fused onto the activation domain (AD), which is capable of activating RNA polymerase. c) Both the bait and prey are inside the same cell, and if the B and ORF protein physically interact, RNA polymerase will be able to transcribe a reporter gene (in this case *His3*) that leads to production of the amino acid histidine.

- **Co-immunoprecipitation** is the technique in which an antibody is raised to a "bait" that interacts and binds with any other "prey" protein. The interacting protein partners are subsequently purified and analysed by Western blotting or mass spectrometry. Interactions detected by this approach are considered to be real. However, this technique has disadvantage that it can verify only interactions between suspected interaction partners.

- **Quantitative immunoprecipitation combined with knock-down (QUICK)** method relies on co-immunoprecipitation, quantitative mass spectrometry and RNA interference (RNAi). Interactions among endogenous non-tagged proteins can be predicted by this technique. This method requires suitable antibodies for co-immunoprecipitation.

▣ **Chemical cross-linking** fixes interacting partners in the complex to make their isolation easier. Subsequently, they are subjected to proteolytic digestion and high-mass MALDI mass spectrometry so that interacting proteins are identified.

▣ **Phase display** is the technique in which a phage vector (f1,M13 or λ) is designed. Genes for a large number of proteins are individually fused to the gene for a phage-coat protein. This allows the display of foreign protein on phage coat, hence permits its possible interaction with other proteins that the phage encounters. A phage display library consisting of a collection of clones displaying a range of proteins on phage coat can also be used to identify the proteins that interact with a protein under investigation.

▣ **Tandem affinity purification (TAP)** approach allows high-throughput identification of interprotein interactions. In contrast to Y2H approach, accuracy of the method can be compared to those of small-scale experiments and the interactions are detected within the correct cellular environment as by co-immunoprecipitation. In 2006, Krogan *et al.* and Gavin *et al.* performed genome-based TAP experiments and provided updated protein interaction data for yeast organism. Figure 9.22 shows different kinds of results when yeast proteins of unknown function were tested by TAP isolation. Two successive affinity purifications are done using two tags fused to the protein of interest; the first step involves binding of the protein A tag to an IgG column. Under mild conditions, the putative complex is released from the column by a specific viral protease and rebound to a second column, where calcium-dependent interaction of calmodulin with a Calmodulin Binding Peptide (CBP) will allow the enrichment of the complex.

It is a good idea to test whether the depletion of the studied protein gives rise to a particular phenotype. If the tagged protein used for complex purification is functional, this phenotype should not be observed. That's why it is an advantage to study one particular protein by the affinity purification of the associated complex in a yeast strain where the functional protein is essential. In this way it can be sure that the tag did not interfere with the protein function and that the purified complex might be the "physiological one".

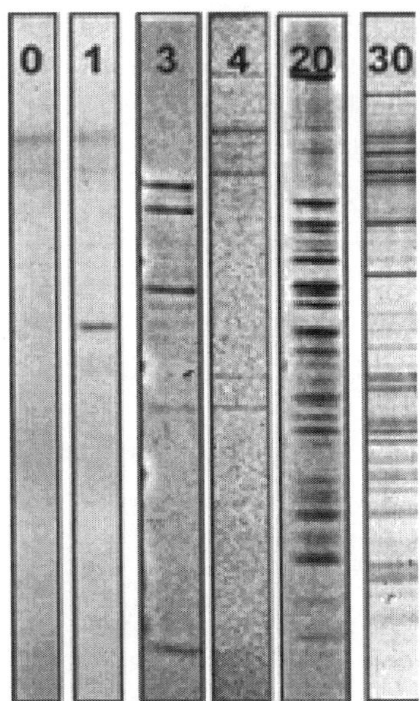

Figure 9.22　Number of proteins isolated by using the TAP strategy in yeast with different unknown proteins as "baits"

Other methods for observing the interacting proteins include Bimolecular Fluorescence Complementation (BiFC), Fluorescence Resonance Energy Transfer (FRET), Dual Polarization Interferometry (DPI), Static Light Scattering (SLS), Fluorescence Correlation Spectroscopy (FCS). Complementary information about protein–protein interaction is provided by following additional approaches:

- **Phylogenetic distribution pattern**　Phylogeny explains the relationship among species, populations, and individuals and even among the proteins or genes. Phylogenetic profiling of a protein is the set of organisms in the protein or its homologues is present. It finds pairs of protein families with similar patterns of presence or absence across large numbers of species. Proteins sharing phylogenetic profile may have functional link or expected to co-evolve.

- **Prediction of co-evolved protein pairs based on similar phylogenetic trees**　This approach involves use of a sequence search tool such as BLAST for finding homologues of a pair of proteins, then building multiple sequence alignments with alignment tools such as ClustalW. From these multiple sequence alignments, phylogenetic distance matrices are calculated for each protein in the hypothesized interacting pair. If the matrices are sufficiently similar they are deemed likely to interact. Aligned using ClustalW and PAM250 matrix, the phylogenetic tree of beta globin is displayed in Figure 9.23. Common ancestor is at the root of this tree in which the implication is that for the given protein, the Carp (fish) and human are not closely related while the amino acid sequences in a beta globin of human and gorilla have great similarities since they have very close taxonomic relationship.

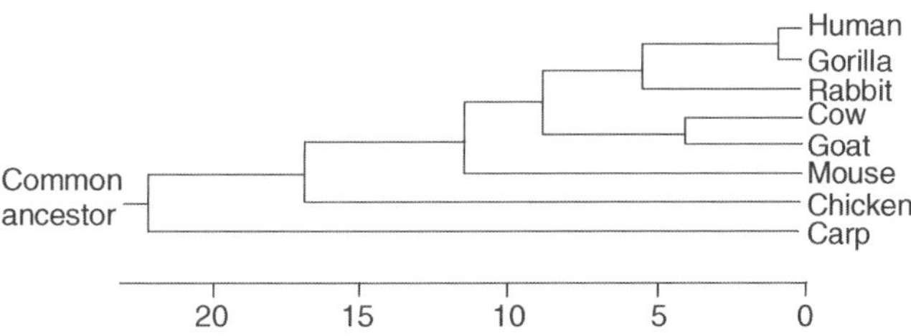

Figure 9.23 Phylogenetic tree of beta globin drawn using PAM250 matrix

Protein–Protein Interaction Databases

- **Human Protein Reference Database** (HPRD; http://hprd.org/), a manually curated database of human protein information with visualization tools.

- **I2D** (I2D; http://ophid.utoronto.ca/ophidv2.201/ppi.jsp) designed to be both a resource for the laboratory scientist to explore known and predicted protein–protein interactions, and to facilitate bioinformatics initiatives exploring protein interaction networks.

- **Interaction Database** (IntAct; http://www.ebi.ac.uk/intact) a public repository for manually curated molecular interaction data from literature.

- **Database of Interacting Proteins** (DIP; http://dip.doe-mbi.ucla.edu/) a manual and automatic catalogue of experimentally determined interactions between proteins.

- **Michigan Molecular Interaction index** (MiMI; http://mimi.ncibi.org/) integrates data from multiple sources.

- **Molecular INTeraction Database** (MINT; http://mint.bio.uniroma2.it/mint/) focuses on experimentally verified protein interactions mined from the scientific literature by expert curators.

- **Mammalian Protein–Protein Interaction Database** (MIPS; http://mips.gsf.de/proj/ppi/) integrates interprotein interactions in mammals.

- ▣ Microbial Protein–Protein Interactions(MiPPI; http://mippi.ornl.gov/) database of interacting proteins in microbes.

- ▣ Biomolecular Interaction Network Database (BIND; http://bond.unleashedinformatics.com/Action?) a manual of interacting network of biomolecues.

- ▣ The Microbial Protein Interaction Database (MPIDB; http://www.jcvi.org/mpidb) integrates microbial interactions and curates microbial interactions from the scientific literature.

Interaction Network Software and Web Servers

- ▣ Agile Protein Interaction Data Analyser (APID; http://bioinfow.dep.usal.es/apid/) is an interactive bioinformatics web tool to explore and analyse in unified and comparative platform mainly currently known information about protein–protein interactions.

- ▣ APID2NET (unified interactome graphic analyser) (http://bioinfow.dep.usal.es/apid/apid2net.html) is an open access tool, included in Cytoscape, which allows surfing unified interactome data by querying APID server and facilitates dynamic analysis of the protein–protein interaction networks.

- ▣ Cytoscape (http://www.cytoscape.org/) is an open source bioinformatics software platform for visualizing molecular interaction networks and integrating these interactions with gene expression profiles and other state data.

- ▣ NetPro (http://www.molecularconnections.com/product.html#netpro) is a comprehensive fully hand-curated knowledgebase of protein–protein, protein–small molecules DNA and RNA interactions.

- ▣ PA 800 Protein Characterization System (http://www.celeader.com/genfiles/CELeadercomPA8_05124356.asp?id=1.2&ln=112) uses Capillary Electrophoresis technology to determine a protein's molecular weight, resolve differences in isoelectric point, generate high-resolution peptide maps and carbohydrate profiles, and provide front-end separation and introduction of these proteins to mass spectrometry.

- ▣ Matrix (http:// orion.icmb.utexas.edu/cgi-bin/matrix/matrix-index.pl)a web server that predicts interactions between two protein families.

- ▣ ADVICE (Automated Detection and Validation of Interaction based on the Co-Evolution) (http://advice.i2r.a.star.edu.sg) a web server providing prediction of interacting proteins using mirror-tree approach.

▣ **STRING** (Search Tool for the Retrieval of Interacting Genes/proteins) (http://www.bork.embl-heidelberg.de/STRING/) a web server predicting gene and protein interactions based on combined evidence of gene linkage, gene fusion and phylogenetic profiles.

PRACTICAL APPLICATIONS OF PROTEOMICS

Drug Discovery

One of the most promising developments evolved from the study of human genes and proteins has been the identification of potential new drugs for the treatment of disease. It relies on genome and proteome information to identify proteins associated with a disease, which computer software can then use it as targets for new drugs. For example, if a certain protein is implicated in a disease, its 3D structure provides the information to design drugs to interfere with the action of the protein. A molecule that fits the active site of an enzyme, but cannot be released by the enzyme, will inactivate the enzyme. This is the basis of new drug-discovery tools, which aim to find new drugs to inactivate proteins involved in disease. As genetic differences among individuals are found, researchers expect to use these techniques to develop personalized drugs that are safe and more effective for the individual.

A computer technique which attempts to fit millions of small molecules to the three-dimensional structure of a protein is called " ". The computer rates the quality of the fit to various sites in the protein, with the goal of either enhancing or disabling the function of the protein, depending on its function in the cell. A good example of this is the identification of new drugs to target and inactivate the HIV-1 protease. The HIV-1 protease is an enzyme that cleaves a very large HIV protein into smaller, functional proteins. The virus cannot survive without this enzyme; therefore, it is one of the most effective protein targets for killing HIV.

Better Diagnosis of Disease

Proteomics has looked at tumour biopsies and blood samples for indicators of cancer, birth defects, and other medical conditions. One diagnostic challenge is distinguishing between disorders with similar symptoms, but requiring different treatments. Proteomics will lead to research breakthroughs allowing doctors to better diagnose and treat diseases. Those pursuing proteomics hope to find biological markers that signal disease, targets for drugs, and detailed understanding of biology on the molecular level. Proteomics may hold the key to personalized medicine. People being people, it is not surprising that most proteomics research so far has been aimed at medical applications. This is not to say that you will see proteomics in your doctor's office. Proteomics is a research discipline that is carried out in laboratories at universities, government institutes, biotech firms and big pharma.

Diagnosis of Cancer

Although proteomics is a relatively new research area, there have already been some promising results in the diagnosis of cancer. For example, proteins have been identified that can be used to diagnose breast cancer, colon cancer and bladder cancer. A protein called stathmin has been identified that is expressed at unusually high levels in cases of childhood leukaemia.

Interestingly, the stathmin protein is phosphorylated in cancer patients —it has an additional phosphate group added to it. Many proteins are modified by phosphorylation after they are synthesized. Such modifications change the chemical properties of the protein allowing the phosphorylated and non-phosphorylated forms to be separated by two-dimensional gel electrophoresis. In the case of stathmin, only the phosphorylated form is associated with childhood leukaemia. This emphasizes the importance of proteomics in disease diagnosis, because a change in protein modification associated with cancer cannot be detected using DNA arrays.

In the case of bladder cancer, proteomics analysis has identified several keratin proteins that are expressed in different amounts as the disease progresses from the early transitional epithelium stage to full blown squamous cell carcinoma. The measurement of keratin levels in bladder cancer biopsies can therefore be used to monitor the progression of the disease. Another protein, psoriasin, is found in the urine of bladder cancer patients and can be used as an early diagnostic marker for the disease. This provides another example of how proteomics, but not DNA arrays, can be used in cancer diagnosis. Urine, in common with most bodily fluids, contains proteins but no RNA.

Monitoring of Disease through Biomarkers

Proteomics can also search for biomarkers that indicate the stage of a disease, or the response of the patient to treatment. Understanding the proteome, the structure and function of each protein and the complexities of protein–protein interactions will be critical for developing the most effective diagnostic techniques and disease treatments in the future. An interesting use of proteomics is using specific protein biomarkers to diagnose disease. A number of techniques allow testing for proteins produced during a particular disease, which helps to diagnose the disease quickly. Techniques include Western blot, immunohistochemical staining, ELISA or mass spectrometry.

Assessment of heart disease commonly use several key protein-based biomarkers. Standard protein biomarkers for CVD (Cardio Vascular Disease) include interleukin-6, interleukin-8, serum amyloid A protein, fibrinogen, and troponins. The cTnI (cardiac troponin I) increases in concentration within 3 to 12 hours of initial cardiac injury and can be found elevated days after an acute myocardial infarction. A number of commercial antibody-based assays as well as other methods are used in hospitals as primary tests for acute myocardial infarction.

In Alzheimer's disease, elevations in β-secretase create amyloid/β-protein, which causes plaque to build up in the patient's brain leading to dementia. Immunohistochemical staining procedure in which antibodies bind to specific antigens or biological tissue of amyloid/β-protein is used to test the increase in amyloid/β-protein. As an effective mode of treatment by targeting the β-secretase enzyme there is decrease in the amyloid/β-protein as a result the progression of the disease slows down.

REVIEW QUESTIONS

1. Define proteomics. Explain different methods of studying proteins.
2. What are the various proteomics databases?
3. Describe the steps involved in construction of ProteinAtlas database.
4. Describe the structure of entries in the UniProtKB/Swiss-Prot database.
5. What is PIR? Explain variety of resources offered by PIR.
6. What is Swiss-Prot? Explain the structure of Swiss-Prot entries.
7. What is Protein Data Bank? What is the latest status of PDB?
8. Describe the details of the PDB file format.
9. What are the points of differences between mmCIF file format and MMDB file format?
10. What types of software are present for visualization of protein structure?
11. Enlist the advantages and disadvantages of RasMol and SwissPdbViewer.
12. Explain the role of CATH and SCOP in protein structure classification.
13. What are the different protein structure bioinformatics software?

14. Describe the significance of protein structure prediction and the software for the prediction of protein structure.

15. What are the different steps involved in homology modelling of a protein.

16. Distinguish between homology modelling and *ab initio* protein modelling method.

17. Would it be possible to predict protein structure with the help of Rosetta? If yes, explain the mechanism.

18. What are the different classes of protein functions? Explain the methods for prediction of protein function.

19. What types of interactions are there in between two different protein molecules?

20. Explain the methods for investigation of protein–protein interaction.

21. Describe the factors affecting protein folding.

22. Enlist the databases of protein–protein interaction.

23. What correlation do you see between proteomics and diagnosis of diseases?

27. Write short notes on:
      i.   Protein microarrays
      ii.   PIR-PSD
      iii. Rosetta@Home
      iv.   PDBsum
      v.   CASP
      vi.   VAST
      vii. DALI
      viii. Chime
      ix.   FSSP
      x.   Diversity in functions of protein
      xi.   Y2H method
      xii. Phylogenetic tree
      xiii. Biomarkers
      xiv. Co-immunoprecipitation
      xv.   Protein-folding disorders
      xvi. Tandem affinity purification

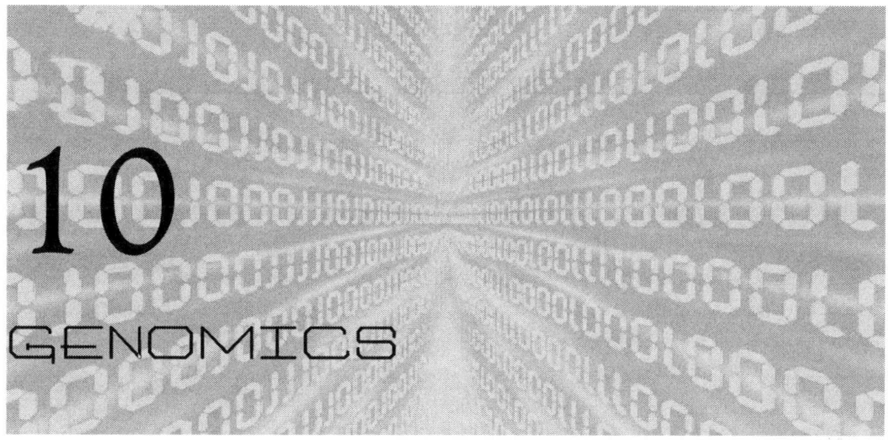

10
GENOMICS

INTRODUCTION

Gene

Gene is the fundamental, physical and functional unit of heredity. It is an ordered sequence of nucleotides located in a particular position on a particular chromosome that encodes a specific functional product (i.e., a protein or RNA molecule). A single gene coding for a particular protein corresponds to a sequence of nucleotides along with one or more regions of a molecule of DNA. The DNA sequence is collinear with the protein sequence. In species for which the genetic material is double-stranded DNA, coding sequence of the gene may appear on either strand. Bacterial genes are continuous regions of DNA. Therefore, the functional unit of genetic sequence information from a bacterium, for example, a string of 300 nucleotides encodes a string of 100 amino acids, or a string of 100 nucleotides encodes a structural RNA molecule of 100 residues.

Whereas in eukaryotes, the relationship between the size of gene and the size of protein encoded is very different from that in bacteria. Eukaryotic genomic DNA that corresponds to the coding part of genes is not continuous, but consists of exons and introns. Exons are the part of the gene that code for proteins and they are interspersed with non-coding introns which must be removed by splicing. The number and size of introns and exons differ considerably between genes and also between species. Only very few genes in yeast have introns, while for human there are about 4 introns per gene on average, and the average size of exons is 150 bp and just above 3400 bp for introns. There is a cellular machinery that splices pre-mRNA, removes introns and forms RNA transcripts, which subsequently translated to synthesize a string of amino acids. Genome size neither correlates with evolutionary status, nor is proportionate to the number of genes. The comparative genome sizes of humans and other organisms being studied is given in Table 10.1.

Table 10.1   List of some organisms showing the number of chromosomes and genes with their average gene density

Organism	Number of chromosomes	Number of genes	Average gene density
H. influenzae (bacterium)	1	1738	1 gene per 1000 bases
Escherichia coli (bacterium)	1	4377	1 gene per 1400 bases
Saccharomyces cerevisiae (yeast)	32	6000	1 gene per 2000 bases
Caenorhabditis elegans (roundworm)	12	19,099	1 gene per 5000 bases
Arabidopsis thaliana (plant)	10	25,498	1 gene per 4000 bases
Drosophila melanogaster (fruit fly)	8	13,601	1 gene per 9000 bases
Mus musculus (mouse)	40	~25,000	1 gene per 100,000 bases
Homo sapiens (humans)	46	~25,000	1 gene per 100,000 bases

Genome

H. Winkler (1920) defined genome as the complete set of chromosomal and extra-chromosomal genes of an organism or virus. Every organism has a definite number of the chromosomes having specific number of base pairs. A typical bacterium has a diameter of about 0.001 mm that contains the genome in the form of 4,639,221 base pairs in a single, circular DNA molecule which can be extended to about 2 mm long. The DNA of higher organisms is organized into chromosomes. Normal human cells contain 23 chromosome pairs that consist of three billion base pairs of DNA. The total amount of genetic information per cell (the sequence of nucleotides of DNA) is very nearly constant for all members of a species, but varies widely between species. Genome also has DNA that do not code for any protein as well as some genes have multiple copies. In such case, the total amount of coding sequences in a cell is difficult to estimate from the genome size. Table 10.2 enlists few organisms with their genome size, number of genes predicted and the name of institutes related to the genomic projects.

GENOMICS

In 1986, Dr. Tom Roderick, a geneticist at the Jackson Laboratory (Bar Harbor, ME) used the term genomics and describes it as a scientific discipline of mapping, sequencing and analysing the genome. In short, it is the study of an organism's entire genome. The field includes intensive efforts to determine the entire DNA sequence of organisms and fine-scale genetic mapping efforts.

Genomics includes the following issues.

- Sequencing the genome
- Identifying genes (transcribed DNA) on the genome
- Interactions between loci and alleles within the genome
- Determining gene function
- Discover how each gene is regulated
- Study natural variations in gene among/between species
- Study variations between healthy/diseased genes

Genomics is applied for

- identification of genes within genomic DNA sequences.
- the alignment and matching the homologous gene sequences in databases to determine function.
- the prediction of the structure and function of gene products.
- describing the interactions between genes and gene products.
- the estimation of phylogenetic relationships.

Table 10.2   List of few organisms with their genome size

Organism	Type	Comment	Genome size	Number of genes predicted	Organization
φX174	Virus	Virus infecting *E. coli*	5.3 kb	10	--
Homo sapiens mitochondrial DNA	--	Cell organelle	1.67 kb	--	- -
Escherichia coli	Bacteria	Pioneer bacteria in molecular biology research	4.6 Mb	4,377	--
Plasmodium falciparum	Parasitic protozoan	Human pathogen (malaria)	22.9 Mb	5,268	Malaria Genome Proje Consortium
Leishmania major	Parasitic protozoan	Human pathogen	32.8 Mb	8,272	Sanger Institute
Trypanosoma brucei	Parasitic protozoan	Human pathogen (sleeping sickness)	26 Mb	9,068	Sanger Institute and TIGR
Entamoeba histolytica	Parasitic protozoan	Human pathogen (amoebic dysentery)	23.8 Mb	9,938	TIGR, Sanger Institute and the London Schoo of Hygiene and Tropica Medicine

Arabidopsis thaliana	Wild mustard	Model plant	120 Mb	25,418	*Arabidopsis* Genome Initiative
Oryza sativa	Rice	Crop and model organism	420 Mb	32–50,000	Beijing Genomics Institute, Zhejiang University and the Chinese Academy of Sciences
Vitis vinifera	Grapevine PN40024	Fruit crop	490 Mb	30,434	The French-Italian Public Consortium for Grapevine Genome Characterization
Aspergillus niger	Fungus	Biotechnology fermentation	33.9 Mb	14,165	--
Aspergillus oryzae	Fungus	Used to ferment soy	37 Mb	12,074	National Institute of Technology and Evaluation
Candida glabrata	Fungus	Human pathogen	12.3 Mb	5,283	Génolevures Consortium
Anopheles gambiae	Mosquito	Vector of malaria	27.8 Mb	13,683	Celera Genomics and Genoscope
Apis mellifera	Honeybee	Social insects	1.8 Gb	10,157	The Honeybee Genome Sequencing Consortium
Bombyx mori	Moth (domestic silk worm)	Silk production	530 Mb	--	University of Tokyo and National Institute of Agrobiological Sciences

(*Contd.*)

Table 10.2 (Continued)

Organism	Type	Comment	Genome size	Number of genes predicted	Organization
Caenorhabditis elegans	Nematode worm	Model animal	97 Mb	19,000	Washington University and the Sanger Institute
Bos taurus	Cow	6*	3.0 Gb	--	Cattle Genome Sequencing International Consortium
Canis lupus familiaris	Dog	7.6*	2.4 Gb	19,300	Broad Institute and Agencourt Bioscience
Equus caballus	Horse	6.8*	2.1 Gb	--	Broad Institute *et al.*
Felis catus	Cat	2*	3 Gb	20,285	The Genome Sequencing Platform, The Genome Assem Team
Homo sapiens	Human	--	3.2 Gb	25,000	Human Genome Project Consortium and Celera Genor
Macaca mulatta	Rhesus Macaque	6*	--	--	Macaque Genome Sequencing Consortium
Mus musculus	Mouse	--	2.5 Gb	24,174	International Collaboration fo Mouse Genome Sequencing
Oryctolagus cuniculus	Rabbit	2*	2.5 Gb	--	Broad Institute *et al.*

Pan troglodytes	Chimpanzee	6*	3.1 Gb	--	Chimpanzee Sequencing and Analysis Consortium
Pongo pygmaeus	Orang-utan	--	3.0 Gb	--	Institute for Molecular Biotechnology
Rattus norvegicus	Rat	1.8* or better	2.8 Gb	21,166	Rat Genome Sequencing Proje Consortium

* Mammals showing shotgun coverage.

Source: http://www.broad.mit.edu/mammals and http://www.ncbi.nlm.nih.gov/sites/entrez

A genome is the sum total of all an individual organism's genes and the genomics is "the study of all the genes of a cell, or tissue, at the DNA (genotype), mRNA (transcriptome), or protein (proteome) levels." Genomics was established by Fred Sanger when he first sequenced the complete genomes of a virus and a mitochondrion. His group established techniques of sequencing, genome mapping, data storage, and bioinformatic analyses in the 1970–1980s. A major branch of genomics is still concerned with sequencing the genomes osf various organisms, but the knowledge of full genomes has created the possibility for the field of functional genomics, mainly concerned with patterns of gene expression during various conditions. Here the most important tools are microarrays and bioinformatics. The study of the full set of proteins in a cell type or tissue, and the changes during various conditions, is called proteomics.

GENOME MAPPING

Genome mapping is the first step to understand a genome structure. It is a process of identifying relative location of genes, mutations or traits on a chromosome. In many eukaryotes, genetic markers (identifiable portion of a chromosome whose inheritance pattern can be followed) represent morphological phenotypes. A cytological map of a chromosome means a banding pattern seen on stained chromosome that can be directly observed under a microscope. The light and dark bands observed on a stained chromosome are the visually distinct markers. However, the banding patterns on a chromosome are not always constant and they are subject to change depending on the extent of chromosomal contraction. Hence, cytological maps are very low resolution and inaccurate physical maps.

The genetic map of a chromosome represents the relative position and the frequency pattern of genetic markers. The closer genetic markers inherited together and are not separated during crossing over because they formed a linkage group, hence genetic map is also called genetic linkage map. The distance between two genetic markers is measured in centiMorgans (cM)— the frequency of recombination of genetic markers. One cM is referred to as one percentage of the total recombination events and it is approximately 1 Mb in humans and 0.5 Mb in *Drosophila*. Physical maps are maps of locations of identifiable landmarks on a genomic DNA regardless of inheritance pattern. Here, the distance between genetic markers is measured directly a kilobases (kb) or megabases (Mb). Physical maps of chromosomes are more accurate and reliable than genetic maps (Figure 10.1).

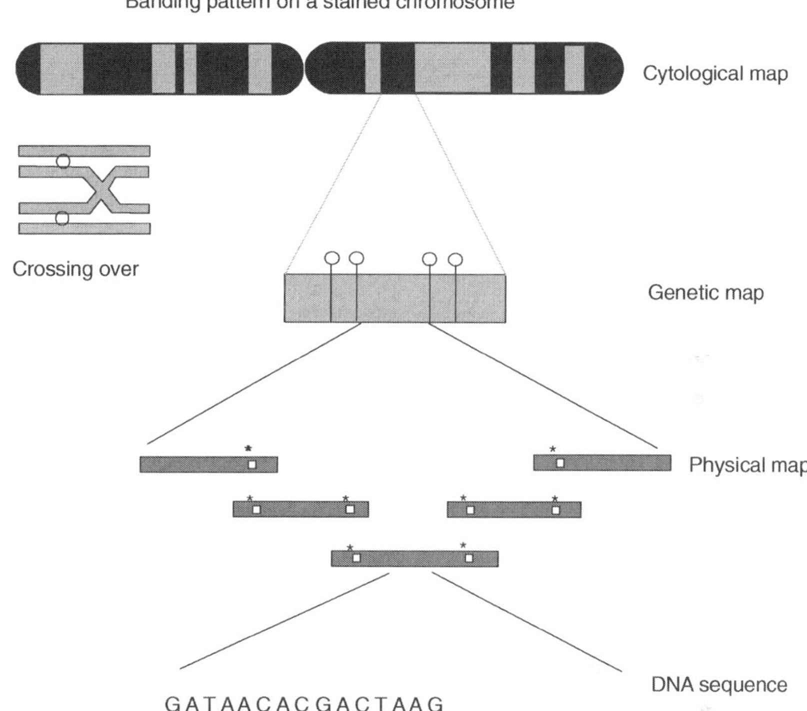

Figure 10.1 Different levels of resolution of genome maps relative to genomic DNA sequence

GENOME PROJECTSH

Biologists wanted to unravel the intricacies of gene structure of prokaryotes and eukaryotes. One of the significant developments in the field of molecular biology is the invention of genome sequencing technique that laid the foundation for the genome projects for various organisms. In 1972, it was the Walter Fiers and his team at the Laboratory of Molecular Biology of the University of Ghent in Belgium for the first time to determine the sequence of a gene for bacteriophage MS2 coat protein. In 1976, the same team determined the complete nucleotide sequence of bacteriophage MS2-RNA. In 1977, Frederick Sanger (two-time Nobel laureate) successfully sequenced the entire DNA-based genome of the bacteriophage ϕ X174 that has 5,368 base pairs.

Since 1995, genomes are being sequenced at a rapid pace. As of September 2007, the complete sequence was known of about 1879 viruses, 577 bacterial species and roughly 23 eukaryotes. Most of the bacteria whose genomes have been completely sequenced are problematic disease-causing agents, such as *Haemophilus influenzae*. Of the other sequenced species, most of them were chosen as model organisms such as yeast *(Saccharomyces cerevisiae)*, the fruit fly *(Drosophila melanogaster)*, the nematode worm *(Caenorhabditis elegans)*, the zebrafish *(Brachydanio rerio)*, the Japanese pufferfish *(Takifugu rubripes)*, the flower plant *(Arabidopsis thaliana)*, the dog *(Canis familiaris)*, the brown rat *(Rattus norvegicus)*, the mouse (*Mus musculus*), and chimpanzee *(Pan troglodytes)*.

Rapid progress in genome science and a glimpse into its potential applications have spurred observers to predict that biology will be the foremost science of the 21st century. Technology and resources generated by the HGP as an international scientific collaboration that seeks to understand the entire genetic blueprint of a human being. It was the landmark project with its primary goal of the sequencing of the three thousand million base pairs and other genomics research are already having a major impact on research across the life sciences. A rough draft of the human genome was completed by the Human Genome Project (HGP) in early 2001. The human (haploid) genome contains 22 pairs of autosomes and 2 sex chromosomes X or Y, requires 24 separate chromosome sequences in order to represent the complete genome size that ranges from 45–279 Mb of DNA, making a total genome content of 3,286 Mb ($\sim$3.3 $\times$ 10^{9}bp). In addition to the human genome, the genomes of about 800 organisms including many microbes have been sequenced in recent years. The web resources for information on sequenced genomes are:

- ▣ DOE Joint Genome Institute (http://jgi.doe.gov/)—Human, plant, animal, and microbial sequencing.
- ▣ GOLD (http://www.genomesonline.org/)—Genomes Online Database provides comprehensive access to information regarding complete and ongoing genome projects around the world.
- ▣ Comprehensive Microbial Resource (http://www.tigr.org/tigr-scripts/CMR2/CMRHomePage.spl)—A tool that allows the researcher to access all of the bacterial genome sequences completed to date.
- ▣ Entrez Genome (http://www.ncbi.nlm.nih.gov/entrez/query.fcgi?db=Genome)—A resource from the NCBI for accessing information about completed and in-progress genomes.

METHODS FOR GENE SEQUENCE ANALYSIS

Since the bacteriophage ϕX174 sequenced in 1977, the DNA sequences of hundreds of organisms have been decoded and stored in databases. The information is analysed to determine genes that encode polypeptides, as well as regulatory sequences. A comparison of genes within a species or between different species can show similarities between protein functions studied through protein–protein interactions, or such a comparison can also be useful to establish relations between species and construction of phylogenetic trees.

Considering the growing amount of data, it long ago became impractical to analyse DNA sequences manually. Hence computer programs are used to search the genome of thousands of organisms, containing billions of base pairs. These programs would compensate for mutations (where a nitrogen base is exchanged, deleted or inserted) in the DNA sequence, in order to identify sequences that are related, but not identical. A variant of this sequence alignment is used in the sequencing process itself. The shotgun sequencing approach is one of the attempts, which was used by J. Craig Venter and H. Smith at The Institute for Genomic Research (TIGR) to sequence the first bacterial genome, *Haemophilus influenzae.*

Shotgun sequencing approach   The steps involved in the whole genome shotgun sequencing technique are shown in Figure 10.2 and are briefly explained as follows:

1. The chromosome isolated from the cells is randomly fragmented into small pieces using ultrasonic waves.

2. The fragments are purified and attached to plasmid vector. Plasmids with single insert are isolated and library of plasmid clones are prepared by transforming *E.coli* with plasmid that lacked restriction enzymes.

3. DNA purified from plasmid and thousands of DNA fragments are sequenced using automated sequencer by employing primers labelled with special dyes. These recognize the plasmid DNA sequences next to bacterial DNA insert. The process is repeated several times for accurate final results.

4. Using special computer program, the sequenced DNA fragments are clustered and assembled into longer stretches of sequence by comparing nucleotide sequence overlaps between fragments. Two fragments are joined to form a large stretch of DNA (contigs), if the

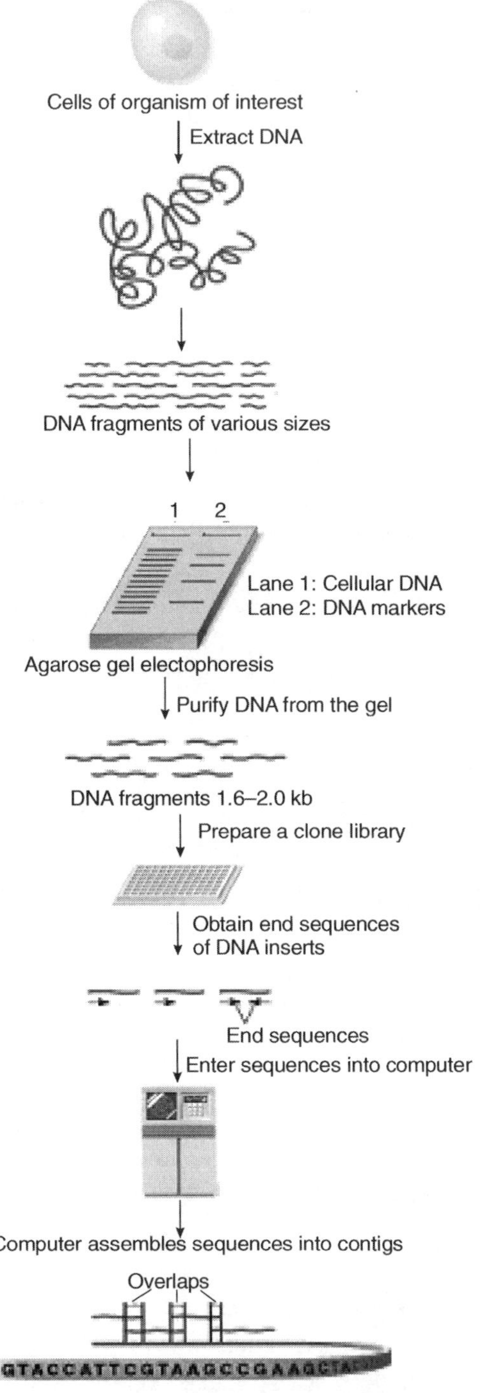

Figure 10.2   Steps in shotgun sequencing genome approach

overlapping sequences are matched. The fragments having overlaps with two contigs gaps are filled with oligonucleotides probes that allow them to place side-by-side and fill in the gap between them.

5. Finally, proofreading of the sequence is done to resolve ambiguities in the sequence and the frameshift mutation is corrected if its presence is detected during the process. For completing the sequencing of genome with the 5,00,000 bp the process requires at least four months.

The process does not give a sequential list of nucleotides, but instead the sequences of thousands of small DNA fragments (each about 600–800 nucleotides long). Shotgun sequencing yields sequence data quickly, but the task of assembling the fragments can be quite complicated for larger genomes. In the case of the Human Genome Project, it took several days of CPU time (on one hundred Pentium III desktop machines clustered specifically for the purpose) to assemble the fragments. Hundreds of thousands of overlapping ~500 bp sequences assembled with fast computers operating in parallel (supercomputer). Shotgun sequencing is the method of choice for virtually all genomes sequenced that combines robotics, lasers, and computers to measure the distance between genetic markers and genome assembly algorithms are a critical area of bioinformatics research.

Role of BAC and YAC in genome sequencing   The physical map consists of contigs of DNA clones that can be correlated with the mapped markers. Currently, Bacterial Artificial Chromosome (BAC), and phage P1-derived Artificial Chromosome (PAC) and Yeast Artificial Chromosomes (YAC) contig maps that cover the entire genome are available.

- ▣ Bacterial artificial chromosome (BAC) is cloning vector that can accommodate large pieces of DNA up to 1 million base pairs. BACs are based on the single-copy F factor plasmid and accept >300 kbp DNA fragments. They can remain for 100 cycles with high degree of stability.

- ▣ P1-derived artificial chromosome (PAC) has the capacity to insert DNA fragments of about 100 kbp. The carrying capacity is twice that of the cosmid (a plasmid with *cos* site) cloning vectors. It has the combined features of P1 vector and BACs and has packaging site (*pac*). It has been used to construct genomic libraries of *Drosophila*, mouse, yeast and human DNA.

- ▣ Another important system for cloning very large DNA pieces is yeast artificial chromosomes (YAC). YACs accept genomic inserts of 100 to 2000 kbp and are used in genome sequencing and positional cloning of genes. These vectors are propagated in *Saccharomyces cerevisiae* and are based on chromosome. Each chromosome has three components, *viz.*, centromere, telomere and origin of replication. YACs form high capacity cloning system in a simple genetic background. Although yeast is not as easy to work with as *E.coli*, it is useful for studying eukaryotic genes that cannot be expressed in *E. coli.*

In addition to above-mentioned approaches, the expressed sequence tag (EST; that has already been discussed in Chapter 8) approach was pioneered by J. Craig Venter and co-workers at National Institute of Health (USA) in the early 1990s. They developed a method of investigating the genes by isolating mRNA molecules instead of genomic DNA and constructed cDNA from the mRNA by the process of reverse transcription. They treated DNA as a part of chromosomal DNA and sequenced to create ESTs that were used as handles for isolating the complete genes. Using this strategy, plenty of database of nucleotide sequences (dbEST—a division of GenBank) of several organisms were generated (up to October 2008 there were 57,425,912 entries in dbEST).

GenBank

GenBank is the NIH genetic sequence database, an annotated collection of all publicly available DNA sequences for almost every organism. The content includes genomic DNA, mRNA, cDNA, ESTs, high-throughput raw data, and sequence polymorphism. GenBank is part of the International Nucleotide Sequence Database Collaboration, which is comprised of DDBJ, EMBL, and GenBank at the NCBI. These three organizations exchange data on a daily basis. GenBank and its collaborators receive DNA sequences produced in laboratories throughout the world more than 260,000 distinct organisms. GenBank continues to grow at an exponential rate (Figure 10.3), doubling every 18 months and a new release is made every two months. As per the Release 155 produced in August 2006, there are over 65 billion nucleotide bases in more than 61 million sequences. There are approximately 108,635,736,141 bases in 27,439,206 sequence records in the GenBank divisions as of February 2008. The complete release notes for the current version of GenBank are available on the NCBI file transfer protocol site.

Each GenBank entry includes a concise description of the sequence, the scientific name and taxonomy of the source organism, and a table of features that identifies coding regions and other sites of biological significance, such as transcription units, sites of mutations or modifications, and repeats.

Bibliographic references are included along with a link to the MedLine unique identifier for all published sequences. Most sequence analysis (http://www.psc.edu/general/software/categories/sequence_analysis.html) programs on PSC supercomputers (http://www.psc.edu/resources.php#hardware) are capable of reading in GenBank data in the GenBank flat file format. The location of the data in the flat file format is built into the MAKSEQ program

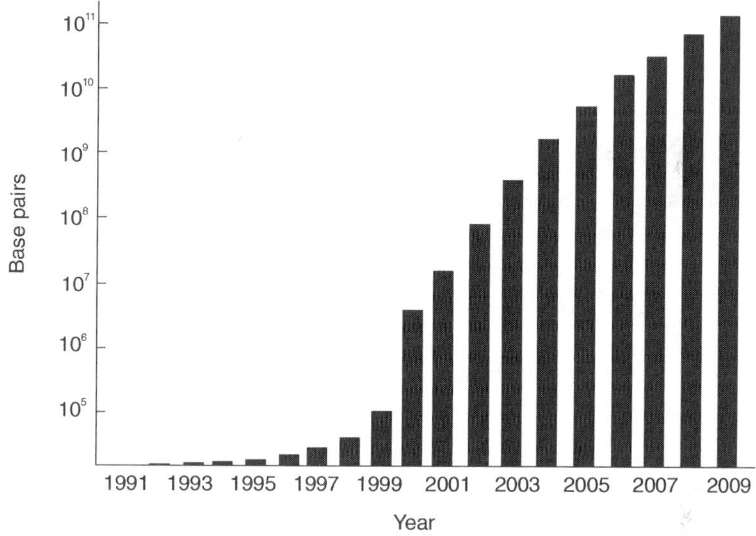

Figure 10.3 Year-wise growth in GenBank

at http://www.psc.edu/general/software/packages/makseq/. However, if you find it necessary to view the GenBank files in the flat file format, they can be found in the directory /biomed/db/genbank.

Sequence submissions to GenBank The information in relation to DNA sequence of a particular organism can be directly submitted to GenBank using BankIt, which is a web-based form, or the stand-alone submission program, Sequin. Upon receipt of a sequence submission, the GenBank staff assigns an accession number to the sequence and performs quality assurance checks. The submissions are then released to the public database, where the entries are retrievable by Entrez or downloadable by FTP.

Many journals require submission of sequence information (http://www.ncbi.nlm.nih.gov/Genbank/submit.html) to a database prior to publication so that an accession number may appear in the paper. There are several options for submitting data to GenBank:

- **BankIt** (http://www.ncbi.nlm.nih.gov/BankIt/index.html.), a WWW-based submission tool for convenient and quick submission of sequence data.

- **Sequin** (http://www.ncbi.nlm.nih.gov/Sequin/), NCBI's stand-alone submission software for MAC, PC, and UNIX platforms, is available by FTP. When using Sequin, the output files for direct submission should be sent to GenBank by e-mail. Sequin automatically performs a number of functions that must be carried out manually in other sequence submissions programs such as BankIt. Sequin obtains the proper genetic code from the name of the organism and automatically determines coding region intervals on the nucleotide sequence by back-translation of the protein sequence.

- **tbl2asn** (http://www.ncbi.nlm.nih.gov/Genbank/tbl2asn2.html), a command-line program, automates the creation of sequence records for submission to GenBank using many of the same functions as Sequin. It is used primarily for submission of complete genomes and large batches of sequences. tbl2asn generates .sqn files for submission to GenBank. Additional manual editing is not required before submission.

- **Barcode Submission Tool** (http://www.ncbi.nlm.nih.gov/WebSub/?tool=barcode), a WWW-based tool for the submission of GenBank sequences and trace data for Barcode of Life (http://www.ncbi.nlm.nih.gov/Genbank/barcode.html) projects. Barcodes are short nucleotide sequences from a standard genetic locus for use in species identification. Currently, the Barcode sequence being accepted for animals is a 5,650 base pair region of the mitochondrial cytochrome oxidase subunit I (COI) gene.

There are specialized, streamlined procedures for bulk submissions of sequences, such as EST (Expressed Sequence Tags) submitted to through dbEST(http://www.ncbi.nlm.nih.gov/dbEST/how_to_submit.html), STS (Sequenced Tagged Site) submitted to through dbSTS, and Genome Survey Sequence (GSS) should be submitted through the dbGSS system. Similarly, High-Throughput Genome Sequence (HTGS) data are most often submitted by large-scale sequencing centres. The GenBank direct submissions group also processes complete microbial genome sequences. GenProtEC is the genome and proteome database of *Escherichia coli* K-12 (strain MG1655).

GenProtEC is dedicated to the functions encoded by the *Escherichia coli* K-12 (strain MG1655) genome defined in the GenBank Accession No.

NC_000913.2 deposit. The annotation work includes multiple types of information:

- Sequence similarity to orthologues as defined by Darwin (start and end of aligned region, identity, and PAM distance).
- Resolution of fused proteins into modular units with independent functions.
- Identification of sequence similar protein groups within *E. coli* that are clustered by transitive relationships. The sequence similarity is limited to PAM 200 and an alignment of at least 83 amino acids.
- Updated literature references.
- Classification of gene products by their gene type and by their cellular role(s). The MultiFun classification system for cellular roles is used to assign gene products to one or more roles. MultiFun has been converted to Gene Ontology terms.
- Families of proteins related by structure and biochemical reaction mechanisms (work in progress).
- SCOP superfamily identification and location (e.g. binding site domains) for *E. coli* proteins.

Genome Assembly

Genome assembly is the process of taking a large number of short DNA sequences, all of which were generated by a shotgun sequencing project, and putting them back together to create a representation of the original chromosomes from which the DNA originated. In a shotgun sequencing project, the entire DNA from a source (usually a single organism, anything from a bacterium to a mammal) is first fractured into millions of small pieces. These pieces are then "read" by automated sequencing machines, which can read up to 900 nucleotides or bases at a time. (The four bases are adenine, guanine, cytosine, and thymine, represented as AGCT.) A genome assembly algorithm works by taking all the pieces and aligning them to one another, and detecting all the places where two of the short sequences, or reads, overlap. These overlapping reads can be merged together, and the process continues.

Genome assembly is a very difficult computational problem, made more difficult because genomes contain large numbers of identical sequences, known as repeats. These repeats can be thousands of nucleotides long, and some occur in thousands of different locations, especially in the large genomes of plants and animals.

Assembly software   Originally, most large-scale DNA sequencing centres developed their own software for assembling the sequences that they produced. However, this has changed as the software has grown more complex and as the number of sequencing centres has increased. Some well known assembly programs include:

- **Phred/Phrap** by Phil Green was one of the first successful assemblers, widely used in the 1990s and early 2000s, especially for smaller genomes.
- **AMOS** (A Modular, Open-Source assembler) is a well-known open source effort to bring together the efforts of leading genome assembly code developers. The home of AMOS is currently http://amos.sourceforge.net/. AMOS was initiated at The Institute for Genomic Research by Steven Salzberg, Mihai Pop, and Art Delcher, who are now at The University of Maryland.
- The **Celera Assembler** was the assembler developed by Gene Myers, Granger Sutton, Art Delcher, and others at Celera Genomics from 1998 until approximately 2002. It was moved to SourceForge and continues to be developed by the original scientists and others, at http://sourceforge.net/projects/wgs-assembler.
- The **Arachne** assembler began in 2000 as the doctoral thesis of Serafim Batzoglou, now at Stanford University. Since that time, it has been developed by a team lead by David B. Jaffe at the Broad Institute, formerly part of the Whitehead Institute. It is available for download at http://www.broad.mit.edu/wga/arachne.

Genome Annotation

If one has to deposit the assembled sequence into a genomic database, it should be analysed first for useful biological functions. After availability of the whole genome sequence for any organism, subsequently the genomic researcher wants to assign meaning or function to different regions of the genome. Thus genome annotation is the process of marking the genes and other biological features in a DNA sequence. In 1995, Dr. Owen White for the first time designed the software system for genome annotation that finds the nucleotide sequences in the genes responsible for synthesis of a protein, the tRNA, and other features, and to make initial assignments of function to those genes. Currently there are several genome annotation software systems working similarly, but the programs available for analysis of genomic DNA are constantly changing and improving.

Genome annotation process provides comments for the features in two steps:

- ▣ **Structural annotation** is the identification of elements on the genome; a process also known as gene prediction that involves determination of open reading frames, coding regions within the genome.
- ▣ **Functional annotation** is attaching biological information to above elements; a process also called functional assignment in which the biochemical and biological function as well as elements involved in regulation and different interactions are searched out.

PHAT   PHAT is a Pretty Handy Annotation Tool developed by Pittsburgh Supercomputing Centre for finding genes in eukaryotic organisms. It was originally developed with a view to annotating *Plasmodium falciparum* but now comes with code to retrain it for other organisms (requires annotated GenBank files for input).

The latest version comes with ready-made parameter sets for the analysis of

- ▣ *Homo sapiens*
- ▣ *Plasmodium falciparum*
- ▣ *Plasmodium vivax*

PHAT supports most of the input sequence formats and has support for analysing sequences in parallel on multiple-processor machines.

Usage   PHAT takes its parameters from a bunch of files all that are located in the same directory. This directory is /home/biomed/src/Phat/Phat/Pars. There are two ways to tell the program which set of parameters to use:

1. One is to specify the directory on the command line

```
fullphat-p/home/biomed/src/Phat/Phat/Pars myseq.fasta

halfphat-p/home/biomed/src/Phat/Phat/Pars myseq.gbk
```

2. Another way is to set the environment variable PHATPARS

```
setenvPHATPARS/home/biomed/src/Phat/Phat/Pars

fullphat myseq.fasta
```

The -p option will override the environment variable if it is already set. Input sequences can be in FASTA or GenBank format. If a multiple FASTA file is passed only in the first sequence in the file will be analysed.

Genome annotation was the major challenge for the human genome project. But annotators completed gene annotation of the human genome by using a combination of theoretical gene prediction and its verification by experiments. Gene prediction program such as GenScan helps in gene prediction by *ab initio*-exon prediction. It is then verified by BLAST searches against a sequence database. The gene predicted in this way is then compared with experimentally determined cDNA and EST sequences using pairwise alignment programs. All the predictions are manually checked by human curator. Once open reading frames are localized, functional assignment of the encoded proteins is worked out by homology searching using BLAST searches against a protein database. The functional annotations to human genome are added by searching protein motifs and domain databases such as Pfam and InterPro.

Nowadays more and more additional information is added to the annotation platform. The additional information allows manual annotators to deconvolute discrepancies between genes that are given the same annotation. For example, the SEED (http://www.theseed.org) database uses genome context information, similarity scores, experimental data, and integrations of other resources to provide the most accurate genome annotations through their subsystems approach. The Ensembl database relies on both curated data sources as well as on a range of different software tools in their automated genome annotation pipeline. Genome annotation is an active area of investigation and involves a number of different organizations in the life science community which publish the results of their efforts in publicly available biological databases accessible via the web and other electronic means.

Genome Similarity

When the human genome project was in progress and those who are learning about genome sequencing for the first time often ask the question—whose genome is being sequenced? Is it a particular individual? This is not an unreasonable question; in fact several unknown samples were collected and pooled for the human genome project. But this does not really matter as all human genomes are deemed to be roughly 99.9% equivalent and on an average, only one in a thousand nucleotides are different in genomes of two different individuals. Therefore researchers can talk in terms of the consensus human genome. The 0.1% difference is exploited in DNA fingerprinting that is used in criminology. Variations in non-coding parts of the genome are analysed to produce patterns that can reliably distinguish individuals, an exception here is identical twins which are much harder to distinguish with DNA fingerprinting.

Single nucleotide polymorphism (SNP, pronounced snip)   Particularly important variations in individual genomes are the single nucleotide polymorphisms or SNPs, which can occur both in coding and non-coding parts of the genome. SNPs are the variations in DNA sequence that occur due to change even in a single base (A, C, G, or T), so that different individuals may have different letters in these positions. For example:

First person: ......CTGCTAGTAG.....

Second person: ......CTGCGAGTAG....

Particularly, nucleotides in SNP positions within or close to genes can influence the gene's protein product. In human genome, SNP's occur at 1.6 to 3.2 million sites that forms the basis for genetic variation and helps in DNA fingerprinting of individuals in search of criminals, rapists, solving parental disputes, confirming identity of individuals, etc. Some abnormal protein variants (mutant variants) are the cause of genetic diseases. SNPs may be responsible for many inheritable differences between individuals, and SNP variation may indicate the predisposition to a genetic disease. For example, there is a evidence that certain combinations of SNPs occur in individuals with Alzheimers disease, cystic fibrosis, Huntington's disease. Alzheimers disease occurs due to difference in a single base in *apoE* gene on chromosome 19 (19q13). Cystic fibrosis is caused due to single gene mutation (Figure 10.4) that occurs due to deletion of 3 bp at codon 508 on chromosome 7 (7q31). Huntington's disease is also caused by autosomal dominance that occurs due to single mutation of A/G/C mutation at chromosome 4 (4q16.3). SNP analysis is therefore important for diagnostics and a SNP database (http://www.ebi.ac.uk/mutations/) is being built with the participation of the EBI. SNP analysis can also be used in population genetics, as some SNPs vary in frequency between populations. There are about 3 million SNPs collected in public SNP databases.

The sequencing projects have revealed that the genomes of what look like very different organisms may be quite similar. It is estimated that the difference between human and chimpanzee genomes is only 1–3%, while between human and mouse only about 5–15%, depending on how the similarity is defined and measured (before the sequencing of chimp and mouse are well under way, these are only estimates). These similarities indicate close evolutionary relationships between these mammalian organisms. It is possible to build a phylogenetic tree of the evolution of proteins, genes and organisms based on sequence comparisons.

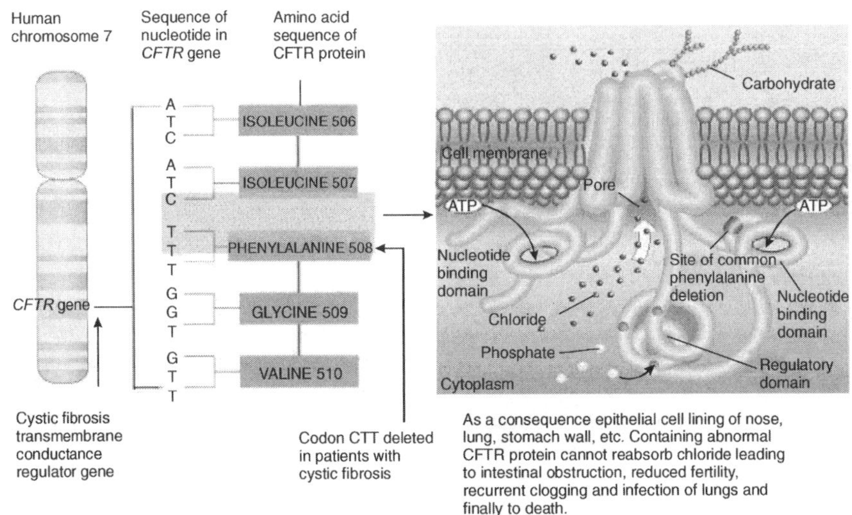

Figure 10.4   *CFTR* gene mutation (deletion of a codon) causing cystic fibrosis

TYPES OF GENOMICS

Comparative Genomics

It is the analysis and comparison of genomes from different species. The purpose is to gain a better understanding of how species have evolved and to determine the function of genes and non-coding regions of the genome. Comparative genomics involves the use of computer programs that can line-up multiple genomes and look for regions of similarity among them. Some of these sequence similarity tools are accessible to the public over the Internet. One of the most widely used homology searching program is BLAST available from the NCBI. BLAST is a set of programs designed to perform similarity searches on all available sequence data.

Researchers have learned a great deal about the function of human genes by examining their counterparts in simpler model organisms such as the mouse (Figure 10.5). Genome researchers look at many different features when comparing the genomes: sequence similarity, gene location, the length and number of coding regions (called exons) within genes, the amount of non-coding gene in each genome, and highly conserved regions maintained in organisms as simple as bacteria and as complex as humans.

Mice and humans (indeed, most or all mammals including dogs, cats, rabbits, monkeys, and apes) have roughly the same number of nucleotides in their genomes, about 3 billion base pairs. This comparable DNA content implies that all mammals contain more or less the same number of genes, and indeed the work of many researchers have provided evidence to confirm that notion.

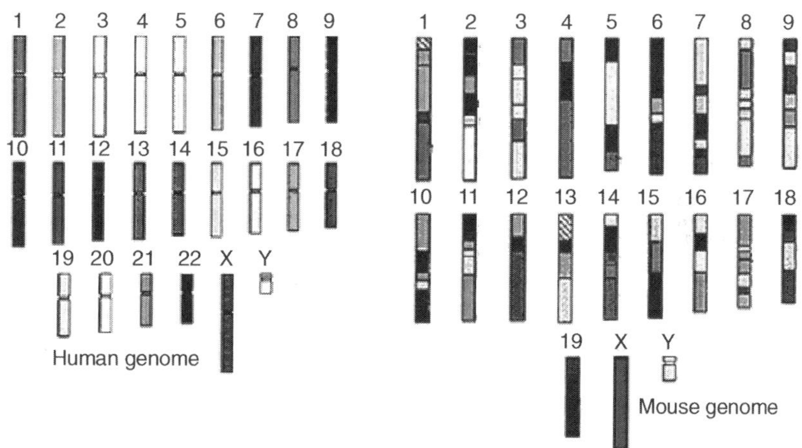

Figure 10.5   Human and mouse genomes sharing 85% similarities

There is no mouse counterpart found for a particular human gene during comparison, and for the most part there is essentially a one-to-one correspondence between genes in the two species. The exceptions generally appear to be of a particular type, i.e., genes that arise when an existing sequence is duplicated.

Gene duplication occurs frequently in complex genomes; sometimes the duplicated copies degenerate to the point where they no longer are capable of encoding a protein. However, many duplicated genes remain active and over time may change enough to perform a new function. Since gene duplication is an ongoing process, mice may have active duplicates that humans do not possess, and vice versa. These appear to make up a small percentage of the total genes. It is believed that the number of human genes without a clear mouse counterpart, and vice versa, won't be significantly larger than 1% of the total. Nevertheless, these novel genes may play an important role in determining species-specific traits and functions.

However, the most significant differences between mice and humans are not in the number of genes each carries, but in the structure of genes and the activities of their protein products. So far gene for a gene is considered, humans are very similar to mice. Really a matter is that subtle changes accumulated in each of the approximately 25,000 genes add together to make quite different organisms. Further, genes and proteins interact in complex ways that multiply the functions of each. In addition, a gene can produce more than one protein product through alternative splicing or post-translational modification; these events do not always occur in an identical way in the two species. A gene can produce more or less protein in different cells at various times in response to developmental or environmental conditions, and many proteins can express disparate functions in various biological contexts. Thus, subtle distinctions are multiplied by more than 30,000 estimated genes.

The often-quoted statement that humans share over 98% of genes with apes (chimpanzees, gorillas, and orang-utans) actually should be quoted in another way, that is, "there is more than 95% to 98% similarity between related genes in humans and apes in general". (Just as in the mouse, quite a few genes probably are not common to humans and apes, and these may influence uniquely human or ape traits.) Similarities between mouse and human genes range from about 70% to 90%, with an average of 85% similarity but a lot of variation from gene to gene (e.g. some mouse and human gene products are almost identical, while others are nearly unrecognizable as close relatives). Some nucleotide changes are neutral and do not yield a significantly altered protein. Others, but probably only a relatively small percentage, would introduce changes that could substantially alter what the protein does.

Put these alterations in the context of known inherited human diseases—a single nucleotide change (point mutation) can lead to inheritance of sickle-cell disease, cystic fibrosis, or breast cancer. A single nucleotide difference can alter protein function in such a way that it causes a terrible tissue malfunction. Single nucleotide changes have been linked to hereditary differences in height, brain development, facial structure, pigmentation, and many other striking morphological differences; due to single nucleotide changes, hands can develop structures that look like toes instead of fingers, and a mouse's tail can disappear completely. Single-nucleotide changes in the same genes but in different positions in the coding sequence might do nothing harmful at all. Evolutionary changes are the same as these sequence differences that are linked to person-to-person variation: many of the average 15% nucleotide changes that distinguish human's and mouse genes are neutral; some lead to subtle changes, whereas others are associated with dramatic differences. Add them all together, and they can make quite an impact, as evidenced by the huge range of metabolic, morphological, and behavioural differences one can see among organisms.

The core of comparative genome analysis is the establishment of the correspondence between genes (orthology analysis) or other genomic features in different organisms. It is intergenomic maps that make it possible to trace the evolutionary processes responsible for the divergence of two genomes. For example, about 2000 genes of yeast are functionally similar to *C. elegans.* Such similarity between organisms suggests that in spite of evolution of organisms about 100 million years ago, their genomes have not changed much. A multitude of evolutionary events acting at various organizational levels, shape the genome evolution. At the lowest level, point mutations affect individual nucleotides. At the highest level, large chromosomal segments undergo duplication, lateral transfer, inversion, transposition, deletion and insertion. Ultimately, whole genomes are involved in processes of hybridization, polyploidization and endosymbiosis, often leading to rapid speciation. The complexity of genome evolution poses many exciting challenges to developers of mathematical models and algorithms—who have recourse to a spectra of algorithmic, statistical and mathematical techniques, ranging from exact, heuristics, fixed parameter and approximation algorithms for problems based on parsimony models to Markov Chain Monte-Carlo algorithms, for Bayesian analysis of problems based on probabilistic models. Many of these studies are based on the homology detection and protein families computation. Specific alignment programs are needed to deal with the unique challenges of whole genome alignment. A number of alignment programs for comparison of large DNA sequences are given below:

- **LAGAN** (Limited Area Global Alignment of Nucleotides; http://lagan.stanford.edu/) is a web-based program designed for pairwise alignment of large genome.

- **MUMmer** (Maximal Unique Match; www.tigr.org/tigr-script/CMR2/webmum/mumplot) is a UNIX program, essentially a modified BLAST, form TIGR for alignment of two entire genome sequences and comparison of location of orthologs.

- **BLASTZ** (http://bio.cse.psu.edu/) is also a UNIX program modified from BLAST useful for pairwise alignment of large genomic DNA sequences.

- **MAVID** (http://baboon.math.berkeley.edu/mavid) is a web-based program that is based on progressive alignment similar to Clustal. It is used for aligning multiple large genomic sequences.

- ▣ **GenomeVista** (http://pipeline.lbl.gov/cgi-bin/GenomeVista) is a database searching program that searches against the human, mouse, rat or *Drosophila* genomes using a large piece of DNA as query.

- ▣ **PipMaker** (http://bio.csepsu.edu/cgi-bin/pipmaker?basic) is a web server that uses BLASTZ heuristic method to find similar regions in two DNA sequences.

Structural Genomics

The objective of structural genomics is to deal with the systematics and determination of the three-dimensional structures of all proteins of a given organism, by experimental methods such as X-ray crystallography, NMR spectroscopy or computational approaches such as homology modelling. It has become feasible due to advances in experimental techniques and as well as due to advances in our understanding of protein structure. Structural genomics projects combine results from different organisms but the special attention is given to human proteome since these proteins are unique so far the diseases due to microorganisms are concerned and knowing the structure of these proteins can lead to improved medical treatment.

The major challenge in structural bioinformatics is the determination of protein function from its 3D structure. Structural genomics in its initial stage dealt with genome analysis, construction of genetic and physical maps, identification of gene, its annotation and comparison of genome structures but later on it emphasizes high-throughput determination of protein structures from amino acid sequence because proteins in every group of individuals vary and so there would also be variation in genome sequence. Many times, a protein's role is not fully understood until its 3D shape is known. Protein structural information is collected by the PDB. In the year 2008, PDB contained 51,491 entries of proteins structures. The protein structure determination efforts are also initiated by German Protein Structure Factory and National Institute of General Medical Sciences (NIGMS) to achieve the objectives globally. Other scientific community is the Protein Structure Initiative that gets immediate access to new structures, as well as to reagents such as clones and proteins. Some of the proteins and clones of the Protein Structure Initiative are discussed in the December 2007 and January 2008 issues of the journal *Structure* (http://www.structure.org/).

Functional Genomics

Functional genomics refers to the analysis of global gene expression and gene functions in a genome. If a researcher locates a gene and able to sequence that gene, the next question would be what function it does? The human genome project was just the first step in understanding humans at the molecular level. Though the project is complete, many questions still remain unanswered, including the function of most of the estimated 30,000 human genes. Researchers also don't know the role of single nucleotide polymorphisms (SNPs)—single DNA base changes within the genome—or the role of non-coding regions and repeats in the genome.

Functional genomics can be roughly defined as using the emerging knowledge about genomes to understand the gene and their product functions and interactions, and most importantly of all, how all this makes organisms to function the way they do. Thus, the functional genomics provides insight into the biological functions of the whole genome through automated high-throughput expression analysis.

GENE FUNCTIONS

As already mentioned, proteins play a variety of roles in a cell. Biologists think that there is likely to be a limited universe of genes and their respective proteins, from the functional point of view, many of which are present in most or all genomes. For instance, about 12% of the ~18,000 nematode worm genes encode proteins whose biological roles could be inferred from their sequence similarity to yeast genes, and vice versa. Almost one-third of the ~6000 yeast genes have functional equivalents in worm genes. The recognition of the limited universe of genes has prompted unification of biology under the name Gene Ontology (GO).

Gene Ontology (GO) Project

The aim of the GO project is to create a coherent nomenclature to describe genes and their products. GO decided that three hierarchical terms were needed to describe different aspects of every protein: biological process, molecular function, and cellular component. The vocabulary is organized in three independent-directed acyclic graphs (DAG's) characterizing the genes at each of these three levels. The use of GO terms allows automated searching of gene function in databases. Proteins in Swiss-Prot and the Interpro databases (another project at the EBI that classifies proteins by a variety of computational methods) are currently being assigned GO terms (http://www.ebi.ac.uk/ego/). GO project is a collaborative effort to address the need for consistent descriptions of gene products in different databases. The project began as collaboration between three model organism databases,

FlyBase (for *Drosophila*), the *Saccharomyces* Genome Database (SGD) and the Mouse Genome Database (MGD), in 1998. Since then, the GO Consortium has grown to include many databases, including several of the world's major repositories for plant, animal and microbial genomes. A good introduction of GO can be found at www.geneontology.org.

As per GO terminology, a gene product might be associated with or located in one or more cellular components; it is active in one or more biological processes, during which it performs one or more molecular functions. Let's take one simple example. The cytochrome *c* oxidase is the gene product that can be described by its 1) cellular component (Where?)—it is located in the mitochondrial matrix and mitochondrial inner membrane, 2) biological process (Why?)—it is present to carry out oxidative phosphorylation and induction of cell death, and 3) molecular function (What?)—being an enzyme it brings about oxidation and reduction of substrate (Figure 10.6).

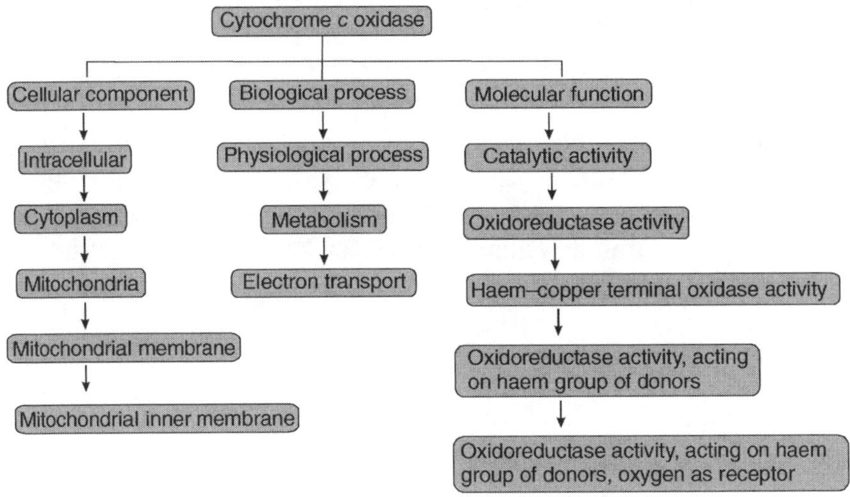

Figure 10.6   Cytochrome *c* oxidase—an example of GO annotation

GO allows to annotate genes and their products with a limited set of attributes. For example, GO does not allow to describe genes in terms of which cells or tissues they're expressed in, which developmental stages they're expressed at, or their involvement in disease. It is not necessary for GO to do these things because other ontologies are being developed for these purposes. The GO consortium supports the development of other ontologies and makes its tools for editing and curating ontologies that are freely available.

ANALYSIS OF GENE EXPRESSION

As mentioned earlier that although the human genome project is completed in 2003, still researchers don't know the functions of each gene out of the estimated 30,000 human genes. It is well-known fact that within the cell always all the genes do not express. Only those genes express whose products are required by the cell. The expression of many genes can be determined by measuring mRNA levels with multiple techniques including microarrays, expressed sequence tags, Serial Analysis of Gene Expression (SAGE), Massively Parallel Signature Sequencing (MPSS), or various applications of multiplexed *in situ* hybridization. All of these techniques are extremely noise-prone and/or subject to bias in the biological measurement, and a major research area in computational biology involves developing statistical tools to separate signal from noise in high-throughput gene expression studies. Such studies are often used to determine the genes implicated in a disorder. One might compare microarray data from cancerous epithelial cells to data from non-cancerous cells to determine the transcripts that are up-regulated and down-regulated in a particular population of cancer cells.

DNA Microarray or DNA Chip

DNA microarrays or DNA chips are small, solid supports onto which the sequences from thousands of different genes are immobilized, or attached at fixed locations. The supports themselves are usually glass microscope slides, silicon chips or nylon membranes. The DNA is printed, spotted, or actually synthesized directly onto the support. The American Heritage Dictionary defines "array" as "to place in an orderly arrangement". It is important that the gene sequences in a microarray are attached to their support in an orderly or fixed way, as a researcher uses the location of each spot in the array to identify a particular gene sequence. Following two types of DNA chips are generally available:

1. Oligonucleotides-based chips contain high density of short oligonucleotides of known sequences, generally 20 to 25 nucleotide long (100,000 to 400,000 oligonucleotides immobilized on the glass slide within an area of 1.3×1.3 cm^2).
2. cDNA-based chips contain high-density cDNA samples derived as PCR amplified DNA fragments.

The former type is often used for Sequencing By Hybridization (SBH) for detection of SNPs and the latter type is often used for studying gene expression patterns in time and space.

Steps involved in a DNA microarray experiment are as follow:

- ▣ Prepare the DNA chip using the chosen target DNAs.
- ▣ Generate a hybridization solution containing mixture of fluorescently labelled cDNAs.
- ▣ Incubate the hybridization mixture containing fluorescently-labelled cDNAs with the DNA chip.
- ▣ Detect bound cDNA using laser technology and store the data in a computer.
- ▣ Analyse the data using computational methods.

One might ask, how does a scientist extract information about a disease condition from a dime-sized glass or silicon chip containing thousands of individual gene sequences? The whole process is based on hybridization probing, a technique that uses fluorescently labelled nucleic acid molecules as "mobile probes" to identify complementary molecules—sequences that are able to base pair with one another. Each single-stranded DNA fragment is made up of four different nucleotides, adenine (A), thymine (T), guanine (G) and cytosine (C), which are linked end-to-end. Adenine is the complement of, or will always pair with, thymine, and guanine is the complement of cytosine. So, the complementary sequence to G-T-C-C-T-A will be C-A-G-G-A-T. When two complementary sequences find each other—such as the immobilized target DNA and the mobile probe DNA, cDNA, or mRNA—they will lock together, or hybridize.

Let us consider two cells: cell type 1 (a healthy cell), and cell type 2 (a diseased cell). Both contain an identical set of four genes A, B, C, and D. Scientists are interested in determining the expression profile of these four genes in the two cell types. To do this, scientists isolated mRNA from each cell type and used this mRNA as templates to generate cDNA with a "fluorescent tag" attached. Different tags (red and green) are used so that the samples can be differentiated in subsequent steps. The two-labelled samples are then mixed and incubated with a microarray containing the immobilized genes A, B, C, and D. The labelled molecules bind to the sites on the array that correspond to the genes expressed in each cell. Each spot on an array is associated with a particular gene. Each colour in an array represents either healthy (control) or diseased (sample) tissue (Figure 10.7). Depending on the type of array used, the location and intensity of a colour informs us whether the gene, or mutation, is present either in the control and/or sample DNA. It also tells us about the expression level of the gene(s) in the sample and control DNA.

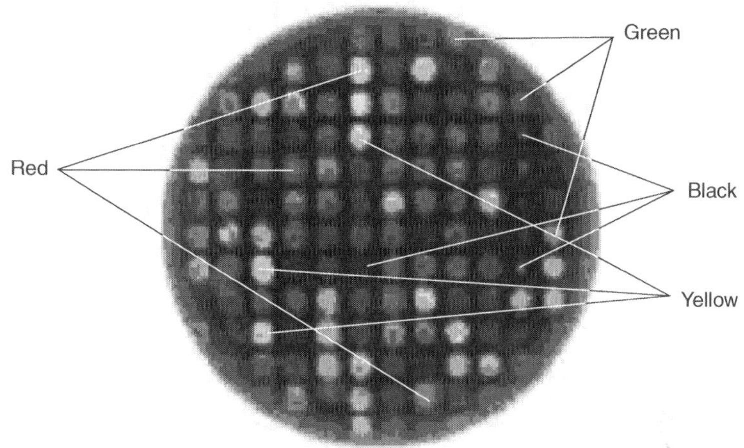

Figure 10.7 The colours of a microarray

In the above Figure 10.7 **GREEN** represents **control DNA** where either DNA or cDNA derived from normal tissue is hybridized to the target DNA. **RED** represents **sample DNA** where either DNA or cDNA is derived from diseased tissue hybridized to the target DNA. **YELLOW** represents a **combination of control and sample DNA** where both hybridized equally to the target DNA. **BLACK** represents areas where **neither the control nor the sample DNA** hybridized to the target DNA.

Usage of different types of microarray   There are three basic types of samples that can be used to construct DNA microarrays—two are genomic and the other is "transcriptomic," i.e., it measures mRNA levels. What makes them different from each other is the kind of immobilized DNA used to generate the array and ultimately, the kind of information that is derived from the chip. The target DNA used will also determine the type of control and the sample DNA that is used in the hybridization solution.

I. *To determine changes in gene expression levels*   Determining the level, or volume, at which a certain gene is expressed is called microarray expression analysis, and the arrays used in this kind of analysis are called as "expression chips". The immobilized DNA is cDNA derived from the mRNA of known genes, and, in otherwords, at least in some experiments, the control and the sample DNA hybridized to the chip is cDNA derived from the mRNA of normal and diseased tissue, respectively. If a gene is over expressed in a certain disease state, then more sample cDNA, as compared to control cDNA, will hybridize to the spot representing the expressed gene. In turn, the spot will fluoresce red with greater intensity than it will fluoresce green. Once the researchers have characterized the expression patterns of various genes involved in many diseases, cDNA derived from diseased tissue of any individual can be hybridized to determine whether the expression pattern of the gene from the individual matches the expression pattern of a known disease. If this is the case, treatment appropriate for that disease can be initiated.

As researchers use expression chips to detect expression patterns—whether or not a particular gene(s) is being expressed more or less under certain circumstances—expression chips may also be used to examine changes in gene expression over a given period of time, such as within the cell cycle. The cell cycle is a molecular network that determines, in the normal cell, whether the cell should pass through its life cycle or not. There are a variety of genes involved in regulating the stages of the cell cycle. Also built into this network are mechanisms designed to protect the body when this regulation system fails or breaks down due to mutations within one of the "control genes," as is the case with cancerous cell growth. An expression microarray experiment could be designed where cell cycle data is generated in multiple arrays and referenced to time "zero." Analysis of the collected data could further elucidate details of the cell cycle and its "clock," providing much needed data on the points at which gene mutation leads to cancerous growth as well as sources of therapeutic intervention.

In the same way, expression chips can be used to develop new drugs. For instance, if a certain gene is over expressed in a particular form of cancer, researchers can use expression chips to see if a new drug will reduce over expression and force the cancer into remission. Expression chips could also be used in disease diagnosis as well. For example, in the identification of new genes involved in environmentally triggered diseases such as those diseases affecting the immune, nervous, and pulmonary/respiratory systems.

II. *To determine genomic gains and losses*   DNA repair genes are thought to be the body's frontline defence against mutations, and, as such, play a major role in cancer. Mutations within these genes often manifest themselves as lost or broken chromosomes. It has been hypothesized that certain chromosomal gains and losses are related to cancer progression and that the patterns of these changes are relevant to clinical prognosis. Using different laboratory methods, researchers can measure gains and losses in the copy number of chromosomal regions in tumour cells. Then, using mathematical models to analyse this data, they can predict which chromosomal regions are most likely to harbour important genes for tumour initiation and disease progression. The results of such an analysis may be depicted as a hierarchical tree-like branching diagram, referred to as a "tree model of tumour progression."

Researchers use a technique called microarray Comparative Genomic Hybridization (microarray CGH) to look for genomic gains and losses, or for a change in the number of copies of a particular gene involved in a disease state. In microarray CGH, large pieces of genomic DNA serve as the target DNA, and each spot of the target DNA in the array has a known chromosomal location. The hybridization mixture will contain fluorescently labelled genomic DNA harvested from both normal (control) and diseased (sample) tissue.

So, if the number of copies of a particular target gene has increased, a large amount of sample DNA will hybridize to those spots on the microarray that represent the gene involved in that disease, whereas comparatively small amounts of control DNA will hybridize to those same spots. As a result, those spots containing the diseased gene will fluoresce red with greater intensity than those spots containing control DNA which they will fluoresce green, indicating that the number of copies of the gene involved in the disease has gone up.

III. *To determine mutations in DNA*   When researchers use microarrays to detect mutations or polymorphisms in a gene sequence, the target, or immobilized DNA, is usually that of a single gene. In this case, though the target sequence placed on any given spot within the array will differ from that of other spots in the same microarray, sometimes by only one or a few specific nucleotides. One type of sequence commonly used in this type of analysis is a SNP—a small genetic change, or variation, that can occur within a person's DNA sequence. Another difference in mutation microarray analysis, as compared to expression or CGH microarrays , is that this type of experiment requires only genomic DNA derived from a normal sample for use in the hybridization mixture.

As discussed earlier, a SNP is a DNA sequence variation occurring when a single nucleotide (A, T, C, or G) in the genome (or other shared sequence) differs between members of a species or between paired chromosomes in an individual. For example, two sequenced DNA fragments from different individuals, AAGCCTA to AAGCTTA, contain a difference in a single nucleotide. In this case we say that there are two alleles: C and T. Almost all common SNPs have only two alleles. For a variation to be considered a SNP, it must occur in at least 1% of the population. There are certain combinations of SNPs occuring in human individuals that are affected with heritable diseases and their analysis helps in diagnosis of disease (Figure 10.8).

Within a population, SNPs can be assigned a minor allele frequency—the lowest allele frequency at a locus that is observed in a particular population.

This is simply the lesser of the two allele frequencies for SNPs. It is important to note that there are variations between human populations, so a SNP allele that is common in one geographical or ethnic group may be much rarer in another. In the past, SNPs with a minor allele frequency of less than or equal to 1% (or 0.5%, etc.) were given the title "SNP," an unwieldy definition.

Once researchers have established that a SNP pattern is associated with a particular disease, they can use SNP microarray technology to test an individual for that disease expression pattern in order to determine if he or she is susceptible to—at risk of developing—that disease. When genomic DNA from an individual is hybridized to an array loaded with various SNPs, the sample DNA will hybridize with greater frequency only to specific SNPs associated with that person. Those spots on the microarray will then fluoresce with greater intensity, demonstrating that the individual being tested may have, or is at risk for developing disease.

As mentioned earlier, SNPs may also be associated with the absorbance and clearance of therapeutic agents. Currently, there is no simple way to determine how a patient will respond to a particular medication. A treatment proven effective in one patient may be ineffective in others. Worse yet, some patients may experience an adverse immunological reaction to a particular drug. Today, pharmaceutical companies are limited to developing agents to which the "average" patient will respond. As a result, many drugs that might benefit a small number of patients never make it to market.

In the future, the most appropriate drug for an individual could be determined in advance of treatment by analysing a patient's SNP profile. Since different people have different responses to treatment. Some are cured, some don't respond while some experience a bad side effect. The ability to target a drug to those individuals most likely to benefit, referred to as "personalized medicine," would allow pharmaceutical companies to bring many more drugs to market and allow doctors to prescribe individualized therapies specific to a patient's needs. Most SNPs are not responsible for a disease state. Instead, they serve as biological markers for pinpointing a disease on the human genome map. DNA microarray also helps in analysing expression patterns of agricultural crops or transgenic plants under the different environmental conditions. Table 10.3 enlists the types of microarray with their particular applications.

DNA from different individuals sequenced

Variation at a single nucleotide

Some individuals will have one
version of SNP, some the other

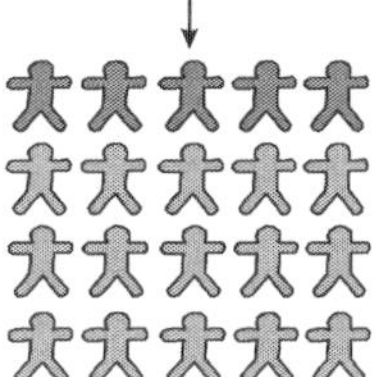

Normal population
In a population, a certain percentage
will have one version, the rest the other

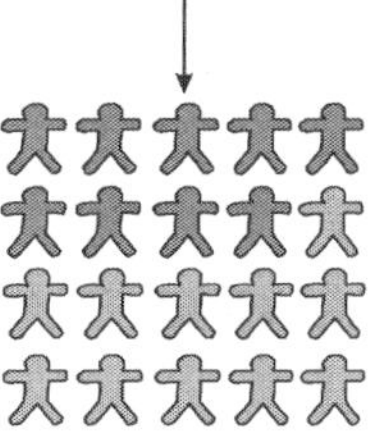

Sample with disease

A higher than expected incidence in
a disease group suggests SNPIG is
associated with a disease (or SNPIA
is protective)

Figure 10.8 SNP analysis helps in diagnosis of disease. SNPIG (Human Individual SNP with G) and SNPIA ((Human Individual SNP with A)

Table 10.3   Microarray types with applications

Microarray type	Application
Comparative genomic hybridization	Tumour classification, risk assessment, and prognosis prediction.
Expression analysis	Drug development, drug response, and therapy development.
Mutation/polymorphism analysis	Drug development, therapy development, and tracking disease progression.

Serial Analysis of Gene Expression (SAGE)

It is a high-throughput, sequence-based approach used by molecular biologists for analysis of gene expression profile. The original technique was developed by Dr. Victor Velculescu at the Oncology Center of Johns Hopkins University and published in 1995. In this method, the analysis of gene is based on the gene sequencing that involves Polymerase Chain Reaction (PCR) to identify differentially expressed mRNA level in a sample. In a particular cDNA pool the relative proportion of the gene, specific ESTs show the relative abundance of the corresponding mRNA transcript. Briefly, SAGE experiments proceed as follows:

- Isolate the mRNA of an input sample (e.g. tumour)
- Extract a small chunk of sequence from a defined position of each mRNA molecule
- Link these small pieces of sequence together to form a long chain (or concatemer)
- Clone these chains into a vector which can be taken up by bacteria
- Sequence these chains using modern high-throughput DNA sequencers
- Process this data with a computer to count the small sequence tags

Several variants have been developed since, most notably a more robust version, that enables very precise annotation of existing genes and discovery of new genes within genomes because of an increased tag-length of 25–27 bp. Useful SAGE modifications (www.sagenet.org) include:

- SAGE on a single cell (MicroSAGE)
- MmeI tagging enzyme (LongSAGE-21 bp)
- EcoP15I tagging enzyme (SuperSAGE-26 bp)

- ▣ Improved efficiencies in development (RL-SAGE)
- ▣ Polyadenylation of total RNA (Prokaryotic SAGE)

The salient features of SAGE are as given below:

- ▣ Exploits large-scale sequencing, but very effective for small-scale sequencing
- ▣ Provides expression data for thousands of gene products without a priori knowledge
- ▣ Identifies expressed genes
- ▣ Determines relative measures of gene expression
- ▣ Affordable and fast comparison of many experiments, stages, etc.
- ▣ Small amount of starting material (single cell studies are possible)
- ▣ Simultaneous gene identification and genome-wide gene expression profiling
- ▣ Efficient, simple, low cost (one clone-many tags)
- ▣ Utilizes common tools of molecular biology
- ▣ Detection of unpredicted ORFs, anti-sense transcripts
- ▣ Reduced measurement error, understandable and tractable artifacts
- ▣ Very quantitative, understandable statistical properties
- ▣ Detection level can be improved with more sequencing.

The detailed SAGE procedure involves the generation of short unique sequence tags by cleaving cDNA that is synthesised from mRNA. The steps are diagrammatically shown in Figure 10.9. The process starts from isolation of mRNA, which is then used to synthesize the cDNA with the help of enzyme reverse transcriptase and biotin-labelled oligo-dT primers. The cDNA molecules are then cleaved with restriction enzyme *Nla*III with a restriction site →CATG and that has a relatively high cutting frequency (*Nla*III cuts every 256 bp on average 4^4 and produces a 4-bp overhangs). Cleaved cDNA is allowed to bind with the streptovidin-coated magnetic beads and divided into two pools that are ligated with different linkers, which have complementary 4-bp overhangs. Another restriction enzyme *Bsm*F1 is also used to delineate the tag and cleaves 10 bp downstream of GGGAC sequence. The restriction fragments are then ligated and amplified with the help of polymerase chain reaction with primers specific to each linker. As a result large number of short ESTs (10–15 bp long) are formed. Several such ESTs may be linked into single array which is finally subjected to automated DNA sequencing. When a large number of clones with linked tags are sequenced, the frequency of occurrence of each tag is counted to obtain an accurate picture of gene expression. This technique is applied to the genomic sequencing of yeast and human. In yeast, 95% of genes are predicted and detected on the basis of the presence of *Nla*III sites.

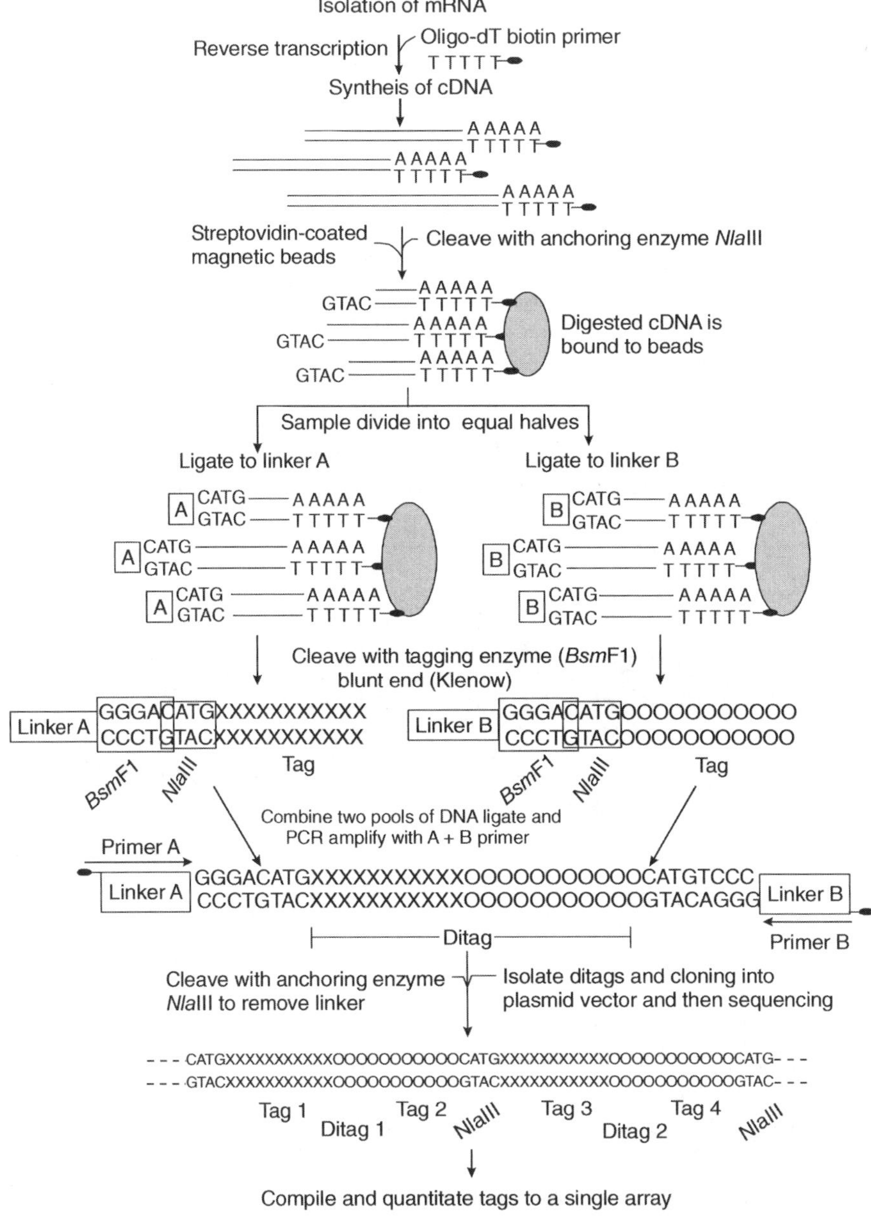

Figure 10.9   Outline of SAGE experimental procedure

SuperSAGE It is the most advanced derivate of the SAGE technology for the analysis of expressed genes in eukaryotic organisms (gene expression profiling). Like in SAGE, a specific tag from each transcribed gene is recovered. By sequencing and counting as many tags as possible, the transcription profile, stating what gene is described and how often, becomes apparent. SuperSAGE uses the type III-endonuclease EcoP15I of phage P1, to cut 26-bp long sequence tags from each transcript's cDNA, expanding the tag size by at least 6 bp as compared to the predecessor techniques SAGE and LongSAGE. The longer tag size allows for a more precise allocation of the tag to the corresponding transcript, because each additional base increases the precision of the annotation considerably.

Like in the original SAGE protocol, so-called ditags are formed, using blunt-ended tags. However, SuperSAGE avoids the bias observed during the less random LongSAGE 20 bp ditag ligation. By direct sequencing with modern high-throughput sequencing techniques, hundred thousands or millions of tags can be analysed simultaneously, producing very precise and quantitative gene expression profiles. Therefore, tag-based gene expression profiling also called "digital gene expression profiling" (DGE) can today provide most accurate transcription profiles that overcome the limitations of microarrays.

The 26-bp tags have a number of advantages over the smaller tags:

- Most notably due to the exact annotation of SuperSAGE tags (at least 10,000 times more accurate than LongSAGE tags), a substantially increased number of transcripts can be differentiated.
- Many different transcript isoforms (representing alternatively spliced transcripts) can be found.
- Novel genes can be discovered, that escape detection on microarrays.
- Sense and anti-sense transcripts and their different regulation can be detected.
- Due to the exact annotation of the 26-bp tags, two or more interacting organisms (parasite–host, pathogen–host) can be analysed simultaneously.
- The 26-bp tags can directly be spotted onto microarrays, and candidate transcripts be combined to produce focused microarrays (i.e., microarrays loaded only with genes, that are relevant for a specific process).
- The 26-bp tag allows the design of highly specific primers for downstream PCR (like for 3´- or 5´-RACE) or of specific probes for the identification of clones from a cDNA library.

Very precise and comprehensive gene expression profiles of any eukaryotic organism can therefore be established, which in many regards are superior to microarrays. Each and every transcript can be quantified by counting the tags in a SuperSAGE library such that quantitative genetics is readily possible with SuperSAGE.

DNA DATA BANK OF JAPAN (DDBJ)

The DNA Data Bank of Japan is a DNA databank located at the National Institute of Genetics of Japan (http://www.ddbj.nig.ac.jp). It is also a member of the International Nucleotide Sequence Database Collaboration (INSDC). It shares its data with EMBL at the EBI and with GenBank at the NCBI. DDBJ began DNA databank activities in earnest in 1986 at the National Institute of Genetics (NIG). The Center for Information Biology at NIG was reorganized as the Center for Information Biology and DNA Data Bank of Japan (CIB-DDBJ) in 2001. The new centre is to play a major role in carrying out research in information biology and to run DDBJ operation in the world. It is generally accepted that research in biology today requires both computer and experimental equipment equally well. In particular, researchers rely on computers to analyse DNA sequence data accumulating at a remarkably rapid rate. Actually, this triggered the birth and development of information biology.

DDBJ is the sole DNA databank in Japan, which is officially certified to collect DNA sequences from researchers and to issue the internationally recognized accession number to data submitters. DDBJ collects data mainly from Japanese researchers, but of course accept data and issue the accession number to researchers in any other countries. Since there is exchange of the collected data with EMBL/EBI and GenBank/NCBI on a daily basis, the three databanks share virtually the same data at any given time (Figure 10.10).

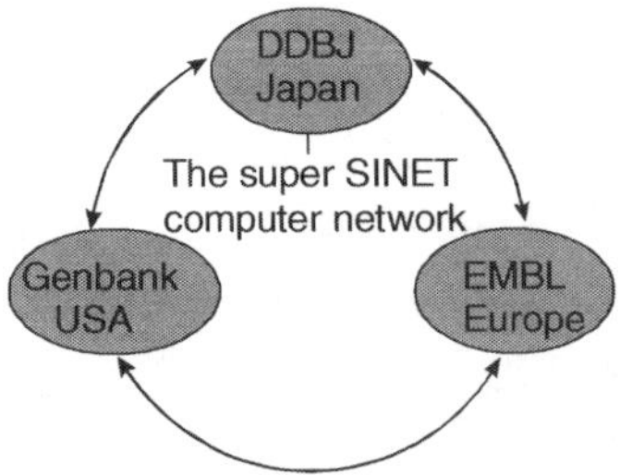

Figure 10.10   DDBJ functions in collaboration with EBI/EMBL (Europe) and NCBI/GenBank(USA)

The increase rates of the data collected, annotated and released to the public in the past year are 43% for the number of entries and 52% for the number of bases. The increase rates are accelerated even after the human genome was sequenced, because sequencing technology has been remarkably advanced and simplified, and research in life science has been shifted from the gene scale to the genome scale. In addition, DDBJ developed the Genome Information Broker (GIB, http://gib.genes.nig.ac.jp) that now includes more than 50 complete microbial genome and *Arabidopsis* genome data. DDBJ also developed a database of the human genome, the Human Genomics Studio (HGS, http://studio.nig.ac.jp). HGS provides one with a set of sequences being as continuous as possible in any one of the 24 chromosomes. Both GIB and HGS have been updated incorporating newly available data and retrieval tools.

THE INSTITUTE FOR GENOMIC RESEARCH (TIGR)

The new J. Craig Venter Institute (JCVI) was formed on October 2006 through the merger of several affiliated and legacy organizations—The Institute for Genomic Research (TIGR) and The Center for the Advancement of Genomics (TCAG), The J. Craig Venter Science Foundation, The Joint Technology Center, and the Institute for Biological Energy Alternatives (IBEA). Today all these organizations have become one large multidisciplinary genomic-focused organization. With more than 500 scientists and staff, more than 250, 000 square feet of laboratory space, and locations in Rockville, Maryland and La Jolla, California, the new JCVI is a world leader in genomic research. Different categories of JCVI software are given in Table 10.4.

Table 10.4 Different categories of JCVI software

JCVI software	Significance and its URL
Sequencing/Assembly	
AMOS	http://amos.sourceforge.net/ Manipulation of input and output files related to whole-genome shotgun assembly. Currently maintained at CBCB at the University of Maryland.
Celera-Assembler	http://www.jcvi.org/cms/research/software/packages/celera-assembler/overview/ Celera-Assembler can reconstruct long sequences of genomic DNA given the fragmentary data produced by whole-genome shotgun sequencing.

(Contd.)

Table 10.4 (Continued)

JCVI software	Significance and its URL
Cloe	http://cloe.sourceforge.net/ Viewer/editor The finishing stage of Sanger sequencing projects.
TIGR-Assembler	http://www.jcvi.org/cms/publications/listing/abstract/article/tigr-assembler-a-new-tool-for-assembling-large-shotgun-sequencing-projects/Enabled the first published whole-genome assembly of a free-living organism in 1995. Last revised in 2003.
TraceTuner	http://www.jcvi.org/cms/research/software/packages/tracetuner/overview/ It is a DNA sequencing quality value, base calling and trace processing software.
Gene Finding	
Glimmer	http://cbcb.umd.edu/software/glimmer/ A microbial gene-finding system, especially in bacteria, archaea, and viruses. Currently maintained at CBCB at the University of Maryland.
PhageFinder	http://phage-finder.sourceforge.net/ A tool to identify prophage regions within bacterial genomes.
Annotation	
AAT	ftp://ftp.tigr.org/pub/software/AAT/ A tool for analysing and annotating eukaryotic genomic sequences.
Manatee	http://manatee.sourceforge.net/ Web-based gene evaluation and genome annotation tool to view, modify, and store annotation for prokaryotic and eukaryotic genomes.
PASA	http://pasa.sourceforge.net/ An eukaryotic genome annotation tool that exploits spliced alignments of expressed transcript sequences to automatically model gene structure.
Genome Analysis	
DAGchainer	http://dagchainer.sourceforge.net/ The DAGchainer software computes chains of synthetic genes found within complete genome sequences.
k-mer Tools	http://kmer.sourceforge.net/ DNA sequence analysis tools including ESTmapper, Snapper (for mapping reads), and ATAC (for aligning genomes).

(Contd.)

Table 10.4 (Continued)

JCVI software	Significance and its URL
MUMmer	http://mummer.sourceforge.net/ A system for rapidly aligning entire genomes (draft or complete). Currently maintained at CBCB at the University of Maryland.
Pathema	http://www.jcvi.org/cms/research/projects/pathema/overview/ A NIAID Bioinformatics Resource Center (BRC) containing in depth curatorial analysis of six target organisms from the list of NIAID category.
Sybil	http://sybil.sourceforge.net/ A web-based software package for comparative genomics.
Microarray Tools	
Magnolia	http://pfgrc.jcvi.org/index.php/bioinformatics/magnolia.html A microarray data management and export system for researchers who use PFGRC microarrays.
SNP Filter Scripts	http://pfgrc.jcvi.org/index.php/compare_genomics/snp_scripts.html Scripts to identify and filter out false-positive SNP calls that are present in raw data from Affymetrix GeneChip® resequencing arrays.
TM4	http://www.tm4.org/ A suite of open source software tools that manage and analyse data from microarray experiments.
Proteomics Tools	
APEX	http://pfgrc.jcvi.org/index.php/bioinformatics/apex.html The APEX tool is an implementation of the absolute protein expression quantization technique, computing protein abundance values for LC-MS/M.
Computational Management Tools	
Ergatis	http://ergatis.sourceforge.net/ A web-based utility that is used to create, run, and monitor reusable computational analysis pipelines.
Workflow	http://sourceforge.net/projects/tigr-workflow/ A Java based, XML driven Workflow Engine suite, which can be used to build, execute and monitor complex process pipelines.

KYOTO ENCYCLOPEDIA OF GENES AND GENOMES (KEGG)

KEGG is a collection of online database dealing with genomes, enzymatic pathways, and biological chemicals. The PATHWAY database records networks of molecular interactions in the cells, and variants of them specific to particular organisms.

The KEGG was initiated by the Japanese human genome programme in 1995. According to the developers, they consider KEGG to be a "computer representation" of the biological system. The KEGG database can be utilized for modelling and simulation, browsing and retrieval of data. It is a part of the systems biology approach. KEGG maintains five main databases, *viz.*, KEGG Atlas, KEGG Pathway, KEGG Genes, KEGG Ligand, and KEGG BRITE.

KEGG Databases

KEGG connects known information on molecular interaction networks, such as pathways and complexes (this is the Pathway Database), information about genes and proteins generated by genome projects (including the gene database), and information about biochemical compounds and reactions (including compound and reaction databases). These databases are different networks, known as the protein network, the gene universe and the chemical universe respectively. There are efforts in progress to add to the knowledge of KEGG, including information regarding ortholog clusters in the KO (KEGG Orthology) database.

KEGG Pathway

KEGG Pathway is a collection of manually drawn pathway maps representing our knowledge on the molecular activities, both metabolic and regulatory interaction and reaction networks for:

1. metabolism—carbohydrate, energy, lipid, nucleotide, amino acid, other amino acid, glycan, PK/NRP, cofactor/vitamin, secondary metabolite, and xenobiotics
2. genetic information processing
3. environmental information processing
4. cellular processes
5. human diseases and also on the structure relationships (KEGG drug structure maps) in drug development

A metabolic pathway in KEGG is an idealization corresponding to a large number of possible metabolic cascades. It can generate a real metabolic pathway of a particular organism, by matching the proteins of that organism to enzymes with the reference pathway. For example, the reference pathway of cytochrome P450 to metabolize the drugs like morphine and codeine is shown in Figure 10.11. In the given pathway, Enzyme commission number such as 2.4.1.17 is shown and that is for a systemic way to identify enzyme regardless of species of origin or language used by investigator.

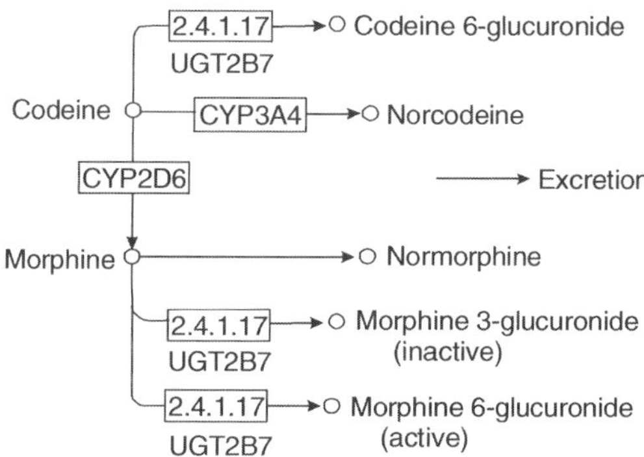

Figure 10.11 A part of the reference pathway of drug metabolism—cytochrome P450 for morphine and codeine (*Source*: http://www.genome.jp/kegg/pathway.html (Kanehisa Laboratories))

The **basic data item in KEGG** is a gene, a gene product, a metabolic compound, or any other molecule in a cell, which may be identified in the form:

database: entry

or

organism: gene

Where "database" is the database name such as GenBank and Swiss-Prot, "entry" is the entry name or the accession number, "organism" is the organism name and "gene" is the gene name. For example,

EC: 6.3.2.3

and

cpd: C00051

represent enzyme glutathione synthase and substrate glutathione, respectively, in the LIGAND database, and *E.coli*: tpiA represents triosephosphate isomerase gene in *E. coli*.

The metabolic pathway diagram in KEGG database represents a consensus view of known metabolic pathways for humans to understand. Each diagram contains enzyme objects that can be manipulated together with the relational operations. The diagram is intended as a drawing of all chemically feasible pathways, rather than a consensus of known pathways. Thus, although the reference diagram has to be drawn and continuously updated manually, the organism-specific pathways are automatically generated. KEGG is actively involved in selected research areas that are given in Table 10.5.

Table 10.5　KEGG for selected research areas

Category	Entry point	Pathway	Brite
Disease	KEGG DISEASE	Human diseases	Pathogens and infectious diseases Human diseases ICD-10 disease classification
Drug	KEGG DRUG	Drug development	Therapeutic category of drugs (Japan) USP drug classification (USA) ATC classification (WHO) TCM drugs in Japan
Glycan	KEGG GLYCAN	Glycan metabolism	Glycosyltransferases Glycan-binding proteins
Lipid PK/NRP Peptide	KEGG COMPOUND	Lipid metabolism PK/NRP biosynthesis	Lipids Polypetides and non-ribosomal peptides Bioactive peptides
Biodegradation	KEGG REACTION	Xenobiotics metabolism	--

Source: http://www.genome.ad.jp/kegg/

New KEGG pathway maps developed up to August 2008

Date	Category	Pathway
2008/08/04	Human diseases	Systemic lupus erythematosus
2008/07/29	Replication and repair	Nonhomologous end joining
2008/07/14	Human diseases	Primary immunodeficiency
2008/07/08	Replication and repair	Homologous recombination
2008/07/07	Human diseases	*Vibrio cholerae* infection
2008/07/07	Human diseases	*Vibrio cholerae* pathogenic cycle
2008/07/04	Xenobiotics metabolism	Drug metabolism (other enzymes)
2008/07/04	Xenobiotics metabolism	Drug metabolism (cytochrome P450)
2008/06/23	Human diseases	Graft-versus-host disease
2008/06/05	Replication and repair	Mismatch repair
2008/06/02	Human diseases	Allograft rejection
2008/05/28	Replication and repair	Nucleotide excision repair
2008/05/14	Replication and repair	Base excision repair
2008/04/28	Human diseases	Autoimmune thyroid disease
2008/03/08	Human diseases	Asthma
2008/03/06	Secondary metabolites	Zeatin biosynthesis
2008/02/27	Secondary metabolites	Insect hormone biosynthesis
2008/01/25	Xenobiotics metabolism	Geraniol degradation
2008/01/25	Xenobiotics metabolism	Fluorobenzoate degradation

Source: http://www.genome.jp/kegg/pathway.html for category and http://www.genome.jp/kegg/pathway/map/ for pathway

New BRITE Hierarchies Added up to August 2008

Date	Category	Hierarchy
2008/08/29	Protein families	Bacterial toxins
2008/07/23	Protein families	DNA repair and recombination proteins
2008/05/16	Drugs	TCM drugs in Japan
2008/03/25	Diseases	ICD-10 disease classification
2008/03/18	Protein families	DNA replication proteins
2008/01/08	Protein families	Ubiquitin system

Source: http://www.genome.jp/kegg-bin/get_htext?

New KEGG Drug Structure Maps Developed During Year 2007-2008

Date	Category	Pathway
2008/02/26	Drug development	Nicotinic cholinergic receptor antagonists
2008/02/12	Drug development	Cholinergic and anticolinergic drugs
2007/12/05	Drug development	Immunosuppressive agents
2007/11/09	Drug development	Antineoplastics—protein kinases inhibitors
2007/10/22	Drug development	Cyclooxygenase inhibitors
2007/09/14	Drug development	Antiviral
2007/08/21	Drug development	Antineoplastics—hormones
2007/08/07	Drug development	Antineoplastics—alkylating agents
2007/08/07	Drug development	Antineoplastics—antimetabolic agents
2007/08/07	Drug development	Antineoplastics—agents from natural products

Source: http://www.genome.jp/kegg/pathway.html#drug for category and http://www.genome.jp/kegg/pathway/map/ for pathway

New KEGG Organisms

Date	Category	Code	Organism
2008/09/24	Bacteria	seg	*Salmonella enterica* subsp. *enterica* serovar *gallinarum*
2008/09/22	Bacteria	bdu	*Borrelia duttonii*
2008/09/22	Bacteria	bre	*Borrelia recurrentis*
2008/09/17	Bacteria	xop	*Xanthomonas oryzae* PXO99A
2008/09/17	Bacteria	cli	*Chlorobium limicola*
2008/09/17	Bacteria	cbt	*Clostridium botulinum* E3
2008/09/17	Bacteria	mpo	*Methylobacterium populi*
2008/09/17	Bacteria	rpi	*Ralstonia pickettii*
2008/09/17	Bacteria	bpy	*Burkholderia phytofirmans*
2008/09/17	Bacteria	nth	*Natranaerobius thermophilus*

Date	Category	Code	Organism
2008/09/17	Bacteria	bhr	*Borrelia hermsii*
2008/09/11	Bacteria	afe	*Acidithiobacillus ferrooxidans*
2008/09/11	Bacteria	xca	*Xanthomonas campestris* pv. *campestris* B100
2008/09/09	Bacteria	sed	*Salmonella enterica* subsp. enterica serovar dublin
2008/09/08	Bacteria	sek	*Salmonella enterica* subsp. *enterica* serovar *Paratyphi* A AKU12601
2008/09/02	Bacteria	vfm	*Vibrio fischeri* MJ11
2008/09/02	Bacteria	rsd	Uncultured Termite group 1 bacterium phylotype Rs-D17
2008/09/01	Bacteria	pmr	*Proteus mirabilis*
2008/09/01	Bacteria	pal	Candidatus *Phytoplasma australiense*

Source: http://www.genome.jp/kegg-bin/show_organism?http://www.genome.jp/kegg/docs/upd_pathway.html

NATIONAL INSTITUTE OF AGROBIOLOGICAL SCIENCES (JAPAN) DNA BANK

The NIAS DNA Bank was established in 1994 as a section of the MAFF Genebank System in-charge of the management of DNA materials and information. The major thrust of its activities focuses on the collection and distribution of DNA materials from agricultural organisms (both plants and animals), materials derived from researches in molecular genetics, and associated genomics information. The DNA bank takes an active role in various genomics projects of the institute. The collaborators from such projects provide the biological resources as well as the necessary input in the construction of various informatics infrastructure. With the re-organization of the institute into an independent body in April 2001, the DNA Bank retains its original function and continues to provide support to the scientific community.

Major Activities of NIAS DNA Bank

1. *Collection and distribution of DNA material* Currently, the DNA Bank is maintaining DNA materials and information that has been

accumulated as part of the genome projects of the Ministry of Agriculture, Forestry and Fisheries such as the Rice Genome Research Program (RGP) and the Animal Genome Research Program (AGP). The biological materials available for distribution include cDNA clones, RFLP markers, PAC/BAC clones and YAC filters.

2. *Information management and other services*   The DNA bank provides access to all genomics data derived from the rice, animal and silkworm genome projects including databases and sequence analysis tools. In addition, it also collects and manages all publicly available nucleotide sequence data (from DDBJ, NCBI/GenBank, EBI/EMBL) and amino acid sequence/proteins (from PIR, Swiss-Prot, PDB). Various retrieval/homology search systems (BLAST, FASTA, WAIS, SRS) using these databases have been constructed to facilitate efficient analysis of sequence information.

SIGNIFICANCE OF GENOME SEQUENCING

The most significant inferences drawn for the genome sequencing study are:

- Complex organisms have more DNA than do simpler ones
- Gene duplication can increase genome size and complexity

Complex Organisms have more DNA than do Simpler Ones

It is now a well-known fact that the genome size varies tremendously among organisms. The first pattern to be detected was that genome sizes are generally correlated with organisms' complexity. The genome of *Mycoplasma genitalium*, the simplest known prokaryote, has only 470 genes. *Rickettsia prowazekii*, the prokaryote that causes typhus, has 634 genes. On the other hand, *Homo sapiens* has about 20,000 protein-coding genes. Figure 10.12 shows the relative sizes of several prokaryotic and eukaryotic genomes. It is not surprising that more complex genetic instructions are needed for building and maintaining a large, complex organism than a small, simple one. What is surprising is that some organisms, such as lungfishes, some salamanders, and lilies, have about 40 times as much DNA as humans do. Clearly, a lungfish or a lily is not 40 times as complex as a human. Why does genome size vary so much?

Some of the apparent variation in genome size disappears when we compare the portion of DNA that actually encodes RNAs or proteins. The size of the coding genome of organisms varies in a way that makes sense. Eukaryotes have more coding DNA than prokaryotes; plants have more coding DNA than single-celled organisms; invertebrates with wings, legs, and eyes have more coding DNA than nematodes; and vertebrates have more coding DNA than invertebrates.

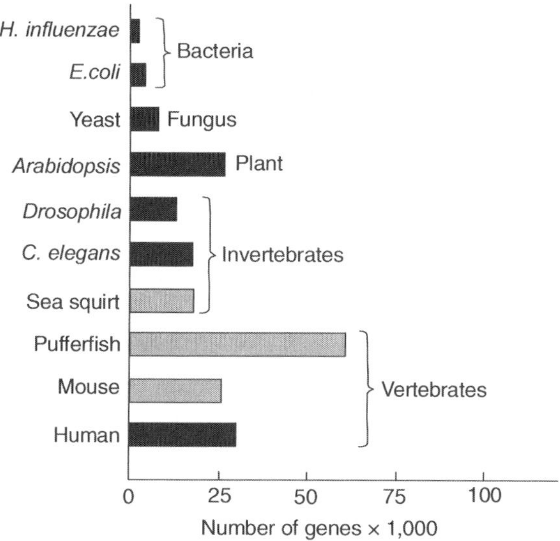

Figure 10.12 Comparative genome sizes of various organisms

The organisms with the largest amount of nuclear DNA (some ferns and flowering plants) have 80,000 times as much as the simplest organisms, but no species has more than 20 times as many protein-coding genes as a bacterium. Therefore, most of the variation in genome size lies not in the number of functional genes, but in the amount of non-coding DNA. Figure 10.13 shows that a large proportion of DNA is non-coding. Most of the DNA of bacteria and yeasts encodes RNAs or proteins, but most of the DNA of more complex organisms is non-coding. Most of the non-coding DNA is probably non-functional.

What maintains such large quantities of non-coding DNA in the cells of most organisms? Does this non-coding DNA have a function, or is it "junk?" Most of this DNA appears to be non-functional. Much of it may consist of pseudogenes that are simply carried in the genome because the cost of doing so is very small. Some of it consists of parasitic transposable elements that spread through populations because they reproduce faster than the host genome. Investigators can use one type of transposable element, retrotransposons, to estimate the rates at which species lose DNA. Retrotransposons copy themselves with the aid of RNA. The most common type of retrotransposon carries duplicated sequences at each end, called Long Terminal repeats (LTRs).

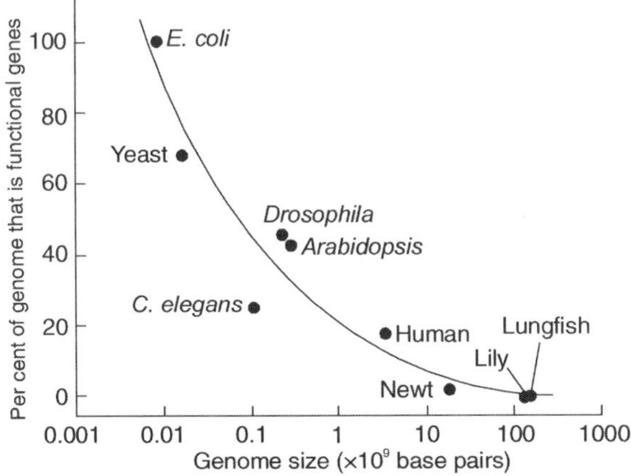

Figure 10.13   Graphical representation of percentage of functional genes in various organisms

Occasionally, LTRs join together in the host genome, at which time the DNA between them is excised. When this happens, one of the LTRs is left behind. The number of such "orphaned" LTRs in a genome is a measure of how many retrotransposons have been lost. By comparing the number of LTRs in the genomes of Hawaiian crickets of the genus *Laupala* and those of fruit flies (*Drosophila*), investigators found that *Laupala* loses DNA more than 40 times more slowly than *Drosophila*. As a result, the genome of *Laupala* is 11 times larger than that of *Drosophila*. Why species differ so greatly in the rate at which they lose DNA is not understood.

Gene Duplication can Increase Genome Size and Complexity

The identical copies of a duplicated gene can have any one of three different fates:

- Both copies of the gene may retain their original function, with the result that the organism produces larger quantities of the gene's RNA or protein product.

- One copy of the gene may be incapacitated by the accumulation of deleterious mutations and become a functionless pseudogene.

- One copy of the gene may retain its original function while the second copy accumulates enough mutations that it can perform a different function.

It is the third of the above fates that is most significant for evolution. How often do gene duplications arise, and which of the three outcomes described above is most likely? These questions can be addressed by counting the number of synonymous nucleotide base changes in the genome of an organism. This number is then compared with the number of base changes that caused protein alterations, to see which number changed faster. Investigators have found that rates of gene duplication are fast enough for a yeast or *Drosophila* population to acquire several hundred duplicate genes over the course of a million years. They also found that most of the duplicated genes in these organisms are very young. Extra genes typically are lost from a genome within 10 million years, which is rapid on an evolutionary time scale.

GENOMICS OF PROKARYOTES AND EUKARYOTES

The biological universe consists of two types of cells—prokaryotic and eukaryotic. Prokaryotic cells consist of a single closed compartment that is surrounded by the plasma membrane, lacks a defined nucleus, and has a relatively simple internal organization. All prokaryotes have cells of this type. Bacteria, the most numerous prokaryotes, are single-celled organisms; the cyanobacteria, or blue-green algae, can be unicellular or filamentous chains of cells. A single *Escherichia coli* bacterium has a dry weight of about 25×10^{-14} g. Eukaryotic cells, unlike prokaryotic cells, contain a defined membrane-bound nucleus and extensive internal membranes that enclose other compartments, called cell organelles. Eukaryotes comprise all members of the plant and animal kingdoms, including the fungi, which exist in both multicellular forms (molds) and unicellular forms (yeasts), and the protozoans. Biochemical and genetical approaches applied to study prokaryotes and eukaryotes generally focus on one gene and its encoded protein at a time. These traditional methods do not give a comprehensive view of the structure and activity of an organism's genome. The field of genomics does just that, encompassing the molecular characterization of whole genomes and the determination of global patterns of gene expression.

In other words, genomics reveals differences in the structure and expression of entire genomes. Recently, the genome sequencing for more than 150 species of prokaryotes and several eukaryotes have been completed. These studies permit comparisons of entire genomes from different species. Although scientists continue to use the classical genetic approach to dissect fundamental cellular processes and biochemical pathways; the availability of complete genomic sequence information for most of the common prokaryotic and eukaryotic experimental organisms has fundamentally changed the way genetic experiments are conducted. Using various computational methods, scientists have identified most of the protein-coding gene sequences in *E. coli,* yeast, *Drosophila, Arabidopsis,* mouse, and humans. The gene sequences, in turn, reveal the primary amino acid sequence of the encoded protein products, providing us with a nearly complete list of the proteins found in each of the major experimental organisms. The results provide overwhelming evidence of the molecular unity of life and the evolutionary processes that made us what we are. By mutating individual genes in a small genome, as of prokaryotes, scientists can determine the minimal genome required for cellular life. Genomics-based methods for comparing thousands of pieces of DNA from different individuals all at the same time are proving useful in tracing the history and migrations of plants and animals and in following the inheritance of diseases in human families. Prokaryotic genome sequencing has important ramifications for the study of organisms that cause human diseases and also helps in development of potential vaccines. New insights arising from eukaryotic genomics are leading to better understanding of evolution, complexity of life and natural selection of the most efficient and precise designs.

GENOME OF BACTERIOPHAGE

Bacterial viruses (bacteriophages) are not free-living organisms; rather, they are infectious parasites that use the resources of a host cell to carry out many of the processes they require to propagate. Many viral particles consist of no more than a genome (usually a single RNA or DNA molecule) surrounded by a protein coat. Almost all plant viruses and some bacterial and animal viruses have RNA genomes. The genomes of DNA viruses vary greatly in size (Table 10.6). Many viral DNAs are circular for at least part of their life cycle. During viral replication within a host cell, specific types of viral DNA called replicative forms may appear; for example, many linear DNAs become circular and all single-stranded DNAs become double-stranded. A typical medium-sized DNA virus is bacteriophage λ (lambda), which infects *E. coli.* In its replicative form inside cells, λ DNA is a circular double helix. This double-stranded DNA contains 48,502 bp and has a contour length of 17.5 μm. Bacteriophage φX174 is a much smaller DNA virus; the DNA in the viral particle is a single-stranded circle, and the double-stranded replicative form contains 5,386 bp. Although viral genomes are small, the contour lengths of their DNAs are much greater than the long dimensions of the viral particles that contain them. The DNA of bacteriophage T4, for example, is about 290 times longer than the viral particle itself. The genomes of RNA viruses, namely, mammalian retroviruses (e.g. HIV) are about 9,000 nucleotides long, and that of the bacteriophage Qβ has 4,220 nucleotides.

The ability of viruses to exchange genetic information through recombination is the basis for virus-based vectors in recombinant DNA technology and hold great promises in the development of gene therapy. Thus, genomics of bacteriophages, plant viruses and animal viruses play an important role as tools in molecular and cellular biology research as well as in bacterial genetics and genetical basis of cancer.

Table 10.6 The length and size of DNA of some bacteriophages

Virus	Length of viral DNA (nm)	Size of viral DNA (bp)
ϕ X174	1,939	5,386
T7	14,377	39,936
λ (lambda)	17,460	48,502
T4	60,800	168,889

GENOME OF CYANOBACTERIA

Cyanobacteria, sometimes called blue-green bacteria because of their photosynthetic pigments, such as chlorophyll *a*, chlorophyll *c*, phycocyanin and phycoerythrin. These are photoautotrophs that require only water, nitrogen gas, oxygen, a few mineral elements, light, and carbon dioxide to survive. They use chlorophyll *a* for photosynthesis and release oxygen gas; many species also fix nitrogen. Their photosynthesis was the basis of the "oxygen revolution" that transformed earth's atmosphere. The total genome sequence of 24 cyanobacteria is completed and made available on webpages, of which 15 of the cyanobacteria come from the marine environment. These are six *Prochlorococcus* strains, seven marine *Synechococcus* strains, *Trichodesmium erythraeum* and *Crocosphaera watsonii*. Studies on genomics of cyanobacteria have demonstrated that these sequences could be used very successfully to infer important ecological and physiological characteristics of marine cyanobacteria. However, there are many more genome projects currently in progress, amongst those there are further *Prochlorococcus* and marine *Synechococcus* isolates, *Acaryochloris* and *Prochloron*, the N_2-fixing filamentous cyanobacteria *Nodularia spumigena*, *Lyngbya aestuarii* and *Lyngbya majuscula*, as well as bacteriophages infecting marine cyanobaceria. Genome sequencing information of cyanobacteria provides insight into the evolutionary origin of photosynthetic pathways.

Genome of *Escherichia coli*

E. coli is a usually harmless inhabitant of the human intestinal tract and also a favourite experimental organism. A single *E. coli* cell contains almost 100 times as much DNA as a bacteriophage λ particle. Different strains of *E. coli* vary from 4300 to 5400 genes. The average bacterial gene is ~1000 bp long and is separated from the next gene by a space of ~100 bp. The chromosome of *E. coli* cell is a single, double-stranded, and circular DNA molecule having 4.2×10^6 bp with a length of about 1.7 mm, some 850 times the length of the *E. coli* cell. It is well-characterized model organism, which has been used for pioneering studies of gene structure and regulation. In *E. coli,* about half the genes are clustered into operons, each of which encodes enzymes involved in a particular metabolic pathway or proteins that interact to form one multi-subunit protein. In 1997, F. Blattner *et al.* of the University of Wisconsin published genome of strain K-12 MG1655, which contains 4639675 bp in a single, circular DNA molecule. *E. coli* strain O157:H7 in hamburger can cause severe illness when ingested. Its genome has 5,416 genes, of which 1,387 are different from those in the familiar (and harmless) laboratory strains of this bacterium. Remarkably, many of these unique genes are also present in other pathogenic species, such as *Salmonella* and *Shigella*. This finding suggests that there is extensive genetic exchange between these species, and that "superbugs" are on the horizon.

The cytoplasm of *E. coli* bacterium has ~20,000 ribosomes and ~200,000 tRNAs, mostly in the form of aminoacyl-tRNA. There are ~1500 mRNA molecules, representing 2–3 copies of each of 600 different messengers. About 7000 RNA polymerase molecules, about 1000 different enzymes, numerous metabolites and cofactors, and a variety of inorganic ions are present in an *E. coli* cell. Many of them are engaged in transcription; probably 2000–5000 enzymes are synthesizing RNA at any one time, the number depends on the growth conditions. In bacteria, regulator RNAs are short molecules, collectively known as sRNAs; *E.coli* contains at least 17 different sRNAs. Some of the sRNAs are general regulators that affect many target genes. They perform function by base pairing with target RNAs (typically mRNAs) to control either their stability or function. *E. coli* has 55 regulatory genes coding for transcriptional activators and 58 for repressor proteins. Genes that encode the proteins and RNA required for protein synthesis make up 3% to 4% of the *E. coli* genome. There are twenty two functional groups of proteins found in *E. coli* (Table 10.7).

Plasmids, viruses, and even phage capsids (in the case of transduction) can transport genes from one bacterial cell to another. There is another type of "gene transport" that occurs within the individual cell. It relies on segments of DNA that can be inserted either at a new location on the same chromosome or into another chromosome. These DNA sequences are called transposable elements. Their insertion often produces phenotypic effects by disrupting the genes into which they are inserted. The first transposable elements typically 1000 to 2000 base pairs long, found at many sites on the *E. coli* main chromosome. The transposable element replicates independently of the rest of the chromosome.

Table 10.7 Twenty two functional groups of proteins found in *E. coli*

Functional class	Number	%
Regulatory function	45	1.05
Putative regulatory proteins	133	3.10
Cell structure	182	4.24
Putative membrane proteins	13	0.30
Putative structural proteins	42	0.98
Phage, transposons, plasmids	87	2.03
Transport and binding proteins	281	6.55
Putative transport proteins	146	3.40
Energy metabolism	243	5.67
DNA replication, recombination, modification, and repair	115	2.68
Transcription, RNA synthesis, metabolism, and modification	55	1.28
Translation, post-translational protein modification	182	4.24
Biosynthesis of cofactors, prosthetic groups, and carriers	103	2.40
Putative chaperones	9	0.21
Nucleotide biosynthesis and metabolism	58	1.35
Amino acid biosynthesis and metabolism	131	3.06
Fatty acid and phospholipid metabolism	48	1.12
Carbon compound catabolism	130	3.03
Central intermediary metabolism	188	4.38
Putative enzymes	251	5.85
Other known genes (gene product or phenotype known)	26	0.61
Hypothetical, unclassified, unknown	1632	38.06

Source: Blattner *et al.* (1997). "The complete genome sequence of *Escherichia coli* K12." *Science.* 277: 1453–74.

The molecular weight of an *E. coli* DNA molecule is about 3.1×10^9 g/mol. The average molecular weight of a nucleotide pair is 660 g/mol, and each nucleotide pair contributes 0.34 nm to the length of DNA. The average size of an ORF is 317 amino acids. If the genes were evenly distributed, the average intergenic region would be 130 bp; the observed average distance between genes is 118 bp. However, the sizes of intergenic regions vary considerably. Some intergenic regions are large. These contain sites of regulatory function, and repeated sequences. The longest intergenic region, 1730 bp, contains non-coding repeat sequences.

Genome information of *Escherichia coli* K-12 MG1655

Organism	*E.coli*
Name	*E.coli*, ECOLI, 511145
Full name	*Escherichia coli* K-12 MG1655
Definition	*Escherichia coli* K-12 MG1655
Annotation	manual
Taxonomy	TAX: 511145
Lineage	Bacteria; Proteobacteria; Gammaproteobacteria; Enterobacteriales; Enterobacteriaceae; Escherichia
Data source	RefSeq
Original DB	Wisconsin, Pasteur, RegulonDB, EcoGene, TIGR, ECOCYC
Chromosome	Circular
Sequence	RS: NC_000913
Length	4639675
Statistics	Number of nucleotides: 4639675 Number of protein genes: 4132 Number of RNA genes: 172
Reference	PMID: 9278503
Authors	Blattner FR, *et al.*
Title	The complete genome sequence of *Escherichia coli* K-12.
Journal	*Science.* 277:1453–74 (1997).

Genome information of *Escherichia coli* APEC O1

Organism	*E.coli*
Name	*E.coli*_APEC, ECOK1, 405955
Full name	*Escherichia coli* APEC O1
Definition	*Escherichia coli* APEC O1 (avian pathogenic *Escherichia coli*)
Annotation	manual
Taxonomy	TAX: 405955
Lineage	Bacteria; Proteobacteria; Gammaproteobacteria; Enterobacteriales; Enterobacteriaceae; Escherichia
Data source	RefSeq
Chromosome	Circular
Sequence	RS: NC_008563
Length	5082025
Plasmid	pAPEC-O1-R; Circular
Sequence	RS: NC_009838
Length	241387
Plasmid	pAPEC-O1-ColBM; Circular
Sequence	RS: NC_009837
Length	174241
Statistics	Number of nucleotides: 5497653 Number of protein genes: 4851 Number of RNA genes: 114
Reference	PMID: 17293413
Authors	Johnson T.J., *et al.*
Title	The genome sequence of avian pathogenic *Escherichia coli* strain O1:K1:H7 shares strong similarities with human extraintestinal pathogenic *E. coli* genomes.
Journal	*J. Bacteriol.* 189:3228–36 (2007).

Genome of *Saccharomyces cerevisiae*

Saccharomyces cerevisiae (baker's yeast) is one of the simplest known eukaryotic organisms containing a nucleus and other specialized intracellular compartments. More than 600 scientists around the world collaborated in mapping and sequencing the yeast genome and it was completed in 1992.

Yeasts *S. cerevisiae* and *S. pombe* genomes of 13.5 Mb and 12.5 Mb have ~6000 and ~5000 genes, respectively. The average open reading frame is ~1.4 kb, so that ~70% of the genome is occupied by coding regions. The major difference between them is that only 5% of *S. cerevisiae* genes have introns, compared to 43% in *S. pombe*. The density of genes is high; organization is generally similar, although the spaces between genes are bit shorter in *S. cerevisiae*. The *S. cerevisiae* genome is distributed over 16 chromosomes. The chromosomes range in size over an order of magnitude, from the 1352 kbp chromosome IV to the 230 kbp chromosome I. The average yeast gene is 1.4 kb long, and very few are longer than 5 kb. In yeast, there are some longer exons that represent uninterrupted genes where the coding sequence is intact. About half of the genes identified by sequence were either known previously or related to known genes.

The yeast genome contains 5885 predicted protein-coding genes, ~140 genes for ribosomal RNAs, 40 genes for small nuclear RNAs, and 275 transfer RNA genes (Figure 10.14).

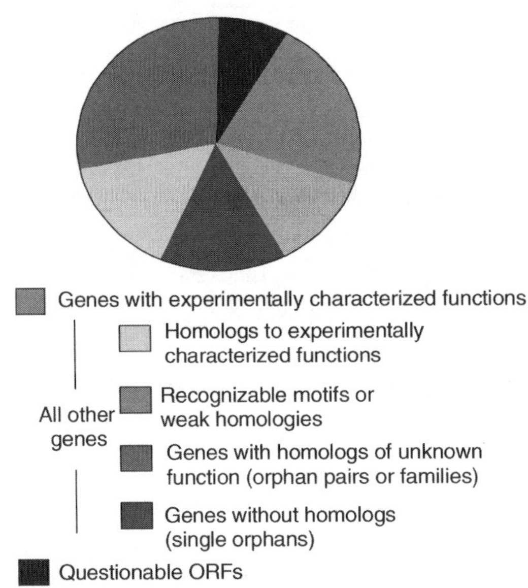

Figure 10.14   Summary of genes in the yeast genome

The known and predicted proteins encoded by the yeast genome have an average molecular weight of 52,728 and contain, on average, 466 amino acid residues. The average molecular weight of amino acids in proteins is 113, taking into account their average relative abundance. This value can be used to estimate the number of residues in a protein from its molecular weight or, conversely, its molecular weight from the number of residues. There is a striking difference between yeast and the higher eukaryotes. In two respects, the yeast genome is denser in coding regions than the known genomes of the more complex eukaryotes, *Caenorhabditis elegans, Drosophila melanogaster,* and human: (1) Introns are relatively rare, and relatively small (only 231 genes in yeast contain introns), (2) There are fewer repeat sequences compared with more complex eukaryotes. The genes of yeast and mammalian mitochondria code for virtually identical mitochondrial proteins, in spite of a considerable difference in gene organization. Vertebrate mitochondrial genomes are very small, with an extremely compact organization of continuous genes, whereas yeast mitochondrial genomes are larger and have some complex interrupted genes. Mammalian mitochondria use their 16 kb genomes to code for 13 proteins, whereas yeast mitochondria use their 60–80 kb genomes to code for as few as 8 proteins.

Table 10.8 Comparison of genomes of *E.coli* and Yeast

Features	*E.coli*	Yeast
Genome length (base pairs)	4.6 Mb	13.6 Mb
Number of proteins	4300	6000
Proteins with roles in:		
Metabolism	650	650
Energy production/storage	240	175
Membrane transporters	280	250
DNA replication/repair/recombination	120	175
Transcription	230	400
Translation	180	350
Protein targeting/secretion	35	430
Cell structure	180	250

The 6000 proteins of the yeast proteome include 5000 soluble proteins and 1000 transmembrane proteins. About half of the proteins are cytoplasmic, a quarter is in the nucleolus, and the remainder are split between the mitochondrion and the ER/Golgi system. Of the 5885 potential protein-coding genes, 3408 correspond to known proteins. The most striking difference between the yeast genome and that of *E. coli* is in the genes for protein targeting (Table 10.8). Both of these single-celled organisms appear to use about the same numbers of genes to perform the basic functions of cell survival. It is the compartmentalization of the eukaryotic yeast cell into organelles that requires it to have so many more genes. Genes encoding several other types of proteins that include histones, cytoskeleton elements, cyclin-dependent kinases, proteins involved in processing of RNA are present in the yeast and other eukaryotic genomes, but have no homologs in prokaryotes.

Genome of *Caenorhabditis elegans*

During the mid-1960s, *Caenorhabditis elegans,* a nematode worm, was selected as a model system for the medical research related to nervous system, because it was perhaps the smallest multicellular animal. The adult is having only 959 cells including about 300 neurons. *C. elegans* has a 1-mm long transparent body and it normally lives in the soil. The roundworm *C. elegans* has proven to be a valuable model organism for studies of cell birth, cell lineage, and cell death. The genome sequencing of *C. elegans* was completed in 1998 due to the efforts of two groups, one located at Washington University Genome Sequencing Centre in USA and the other located at Sanger Centre in UK. It was the first full DNA sequence of a multicellular organism in which 80% of the genome was sequenced using clones from the genomic library that could be multiplied in bacteria and the remaining 20% of the genome was sequenced using YACs that are multiplied in yeast. The *C. elegans* genome contains ~97 Mb of DNA distributed on paired chromosomes I, II, III, IV, V and X (Table 10.9).

Table 10.9   Distribution of *C. elegans* genes

Chromosome	Size (Mb)	Number of protein genes	Density of protein genes (kb/gene)	Number of tRNA genes
I	13.9	2803	5.06	13
II	14.7	3259	3.65	6
III	12.8	2508	5.40	9
IV	16.1	3094	5.17	7
V	20.8	4082	4.15	5
X	17.2	2631	6.54	3

There is no Y chromosome. Different genders in *C. elegans* appear in the XX genotype, a self-fertilizing hermaphrodite; and the XO genotype, a male.

In mammals, one of the two female X chromosomes is inactivated completely. The result is that females have only one active X chromosome, which is the same situation found in males. The active X chromosome of females and the single X chromosome of males are expressed at the same level. In *C. elegans*, the expression of each female X chromosome is halved relative to the expression of the single male X chromosome.

The *C. elegans* genome is eight times larger than that of yeast (97 Mb) and has four times as many protein-coding genes (19,099). Once again, sequencing revealed far more genes than expected. When the sequencing effort began, researchers estimated that the worm would have about 6,000 genes and about that many proteins. Clearly, it has far more. About 3,000 genes in the worm have direct homologs in yeast; these genes code for basic eukaryotic cell functions. The gene density is relatively low for a eukaryote, with ~1 gene/5 kb of DNA. Exons cover ~27% of the genome; the genes contain an average of 5 introns each. Approximately 25% of the genes are in clusters of related genes. Many *C. elegans* proteins are common to other life forms. Others are apparently specific to nematodes.

Eukaryotic genes are transcribed individually, each gene producing a monocistronic messenger. There is only one general exception to this rule; in the genome of *C.elegans*, ~15% of the genes are organized into polycistronic units which is associated with the use of *trans*-splicing to allow expression of the downstream genes in these units. The genome of *C. elegans* DNA varies between regions rich in genes and regions in which genes are more sparsely organized. Only ~42% of the genes have putative counterparts outside the phylum Nematoda, 34% are homologous to proteins of other nematodes and 24% have no known homologues outside *C. elegans* itself. Many of the proteins have been classified according to structure and function. Twenty commonest protein domains found in *C. elegans* are given in Table 10.10.

In recent experiments, microarrays representing about 94 per cent of the *C. elegans* genes were used to monitor transcription at different stages of development and in both sexes. The results showed that expression of about 58 per cent of the monitored genes changes more than two-fold during development, and another 12 per cent are transcribed in sex-specific patterns. The *C. elegans* genome contains 659 genes for tRNA, almost 44% of them are present on the X chromosome and rRNAs appear in a long tandem array at the end of chromosome I. The 5S RNAs appear in a tandem array on chromosome V.

Table 10.10   Twenty common protein domains found in *C. elegans*

Type of domain	Number
Seven-transmembrane spanning chemoreceptor	650
Eukaryotic protein kinase domain	410
Two domain, C4-type zinc finger	240
Collagen	170
Seven-transmembrane spanning receptor (rhodopsin-family)	140
C2H2-type zinc finger	130
C-type lectin	120
RNA recognition motif	100
C3HC4-type (RING finger) zinc fingers	90
Protein tyrosine phosphatase	90
Ankyrin repeat	90
WD domain G-beta repeat	90
Homeobox domain	80
Neurotransmitter gated ion channel	80
Cytochrome P450	80
Conserved C-terminal helicase	80
Short chain and alcohol dehydrogenases	80
UDP-glucoronosyl and UDP-glucosyl transferases	70
EGF-like domain	70
Immunoglobulin superfamily	70

Source: *C. elegans* Genome Consortium paper in the 11 December 1998. *Science.*

Genome of *Drosophila melanogaster*

The fruit fly, *Drosophila melanogaster* is a much larger organism than *C. elegans*, both in size (the fly has 10 times more cells) and complexity. *Drosophila* is used in classical genetics and development studies. In 1999, the complete genome sequencing of *Drosophila* was possible due to the collaboration between Celera Genomics and the Berkeley Drosophila Genome Project. The total chromosomal DNA of *D. melanogaster* is about ~180 Mb. It contains more than 35 times as much DNA as *E. coli*. The fly's 13,601 genes are approximately double the number in yeast, but are fewer than in *C. elegans*. The *Drosophila* genome consists of is five chromosomes:

three large autosomes, a Y chromosome, and a fifth tiny chromosome containing only ~1 Mb of euchromatin. Approximately one-third (60 Mb) of the genome is contained in heterochromatin, highly coiled, compact and densely stained regions flanking the centromeres. The other two-thirds (120 Mb) of the genome is euchromatin in the form relatively uncoiled, less-compact and fairly stained region that contains active genes. Berkeley Drosophila Genome Project (BDGP), European Drosophila Genome Project (EDGP), and Celera Genomics jointly sequenced euchromatic part of the genome. It was for the first time that instead of "clone by clone" method the "Whole Genome Shotgun" (WGS) approach used to sequence the *Drosophila* genome.

The heterochromatin in *D. melanogaster* contains many tandem repeats of the sequence AATAACATAG, and relatively few genes. In *Drosophila*, the average density of genes in the euchromatin sequence is 1 gene/9 kb; much lower than the typical 1 gene/kb densities of prokaryotic genomes. The functional distribution of 13,601 genes in *Drosophila* is shown in Figure 10.15. In *Drosophila*, the expression of the single male X chromosome is doubled relative to the expression of each female X chromosome. About 20% of *Drosophila* genes code for proteins concerned with maintaining or expressing genes, ~20% for enzymes, <10% for proteins concerned with the cell cycle or signal transduction. Half of the genes of *Drosophila* code for products of unknown function.

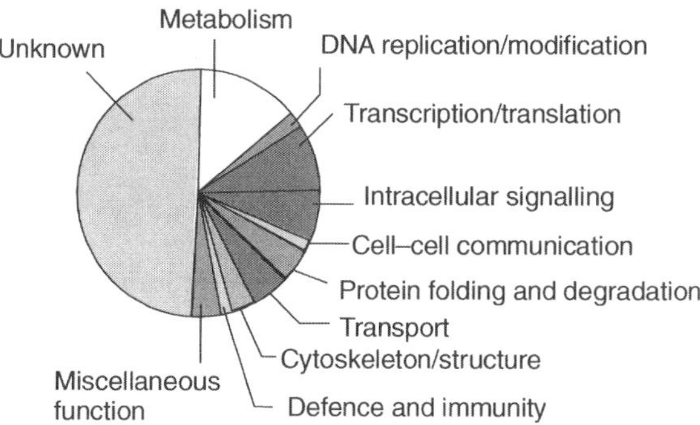

Figure 10.15 Pie chart showing the functional distribution genes in *Drosophila*

Many genes that are present in the worm genome have homologs with similar sequences in fly DNA. Although arthropods are not closely related to mammals, about half of the fly's genes have mammalian homologs. Comparative genomics has made an important contribution to medicine through the discovery of homologs in other organisms of genes that are implicated in human diseases including cancer, and cardiovascular, neurological, endocrinological, renal, metabolic, and haematological diseases. The *Drosophila* genome contains 177 genes with sequences similar to genes that also occur in the human genome and are involved in human diseases. Human disease-associated genes can be introduced into the fly and its expression pattern can be studied. For example, the expression of human gene in the fly responsible for spinocerebellar ataxia type 3 produces similar neuronal cell degeneration. Hence, the *Drosophila* models are used in medical research to provide an insight for Parkinson disease and malaria.

GENOME OF MOSQUITO [http://www.cvc.ucr.edu/]

Mosquitoes are vectors of human diseases, such as malaria and dengue that remain the scourge of developing countries located in the tropical and sub-tropical regions of the earth. A unique aspect of research at the Institute for Integrative Genome Biology is a concentration on mosquito genomics. Researchers associated with this institute want to utilize the tools of genomics, proteomics and bioinformatics to develop new strategies for the control of diseases spread by mosquitoes. New strategies are urgently needed due to the impact of mosquito-borne disease on human welfare and development, the entrenchment of drug resistance within parasite populations, the development of insecticide resistance by mosquitoes and by the increasing urbanization in tropical and sub-tropical countries. The Institute for Integrative Genome Biology is involved in the following large projects:

- The *Culex pipiens quinquefascaitus* (Southern house mosquito) genome project. This mosquito species is a major vector of West Nile virus and is a primary global vector of filariasis. This genome project will follow the successful *Anopheles gambiae* genome project in facilitating the identification of mosquito genes involved in the active transmission of pathogens through the mosquito.

- The global Mosquito Microchip Consortium manufactures uses and distributes DNA microchips containing the expressed genes of *Anopheles gambiae*. Analysis of the expression profile of these genes helps the identification of mosquito genes that are up or down regulated in response to pathogen infection. Researchers of this Consortium attempting for the development, distribution use similar DNA microarray chips for *Aedes aegpti*—the principal vector of dengue fever and yellow fever, and *Culex pipiens quinquefasciatus* over the next few years.

Genome of *Arabidopsis thaliana*

A small sized, easily manipulating flowering plant *Arabidopsis thaliana* is an annual crucifer weed called "thale cress", the member of mustard family that has been a favourite material for plant scientists and has been compared with *Drosophila* by animal geneticist. *A. thaliana* is noted for its adaptations to diverse climatic conditions and hence plant researchers strive for the molecular mechanism underlying this property. The *Arabidopsis thaliana* has five pairs of chromosomes with the genome of ~125 Mb, of which 115.4 Mb was reported in 2000 by an international collaboration, called The *Arabidopsis* Genome Initiative (TAGI). Almost entire genome of *A. thaliana* was completed and published in 14th December, 2000 issue of journal *Nature*. The number of genes was higher than those predicted for either *Drosophila* or those for *C. elegans*. The genes of *Arabidopsis* are relatively small and its DNA sequence reveals about 25,498 protein-coding genes (Table 10.11), but remarkably, many of these are duplicates of other genes and have probably originated by chromosomal rearrangements. When these duplicate genes are subtracted from the total, about 15,000 unique genes are left, a number not too dissimilar from the fruit fly and roundworm. Indeed, many of the genes found in these invertebrate animals have homologs in the plant, suggesting that plants and animals have a common ancestor.

Table 10.11   The genomic information of *Arabidopsis thaliana*

Chromosome	Number of genes	Length (bp)	Density (kb/gene)	Mean gene length
1	6543	29105111	4.0	2078
2	4036	19646945	4.9	1949
3	5220	23172617	4.5	1925
4	3825	17549867	4.6	2138
5	5874	25353409	4.4	1974
Total	25498	115409949		

In general, *Arabidopsis* genes are relatively compact, with 1 gene/4.6 kb on the average, each having many exons, each having the length of about 200 base pairs and interrupted by short non-coding introns of about 160–185 base pairs long. Small intergenic regions suggest small regulatory sequences for the genes, which is in sharp contrast to long regulatory sequences known in many animal genes. Exons in *Arabidopsis* are GC rich (44%) as compared to GC content of introns (32%), which is considered to be a distinctive feature of plant genes.

A. thaliana has some genes that distinguish it as a plant. For example, 420 genes for the assembly of the cell wall and growth, 139 genes involved in photosynthesis, 300 genes for the transport of water into the root and throughout the plant, 94 genes for the uptake and metabolism of inorganic substances from the environment and in the synthesis of specific molecules used for defense against plant predators. Therefore, the plant has been used for genetic screens to identify genes involved in nearly every aspect of plant life. Fifty-eight per cent of the genome contains 24 duplicated segments, = 100 kb in length. As mentioned earlier, the genes in the duplicated segments are usually not identifiable homologues. However, 4140 genes (17% of all genes) appear in 1528 tandem arrays of gene families, containing up to 23 members. The pattern of duplications suggests that an ancestor of *Arabidopsis* underwent a whole-genome duplication, to form a tetraploid plant species.

The Arabidopsis Information Resource (TAIR) maintains a database of genetic and molecular biology data for the model higher plant *Arabidopsis thaliana* (http://www.arabidopsis.org/portals/education/ aboutarabidopsis.jsp). Data available from TAIR includes the complete genome sequence along with gene structure, gene product information, metabolism, gene expression, DNA and seed stocks, genome maps, genetic and physical markers, publications, and information about the Arabidopsis research community. Gene product function data is updated every two weeks from the latest published research literature and community data submissions. Gene structures are updated 1–2 times per year using computational and manual methods as well as community submissions of new and updated genes. TAIR also provides extensive linkouts from the data pages to other Arabidopsis resources. Similarly, The Arabidopsis Biological Resource Center at The Ohio State University collects, reproduces, preserves and distributes seed and DNA resources of *Arabidopsis thaliana* and related species. Stock information and ordering for the ABRC are fully integrated into TAIR.

Genome of Mouse (*Mus musculus*)

The combined efforts of four research institutions including Sanger Centre (UK), the Genome Sequence Centre in British Columbia (Canada), Washington University in St. Louis (USA), and the Institute Genome Research (TIGR) in Rockville (USA) lead to publish a physical map of mouse genome during the year 2001. Later, the genome sequence of the mouse (*Mus musculus*) was generated by Mouse Genome Sequencing Consortium

(Sanger Centre, Washington University, and Whitehead Institute) and uploaded the assembled genome sequence from the C57BL/6J strain, and primary gene annotation at http://www.ensembl.org/Mus_musculus/ resources.html. The genomic map was constructed using 290,000 bacterial clones assembled into just 3,080 Megabases, representing 98% of the entire genome. Such a small number of contigs made it easy to assemble the sequence of whole genome of the mouse using shotgun sequence approach. There are 20 chromosomes in the mouse haploid genome having ~30,000 protein-coding genes, including ~4000 pseudogenes. There are ~800 RNA-coding genes constituting ~350 tRNA genes and 150 pseudogenes, and ~450 other non-coding RNA genes.

Synteny is the term to describe a relationship between chromosomal regions of different species where homologous genes occur in the same order. Syntenic relationships are extensive between mouse and human genomes, and most active genes are in a syntenic region. Mouse chromosome 1 has 21 segments of 1–25 Mb that are syntenic with regions corresponding to parts of 6 human chromosomes. 99% of mouse genes have a homologue in the human genome; and for 96% that homologue is in a syntenic region. Comparing the genomes provides interesting information about the evolution of species. The number of gene families in the mouse and human genomes is the same, and a major difference between the species is the differential expansion of particular families in one of the genomes. Of 25 families, where the size has been expanded in mouse, 14 contain genes specifically involved in rodent reproduction, and 5 contain genes specific to the immune system. There are several known active transposons in the mouse genome indicating that a transposition can introduce several types of collateral damage as well as inserting into a new site may have more chances of chromosomal rearrangements and deletions. There are mouse models available to study molecular defects leading to human genetic diseases. New treatments for diseases can be tested on such mouse models by minimizing human exposure to untested treatments.

HUMAN GENOME PROJECT

The Human Genome Project (HGP) represents biology's first big science project. HGP initiative was put into place to acquire fundamental information needed to further our basic scientific understanding of human genetics and the role of various genes in human health and disease. The project was started in 1990 and was funded by the US Department of Energy. The complete DNA sequence of the human genome is only one of the goals of the HGP.

Other goals include i) construction of a high resolution map of the human genome, ii) Production of physical maps of all human chromosomes and other selected model organisms, iii) development of capabilities for collecting, storing, distributing and analysing data produced, iv) creation of appropriate technologies necessary to achieve these goals.

Milestones of HGP

- 1990—Project initiated as joint effort of U.S. Department of Energy and the National Institutes of Health.
- June 2000—Completion of a working draft of the entire human genome
- February 2001—Analyses of the working draft are published
- April 2003—HGP sequencing is completed and Project is declared and finished two years ahead of schedule.

The imminent announcement of the completion of the HGP (in the year 2003), fifty years after Francis Crick and James Watson historic breakthrough (in the year 1953), marks both an end and a new beginning. It was the triumphant culmination of an extraordinary scientific collaboration that involved 20 research groups from six countries—US, UK, Japan, France, Germany and China. The US Department of Energy and National Institute of Health, US; Centre d'Etudes Polymorphism Humaine (CEPH) and Genethon (a factory of human genetics) in France and Welcome Trust Sanger Institute, England were the major contributory research institutions where this task was achieved. In February 2001, the International Human Genome Sequencing Consortium and Celera Genomics published, separately, drafts of the human genome.

Before the sequencing of the human genome began, most scientists estimated that they would find between 100,000 and 150,000 genes. It was surprising when the actual sequence revealed that there are only few genes predictable than the number of genes estimated. In April 2003, estimates from gene-prediction programs suggested there might be 24,500 or fewer protein-coding genes in human genome, whereas the Ensembl genome-annotation system estimates them at 23,299. In fact, there are many more human mRNAs than there are human genes, and most of this variation comes from alternative splicing. Alternative splicing can be a deliberate mechanism for generating a family of different proteins from a single gene. It is estimated that ~35% of genes have alternative splicing patterns. The human genome is distributed over 22 autosomal pairs plus the X and Y sex-chromosomes. The average gene consists of 3000 bases, but sizes vary with the largest known genes containing 2.4 million bases. Almost 99.9 per cent of nucleotide bases are same in all the people. Chromosome 1 has the most genes (2968) and the Y (male) chromosome has the fewest (231). The DNA contents of the autosomes range from 279 Mbp down to 48 Mbp. The X chromosome contains 163 Mbp and the Y chromosome only 51 Mbp. The human genome sequence amounted to ~3.2 × 10^9 bp, thirty times larger than the genomes of *C. elegans* or *D. melanogaster.* Comparatively, the human genome has a much greater portion (50%) of repeat sequences than the mustard weed *A. thaliana* (11%), the nematode worm *C. elegans* (7%), and the fly *D. melanogaster* (3%).

Types of Sequences in the Human Genome

The exons of human protein-coding genes are relatively small compared to those in other known eukaryotic genomes. The introns are relatively long. As a result many protein-coding genes span long stretches of DNA. For instance, the dystrophin gene, coding for a 3,685 amino acid protein, is >2.4 Mbp long. The pie chart (Figure 10.16) divides the genome into transposons (transposable elements), genes, and miscellaneous sequences. There are four main classes of transposons. Long Interspersed Elements (LINEs), 6 to 8 kbp long, typically include a few genes encoding proteins that catalyse transposition. There are about 850,000 LINEs that constitute about 21% of the human genome. Some of the LINESs are transcribed and translated into proteins. Short Interspersed Elements (SINEs) are about 100 to 300 bp long and are transcribed, but not translated. SINEs constitute about 13% of the total human genome.

Of the 1.5 million in the human genome more than 1 million are *Alu* elements, so called because they generally include one copy of the recognition sequence for *Alu* I, a restriction endonuclease. The genome also contains 450,000 copies of retrovirus-like transposons, 1.5 to 11 kbp long that accounts for 8% of the total human genome. Although these are "trapped" in the genome and cannot move from one cell to another, they are evolutionarily related to the retroviruses, which include HIV. A final class of transposons (making up <1% and not shown in the pie chart) consists of a variety of transposon remnants that differ greatly in length. About 30% of the genome consists of sequences included in genes for proteins, but only a small fraction of this DNA is in exons (coding sequences). About 25% of the total human genome are miscellaneous sequences that include Simple Sequence Repeats (SSR) and large Segmental Duplications (SD), the latter being segments that appear more than once in different locations. Among the unlisted sequence elements (denoted by a question mark) are genes encoding RNAs (which can be harder to identify than genes for proteins) and remnants of transposons that have been evolutionarily altered so that they are now hard to identify.

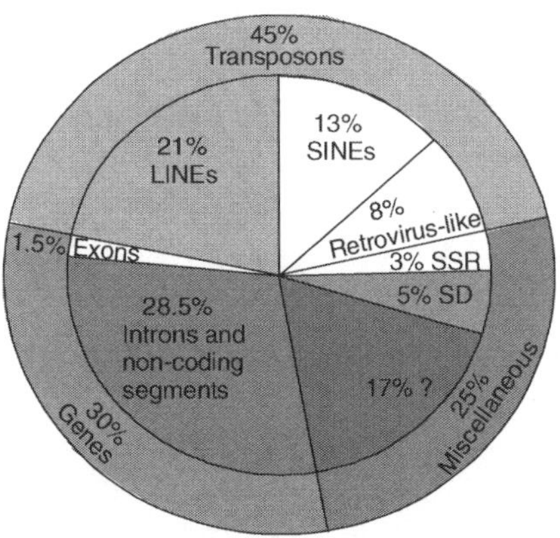

Figure 10.16   Composition of different sequences in human genome

Protein-coding genes   Consortium researchers have confirmed the existence of 19,599 protein-coding genes in the human genome and identified another 2,188 DNA segments that are predicted to be protein-coding genes. Of the 94 per cent of human genomic DNA that has been sequenced, only ~1.5 per cent corresponds to protein-coding sequences (exons). Most human exons contain 50–200 base pairs, although the 3′ exon in many transcription units is much longer. Human introns vary in length considerably. Although many are ~90 bp long, some are much longer; their median length is 3.3 kb. Approximately one-third of human genomic DNA is thought to be transcribed into pre-mRNA precursors, but some 95 per cent of these sequences are in introns, which are removed by RNA splicing. Alternative splicing of pre-mRNA can be used to produce different proteins. The transcripts of over half the genes in the human genome are alternatively spliced, which increases the number of proteins that can be encoded by a single gene. Analysis of the human protein repertoire implied by the genome sequence has proved difficult because of the problems in reliably detecting genes, and alternative splicing patterns. The International Human Genome Sequencing Consortium has estimated that the total count will be ~32,000 genes in all. In addition to protein-coding genes, the human genome contains thousands of RNA genes, including tRNA, rRNA and microRNA, and other non-coding RNA genes.

Genome information of *Homo sapiens* (human)

Organism	Human (has)
Name	H.sapiens, HUMAN, 9606
Full name	Homo sapiens (human)
Definition	Homo sapiens (human)
Annotation	manual
Taxonomy	TAX: 9606
Lineage	Eukaryota; Metazoa; Chordata; Craniata; Vertebrata; Euteleostomi; Mammalia; Eutheria; Euarchontoglires; Primates; Haplorrhini; Catarrhini; Hominidae; Homo
Data source	RefSeq
Original DB	NCBI, HGNC, HPRD, Ensembl
Statistics	Number of protein genes: 24246 Number of RNA genes: 24
Reference	PMID: 11237011
Authors	Lander ES, *et al.*
Title	Initial sequencing and analysis of the human genome.
Journal	*Nature.* 409:860–921 (2001).
Reference	PMID: 11181995
Authors	Venter J.C., *et al.*
Title	The sequence of the human genome.
Journal	*Science.* 291:1304–51 (2001)
Reference	PMID: 15496913
Authors	International Human Genome Sequencing Consortium.
Title	Finishing the euchromatic sequence of the human genome.
Journal	*Nature.* 431:931–45 (2004).
Reference	PMID: 16710414 (chromosome 1)
Authors	Gregory S.G., *et al.*
Title	The DNA sequence and biological annotation of human chromosome 1.
Journal	*Nature.* 441:315–321 (2006).
	(Similar entries are made for chromosome 2 to 22 and for both sex chromosomes, X and Y)

Impact of HGP

When the sequence is known what will it mean and was it worth the effort? It is expected that there will be many medical benefits from knowing the structure of all genes in the human genome. This will provide more comprehensive pre-natal and pre-symptomatic diagnosis in individuals judged to be at risk of carrying a disease gene. In addition, information about gene structure and regulation can be used to understand the biological process in humans. It is also expected that this information will provide a framework for developing new therapies for disease, which include gene therapy approaches. While most single gene disorders are rare, they are the easiest targets for developing medical therapies. The more common disorders are multifactorial, such as heart disease and adult onset diabetes, and will prove more problematic to identify the genes involved. However, ultimately the information from the HGP will be a benefit to the medical community; it just may take some time to realize them.

Potential Benefits of HGP

Technology and resources generated by the HGP and other genomics research are already having a major impact on research across the life sciences. This accomplishment has been hailed as one of the greatest intellectual achievements of humankind. The potential for commercial development of genomics research presents U.S. industry with a wealth of opportunities, and sales of DNA-based products and technologies in the biotechnology industry are projected to exceed $45 billion by 2009 (Consulting Resources Corporation *Newsletter*, Spring 1999). Rapid progress in genomics and its potential applications are applied for human welfare. Some of the current and potential applications of genome research include:

- Molecular medicine
- Microbial genomics
- Risk assessment
- Bioarchaeology, anthropology, evolution, and human migration
- DNA forensics (identification)
- Agriculture, livestock breeding, and bioprocessing

Molecular medicine   It helps in

- improved diagnosis of disease
- earlier detection of genetic pre-dispositions to disease
- rational drug design

□ gene therapy and control systems for drugs

□ pharmacogenomics "custom drugs"

Technology and resources promoted by the Human Genome Project are starting to have profound impacts on biomedical research and promise to revolutionize the wider spectrum of biological research and clinical medicine. Increasingly detailed genome maps have aided researchers seeking genes associated with dozens of genetic conditions, including myotonic dystrophy, fragile X syndrome, neurofibromatosis types 1 and 2, inherited colon cancer, Alzheimer's disease, and familial breast cancer.

On the horizon is a new era of molecular medicine characterized less by treating symptoms and more by looking to the most fundamental causes of disease. Rapid and more specific diagnostic tests will make possible earlier treatment of countless maladies. Medical researchers also will be able to devise novel therapeutic regimens based on new classes of drugs, immunotherapy techniques, avoidance of environmental conditions that may trigger disease, and possible augmentation or even replacement of defective genes through gene therapy.

Microbial genomics   This can be used in the following ways.

□ to create new energy sources (biofuels).

□ to develop environmental monitoring techniques to detect pollutants.

□ for safe, efficient environmental remediation.

□ for carbon sequestration

□ to have protection from biological and chemical warfare

□ to understand disease vulnerabilities and revealing drug targets

In 1994, taking advantage of new capabilities developed by the genome project, DOE initiated the Microbial Genome Program (MGP) to sequence the genomes of bacteria useful in energy production, environmental remediation, toxic waste reduction, and industrial processing. Despite our reliance on the inhabitants of the microbial world, we know little of their number or their nature, estimates are t hat less than 0.01% of all microbes have been cultivated and characterized. Programs like the DOE MGP help to lay a foundation for knowledge that will ultimately benefit human health and the environment. The economy will benefit from further industrial applications of microbial capabilities.

Information generated from the characterization of complete genomes in MGP will lead to insights into the development of such new energy-related biotechnologies as photosynthetic systems, microbial systems that function in extreme environments and organisms that can metabolize readily available renewable resources and waste material with equal facility. Expected benefits also include development of diverse new products, processes, and test methods that will open the door to a cleaner environment. Biomanufacturing will use non-toxic chemicals and enzymes to reduce the cost and improve the efficiency of industrial processes. Already, microbial enzymes are being used to bleach paper pulp, stone wash denim, remove lipstick from glassware, break down starch in brewing, and coagulate milk protein for cheese production. In the health arena, microbial sequences may help researchers find new human genes and shed light on the disease-producing properties of pathogens.

Microbial genomics will also help pharmaceutical researchers gain a better understanding of how pathogenic microbes cause disease. Sequencing these microbes will help reveal vulnerabilities and identify new drug targets. Gaining a deeper understanding of the microbial world also will provide insights into the strategies and limits of life on this planet. Data generated in this young program already have helped scientists to identify the minimum number of genes necessary for life and confirm the existence of a third major kingdom of life. Additionally, the new genetic techniques now allow us to establish more precisely the diversity of microorganisms and identify those critical to maintaining or restoring the function and integrity of large and small ecosystems; this knowledge also can be useful in monitoring and predicting environmental change. Finally, studies on microbial communities provide models for understanding biological interactions and evolutionary history.

Risk assessment   The risk assessment is done to

- assess health damage and risks caused by radiation exposure, including low-dose exposures
- assess health damage and risks caused by exposure to mutagenic chemicals and cancer-causing toxins
- reduce the likelihood of heritable mutations

Understanding the human genome will have an enormous impact on the ability to assess risks posed to individuals by exposure to toxic agents. Scientists know that genetic differences make some people more susceptible and others more resistant to such agents. Far more work must be done to determine the genetic basis of such variability. This knowledge will directly address DOE's long-term mission to understand the effects of low-level exposures to radiation and other energy-related agents, especially in terms of cancer risk.

Bioarchaeology, anthropology, evolution, and human migration These fields are used to:

▣ study the evolution through germ-line mutations in lineages

▣ study the migration of different population groups based on female genetic inheritance

▣ study the mutations on the Y chromosome to trace lineage and migration of males

▣ compare breakpoints in the evolution of mutations with ages of populations and historical events

Understanding genomics will help us to understand human evolution and the common biology we share with all of life. Comparative genomics between humans and other organisms such as mice already has led to similar genes associated with diseases and traits. Further comparative studies will help to determine the yet-unknown function of thousands of other genes.

Comparing the DNA sequences of entire genomes of different microbes will provide new insights about relationships among the three kingdoms of life: archaebacteria, eukaryotes, and prokaryotes.

DNA forensics (identification) DNA forensics are applied to:

▣ identify potential suspects whose DNA may match evidence left at crime scenes

▣ exonerate persons wrongly accused of crimes

▣ identify crime and catastrophe victims

▣ establish paternity and other family relationships

▣ identify endangered and protected species as an aid to wildlife officials (could be used for prosecuting poachers)

▣ detect bacteria and other organisms that may pollute air, water, soil, and food

▣ match organ donors with recipients in transplant programs

▣ determine pedigree for seed or livestock breeds

▣ authenticate consumables such as caviar and wine

Any type of organism can be identified by examination of DNA sequences unique to that species. Identifying individuals is less precise at this time, although when DNA sequencing technologies progress further, direct characterization of very large DNA segments, and possibly even whole genomes, will become feasible and practical and will allow precise individual identification.

To identify individuals, forensic scientists scan about 10 DNA regions that vary from person to person and use the data to create a DNA profile of that individual (sometimes called a DNA fingerprint). There is an extremely small chance that another person has the same DNA profile for a particular set of regions.

Agriculture, livestock breeding and bioprocessing These fields are used in production of :

▣ disease-, insect-, and drought-resistant crops
▣ healthier, more productive, disease-resistant farm animals
▣ more nutritious produce
▣ biopesticides
▣ edible vaccines incorporated into food products
▣ new environmental cleanup uses for plants like tobacco

Understanding plant and animal genomes will allow us to create stronger, more disease-resistant plants and animals —reducing the costs of agriculture and providing consumers with more nutritious, pesticide-free foods. Already growers are using bioengineered seeds to grow insect- and drought-resistant crops that require little or no pesticide. Farmers have been able to increase outputs and reduce waste because their crops and herds are healthier.

Alternate uses for crops such as tobacco have been found. One researcher has genetically engineered tobacco plants in his laboratory to produce a bacterial enzyme that breaks down explosives such as TNT and dinitroglycerin. Waste that would take centuries to break down in the soil can be cleaned up by simply growing these special plants in the polluted area.

Ethical, Legal, and Social Issues of the HGP

As mentioned earlier, HGP was completed in 2003. One of the key research areas was ethical, legal, and social issues research (ELSI). The U.S. Department of Energy and the National Institutes of Health devoted 3% to 5% of their annual Human Genome Program budgets toward studying the ethical, legal, and social issues surrounding availability of genetic information. This represents the world's largest bioethics program. It has become a model for ELSI programs around the world.

Rapid advances in the science of genetics and its applications have presented new and complex ethical and policy issues for individuals and society. ELSI programs that identify and address these implications have been an integral part of the U.S. HGP since its inception. These programs have resulted in a body of work that promotes education and helps to guide the conduct of genetic research and the development of related medical and public policies. A continuing challenge is to safeguard the privacy of individuals and groups who contribute DNA samples for large-scale sequence-variation studies. Other concerns have been to anticipate how the resulting data may affect concepts of race and ethnicity; identify potential uses (or misuses) of genetic data in workplaces, schools, and courts; identify commercial uses; and foresee impacts of genetic advances on the concepts of humanity and personal responsibility.

IDENTIFYING GENES INVOLVED IN HUMAN DISEASE

Most genes are identified by mapping the gene location to the human genome and subsequent identification of the mutations causing the condition. This is done through studies of affected families. The mutation in the genome causing the disease is genetically mapped to a location in the human genome. This is done using genetic mapping techniques, which rely on the naturally occurring sequence variation in the human genome. A genetic disease locus sequence variation can be mapped to a specific location in the genome by looking for its association to a closely linked DNA polymorphism. These DNA polymorphisms can be restriction site length Polymorphism (RFLPs), or Variable Number Tandem Repeat Sequences (VNTR) as well as other types of variable sequences such as CA repeats. A genetic linkage map is constructed by observing how frequently two DNA markers are inherited together. Figure 10.17 shows that marker (M) and Huntington disease (HD) loci are preferentially inherited by the offspring. The one child that inherits only marker (M) shows that recombination has separated the two markers during the production of the gametes from the father. The frequency of this event helps to determine the distance between the two DNA sequences on a genetic map.

Once the genetic location of the disease locus is found the positional cloning of the specific disease gene is accomplished. It involves the use of a low and high resolution physical map of the DNA around the disease gene. HGP has accomplished this for the entire genome. The production of a complete high resolution physical DNA map including the complete nucleotide sequence allows the researcher to utilize this data to clone more

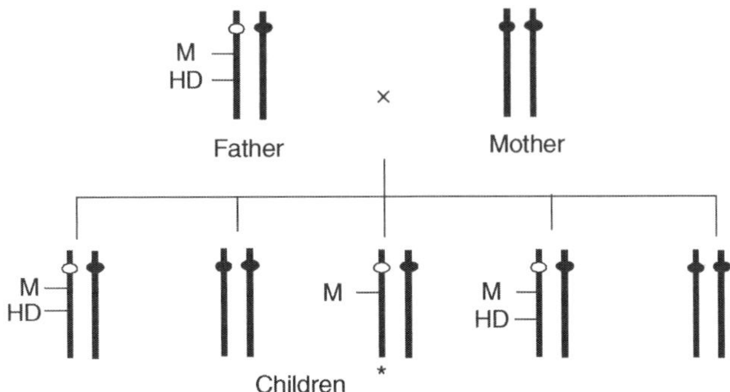

Figure 10.17   Inheritance of the Marker (M) and Huntington (HD) gene from parents. * Recombination frequency of this event reflects the distance between the marker M and HD.

easily the genes that may be involved in the disease studied. The completed sequencing of the human genome provides all the information on the physical map and the nucleotide sequence of all human genes. Once a genetic location is established the DNA sequence of that region can be easily obtained from the database. The potential genes can be identified and subsequently screened for mutations in individuals showing the disease phenotype. This will allow researchers to more easily identify genes involved in diseases such as diabetes and heart disease as well as a host of other complex human disease conditions.

Most aspects of human biology involve both genetic (inherited) and non-genetic (environmental) factors. Some inherited variation influences aspects of our biology that are not medical in nature (height, eye colour, ability to taste or smell certain compounds, etc.). Moreover, some genetic disorders only cause disease in combination with the appropriate environmental factors (such as diet). With these caveats, genetic disorders may be described as clinically defined diseases caused by genomic DNA sequence variation. In the most straightforward cases, the disorder can be associated with variation in a single gene. For example, cystic fibrosis is caused by mutations in the *CFTR* gene, and is the most common recessive disorder in caucasian populations with over 1300 different mutations known. Disease-causing mutations in specific genes are usually severe in terms of gene function, and are fortunately rare, thus genetic disorders are similarly individually rare. However, since there are many genes that can vary to cause genetic disorders, in aggregate they comprise a significant component of known medical conditions, especially in pediatric medicine. Molecularly characterized genetic disorders are those for which the underlying causal gene has been identified, currently there are approximately 2200 such disorders annotated in the OMIM database.

Diagnosis and treatment of genetic disorders are usually performed by a geneticist-physician trained in clinical/medical genetics. The results of the HGP are likely to provide increased availability of genetic testing for gene-related disorders, and eventually improved treatment. Parents can be screened for hereditary conditions and counselled on the consequences, the probability it will be inherited, and how to avoid or ameliorate it in their offspring.

As noted above, there are many different kinds of DNA sequence variation, ranging from complete extra or missing chromosomes down to single nucleotide changes. It is generally presumed that much naturally occurring genetic variation in human populations is phenotypically neutral, i.e., has little or no detectable effect on the physiology of the individual (although there may be fractional differences in fitness defined over evolutionary time frames). Genetic disorders can be caused by any or all known types of sequence variation. To molecularly characterize a new genetic disorder, it is necessary to establish a causal link between a particular genomic sequence variant and the clinical disease under investigation. Such studies constitute the realm of human molecular genetics.

With the advent of the Human Genome and International HapMap Project, it has become feasible to explore subtle genetic influences on many common disease conditions such as diabetes, asthma, migraine, schizophrenia, etc. Although some causal links have been made between genomic sequence variants in particular genes and some of these diseases, often with much publicity in the general media, these are usually not considered to be genetic disorders per se as their causes are complex, involving many different genetic and environmental factors. Thus there may be disagreement in particular cases whether a specific medical condition should be termed a genetic disorder.

Knockout Mice Helping to Determine Human Gene Function

One of the transgenic animal models is knockout mice, which can provide a unique insight into genetic basis of a human disease as well as a mechanism for studying the various cellular activities in which the product of a particular gene might be engaged. In knockout mice, genetic code has been altered by the insertion of foreign genetic material into their DNA. Using this approach, genetic engineers target specific genes, causing them to be expressed or inactivated. These mice are then bred, creating a population of offspring with the trait. When researchers isolate human genes with unknown functions, they can create knockout mice with these genes and observe the results. Instead of creating merely the mouse equivalent of the human gene, researchers are able to reproduce and express actual human genes and their corresponding proteins in mice. Subsequent offspring will inherit not only the instructions coded by their original mouse genome, but also the traits coded for by the inserted human DNA. This helps researchers to understand health and disease by observing how genes work in cells.

Development of knockout mice that mimic certain human diseases can be illustrated by cystic fibrosis. The recessive mutation that causes this disease eventually was shown to be located in a gene known as *CFTR*, which encodes a chloride channel. Using the cloned wild-type human *CFTR* gene, researchers isolated the homologous mouse gene and subsequently introduced mutations in it. The gene-knockout (i.e., both alleles of a gene are deleted) technique was then used to produce homozygous mutant mice, which showed symptoms (i.e., a phenotype in the form of defective chloride channel in epithelial cells leading to excessive mucus in lungs), similar to those of humans with cystic fibrosis. These knockout mice are currently being used as a model system for studying this genetic disease and developing effective therapies. Knockout mice have many benefits. They not only allow researchers to determine gene function and understand diseases at the molecular level, but they also help scientists in testing new drugs and devising novel therapies.

GENE THERAPY

Chromosomes carry genes, which are the basic physical and functional units of heredity. Genes are specific sequences of bases that encode instructions to make proteins. Alteration in gene sequence leads to synthesis of abnormal proteins, which are unable to carryout their normal functions and hence genetic disorders can result. Sickle-cell anaemia was the first genetic disease detected, although the gene responsible for this disease was not cloned until much later. The first "disease-causing gene" to be discovered was the dystrophin gene that causes muscular dystrophy. The list of "disease genes" grows everyday. Diseases caused by single gene defects include cystic fibrosis, diabetes mellitus, pituitary dwarfism, emphysema, familial hyper-cholesterolemia, thalassemia major, haemophilia, lysosomal disease, Gaucher's disease, glycogen storage disease, phenylketonuria, Severe Combined Immunodeficiency disease (SCID) , muscular dystrophy, etc. Acquired diseases include diseases which are caused mainly due to somatic mutations in some genes. Majority of cancers, cardiovascular diseases (e.g. coronary artery disease, myocardial infarction), neurodegenerative diseases (e.g. Alzheimer's disease, Parkinson's disease), rheumatoid arthritis, and AIDS belong to this class. Gene therapy is the technique for correcting diseased genes responsible for disease development. Researchers may use one of following approaches for correcting faulty genes:

- A normal gene may be inserted into a non-specific location within the genome to replace a non-functional gene. This is the most common approach.

- An abnormal gene could be swapped for a normal gene through homologous recombination.

- An abnormal gene could be repaired through selective reverse mutation, which returns the gene to its normal function.

- The regulation (the degree to which a gene is turned on or off) of a particular gene could be altered.

Working Principle of Gene Therapy

Gene therapy is the technique involving replacement of defective gene with a functional one. A carrier molecule called a vector must be used to deliver the therapeutic gene to the patient's target cells. Currently, the most common vector is a virus that has been genetically altered to carry normal human DNA. Viruses have evolved a way of encapsulating and delivering their genes to human cells in a pathogenic manner. Scientists have tried to take advantage of this capability and manipulate the virus genome to remove disease-causing genes and insert therapeutic genes.

Target cells such as the patient's liver or lung cells are infected with the viral vector. The vector then unloads its genetic material containing the therapeutic human gene into the target cell. The generation of a functional protein product from the therapeutic gene restores the target cell to a normal state.

Some of the different types of viruses used as gene therapy vectors are as follows:

- **Retroviruses** A class of viruses that can create double-stranded DNA copies of their RNA genomes. These copies of its genome can be integrated into the chromosomes of host cells, e.g. Human immunodeficiency virus (HIV).

◙ **Adenoviruses** A class of viruses with double-stranded DNA genomes that cause respiratory, intestinal, and eye infections in humans. The virus that causes the common cold is an adenovirus.

◙ **Adeno-associated viruses** A class of small, single-stranded DNA viruses that can insert their genetic material at a specific site on chromosome 19.

◙ **Herpes simplex viruses** A class of double-stranded DNA viruses that infect a particular cell type, neurons. Herpes simplex virus type 1 is a common human pathogen that causes cold sores.

Besides virus-mediated gene-delivery systems, there are several non-viral options for gene delivery. The simplest method is the direct introduction of therapeutic DNA into target cells. The mechanical or electrical methods include low-voltage electroporation or microinjection, particle bombardment, and pressure-mediated DNA delivery. These approaches have limitations because they can be used only with certain tissues and require large amounts of DNA.

Another non-viral approach involves the use of chemical to enhance DNA uptake by the cell. Liposomes (artificial lipid sphere with an aqueous core) carry the therapeutic DNA and are capable of passing the DNA through the target cell's membrane. Therapeutic DNA complexed with 2-diethylamino ether (DEAE) (dextran), or calcium phosphate can enter the cell. Chemically linked therapeutic DNA will bind to special cell receptors. Once bound to these receptors, the therapeutic DNA constructs are engulfed by the cell membrane and passed into the interior of the target cell. This delivery system tends to be less effective than other options in addition to the limitation due to toxicity of the chemicals upon systemic administration.

Researchers are also experimenting with introducing a 47th (artificial human) chromosome into target cells. This chromosome would exist autonomously alongside the standard 46 without affecting their workings or causing any mutations. It would be a large vector capable of carrying substantial amounts of genetic code, and scientists anticipate that, because of its construction and autonomy, the body's immune systems would not attack it. A problem with this potential method is the difficulty in delivering such a large molecule to the nucleus of a target cell.

Current Status of Gene Therapy Research

The first human gene therapy trial was carried out at the NIH in Bethesda, Maryland, in 1990. The patient was a four-year-old girl crippled by ADA deficiency (often called "the boy in the bubble disease")—one form of SCID results from genetically inherited defects in the gene encoding adenosine deaminase (ADA), an enzyme involved in nucleotide biosynthesis. Bone marrow cells from the child were transformed with an engineered retrovirus containing a functional ADA gene.

The Food and Drug Administration (FDA) has not yet approved any human gene therapy product for sale. Current gene therapy is experimental and has not proven very successful in clinical trials. Scientific and medical research related to gene therapy sometimes lead to stories with sad endings and thus hampered the progress in the field. In September 2000, gene therapy suffered a major setback with the death of 18-year-old Jesse Gelsinger, who was participating in a gene therapy trial for ornithine transcarboxylase deficiency (OTCD). He died from multiple organ failures after 4 days starting the treatment. His death is believed to have been triggered by a severe immune response to the adenovirus carrier since he was given too many adenoviruses.

A gene therapy trial initiated in 1999 was successful in correcting a form of SCID caused by defective cytokine receptors (in particular a subunit called γc), as reported in 2000 by physician researchers in Europe. These researchers introduced the corrected gene for the γc cytokine-receptor subunit into CD34$^+$ cells. (The stem cells that give rise to immune system cells have a protein called CD34$^+$ on their surface; these cells can be separated from other bone marrow cells by antibodies to CD34$^+$.) The transformed cells were placed back into the patients' bone marrow. In this trial, introduction of the corrected gene clearly conferred a growth advantage over the untreated cells. A functioning immune system was detected in four of the first five patients within 6 to 12 weeks, and levels of mature immune system T lymphocytes reached the levels found in age-matched control subjects (who did not have SCID) within 6 to 8 months. Immune system function was restored and nearly 4 years later (mid-2003) most of the children are leading normal lives. Similar results have been obtained with four additional patients. This provided dramatic confirmation that human gene therapy could cure a serious genetic disease.

Another major blow came in January 2003, when the FDA placed a temporary halt on all gene therapy trials using retroviral vectors in blood stem cells. FDA took this action after it learned that a second child treated in a French gene therapy trial had developed a leukaemia-like condition. Both this child and another who had developed a similar condition in August 2002 had been successfully treated by gene therapy for X-linked SCID (X-SCID), also known as "bubble baby syndrome." FDA's Biological Response Modifiers

Advisory Committee (BRMAC) met at the end of February 2003 to discuss possible measures that could allow a number of retroviral gene therapy trials for treatment of life-threatening diseases to proceed with appropriate safeguards. In April of 2003 the FDA eased the ban on gene therapy trials using retroviral vectors in blood stem cells.

Factors Affecting Gene Therapy

Short-lived nature of gene therapy Before gene therapy can become a permanent cure for any condition, the therapeutic DNA introduced into target cells must remain functional and the cells containing the therapeutic DNA must be long-lived and stable. Problems with integrating therapeutic DNA into the genome and the rapidly dividing nature of many cells prevent gene therapy from achieving any long-term benefits. Patients will have to undergo multiple rounds of gene therapy.

Immune response Anytime a foreign object is introduced into human tissues, the immune system is designed to attack the invader. The risk of stimulating the immune system in a way that reduces gene therapy effectiveness is always a potential risk. Furthermore, the immune system's enhanced response to invaders it has seen before makes it difficult for gene therapy to be repeated in patients.

Problems with viral vectors Viruses, while the carrier of choice in most gene therapy studies, present a variety of potential problems to the patient —toxicity, immune and inflammatory responses, and gene control and targeting issues. In addition, there is always the fear that the viral vector, once inside the patient, may recover its ability to cause disease.

Multigene disorders Conditions or disorders that arise from mutations in a single gene are the best candidates for gene therapy. Unfortunately, some of the most commonly occurring disorders, such as heart disease, high blood pressure, Alzheimer's disease, arthritis, and diabetes, are caused by the combined effects of variations in many genes. Multigene or multifactorial disorders such as these would be especially difficult to treat effectively using gene therapy. For more information on different types of genetic disease, log on to http://www.ornl.gov/sci/techresources/Human_Genome/medicine/assist.shtml.

Recent Developments in Gene Therapy Research

- Results of world's first gene therapy for inherited blindness show sight improvement. In 28 April 2008, UK researchers from the UCL Institute of Ophthalmology and Moorfields Eye Hospital NIHR Biomedical Research Centre have announced results from the world's first clinical trial to test a revolutionary gene therapy treatment for a type of inherited blindness. The results, published in the New England Journal of Medicine (April 2008), show that the experimental treatment is safe and can improve sight. The findings are a landmark for gene therapy technology and could have a significant impact on future treatments for eye disease. (http://www.ucl.ac.uk/media/library/Genetherapyblind).

Previous information on this trial (May 1, 2007): A team of British doctors from Moorfields Eye Hospital and University College in London conduct first human gene therapy trials to treat Leber's congenital amaurosis, a type of inherited childhood blindness caused by a single abnormal gene. The procedure has already been successful at restoring vision for dogs. This is the first trial to use gene therapy in an operation to treat blindness in humans. (http://www.reuters.com/article/scienceNews/idUSL016653620070501?pageNumber=1).

- A combination of two tumour-suppressing genes delivered in lipid-based nanoparticles drastically reduces the number and size of human lung cancer tumours in mice during trials conducted by researchers from The University of Texas M. D. Anderson Cancer Center and the University of Texas Southwestern Medical Center. (http://www.newswise.com/p/articles/view/526526/).

- Researchers at the National Cancer Institute (NCI), part of the National Institutes of Health, successfully re-engineer immune cells, called lymphocytes, to target and attack cancer cells in patients with advanced metastatic melanoma. This is the first time that gene therapy is used successfully to treat cancer in humans. (For more information log on to http://www.cancer.gov/newscenter/pressreleases/MelanomaGeneTherapy).

- Gene therapy is effectively used to treat two adult patients for a disease affecting non-lymphocytic white blood cells called myeloid cells. Myeloid disorders are common and include a variety of bone marrow failure syndromes, such as acute myeloid leukaemia. The study is the first to show that gene therapy can cure diseases of the myeloid system. (http://www.cincinnatichildrens.org/about/news/release/2006/3-gene-therapy.htm).

- Gene Therapy cures deafness in guinea pigs. Each animal had been deafened by destruction of the hair cells in the cochlea that translate sound vibrations into nerve signals. A gene, called *Atoh1*, which stimulates the hair cells' growth, was delivered to the cochlea by an adenovirus. The genes triggered re-growth of the hair cells and many of the animals regained up to 80% of their original hearing thresholds. This study, which many pave the way to human trials of the gene, is the first to show that gene therapy can repair deafness in animals. (http://www.newscientist.com/article.ns?id=dn7003).

▣ University of California, Los Angeles research team gets genes into the brain using liposomes coated in a polymer call polyethylene glycol (PEG). The transfer of genes into the brain is a significant achievement because viral vectors are too big to get across the "blood-brain barrier." This method has potential for treating Parkinson's disease. (http://www.newscientist.com/news/news.jsp?id =ns99993520).

▣ RNA interference (RNAi) or gene silencing may be a new way to treat Huntington's disease. Short pieces of double-stranded RNA (short, interfering RNAs or siRNAs) are used by cells to degrade RNA of a particular sequence. If a siRNA is designed to match the RNA copied from a faulty gene, then the abnormal protein product of that gene will not be produced. (http://www.newscientist.com/news/news.jsp?id=ns99993493).

▣ New gene therapy approach repairs errors in messenger RNA derived from defective genes. This technique has the potential to treat the blood disorder thalassaemia, cystic fibrosis, and some cancers. (NewScientist.com (October 11, 2002)).

▣ Gene therapy for treating children with X-SCID or the "bubble boy" disease has been stopped in France when the treatment caused leukaemia in one of the patients. (NewScientist.com (October 3, 2002)).

▣ Researchers at Case Western Reserve University and Copernicus Therapeutics are able to create tiny liposomes 25 nm across that can carry therapeutic DNA (Nanoballs) through pores in the nuclear membrane. (NewScientist.com (May 12, 2002)).

▣ Sickle cell is successfully treated in mice. (*The Scientist* (March 2002)).

DRUG DESIGNING

An old strategy for drug design and development involves trial and error approach and is laborious. The traditional processes are now supplemented by more direct approaches and are derived from the molecular processes involved in the understanding of the diseases. Advances in genomics and proteomics make major contributions not only to our understanding of cell and tissue function but also to the quality of human health. These types of studies lead to discovery of new drugs in a more sophisticated way. Drug is a chemical entity, when consumed or injected, results in the control or eradication of a particular disease or infection. Drugs work either by stimulating or blocking the activity of their targets. Thus, a drug can be an agonist (produce an action) or an antagonist (opposes an action). For a chemical compound to qualify as a drug, it must be safe, effective, and stable (both chemically and metabolically). In addition to these parameters, drug must be absorbed and distributed in the body as well as it must be either isolated from its natural sources or chemically synthesized and finally drug must be patentable. Drug design is the approach of finding drugs by design, based on their biological targets. Typically a drug target, which is a key molecule involved in a particular metabolic or signalling pathway, is specific to a disease condition or pathology, or to the infectivity or survival of a microbial pathogen.

Objectives of Drug Designing

- Finding a chemical compound that can fit to a specific cavity on a protein target both geometrically and chemically.
- After passing the animal tests and human clinical trials, this compound becomes a drug available to treat the patient.

The structure of the drug molecule that can specifically interact with the biomolecules can be modelled using computational tools that can allow a drug molecule to be constructed within the biomolecule using knowledge of its structure and the nature of its active site. Construction of the drug molecule can be made inside-out or outside-in depending on whether the core or the R-groups are chosen first. However, many of these approaches are plagued by the practical problems of chemical synthesis. Newer approaches have also suggested the use of drug molecules that are large and proteinaceous in nature rather than as small molecules. There have also been suggestions to make these using mRNA. RNA interference (RNAi) or gene silencing may also have therapeutical application.

Rational Drug Design

In the search for new drugs and chemical reagents, chemists have traditionally resorted to laborious trial-and-error testing of individual compounds. More recently, as techniques for analysing molecular structures have improved, many researchers have taken a different approach, often called rational drug design, in which they try to create molecules that will have the useful features they desire. This approach, too, is often slow and difficult. Rational drug design is a modern approach to obtain new pharmaceutical agents in which chemists synthesize novel substrates based on knowledge of substrates and reaction mechanisms. Unlike the historical method of drug discovery, by trial-and-error testing of chemical substances on cultured cells or animals, and matching the apparent effects to treatments, rational drug design begins with a knowledge of specific chemical responses in the body or target organism, and tailoring combinations of these to fit a treatment profile. Due to the complexity of the drug design process, two terms of interest are still serendipity and bounded rationality. Those challenges are caused by the large chemical space describing potential new drugs without side effects.

A particular example of rational drug design involves the use of three-dimensional information about biomolecules obtained from such techniques as X-ray crystallography and NMR spectroscopy. This approach to drug discovery is sometimes referred to as Quantitative Structure Activity Relationships (QSAR). QSAR represent an attempt to correlate structural or property descriptors of compounds with activities. These physico-chemical descriptors, which include parameters to account for hydrophobicity, boiling point, vapour pressure, melting point, refractive index, density, viscosity, topology, electronic properties, and steric effects, are determined empirically or, more recently, by computational methods. Activities used in QSAR include chemical measurements and biological assays. QSAR currently are being applied in many disciplines, with many pertaining to drug design and environmental risk assessment.

The purpose of QSAR is to derive a function that links biological activity (such as sweetness correlations, mutagenicity, general toxicities, etc.) of a group of related chemical compounds with parameters that describe a structural feature of these molecules. This feature also reflects the property of binding cavity on a protein target. The derived function is used as a guide to select the best candidate for drug design.

QSAR is applied for:

1. analysis and modelling of pharmacological molecule activities.
2. evaluation of experimental chemical properties.
3. evaluation of environmental molecule impact.
4. evaluating and modelling of toxicological molecule activities.
5. prediction of properties, activities or behaviours of not yet synthesized molecules.

Researchers have attempted for many years to develop drugs based on QSAR. Easy access to computational resources was not available when these efforts began, so attempts consisted primarily of statistical correlations of structural descriptors with biological activities. However, as access to high-speed computers and graphics workstations became common place, this field has evolved into what is often termed rational drug design or computer-assisted drug.

The first unequivocal example of the application of structure-based drug design leading to an approved drug is the carbonic anhydrase inhibitor dorzolamide, which was approved in 1995. Carbonic anhydrase is strongly inhibited by the drug dorzolamide, which is used as a diuretic (i.e., to increase the production of urine) and to lower excessively high pressure in the eye (due to accumulation of intraocular fluid) in glaucoma. Carbonic anhydrase catalyses the hydration of some aldehydes and ketones, and the hydrolysis of alkyl- and aryl esters.

$$CO_2 + H_2O \rightleftharpoons HCO_3 + H^+$$

It is a zinc-containing enzyme of about 30,000 daltons, and the three-dimensional structure has been characterized by X-ray diffraction. Physiologically, carbonic anhydrase is involved in gastric, urinary, pancreatic, lacrimal, and cerebrospinal secretions.

The activity of a drug at its binding site is one part of the design. Another to take into account is the molecule's drug likeness, which summarizes the necessary physical properties for effective absorption. One way of estimating drug likeness is "Lipinski's Rule of Five", which has been a crucial filter for drug development program. The Lipinski's rule quantifies the properties that are expected out of a poor drug absorption or permeation which is more likely when,

```
The molecular weight of drug ≤500 (opt = ~ 350)

The hydrogen bond acceptors ≤10 (opt = ~5)

The hydrogen bond donors ≤5 (opt = ~2)

-2 cLog P <5 (opt = ~3.0)

rotable bonds ≤5
```

where, P, the slope of the line, is a proportionality constant pertaining to a given equilibrium.

Compounds violating more than one of above mentioned Lipinski's rules are assumed to have problems with bioavailability.

Pharmaceuticals, or drugs, are often special inhibitors specifically targeted at a particular enzyme in order to overcome infection or to alleviate illness. An important case study in rational drug design is **imatinib**, a tyrosine kinase inhibitor designed specifically for the bcr-abl fusion protein that is characteristic for Philadelphia chromosome-positive leukaemias (chronic myelogenous leukaemia and occasionally acute lymphocytic leuka emia). Signalling from receptor tyrosine kinases leads to activation of several cytosolic protein kinases that translocate into the nucleus and regulate the activity of nuclear transcription factors. It further regulates transcription of many genes essential for cell division and for many cell differentiation processes. Imatinib is substantially different from previous drugs for cancer, as most agents of chemotherapy simply target rapidly dividing cells, not differentiating between cancer cells and other issues.

Highly Active AntiRetroviral Therapy (HAART) is a successful example of rational drug design. HIV is arguably the best characterized organism known to science, this knowledge has been aggressively pressed to service in the fight against AIDS. In a matter of a decade, AIDS has changed from a uniformly deadly disease to a manageable disease in many cases, because of the nanoscale design of effective anti-HIV drugs. Several of the enzymes involved in the life cycle of HIV have been characterized and used as targets for drug design. The reverse transcriptase, which copies the viral genome into a form that is read by the infected cell, was the first to be characterized and the first that was subjected to drug therapy. Because it acts on nucleic acids, early drug design efforts focused on modified nucleotides, which are added to a growing nucleic acid strand but which have modifications that prevent further growth. Thus they prematurely terminate the copying of the viral genome, creating defective copies. Drugs such as azidothymidine (AZT, Zidovudine) and dideoxyinosine (DDI) fall into this category. Rational drug design efforts by a number of pharmaceutical companies have yielded several highly effective inhibitors of the protease. When used in combination with inhibitors of reverse transcriptase, they form the powerful triple cocktail of HAART therapy.

Computer-Assisted Drug Design

Computer-Assisted Drug Design (CADD), also called Computer-Assisted Molecular Design (CAMD), represents more recent applications of computers as tools in the drug design process. The use of computers to aid the drug design process has become increasingly widespread over the past decades. These computational methods, often referred to as *in silico* approaches, are attractive to pharmaceutical and biotechnology companies because they are generally less expensive than experiments and can often access quantities inaccessible by other means. Furthermore, the sheer volume of data generated in the drug development industry, potentially millions of molecules and hundreds of millions of data points, necessitates the use of computers as data storage and data manipulation devices. As a result, there now exists, within pharmaceutical and biotechnology companies, established computational chemistry/molecular modelling groups that employ a range of computational techniques in their drug optimization efforts.

CADD is an emerging technology that makes use of knowledge of the steric and electronic aspects of the receptor–ligand, enzyme–substrate interaction to identify pharmacophores or aid in their design or both. Computers analyse the chemical structures of potential drugs and pinpoint the most promising candidates. Existing computer programs check a wide range of chemical features to help distinguish between drug-like and non-drug materials. These programs usually cannot screen for all features at the same time, an approach that risks overlooking promising drug-like substances. In considering this topic, it is important to emphasize that computers cannot substitute for a clear understanding of the system being studied. That is, a computer is only an additional tool to gain better insight into the chemistry and biology of the problem at hand.

In most current applications of CADD, attempts are made to find a ligand (the putative drug) that will interact favourably with a receptor that represents the target site. Binding of ligand to the receptor may include hydrophobic, electrostatic, and hydrogen-bonding interactions. In addition, solvation energies of the ligand and receptor site are also important because partial to complete desolvation must occur prior to binding. This approach to CADD optimizes the fit of a ligand in a receptor site. However, optimum fit in a target site does not guarantee that the desired activity of the drug will be enhanced or that undesired side effects will be diminished. Moreover, this approach does not consider the pharmacokinetics of the drug.

The approach used in CADD is dependent upon the amount of information that is available about the ligand and receptor. Ideally, one would have 3-dimensional structural information for the receptor and the ligand–receptor complex from X-ray diffraction or NMR. The ideal is seldom realized. In the opposite extreme, one may have no experimental data to assist in building models of the ligand and receptor, in which case computational methods must be applied without the constraints that the experimental data would provide.

As per the information available, one can apply either ligand-based or receptor-based molecular design methods. The ligand-based approach is applicable when the structure of the receptor site is unknown, but when a series of compounds have been identified that exert the activity of interest. To be used most effectively, one should have structurally similar compounds with high activity, with no activity, and with a range of intermediate activities. In recognition site mapping, an attempt is made to identify a pharmacophore, which is a template derived from the structures of these compounds. Pharmacophore is represented as a collection of functional groups in three-dimensional space that is complementary to the geometry of the receptor site. Advantages of this technology include a direct analysis of the interaction of a ligand with a receptor (or substrate with an enzyme), the use of small amounts of material, and rapid throughput.

In applying this approach, conformational analysis will be required, the extent of which will be dependent on the flexibility of the compounds under investigation. One strategy is to find the lowest energy conformers of the most rigid compounds and superimpose them. Conformational searching on the more flexible compounds is then done while applying distance constraints derived from the structures of the more rigid compounds. Ultimately, all of the structures are superimposed to generate the pharmacophore. This template may then be used to develop new compounds with functional groups in the desired positions. In applying this strategy, one must recognize that one is assuming that it is the minimum energy conformers that will bind most favourably in the receptor site. In fact, there is no a priori reason to exclude higher energy conformers as the source of activity.

The receptor-based approach to CADD applies when a reliable model of the receptor site is available, as from X-ray diffraction, NMR, or homology modelling. With the availability of the receptor site, the problem is to design ligands that will interact favourably at the site, which is a docking problem. CADD uses computational chemistry to discover, enhance, or study drugs and related biologically active molecules. For example, for the development of aromatic and heterocyclic sulphonamides and other compounds as inhibitors of carbonic anhydrase rely upon the use of both traditional QSAR and computer graphical methods (Figure 10.18). Some of these compounds have found application as diuretics.

The active site of carbonic anhydrase is a cavity approximately 12 Å deep with a zinc atom near the bottom of the cavity. The active site is divided into a hydrophilic half and a hydrophobic half. In the complex, the inhibitor appears to be bound such that the sulphonamide moiety occupies the fourth coordination site of the zinc atom, with the other three sites being occupied by histidine residues.

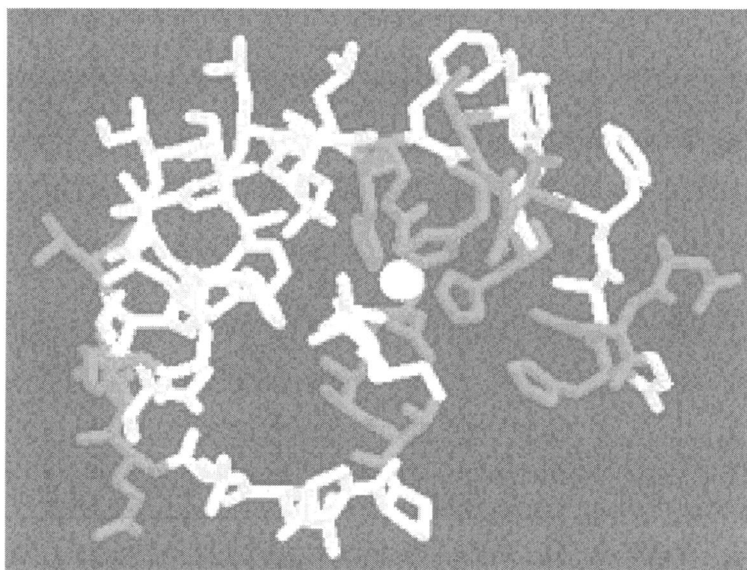

Figure 10.18 Active site of carbonic anhydrase containing the inhibitor MTS [(4S-*trans*) 4-(methylamino) 5, 6-dihydro 6-methyl 4H-thieno (2, 3-B) thiopyran 2-sulphonamide 7,7-dioxide].

Methods for protein-based drug design used can include simple molecular modelling, using molecular mechanics, molecular dynamics that deals with time evolution of a molecular system, semi-empirical quantum chemistry methods, *ab initio* quantum chemistry methods, density functional theory and Monte-Carlo simulation that analyses the conformational space in terms of thermodynamic properties like entropy, enthalpy, etc. In addition, the combinatorial biochemistry also plays a key role in drug designing. It refers to the generation of novel molecules derived from natural products by genetic engineering of biosynthetic pathways in living organism. The purpose is to reduce the number of targets for a good drug that have to be subjected to expensive and time-consuming synthesis and trialling. Thus, drug designing is a highly empirical process that requires intellectual and technological contributions from several scientific disciplines. The final output of their efforts is in the form of lead compound that can be suitably formulated and release for clinical trials (Figure 10.19).

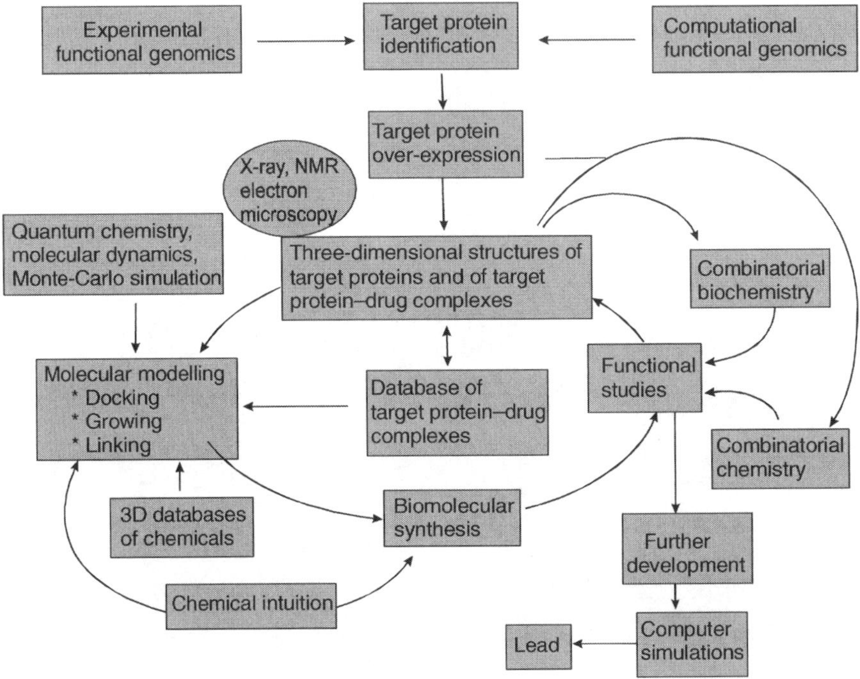

Figure 10.19   Protein-based drug design cycle

Examples of designed drugs

- ▣ Dorzolamide, a carbonic anhydrase inhibitor used to treat glaucoma.

- ▣ Cimetidine, the histamine receptor antagonist that promotes the healing of duodenal ulcers by inhibiting secretion of gastric acid.

- ▣ Many of the atypical antipsychotics that can often significantly alleviate hallucinations and other abnormal behaviours.

- ▣ Selective Serotonin Reuptake inhibitors (SSRIs) such as Prozac and Paxil, block the reuptake of serotonin back into presynaptic neurons, a class of antidepressants that reduce mental depression.

- ▣ Zanamivir, an antiviral drug that is used in the treatment of viral infections.

- ▣ Enfuvirtide, a peptide that prevents HIV entry in the cell.

- ▣ Zolpidem and Zopiclone—the nonbenzodiazepines that used as antianxiety drugs.

- ▣ Selective COX-2 inhibitor—prostaglandins that are potent mediators of inflammation. The first and committed step in the production of prostaglandins from arachidonic acid is the bis-oxygenation of arachidonate to prostaglandin PGG2. This is followed by reduction to PGH2 in a peroxidase reaction. Both these reactions are catalysed by prostaglandin endoperoxide synthase, also known as PGH2 synthase or cyclooxygenase (COX). This enzyme is inhibited by the family of drugs known as Non-Steroidal Anti-Inflammatory Drugs (NSAIDs). COX-2 (Figure 10.20), which is induced by cytokines, mitogens, and endotoxins in inflammatory cells and is responsible for the production of prostaglandins in inflammation.

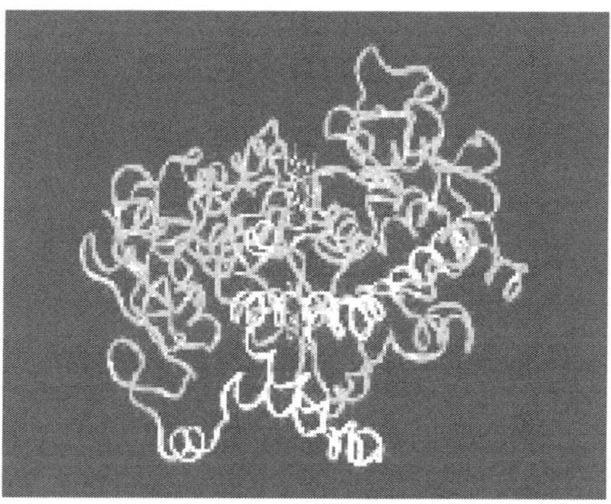

Figure 10.20   The COX-2 enzyme from mouse (PDB ID 6COX)

Drug Development

Drug development is defined in many pharmaceutical companies as the process of taking a new chemical lead through the stages necessary to allow it to be tested in human clinical trials, although a broader definition would encompass the entire process of drug discovery and clinical testing of novel drug candidates. The development of a new therapeutic drug is a complex, lengthy and expensive process (Figure 10.21). It can take from 12–20 years and over $300 to $500 million to bring a drug from concept to market. This includes 2–4 years of pre-clinical development, 3–6 years of clinical development and additional time for dealing with the regulatory authorities. The results of pre-clinical testing are also used to determine how to best formulate the drug for its intended clinical use, e.g. as a pill, aerosol or cream.

The drug discovery process starts from biological screening of thousands of compounds, about 1000 "hits" are selected that have optimum bioactivity, from them "leads" are chosen on the basis of their therapeutic potential. Few drug candidates having desired properties are selected from "leads" and finally a single drug is release for clinical trials.

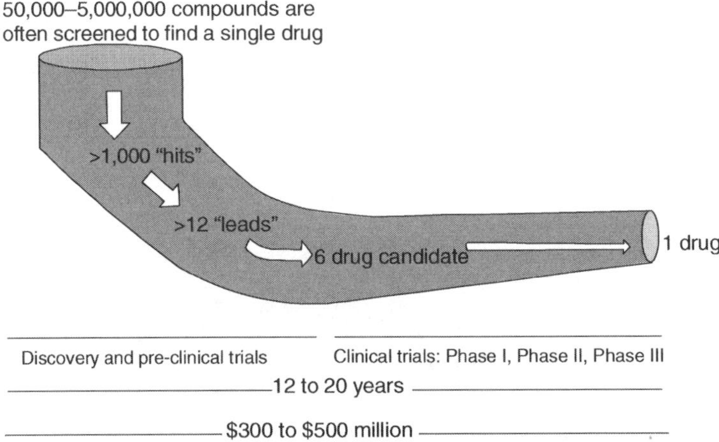

Figure 10.21   Drug discovery pipeline

Once the pre-clinical studies have been completed, the hundreds of lead compounds will have been whittled down to many fewer useful candidate drugs. Some of these may then be advanced to the clinical development stage, which involves testing in humans. Before this can take place, the pre-clinical studies must be submitted to the appropriate regulatory authorities. If the application is successful, the compound can be registered as an "investigational new drug".

Clinical development usually consists of phase I, phase II and phase III clinical trials. These are tests on human volunteers that provide more information on drug safety and activity. By the end of the clinical development phase, most of the investigational new drugs will have been eliminated on safety or efficacy grounds and only a very few compounds will be submitted to the regulatory authorities as a new drug application, which includes permission to market. After approval, pharmaceutical companies have a short period of exclusivity before patents expire and other companies can market the same drugs as generics. This time is used to recoup the massive investment required to develop and launch a new drug. However, the companies must also continue to test their drugs and monitor the feedback from healthcare professionals in order to identify undiscovered side effects, risk factors and interactions.

The summary of drug discovery and development can be put into following steps:

1. *Recognition of drug targets and mechanisms of drug action*

- Enzymes–inhibitors (reversible, irreversible)
- Receptors–agonists and antagonists
- Ion Channels–blockers
- Transporters–uptake inhibitors
- DNA–intercalating agents, minor groove binders, antisense drugs

2. *Identification of target and discovery of lead*

- Identification of target (e.g. enzyme, receptor, ion channel, and transporter)
- Determination of DNA and protein sequence
- Elucidation of structure and function of protein
- Proving therapeutic concept in animals ("knockouts")
- Development of assay for high-throughput molecular screen
- Mass screening and/or directed synthesis program
- Selection of one or more lead structures

3. *Optimization of lead and drug development*

- Determination of 3D structure of target receptor complexed with leads
- Molecular modelling design and refinement of new leads
- Synthesis and biological testing of new leads
- Optimization of selectivity, bioavailability, and pharmacokinetics
- Pharmaceutical formulation
- Pre-clinical and clinical development
- Drug approval and market introduction

PHARMACOGENOMICS

Generally, medicines are prescribed for a disease by the physician without looking into the constitution of the patient, except in some cases where a physician may ask or test allergic responses of the patient to certain drugs or antibiotics. Pharmacogenomics is the study of the relationship between a specific person's genetic make-up and his or her response to drug treatment. Because people differ in their ability to metabolize drugs, different patients with the same condition may require different dosages. The fast-growing field of pharmacogenomics deals with the study of how an individual's genetic inheritance affects the body's response to specific drugs. The term comes from the words pharmacology and genomics and is thus the intersection of pharmaceuticals and genetics. It is based on knowledge of bioinformatics, genome mapping, single nucleotide polymorphisms (SNPs), expressed sequence tags (ESTs), microarray technology and molecular genetics (Figure 10.22). Genetic variation in drug metabolism has been a medical problem for a long time. The emerging field of pharmacogenomics is identifying the genes responsible for this variation and developing tests to predict who will react best to which medications. The end result of all of this knowledge may lead to a new approach to medical care, in which each person's genome will be used to prescribe lifestyle changes and treatments that can maximize that person's genetic potential. Thus, pharmacogenomics is an important area of biotechnology research to examine heterogeneity of disease and individual's responses to medicine.

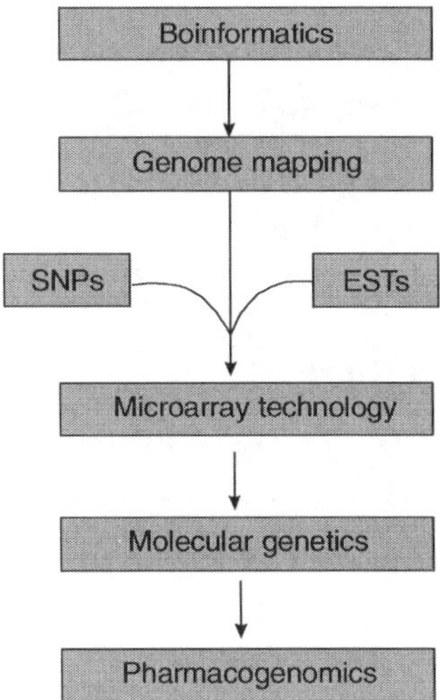

Figure 10.22   Bioinformatics provides input to pharmacogenomics

The efficacy and toxicity of many medications depend on genetic polymorphisms in patients for drug-metabolizing enzymes, transporters, receptors, and other drug targets. These genes may also transcribe and translate into proteins that are involved in absorption, distribution, metabolism, and elimination (ADME) of the drug. Pharmacogenomics holds the promise that drugs might one day be tailor-made for individuals and adapted to each person's own genetic make-up. Environment, diet, age, lifestyle, and state of health all can influence a person's response to medicines, but understanding an individual's genetic make-up is thought to be the key to creating personalized drugs with greater efficacy and safety. This will change the practice and economics of medicine. PharmGKB (http://www.pharmgkb.org/) is a webpage dedicated to services that include integrated resources about the variation in human genes leading to variation in response to a particular drug (Figure 10.23).

Figure 10.23 Home Page of PharmGKB(Source: http://www.pharmgkb.org/)

Prospect of Pharmacogenomics

More powerful medicines Pharmaceutical companies will be able to create drugs based on the proteins, enzymes, and RNA molecules associated with genes and diseases. This will facilitate drug discovery and allow drug makers to produce a therapy more targeted to specific diseases. This accuracy not only will maximize therapeutic effects but also decrease damage to nearby healthy cells.

Better, safer drugs the first time Instead of the standard trial-and-error method of matching patients with the right drugs, doctors will be able to analyse a patient's genetic profile and prescribe the best available drug therapy from the beginning. Not only this will take the guesswork out of finding the right drug, it will speed recovery time and increase safety as the likelihood of adverse reactions is eliminated. Pharmacogenomics has the potential to dramatically reduce the estimated 100,000 deaths and 2 million hospitalizations that occur each year in the US as the result of adverse drug response.

More accurate methods of determining appropriate drug dosages Current methods of basing dosages on weight and age will be replaced with dosages based on a person's genetics —how well the body processes the medicine and the time it takes to metabolize it. This will maximize the therapy's value and decrease the likelihood of overdose.

Advanced screening for disease Knowing one's genetic code will allow a person to make adequate lifestyle and environmental changes at an early age so as to avoid or lessen the severity of a genetic disease. Likewise, advance knowledge of particular disease susceptibility will allow careful monitoring, and treatments can be introduced at the most appropriate stage to maximize their therapy.

Better vaccines Vaccines made of genetic material, either DNA or RNA; promise all the benefits of existing vaccines without all the risks. They will activate the immune system but will be unable to cause infections. They will be inexpensive, stable, easy to store, and capable of being engineered to carry several strains of a pathogen at once.

Improvements in the drug discovery and approval process Pharmaceutical companies will be able to discover potential therapies more easily using genome targets. Previously failed drug candidates may be revived as they are matched with the niche population they serve. The drug approval process should be facilitated as trials are targeted for specific genetic population groups, providing greater degrees of success. The cost and risk of clinical trials will be reduced by targeting only those persons capable of responding to a drug.

Decrease in the overall cost of health care Decreases in the number of adverse drug reactions, the number of failed drug trials, the time it takes to get a drug approved, the length of time patients are on medication, the number of medications patients must take to find an effective therapy, and the effects of a disease on the body (through early detection), and an increase in the range of possible drug targets will promote a net decrease in the cost of health care.

Use of Pharmacogenomics

Based on the experimental results obtained by researchers of several pharmaceutical companies, to a limited degree, the cytochrome P450 (CYP) family of liver enzymes is responsible for breaking down more than 30 different classes of drugs. DNA variations in genes that code for these enzymes can influence their ability to metabolize certain drugs. Less active or inactive forms of CYP enzymes that are unable to break down and efficiently eliminate drugs from the body can cause drug overdose in patients. Today, clinical trials researchers use genetic tests for variations in cytochrome P450 genes to screen and monitor patients. In addition, many pharmaceutical companies screen their chemical compounds to see how well they are broken down by variant forms of CYP enzymes.

There is another enzyme called ThioPurine Methyltransferase (TPMT) that plays an important role in the chemotherapy treatment of a common childhood leukaemia by breaking down a class of therapeutic compounds called thiopurines. A small percentage of Caucasians have genetic variants that prevent them from producing an active form of this protein. As a result, thiopurines elevate to toxic levels in the patient because the inactive form of TPMT is unable to break down the drug. Today, doctors can use a genetic test to screen patients for this deficiency, and the TPMT activity is monitored to determine appropriate thiopurine dosage levels.

Barriers to Pharmacogenomics Progress

Pharmacogenomics is a developing research field that is still in its infancy. Several of the following barriers will have to be overcome before many pharmacogenomics benefits can be realized:

- Complexity of finding gene variations that affect drug response such as Single Nucleotide Polymorphisms (SNPs) are DNA sequence variations that occur when a single nucleotide (A, T, C, or G) in the genome sequence is altered. SNPs occur every 100 to 300 bases along the 3-billion-base human genome, therefore millions of SNPs must be identified and analysed to determine their involvement (if any) in drug response. Further complicating the process is our limited knowledge of which genes are involved with each drug response. Since many genes are likely to influence responses, obtaining the big picture on the impact of gene variations is highly time-consuming and complicated.

- Limited drug alternatives, i.e., only one or two approved drugs may be available for the treatment of a particular condition. If patients have gene variations that prevent them using these drugs, they may be left without any alternatives for treatment.

▣ Disincentives for drug companies to make multiple pharmacogenomic products—most pharmaceutical companies have been successful with their "one size fits all" approach to drug development. Since it costs hundreds of millions of dollars to bring a drug to market, will these companies be willing to develop alternative drugs that serve only a small portion of the population.

▣ Educating health care providers, since introducing multiple pharmacogenomic products to treat the same condition for different population subsets undoubtedly will complicate the process of prescribing and dispensing drugs. Physicians must execute an extra diagnostic step to determine which drug is best suited to each patient. To interpret the diagnostic accurately and recommend the best course of treatment for each patient, all prescribing physicians, regardless of specialty, will need a better understanding of genetics.

REVIEW QUESTIONS

1. Define gene and genome. Enlist the number of genes predicted and genomic size of prokaryotes and eukaryotes.

2. What is genomics and genomic project? Describe various genomics projects with their significant outputs.

3. What are the different web resources available for information on sequenced genomes?

4. Explain the steps involved in the whole genome shotgun sequencing technique.

5. Define BAC and YAC. What is the role of BAC and YAC in genome sequencing?

6. What is GenBank? Describe the role of GenBank in genomic database management.

7. What are the different options for submitting data to GenBank?

8. Explain the goals and growth of GenBank database.

9. What is genome assembly? Describe the various genome assembly software.

10. Define genome annotation. What are points of differences between structural and functional genome annotation?

11. Explain different types of genomics with their significance.

12. What is Gene Ontology project? Describe GO for cytochrome *c* oxidase with all its components.

13. What are the different ways for the analysis of gene expression?

14. Define DNA microarray. Explain different steps involved in a DNA microarray experiment.

15. Explain the experimental steps for Serial Analysis of Gene Expression.

16. What is SNP microarray technology? Describe its role in disease diagnosis.

17. What is DNA Data Bank of Japan? What are the different collaborations and goals of DDBJ?

18. What is TIGR? Describe its role in genomics.

19. What is the significance of Kyoto Encyclopedia of Genes and Genomes? Explain the KEGG pathway.

20. Enlist different categories of JCVI software with their importance.

21. What are the different parameters in genome information of *Escherichia coli* K-12 MG1655?

22. Enlist twenty commonest protein domains found in *C. elegans.*

23. What are the different types of sequences found in the human genome?

24. Is it true that complex organisms have more DNA than do simpler ones? If yes, explain the concept with examples.

25. What is HGP? What is difference between draft sequence and finished sequence of human genome?

26. Describe ethical, legal, and social issues of the HGP.

27. What are goals and potential benefits of HGP?

28. Define gene therapy. Describe the working principle of gene therapy with the recent developments.

29. What are the different factors affecting gene therapy?

30. Define drug. What are objectives of the drug designing?

31. Explain the concept of rational drug design with example.

32. What is computer-assisted drug design?

33. Define drug development. Explain the steps involved in drug development.

34. Is there any relation between new chemical entity development and drug development?

35. What is Quantitative Structure-Activity Relationship? Enlist its applications.

36. Explain the term pharmacogenomics with its prospect.
37. Write short notes on
    i.   Barriers to pharmacogenomics progress
    ii.   Bubble baby syndrome
    iii.   Human clinical trial
    iv.   PharmGKB
    v.   HAART
    vi.   Lipinski's Rule of Five
    vii.   Protein-based drug design cycle
    viii.   Ethical considerations for gene therapy
    ix.   ADA deficiency
    x.   Human *CFTR* gene
    xi.   Human protein-coding genes
    xii.   Knockout mice and human gene function
    xiii.   The Arabidopsis Information Resource
    xiv.   Mosquito genomics
    xv.   Bacteriophage genomics
    xvi.   SuperSAGE
    xvii.   Pretty Handy Annotation Tool
    xvii.   CYP enzymes

GLOSSARY

Ab initio prediction   Computational prediction based on first principles or using the most elementary information.

Accession number   An identifier supplied by the curators of the major biological databases upon submission of a novel entry that uniquely identifies that sequence (or other). RefSeq accession numbers are written in the following format: two letters followed by an underscore and six digits (e.g. NT_123456). The first two letters of the RefSeq accession number indicate the type of sequence included in the record as described below:

```
NT_123456 constructed genomic
contigs

NM_123456 mRNAs (actually the
cDNA sequences constructed from
mRNA)

NP_123456 proteins C_123456
chromosomes
```

Active site   The amino acid residues at the catalytic site of an enzyme. These residues provide the binding and activation energy needed to place the substrate into its transition state and bridge the energy barrier of the reaction undergoing catalysis.

Algorithm   A series of steps defining a procedure or formula for solving a problem, which can be coded into a programming language and executed. Bioinformatics algorithms typically are used to process, store, analyse, visualize and make predictions from biological data.

Alignment   The result of a comparison of two or more gene or protein sequences in order to determine their degree of base or amino acid similarity. Sequence alignments are used to determine the similarity, homology, function or other degree of relatedness between two or more genes or gene products.

Alpha helix (α-helix)   Common folding pattern in proteins in which a linear sequence of amino acids fold into a right-handed helix stabilized by internal hydrogen-bonding between backbone atoms.

Alternate splicing   Removal of introns in eukaryotic pre-mRNAs, resulting in one gene producing several different mRNA and protein products.

***Alu* family**   Intermediately repetitive DNA sequence that is dispersed in human genome. The sequence is more or less 300 bp long. The name *Alu* comes from the restriction endonuclease that cleaves it.

Amino acid sequence   The order of amino acids in a protein molecule.

Annotation   A combination of comments, notations, references, and citations, either in free format or utilizing a controlled vocabulary, which together describe all the experimental and inferred information about a gene or protein.   Annotations can also be applied to the description of other biological systems. Batch, automated annotation of bulk biological sequence is one of the key uses of bioinformatics tools.

Antisense   DNA or RNA composed of the complementary sequence to the target DNA/RNA. Also used to describe a therapeutic strategy that uses antisense DNA or RNA sequences to target a specific gene (DNA sequences or mRNA) implicated in disease, in order to bind and physically inhibit their expression by blocking them.

Applied proteomics   A branch of bioinformatics, the current applications of which seem to be focusing on toxicology and drug target identification.

Assembly   Compilation of overlapping sequences from one or more related genes that have been clustered together based on their degree of sequence identity or similarity. Sequence assembly may be used to piece together "shotgun" sequencing fragments (*See* shotgun sequencing) based upon overlapping restriction enzyme digests, or may be used to identify and index novel genes from "single-pass" cDNA sequencing efforts.

Autoradiography   A method used to locate radioisotope-labelled materials which have been separated in gels or are present in blots. The location of the radio-labelled material is determined by overlaying the test material with a photographic film that is sensitive to the radioisotope.

Bacterial artificial chromosome (BAC)   Cloning vector that can incorporate large fragments of DNA.

Bait   Refers to a protein. The basic format of the **yeast-two hybrid** system involves the creation of two hybrid molecules, one in which the "**bait**" protein is fused with a transcription factor, and the other in which the "**prey**" protein is fused with a related transcription factor. If the bait and prey proteins indeed interact then the two factors fused to these two proteins are also brought into proximity with each other. As a result a specific signal is produced, indicating that an interaction has taken place.

Base pair   A pair of nitrogenous bases (a purine and a pyrimidine), held together by hydrogen bonds, which form the core of DNA and RNA, i.e., the A : T, G : C and A : U interactions.

Beta sheet   A three-dimensional arrangement taken up by polypeptide chains that consist of alternating strands linked by hydrogen bonds.

The alternating strands together form a sheet that is frequently twisted. It is one of the secondary structural characteristics of proteins.

Bioinformatics The field of endeavour that relates to the collection, organization and analysis of large amounts of biological data using networks of computers and databases (usually with reference to the genome project and DNA sequence information). It involves the use or development of techniques in the fields of applied mathematics, informatics, statistics, computer science, artificial intelligence, and chemistry, and finding solutions to biological problems usually at the molecular level. Major research efforts in the field include sequence alignment, gene finding, genome assembly, protein structure alignment, protein structure prediction, prediction of gene expression and protein–protein interactions, and the modelling of evolution.

Biological database A collection of biological data that is organized so that its contents can easily be accessed, managed, and updated.

Bit score Statistical indicator in database sequence similarity searches. It is suitable for comparing search results from different databases. The higher the bit score, the better the match is.

BLAST (Basic Local Alignment Search Tool) Search engine used to perform fast similarity searches of the query for protein or DNA.

Boolean expression Database retrieval method of expression query by connecting query words using the logical operators AND, OR and NOT between the words.

Boot To restart a computer after it has crashed.

Bootstrap analysis Statistical method for assessing the consistency of phylogenetic tree topologies based on the generation of large number of replicates with slight modifications in input data.

CATH A hierarchical classification of protein domain structures based on class (C), architecture (A), topology (T), and homologous superfamily (H).

cDNA (complementary DNA) A DNA strand copied from mRNA using reverse transcriptase.

cDNA library A set of DNA fragments prepared from the total mRNA obtained from a selected cell, tissue or organism. A cDNA library represents the entire expressed DNA in a cell.

Central dogma of molecular biology The principle proposed by Crick and his co-workers that the transfer of genetic information is from gene to protein. The first step in this prcess is known as transcription in which the messenger RNA (ribonucleic acid) is assembled as a complementary copy of one of the two strands of DNA. The second step is referred to as translation in which mRNA is decoded to synthesize protein.

Chaperones A class of proteins which bind to incompletely folded or assembled proteins in order to assist their folding or prevent them from aggregating.

Chemical proteomics Linking the new proteins with known catalytic activities, proteome-scale screens for generic enzyme activities (e.g. protease and phosphatase) should be implemented. Although it is impossible to screen for chemical reactions that are unknown, in theory, identifying small molecules that bind to the new proteins may elucidate clues to new activities. These ligands might be found by screening the new proteins against diverse chemical libraries using existing methods such as NMR spectroscopy, microcalorimetry, or microarrays. The general concept of ascribing function to new proteins by discovering small molecule ligands might be referred to as chemical proteomics.

Chime A web-browser-based molecular visualization application that allows interactive display of graphics of protein structure.

Chromatography A versatile physical technique in which the crude mixture of dissolved components is separated as it moves through some type of porous matrix.

Cloning vector A small DNA molecule, usually derived from a bacteriophage or plasmid, which is used to carry the fragment of DNA to be cloned into the recipient cell, and which enables the DNA fragment to be replicated.

Cluster The grouping of similar objects in a multidimensional space. Clustering is used for constructing new features which are abstractions of the existing features of those objects. The quality of the clustering depends crucially on the metric distance in space. In bioinformatics, clustering is performed on sequences, high-throughput expression and other experimental data. Clusters of partial or complete gene sequences can be used to identify the complete (contiguous) sequence and its function. Clustering expression data enables the researcher to discern patterns of co-regulation in groups of genes.

Coding regions (CDS) The portion of a genomic sequence bounded by start and stop codons that identify the sequence of the protein being coded for by a particular gene.

Common gateway interface programming A writing application that acts as an interface or gateway program between the client browser, web server, and a traditional programming application.

Comparative genomics The establishment of the correspondence between genes (orthology analysis) or other genomic features in different organisms. It is these intergenomic maps that make it possible to trace the evolutionary processes responsible for the divergence of two genomes. A multitude of evolutionary events acting at various organizational levels shape genome evolution.

Computational proteomics Large-scale generation and analysis of 3D and 4D protein structural information and the application of structural knowledge across all life science disciplines.

Conformation The precise three-dimensional arrangement of atoms and bonds in a molecule describing its geometry and hence its molecular function.

Consensus sequence Average or most typical form of a sequence that is reproduced with minor variations in a group of related DNA, RNA, or protein sequences. The consensus sequence shows the nucleotide or amino acid most often found at each position. The preservation of a consensus implies that the sequence is functionally important.

Conservation A process of substitution of one amino acid in a polypeptide by another amino acid without changing the physico-chemical properties. For example, a hydrophobic amino acid residue can be replaced by another hydrophobic residue.

Conserved sequence A base sequence in a DNA molecule (or an amino acid sequence in a protein) that has remained essentially unchanged throughout evolution.

Contig A series of overlapping DNA clones of known order along a chromosome from an organism of interest. Contigs are typically created computationally, by comparing the overlapping ends of several sequencing reads generated by restriction enzyme digestion of a segment of genomic DNA. The creation of contigs in the presence of sequencing errors, ambiguities and the presence of repeats is one of the most computationally challenging aspects of the role of bioinformatics in genome analysis.

Contig map A map depicting the relative order of a linked library of small overlapping clones representing a complete chromosomal segment.

Crash An unexpected failure of a computer program, or of the operating system itself.

Crystal structure Term used to describe the high-resolution molecular structure derived by X-ray crystallographic analysis of protein or other biomolecular crystals.

Database Computed archive used for storage and organization of data that can be retrieved easily via a variety of search methods.

Database management system (DBMS) Computer software designed for the purpose of managing databases based on a variety of data models.

Data mining The ability to query very large databases in order to satisfy a hypothesis ("top-down" data mining); or to interrogate a database in order to generate new hypotheses based on rigorous statistical correlations ("bottom-up" data mining).

Data processing The systematic performance of operations upon data such as handling, merging, sorting, and computing. The semantic content of the original data should not be changed, but the semantic content of the processed data may be changed.

Data warehouses Vast arrays of heterogeneous (biological) data, stored within a single logical data repository, which are accessible to different querying and manipulation methods.

Dendrogram A graphical procedure for representing the output of a hierarchical clustering method. A dendrogram is strictly defined as a binary tree with a distinguished root, that has all the data items at its leaves. Conventionally, all the leaves are shown at the same level of the drawing. The ordering of the leaves is arbitrary, as is their horizontal position. The heights of the internal nodes may be arbitrary, or may be related to the metric information used to form the clustering.

Dihedral angle The angle of rotation about the bond in a molecule.

Discriminator An abstraction of a conserved motif, or motifs (e.g. a regular expression pattern, or a fingerprint), within an alignment used to search either an individual query sequence or a full database for the occurrence of that same, or similar, motif.

DNA methylation Addition of a methyl group to DNA. Extensive methylation of the cytosine base in CG sequences is used in vertebrates to keep genes in an inactive state.

DNA microarrays The deposition of oligonucleotides or cDNAs on to an inert substrate such as glass or silicon. Thousands of molecules may be organized spatially into a high-density matrix. These DNA chips may be probed to allow expression monitoring of many thousands of genes simultaneously. Uses include study of polymorphism in genes, de novo sequencing and molecular diagnosis of disease.

DNA sequencing The technique in which the specific sequence of bases forming a particular DNA region is deciphered.

Domain (protein) A region of special biological interest within a single protein sequence. However, a domain may also be defined as a region within the three-dimensional structure of a protein that may encompass regions of several distinct protein sequences that accomplish a specific function. A domain class is a group of domains that share a common set of well-defined properties or characteristics.

Dot blot technique A variety of Southern or northern blotting technique where DNA/RNA is not subjected to gel electrophoresis. The cloned or pure forms of extracted DNAs to be tested are spotted adjacent to each other on a nitrocellulose filter membrane. DNA blots are immobilized and denatured on the filter as single DNA blot. The filter is then hybridized with single-stranded DNA radiolabelled probe. The dot representing sequence related to the probe will illuminate by autoradiography.

Dotplot method (dot matrix method) A simple graphical picture that gives an overview of the similarities between two sequences in two-dimensional matrix.

Drug An agent that affects a biological process. Specifically, a molecule whose molecular structure can be correlated with its pharmacological activity.

Drug discovery cycle The cycle of events required to develop a new drug. Typically this involves research, preclinical testing and clinical development, and can take from 5 to 12 years.

Dynamic programming A method of solving problems exhibiting the properties of overlapping sub-problems and optimal substructure that can be applied to produce global alignments via the Needleman–Wunsch algorithm, and local alignments via the Smith–Waterman algorithm.

Electrophoresis The use of an external electric field to separate large biomolecules on the basis of their charge by running them through acrylamide or agarose gels.

Ensembl A joint project between EMBL, EBI and the Wellcome Trust Sanger Institute (WTSI) to develop a software system which produces and maintains automatic annotation on selected eukaryotic genomes.

ENTREZ The data retrieval facility developed by NCBI that allows access to molecular biology databases and bibliographic citations from NCBI's integrated databases.

Ethernet System for the connection of computer networks.

European Bioinformatics Institute (EBI) Established as an outstation of EMBL for the development and distribution of EMBL nucleotide sequence database as well as for a number of protein-related databases.

European Molecular Biology Laboratories (EMBL) Collective term for European laboratories that use biocomputing methods in molecular biology research specifically in the study of nucleotide and protein sequencing. It established the network called EMBnet to provide up-to-date information of biological databases, services and training to users dispersed in European laboratories.

E-value   The number of different alignments with a score equal to or better than S that can be expected to occur simply by chance (expectation value).

Evolutionary biology   The study of the origin and descent of species, as well as their change over time. It helps in tracing the evolution of a large number of organisms by measuring changes in their DNA, rather than through physical taxonomy or physiological observations alone, to compare entire genomes, which permits the study of more complex evolutionary events, such as gene duplication, lateral gene transfer, and the prediction of factors important in bacterial speciation, to build complex computational models of populations to predict the outcome of the system over time as well as to track and share information on an increasingly large number of species and organisms.

Exon   The region of DNA within a gene that codes for a polypeptide chain or domain. Typically a mature protein is composed of several domains coded by different exons within a single gene.

ExPASy (Expert protein analysis system) A comprehensive proteomic web server maintained by Swiss Institute of Bioinformatics.

Expressed sequence tags (ESTs)   A small sequence from an expressed gene that can be amplified by PCR. ESTs act as physical markers for cloning and full-length sequencing of the cDNAs of expressed genes. They are typically identified by purifying mRNAs, converting to cDNAs, and then sequencing a portion of the cDNAs.

Expression (gene or protein)   A measure of the presence, amount, and time-course of one or more gene products in a particular cell or tissue. Expression studies are typically performed at the RNA (mRNA) or protein level in order to determine the number, type, and level of genes that may be up-regulated or down-regulated during a cellular process, in response to an external stimulus, or in sickness or disease.   Gene chips and proteomics now allow the study of expression profiles of sets of genes or even entire genomes.

Expression profile   The level and duration of expression of one or more genes selected from a particular cell or tissue type, generally obtained by a variety of high-throughput methods such as sample sequencing, serial analysis, or microarray-based detection.

False positive   A sequence incorrectly identified by a discriminator as possessing a particular motif or pattern.

FASTA   A FAST homology search all sequences. It is a DNA and protein sequence alignment software package.

FASTA format   Sequence format that begins with a single-line description followed by lines of sequence data. This format can be used as query input when searching bioinformatics tools such as BLAST or ClustalW. The description line is distinguished from the sequence data by a "greater than" (">") symbol in the first column. It is recommended that all lines of text be shorter than 80 characters in length. Blank lines are not allowed in the middle of FASTA input.

Fingerprint A set of motifs used to predict the occurrence of similar motifs, in either an individual sequence or in a database. Fingerprints are refined by iterative scanning of a composite protein sequence database. A composite or multiple-motif fingerprint contains a number of aligned motifs taken from different parts of a multiple alignment. True family members are then easy to identify by virtue of possessing all elements of the fingerprint, while subfamily members may be identified by possessing only part of it.

Frameshift A deletion, substitution, or duplication of one or more bases that cause the reading frame of a structural gene to shift from the normal series of triplets.

Functional genomics The use of genomic information to delineate protein structure, function, pathways and networks. Function may be determined by "knocking out" or "knocking in" expressed genes in model organisms such as worm, fruit fly, yeast or mouse.

Fusion protein The protein resulting from the genetic joining and expression of two different genes.

Gap penalties The penalty applied to a similarity score for the introduction of an insertion or deletion gap, the extension of a gap, or both. Gap penalties are usually subtracted from a cumulative score being determined for the comparison of two or more sequences via an optimization algorithm that attempts to maximize that score.

Gaps (affine gaps) A gap is defined as any maximal, consecutive run of spaces in a single string of a given alignment. Gaps help to create alignments that better conform to underlying biological models and more closely fit patterns that one expects to find in meaningful alignment. The idea is to take into account the number of continuous gaps and not the number of spaces alone when calculating an alignment. Affine gaps contain a component for gap insertion and a component for gap extension, where the extension penalty is usually much lower than the insertion penalty. This mimics biological reality as multiple gaps would imply multiple mutations, but a single mutation can lead to a long gap quite easily.

Gel filtration/size-exclusion chromatography The technique used to separate proteins (or nucleic acid) primarily by molecular weight using separation material in the form of tiny beads composed of cross-linked polysaccharides (dextrans or agarose) of different porosity.

GenBank Databank of genetic sequences operated by a division of the National Institute of Health.

Gene The fundamental physical and functional unit of heredity. A gene is an ordered sequence of nucleotides located in a particular position on a particular chromosome that encodes a specific functional product (i.e., a protein or RNA molecule).

Gene chips (also Gene arrays) The covalent attachment of oligonucleotides or cDNA directly onto a small glass or silicon chip in organized arrays. Over 50,000 different DNA fragments can be presented on a single chip providing a high-throughput parallel method of probing gene expression, genotype or gene function.

Gene expression The process by which a gene-coded information is converted into the structures present and operating in the cell. Expressed genes include those that are transcribed into mRNA and then translated into protein and those that are transcribed into RNA but not translated into protein (e.g. tRNAs and RNAs).

Gene families Subsets of genes containing homologous sequences which usually correlate with a common function.

Gene index A listing of the number, type, label and sequence of all the genes identified within the genome of a given organism. Gene indices are usually created by assembling overlapping EST sequences into clusters, and then determining if each cluster corresponds to a unique gene. Methods by which a cluster can be identified as representing a unique gene include identification of long open reading frames (ORFs), comparison to genomic sequence, and detection of SNPs or other features in the cluster that are known to exist in the gene.

Gene library A collection of cloned DNA fragments created by restriction endonuclease digestion that represent part or all of an organism's genome.

Gene mapping Determination of the relative positions of genes on a DNA molecule (chromosome or plasmid) and of the distance, in linkage units or physical units, between them.

Gene ontology (GO) A controlled vocabulary of terms relating to molecular function, biological process, or cellular components developed by the Gene Ontology Consortium. A controlled vocabulary allows scientists to use consistent terminology when describing the roles of genes and proteins in cells.

Gene prediction Predictions of possible genes made by a computer program based on how well a stretch of DNA sequence matches known gene sequences.

Gene product The biochemical material, either RNA or protein, resulting from expression of a gene. The amount of gene product is used to measure how active a gene is; abnormal amounts can be correlated with disease-causing alleles.

Gene therapy The use of genetic material for therapeutic purposes. The therapeutic gene is typically delivered using recombinant virus or liposome- based delivery systems.

Genetic code The mapping of all possible codons into 20 amino acids including the start and stop codons.

Genome All the genetic material in the chromosomes of a particular organism; its size is generally given as its total number of base pairs. The normal human genome consists of three billion base pairs of DNA.

Genome projects Research and technology development efforts aimed at mapping and sequencing some or all of the genome of human beings and other organisms.

Genomic annotation The process of marking the genes and other biological features in a DNA sequence. The first genome annotation software system was designed in 1995 by Dr. Owen White, who sequenced and analysed the first genome of a free-living organism, the bacterium *Haemophilus influenzae*. He built a software system to find the genes (places in the DNA sequence that encode a protein), the transfer RNA, and other features, and to make initial assignments of function to those genes. Most current genome annotation systems work similarly, but the programs available for analysis of genomic DNA are constantly changing and improving.

Genomic DNA sequence DNA sequence typically obtained from mammalian or other higher-order species, which include both intron and exon sequence (coding sequence), as well as non-coding regulatory sequences such as promoter, and enhancer sequences.

Genomics The study of the structure and function of large numbers of genes.

Genotype Strictly, all of the genes possessed by an individual. In practice, the particular alleles present in a specific genetic locus.

GenScan An online program to identify complete gene structures in genomic DNA.

Gibbs sampling Local alignment strategy that matches up two or more sequences over their entire lengths.

Global alignment The alignment of two nucleic acid or protein sequences over their entire length.

GOA A project run by the European Bioinformatics Institute that aims to provide assignments of gene products to the Gene Ontology (GO) resource.

Gophers These are character-oriented approach tools for locating data on Internet that help the users to locate essentially all textual information stored on Internet servers.

Heterodimer Protein composed of two different chains or subunits.

Heteroduplex Hybrid structure formed by annealing of two DNA strands (or an RNA and DNA) that have sufficient complementarity in their sequence to allow hydrogen-bonding.

Heuristics Computational strategy to find a near-optimal solution by using rules of thumb. It saves the computational time.

Hidden Markov Model (HMM) A joint statistical model for an ordered sequence of variables. The result of stochastically perturbing the variables in a Markov chain (the original variables are thus "hidden"), where the Markov chain has discrete variables which select the "state" of the HMM at each step. The perturbed values can be continuous and are the "outputs" of the HMM. A Hidden Markov Model is equivalently a coupled mixture model where the joint distribution over states is a Markov chain. Hidden Markov Models are valuable in bioinformatics because they allow a search or alignment algorithm to be trained using unaligned or unweighted input sequences; and because they allow position-dependent scoring parameters such as gap penalties, thus more accurately modelling the consequences of evolutionary events on sequence families.

High-performance liquid chromatography (HPLC) An enhanced version of the column chromatography techniques in which the column materials are much more finely divided with more interaction sites and thus greater resolving power. As a result, there is high resolution as well as rapid separation of biomolecules.

High-throughput screening The method by which a very large number of compounds are screened against a putative drug target in either cell-free or whole-cell assays. Typically, these screenings are carried out in 96 well plates using automated, robotic station based technologies or in higher-density array ("chip") formats.

Homeodomain A 60 amino acid protein domain coded for by the homeobox region of a homeotic gene.

Homologous Descended from a common ancestor.

Homology Two or more biological species (animals, plants, or microbes), systems or molecules (DNA, RNA or proteins) are related by divergent evolution from a common ancestor.

Homology searches Searches typically performed with a query DNA or protein sequence to identify the known gene or gene product sharing significant similarity that might inform about ancestry and possible function of gene in question.

Homology modelling Method for predicting the three-dimensional structure of a protein based on homology by assigning the structure of an unknown protein using an existing homologous protein structure as a template.

HPRD (Human Protein Reference Database) A centralized platform to visually depict and integrate information pertaining to domain architecture, post-translational modifications, interaction networks and disease association for each protein in the human proteome.

Human Genome Initiative Collective name for several projects begun in 1986 by DOE to create an ordered set of DNA segments from known chromosomal locations, develop new computational methods for analysing genetic map and DNA sequence data, and develop new techniques and instruments for detecting and analysing DNA. This DOE initiative is now known as the Human Genome Program. The national effort, led by DOE and NIH, is known as the Human Genome Project.

In silico (biology) The use of computers to simulate, process, or analyse a biological experiment.

Introns Nucleotide sequences found in the structural genes of eukaryotes that are non-coding and interrupt the sequences containing information that codes for polypeptide chains. Intron sequences are spliced out of their RNA transcripts before maturation and protein synthesis.

Isoelectric focusing One of the separation techniques used to separate proteins according to their isoelectric points (pI)—the pH at which there is zero net charge and no migration of proteins in an electric field.

Iteration A series of steps in an algorithm whereby the processing of data is performed repetitively until the result exceeds a particular threshold. Iteration is often used in multiple sequence alignments whereby each set of pairwise alignments is compared with every other, starting with the most similar pairs and progressing to the least similar, until there are no longer any sequence-pairs remaining to be aligned.

Iterative pairwise alignment The approach that uses pairwise alignment scores to iteratively add one additional string to a growing multiple alignment.

Java An object-oriented programming language tailored for network computing.

Junk DNA The excess DNA that is present in the genome beyond that required to encode proteins. A misleading term since these regions are likely to be involved in gene regulation, and other as yet unidentified functions.

Lab on a chip Term describing microdevices that allow rapid, microanalytical analysis of DNA or protein in a single, fully integrated system. Typically, these devices are miniature surfaces, made of silicon, glass or plastic, which carry the necessary microdevices (pumps, valves, microfluidic controllers, and detectors) that allow sample separation and analysis. These devices are used in drug discovery, genetic testing and separation science.

Lead compound A candidate compound identified as the best "hit" (tight binder) after screening of a combinatorial (or other) compound library, that is then taken into further rounds of screening to determine its suitability as a drug.

Lead optimization The process of converting a putative lead compound ("hit") into a therapeutic drug with maximal activity and minimal side effects, typically using a combination of computer-based drug design, medicinal chemistry and pharmacology.

Leucine zipper Protein motif which binds with DNA in which 4–5 leucines are found at 7 amino acid intervals. This motif is present typically in transcription factors and other proteins that bind DNA.

Ligand Any small molecule that binds to a protein or receptor; the cognate partner of many cellular proteins, enzymes, and receptors.

Local alignment Alignment of portions (rather than the entire sequence length) of two nucleic acid or amino acid sequences.

Machine learning Computational approach to detect pattern by progressive optimization of the internal parameters of an algorithm.

Markov chain Any multivariate probability density whose independence diagram is a chain. The variables are ordered, and each variable "depends" only on its neighbours in the sense of being conditionally independent of the others. Markov chains are an integral component of Hidden Markov Models.

Martinsried Institute for Protein Sequences (MIPS) The European partner of PIR-International Protein Sequence Database. Its central activity is to collect, distribute and maintain up-to-date protein sequence data within Europe.

Mass spectrometry An indispensable tool for genomics and proteomics research. It is a technique of separating ions of different mass and energy under the fixed magnetic and/or electric field.

Matrix assisted laser desorption ionization (MALDI) The method used to measure the mass of a wide range of macromolecules. The charged molecular ions of analyte produced during a gas-phase proton transfer reaction with the matrix molecule, are detected and analysed by MALDI-TOF (time-of-fight) mass spectrometer.

Maxam and Gilbert's method A chemical degradation process that uses chemicals to cleave DNA at specific bases, resulting in the formation of DNA fragments of different lengths to be separated by gel electrophoresis and finally autoradiographed to note the sequence of nucleotides.

Microarray A 2D array, typically on a glass, filter, or silicon wafer, upon which genes or gene fragments are deposited or synthesized in a predetermined spatial order allowing them to be made available as probes in a high-throughput, parallel manner. It includes sets of miniaturized chemical reaction areas that may also be used to test antibodies or proteins.

Missense mutation A point mutation in which one codon (triplet of bases) is changed into another designating a different amino acid.

Monte Carlo procedure Computer algorithm that produces random numbers based on a particular statistical distribution.

Morbid map A diagram showing the chromosomal location of genes associated with disease.

Motif A conserved element of a protein sequence alignment that usually correlates with a particular function. Motifs are generated from a local multiple protein sequence alignment corresponding to a region whose function or structure is known. It is sufficient that it is conserved, and hence likely to be predictive of any subsequent occurrence of such a structural/functional region in any other novel protein sequence.

Multigene family A set of genes derived by duplication of an ancestral gene, followed by independent mutational events resulting in a series of independent genes either clustered together on a chromosome or dispersed throughout the genome.

Multiple (sequence) alignment A sequence alignment of three or more biological sequences, generally protein, DNA, or RNA. Multiple alignment of k sequences is a rectangular array, consisting of characters taken from the alphabet A, that satisfies the following conditions: There are exactly k rows; ignoring the gap character, row number i is exactly the sequence s_i; and each column contains at least one character different from "-". In practice multiple sequence alignments include a cost/weight function, that defines the penalty for the insertion of gaps (the "-" character) and weights identities and conservative substitutions accordingly. Multiple alignment algorithms attempt to create the optimal alignment defined as the one with the lowest cost/weight score.

Multiplex sequencing Approach to high-throughput sequencing that uses several pooled DNA samples run through gels simultaneously and then separated and analysed.

National Centre for Biotechnology information (NCBI) Centre that maintains GenBank, the DNA sequence database and is the foremost repository of publicly available genomic and proteomic data.

Needleman–Wunsch algorithm A global alignment algorithm that applies dynamic programming in a sequence alignment.

Neural net An interconnected assembly of simple processing elements, units or nodes, whose functionality is loosely based on the animal brain. The processing ability of the network is stored in the inter-unit connection strengths, or weights, obtained by a process of adaptation to, or learning from, a set of training patterns. Neural nets are used in bioinformatics to map data and make predictions, such as taking a multiple alignment of a protein family as a training set in order to identify novel members of the family from their sequence data alone.

Northern blotting technique The method used for detection and quantitative estimation of RNA in the sample and is an extension of Southern blotting, since RNA was not found to bind with nitrocellulose filter.

Nuclear magnetic resonance (NMR) spectroscopy The significant method for determining the three-dimensional structures of macromolecules by detecting the spinning patterns of their atomic nuclei when they are subjected to external magnetic field.

OMIM (Online Mendelian Inheritance in Man) A catalogue of human genes and genetic disorders.

Open reading frame (ORF) Any stretch of DNA that potentially encodes a protein. Open reading frames start with a start codon, and end with a termination codon. No termination codons may be present internally. The identification of an ORF is the first indication that a segment of DNA may be a part of a functional gene.

Operating system A program that acts as an interface between user of a computer and computer hardware such as Windows, DOS, Linux, UNIX, etc.

Orthologs Genes in different species that evolved from a common ancestral gene by speciation. Normally, orthologs retain the same function in the course of evolution. Identification of orthologs is critical for reliable prediction of gene function in newly sequenced genomes.

Pairwise sequence alignment The method used to find the best-matching local or global alignments of two query sequences.

Paralogs Genes which are related by duplication within a genome. Orthologs retain the same function in the course of evolution, whereas paralogs evolve new functions, even if these are related to the original one.

Pattern Molecular biological patterns usually occur at the level of the characters making up the gene or protein sequence. A pattern language must be defined in order to apply different criteria to different positions of a sequence. In order to have position-specific comparison done by a computer, a pattern-matching algorithm must allow alternative residues at a given position, repetitions of a residue, exclusion of alternative residues, weighting, and ideally, combinatorial representation.

Pattern-Hit initiated (PHI)-BLAST The program designed to search for proteins that contain a pattern specified by the user and are similar to the query sequence.

Pfam A database of multiple alignments of protein domains or conserved protein regions.

Pharmacodynamics The study of drugs which act at target sites in the body.

Pharmacogenomics The use of (DNA-based) genotyping in order to target pharmaceutical agents to specific patient populations. Genetic differences are known to affect responses to many types of drug therapy, and pharmacogenomics analysis serves to customize the use of pharmaceuticals for specific subgroups of patients. The rationale for this approach is that observed gene expression differences may correlate with, and explain the differences in side-effects and efficacy to drugs in humans.

Pharmacokinetics The study of how the body absorbs, distributes, breaks down, and eliminates the drug.

Pharmacophore The spatial mutual orientation of atoms or groups of atoms assumed to be recognized by and to interact with a receptor or the active site of a receptor.

Phylogeny Evolutionary history of an organism or group of organisms, often presented in chart form as a phylogenetic tree.

Physical map A map of the locations of identifiable landmarks on DNA (e.g. restriction enzyme cutting sites, genes), regardless of inheritance. Distance is measured in base pairs. For the human genome, the lowest-resolution physical map is the banding patterns on the 24 different chromosomes; the highest-resolution map would be the complete nucleotide sequence of the chromosomes. The DNA may be cloned into any one of the available vector systems—YACs, cosmids, phage, or even plasmids. Major advantages of ordered clone maps are that they are of high resolution and directly provide the clones for further study.

Polyacrylamide gel electrophoresis (PAGE) The most widely used electrophoresis technique for separation and characterization of proteins and nucleic acids.

Polymerase chain reaction (PCR) A technique used to amplify or generate large amounts of replica DNA of a segment of any DNA whose "flanking" sequences are known. Oligonucleotide primers which bind these flanking sequences are used by an enzyme (*Taq* polymerase) to copy the sequence in between the primers. Cycles of heat to break apart the DNA strands, cooling to allow the primers to bind, and heating again to allow the enzyme to copy the intervening sequence lead to a doubling of DNA at each cycle. The reactions are typically carried out on a regulated heating block and consist of 30–35 cycles of repeated amplification of the entire DNA present. Single molecules of "target" DNA can be amplified to microgram amounts of DNA. The target DNA can be of any origin.

Position-specific iterated (PSI)-BLAST The most sensitive BLAST program, making it useful for finding very distantly related proteins.

Position-specific scoring matrices (PSSM) The form of statistical table that provides probability information of amino acids or nucleotides at each position of an ungapped multiple sequence alignment.

Post-transcriptional modification Alterations made to pre-mRNA before it leaves the nucleus and becomes mature mRNA.

Post-translational modification Alterations made to a protein after its synthesis at the ribosome. It mainly includes trimming or proteolysis, phosphorylation (addition of phosphate groups to proteins in the presence of protein kinases), glycosylation (addition of sugar coating to proteins) and hydroxylation.

Practical extraction report language (PERL) Designed for working with text, generating reports, and manipulating files of biological databases.

Primary sequence (protein) The linear sequence of a polypeptide or protein.

Probe Single-stranded DNA or RNA molecules of specific base sequence, labelled either radioactively or immunologically, that are used to detect the complementary base sequence by hybridization.

Profile Sequence profiles are usually derived from multiple alignments of sequences with a known relationship, and consist of tables of position-specific scores and gap-penalties. Each position in the profile contains scores for all of the possible amino acids, as well as one penalty score for opening and one for continuing a gap at the specified position. Attempts have been made to further improve the sensitivity of the profile by refining the procedures to construct a profile starting from a given multiple alignment. Other representations for sequence domains or motifs do not necessarily require the presence of a correct and complete multiple alignment, such as Hidden Markov Models.

Progressive alignment method The most widely used method to multiple sequence alignments uses a heuristic search known as progressive technique that speeds up the alignment of multiple sequences through a multistep process.

PROSITE A database of protein families and domains. It consists of entries describing the domains, families and functional sites as well as amino acid patterns, signatures, and profiles in them.

Protein annotation In Swiss-Prot, as in most other sequence databases, two classes of data can be distinguished: the core data and the annotation. For each sequence entry the core data consists of the sequence data, the citation information (bibliographical references), and the taxonomic data (description of the biological source of the protein), while the annotation consists of the description of the following items: Function(s) of the

protein, post-translational modification(s) for example, glycosylation, phosphorylation, acetylation, GPI-anchor, etc. Domains and sites for example, calcium-binding regions, ATP-binding sites, zinc fingers, homeobox, etc. Secondary structure, quaternary structure, similarities to other proteins, disease(s) associated with deficiencies(s) in the protein, sequence conflicts, variants, etc.

Protein Data Bank (PDB) The central repository for 3D structural data obtained by X-ray crystallography or NMR spectroscopy of proteins and nucleic acids.

Protein families Sets of proteins that share a common evolutionary origin reflected by their relatedness in function which is usually reflected by similarities in sequence, or in primary, secondary or tertiary structure. Subsets of proteins with related structure and function.

Protein ID (GenBank) The protein ID is an identification number assigned to the amino acid sequence data included within a sequence record. This sequence identifier uses the accession version format. Each protein ID is made up of three letters followed by five digits, a period, and a version number. For example, in a sequence record M12345, the protein ID for the sequence translation could be AAA35650.1.

Protein information resource (PIR) Established as a resource to assist researchers in the identification and interpretation of protein sequence information.

Proteome The entire protein complement of a given organism.

Proteomics The study of the proteome. The study of the set of proteins produced (expressed) by an organism, tissue or cell, and the changes in protein expression patterns in different environments and conditions. Typically, all the expressed proteins in a particular cell or tissue type, obtained by identifying the proteins from cell extracts using a combination of 2D gel electrophoresis and mass spectrometry and subsequent large-scale analysis of the protein composition and function.

PubMed A centralized Web-based library of scientific articles. It provides access to peer-reviewed literature relevant for biological sciences.

Query (sequence) A DNA, RNA or protein sequence used to search a sequence database in order to identify close or remote family members (homologs) of known function, or sequences with similar active sites or regions (analogs), from whom the function of the query may be deduced.

Ramachandran plot The resulting diagram in which ϕ and ψ angles of amino acids of a particular protein are graphically plotted against each other.

RasMol (Raster display of molecules) Software for visualizing three-dimensional structures of many types of molecules, including proteins and nucleic acids.

Rational drug design (Structure-based drug design) The development of drugs based on the 3-dimensional molecular structure of a particular target protein.

Reading frame A sequence of codons that begins with an initiation codon and ends with a termination codon, typically of at least 150 bases (50 amino acids) coding for a polypeptide or protein chain.

Repeats (repeat sequences) Approximate repeats occur throughout the DNA of higher organisms (mammals). For example, the *Alu* sequences of length about 300 characters appear hundreds of thousands of times in human DNA with about 87% homology to a consensus *Alu* string. Some short substrings such as TATA-boxes, poly(A) and (TG) also appear more often than by chance. Repeat sequences may also occur within genes, as mutations or alterations to those genes. Repetitive sequences, especially mobile elements, have many applications in genetic research. DNA transposons and retroposons are routinely used for insertional mutagenesis, gene mapping, gene tagging, and gene transfer in several model systems.

Restriction enzymes or **molecular scissors** Enzymes that recognize certain DNA sequences which they cleave. They are thought to protect cells from viral infection and are useful in recombinant DNA technology.

Restriction fragment length polymorphisms (RFLPs) Variation within the DNA sequences of organisms of a given species that can be identified by fragmenting the sequences using restriction enzymes, since the variation lies within the restriction site. RFLPs can be used to measure the diversity of a gene in a population.

Reverse position specific BLAST (RPS-BLAST) More sensitive way of identifying conserved domains in proteins than standard BLAST searching.

Reverse transcriptase Enzyme that catalyses synthesis of a complementary DNA (cDNA) strand using RNA as a template. Also called RNA-dependent DNA polymerase. The process is called reverse transcription.

Reverse transcriptase-PCR (RT-PCR) Procedure in which PCR amplification is carried out on DNA that is first generated by the conversion of mRNA to cDNA using reverse transcriptase.

Sanger Center Established by Wellcome Trust and Medical Research Council, UK, to focus especially on mapping and sequencing the human genome.

Sanger's dideoxyribonucleotide synthetic method An approach of genomic sequencing that involves use of enzyme to synthesize DNA of varying length in four different reactions, stopping the DNA replication at positions occupied by one of the four bases, and then determines the resulting fragment length by gel electrophoresis. Based on the size of fragments an order of nucleotides is determined automatically.

SCOP (Structural Classification of Proteins) The database of structural motifs of proteins generated manually based on class, fold, superfamily, and family classification.

Secondary structure (protein) The organization of the peptide backbone of a protein that occurs as a result of hydrogen bonds, e.g. α-helix, β-pleated sheet.

Sequence alignment A way of arranging the primary sequences of DNA, RNA, or protein to identify regions of similarity that may be a consequence of functional, structural, or evolutionary relationships between the sequences.

Sequence logo Graphical representation of a multiple sequence alignment that displays a consensus sequence with frequency information.

Sequence retrieval system (SRS) A network browser for databanks in molecular biology, integrating and linking a number of protein and nucleotide databases as well as for feeding the sequence retrieved into analytical tools such as sequence comparison and alignment programs.

Sequence tagged site (STS) A short DNA sequence that is easily recognizable and occurs only once in a genome. STS is a unique sequence from a known chromosomal location that can be amplified by PCR. STSs act as physical markers for genomic mapping and cloning.

Shotgun cloning The cloning of an entire gene segment or genome by generating a random set of fragments using restriction endonucleases to create a gene library that can be subsequently mapped and sequenced to reconstruct the entire genome.

Similarity To relate one nucleotide or protein sequence to another. The extent of similarity between two sequences is based on the per cent of sequence identity and/or conservation.

Similarity (homology) search Homology methods are the most powerful and are based on the detection of significant extended sequence similarity to a protein of known structure, or of a sequence pattern characteristic of a protein family.

Single nucleotide polymorphisms (SNPs) Variations of single base pairs scattered throughout the human genome that serve as measures of the genetic diversity in humans. About 1 million SNPs are estimated to be present in the human genome, and SNPs are useful markers for gene mapping studies.

Smith–Waterman algorithm Local pairwise alignment algorithm that applies dynamic programming in alignment.

Software tools Tools in bioinformatics range from simple command-line tools, to more complex graphical programs and stand-alone web-services available from various bioinformatics companies or public institutions. The computational biology tool best known among biologists is probably BLAST, an algorithm for determining the similarity of arbitrary sequences against other sequences, possibly from curated databases of protein or DNA sequences. The NCBI provides a popular web-based implementation that searches their databases. BLAST is one of a number of generally available programs for doing sequence alignment.

Southern blotting A procedure for the identification of DNA by transmitting a fragment isolated on an agarose gel to a nitrocellulose filter where it can be hybridized with a complementary "probe" sequence.

Spliceosome Protein–RNA complex that removes introns in eukaryotic nuclear RNAs.

Structural genomics The large-scale analysis of 3D protein structure.

Structure prediction Algorithms that predict the secondary, tertiary and sometimes even quaternary structure of proteins from their sequences. Determining protein structure from sequence has been dubbed "the second half of the genetic code" since it is the folded tertiary structure of a protein that governs how it functions as a gene product. As yet most structure prediction methods are only partially successful, and typically work best for certain well-defined classes of proteins.

Substitution matrix A model of protein evolution at the sequence level resulting in the development of a set of widely used substitution matrices. These are frequently called Dayhoff, MDM (mutation data matrix), BLOSUM or PAM (per cent accepted mutation) matrices. They are derived from global alignments of closely related sequences. Matrices for greater evolutionary distances are extrapolated from those for lesser ones.

Supersecondary structure The arrangement of alpha-helices or beta-strands in a protein sequence into discrete folded structures, e.g. beta-barrels, or beta-alpha-beta-motifs.

Surface enhanced laser desorption-ionization (SELDI) An ultra high-throughput screening (uHTS) system utilized for protein profiling of extracts obtained from the cells, tissues or physiological fluids. The method involves use of patented SELDI protein chip on the surface of which proteins isolated from complex biological mixture is directly applied.

Swiss-Prot A manually curated biological database of protein sequences developed by the Swiss Institute of Bioinformatics and the European Bioinformatics Institute.

Tertiary structure Folding of a protein chain via interactions of its sidechain molecules including formation of disulphide bonds between cysteine residues.

Transcriptome Complete set of mRNA molecules produced by a cell under condition, i.e., as per the necessity of cell, active gene transcribes to produce mRNA (known as transcript). Its study is known as transcriptomics.

Transposase The enzyme encoded by transposable elements that undergo conservative transposition.

Transposition The process by which mobile genetic elements move from one location in genome to another.

Transposon A mobile piece of DNA that is flanked by the terminal repeat sequences and typically bears genes coding for transposition function.

TrEMBL (Translated EMBL) A very large protein database in Swiss-Prot format generated by computer translation of the genetic information from the EMBL nucleotide sequence database.

True positive A sequence correctly identified by a discriminator/ algorithm as possessing a particular motif or pattern.

TwinScan Gene structure prediction system that directly extends the probability model of Genscan, allowing it to exploit the homology between two related genomes.

Two-dimensional gel electrophoresis A technique that helps in study of proteins. A key technology for proteomics.

Unidentified reading frame (URF) An open reading frame encoding a protein of undefined function.

Universal Protein (UniProt) Resource for protein sequences and is the central hub for the collection of functional information on proteins, with accurate, consistent, and rich annotation, the amino acid sequence, protein name or description, taxonomic data.

Variable numbers of tandem repeats (VNTRs) DNA sequence blocks of 2–60 base pairs which are repeated from two to more than 20 times in different individuals. This polymorphism makes VNTRs very useful DNA markers used in genomic mapping, linkage analysis and also DNA fingerprinting.

Virtual libraries The creation and storage of vast collections of molecular structures in an electronic database. These databases may be queried for subsets that exhibit specific physico-chemical features, or may be virtually screened for their ability to bind a drug target. This process may be performed prior to the synthesis and testing of the molecules themselves.

VLSI Very large-scale integration allowing over 100,000 transistors on a chip.

Weight matrix The density of binding sites in a gene or sequence can be used to derive a ratio of density for each element in a pattern of interest. The combined individual density ratios of all elements are then collectively used to build a scoring profile known as a weight matrix. This profile can be used to test the prediction of the identification of the selected pattern and the ability of the algorithm to discriminate them from non-pattern sequences.

Western blotting technique This technique is used to detect the presence of proteins in a given sample. It works on the principle of antigen–antibody reaction, hence also known as immuno detection technique.

Worldwide protein data bank (wwPDB) Organizations that act as deposition, data processing and distribution centres for PDB data.

X-ray crystallography An indispensable tool in elucidation of three-dimensional architecture of matter in crystalline state at molecular and atomic resolution.

Yeast artificial chromosome (YAC) A vector used to clone DNA fragments (up to 400 kb); it is constructed from the telomeric, centromeric, and replication origin sequences needed for replication in yeast cells.

Yeast 2-hybrid system A yeast-based method used to simultaneously identify, and clone the gene for, proteins interacting with a known protein. The basis of this method is a "transcriptional reporter assay" in which reporter gene expression is dependent on two domains. The first domain is linked to the known protein. The second domain is genetically linked to a library. If the library is screened against the known protein the two domains will interact only if a protein from the library binds the known protein, resulting in transcription activation of the reporter gene, and a blue colour. The blue yeast clone will contain the gene encoding the newly identified protein.

Z-DNA A conformation of DNA existing as a left-handed double helix (the phosphate–sugar backbone forms a left-handed zigzag course), which may play a role in gene regulation.

Zinc fingers A protein motif formed by the interaction of repeated cysteine and histidine residues with a zinc ion. The spacing of the repeats results in fingerlike arrangements of the protein loops formed from the interaction, which interact with DNA. These motifs are typically found in transcription factors.

REFERENCES

Achuthsankar, S. Nair. (2007). Computational Biology & Bioinformatics-A Gentle Overview, Communications of Computer Society of India.

Adams, M.D. *et al.* (2000). " The genome sequence of *Drosophila melanogaster.*" *Science.* 287: 2185–95.

Adams, M.D., Kelley, J.rM., Gocayne, J.D., Dubnick, M., Polymeropoulos, M.H., Xiao, H., Merril, C.R., Wu, A., Olde, B. and Moreno, R.F. *et al.* (1991). "Complementary DNA sequencing: expressed sequence tags and human genome project." *Science.* 21: 252(5013): 1651–6.

Alberts, B. (1998). "The cell as a collection of protein machines: preparing the next generation of molecular biologists."*Cell.* 92: 291–294.

Alberts, B., Bray, D., Lewis, J., Raff, M., Roberts, K. and Watson, J.D. (1989). *The Molecular Biology of the Cell,* 2nd. edn. Garland Publishing, New York.

Aloy, P. and Russell, R.B. (2005). "InterPreTS: protein interaction prediction through tertiary structure." *Bioinformatics.* 19(1): 161–162.

Altschul, S.F. and Koonin, E.V. (1998). "Iterated profile searches with PSI-BLAST — A tool for discovery in protein databases." *Trends in Biochemical Sciences.* 23: 444–7.

Altschul, S.F. *et al.* (1990). "Basic local alignment search tool." *J. Mol. Biol.* 215: 403–410.

Altschul, S.F. *et al.* (1997). "Gapped BLAST and PSI-BLAST: A new generation of protein database search programs." *Nucleic Acids Res.* 25: 3389–3402.

Altschul, S.F., Gish, W., Miller, W., Myers, E.W. and Lipman, D.J. (1990). "Basic local alignment search tool." *J. Mol. Biol.* 215(3): 403–410.

Aluru, Srinivas. (2006). *Handbook of Computational Molecular Biology.* Chapman & Hall/CRC Computer and Information Science Series.

Appel, R.D., Barouche, A. and Hochstrasser, D.F. (1994). "A generation of information-retrieval tools for biologist with the example of the ExPASy WWW server." *TiBS.* 21(5): 191.

Arabidopsis Genome Initiative. (2000). In Analysis of the genome sequence of the flowering plant *Arabidopsis thaliana. Nature.* 408: 796–815.

Arun, K. Attri and Allen, P. Minton. (2005). "Composition gradient static light scattering: A new technique for rapid detection and quantitative characterization of reversible macromolecular hetero-associations in solution." *Analytical Biochemistry.* 346: 132–138.

Attwood, T.K., Michie, A.D. and Jones, M.L. (1996). "DbBrowser: Integrated access to database worldwide." *TiBS.* 21(5): 191.

Aytuna, A.S., Keskin, O.and Gursoy, A. (2005). "Prediction of protein-protein interactions by combining structure and sequence conservation in protein interfaces." *Bioinformatics.* 21 (12): 2850–2855.

Bairoch Amos. (2000). "Serendipity in bioinformatics, the tribulations of a Swiss bioinformatician through exciting times!." *Bioinformatics.* 16: 48–64.

Baker, D. and Sali, A. (2001). "Protein structure prediction and structural genomics." *Science.* 294(5540): 93–96.

Baldi, P. and Brunak, S. (2001). *Bioinformatics: The Machine Learning Approach,* 2nd edn. MIT Press.

Barnes, M.R. and Gray, I.C. (eds.). (2003). *Bioinformatics for Geneticists,* 1st edn. Wiley.

Baxevanis, A.D. and Ouellette, B.F.F. (eds.) (2005). *Bioinformatics: A Practical Guide to the Analysis of Genes and Proteins,* 3rd edn. Wiley.

Baxevanis, A.D., Petsko, G.A., Stein, L.D. and Stormo, G.D. (eds.). (2007). *Current Protocols in Bioinformatics.* Wiley.

Benno Schwikowskil, Peter Uetz and Stanley Fields. (2000). "A network of protein-protein interactions in yeast." *Nature Biotechnology.* 18: 1257–1261.

Berman, H.M. *et al.* (2000). "The protein data bank." *Nucleic Acids Res.* 28: 235–242.

Berman, H.M., Henrick, K. and Nakamura, H. (2003). "Announcing the worldwide Protein Data Bank." *Nature Structural Biology.* 10(12): 980.

Berman, H.M., Westbrook, J., Feng, Z., Gilliland, G., Bhat, T.N., Weissig, H., Shindyalov, I.N. and Bourne, P.E. (2000). "The Protein Data Bank." *Nucleic Acids Research.* 28: 235–242.

Bernstein, F.C., Koetzle, T.F., Williams, G.J., Meyer, Jr. E.F., Brice, M.D., Rodgers, J.R., Kennard, O., Shimanouchi, T. and Tasumi, M. (1977). "The Protein Data

Bank: a computer-based archival file for macromolecular structures." *J. Mol. Biol.* 112: 535–542.

Berstein, F.C., Koetzle, T.F. and William, G.J.B. (1977). "A computer based archival file for macromolecular structures." *Journal of Molecular Biology.* 112: 535–542.

Black, D.L. (2000): "Protein diversity from alternative splicing: A challenge for bioinformatics and post-genome biology." *Cell.* 103: 367–370.

Blattner, F.R. *et al.* (1997). "The complete genome sequence of *Escherichia coli* K-12." *Science.* 277: 1453–1474.

Bonvin, A.M. (2006). "Flexible protein-protein docking." *Current Opinion in Structural Biology.* 16: 194–200.

Bowie, J.U., Luthy, R. and Eisenberg, D. (1991). "A method to identify protein sequences that fold into a known three-dimensional structure." *Science.* 253 (5016): 164–170.

Burge, C.B. and Karlin, S. (1997). "Finding the genes in genomic DNA." *J. Mol. Bio.* 268: 78–94.

Burset, M. and Guigo, R. (1996). "Evaluation of gene structure prediction programs." *Genomics.* 34: 353–367.

C. elegans Sequencing Consortium. "Genome sequence of the nematode *C. elegans*: A platform for investigating biology." *Science.* 282: 2012–8.

Campbell, A.M. and Heyer, L.J. (2003). *Discovering Genomics, Proteomics, and Bioinformatics.* Cold Spring Harbor Laboratory Press.

Chandonia, J.M. and Karplus, M. (1996). "The importance of larger data sets for protein secondary structure prediction with neural networks." *Protein Sci.* 5: 768–774.

Chandonia, J.M. and Karplus, M. (1999). "New methods for accurate prediction of protein secondary structure." *Proteins.* 35: 293–306.

Chen, L. *et al.* (2004). "TargetDB: a target registration database for structural genomics projects." *Bioinformatics.* 20: 2860–2862.

Claverie, J.M. and Notredame, C. (2003). *Bioinformatics for Dummies.* Wiley.

Craig, R. and Beavis, R.C. (2004). "TANDEM: matching proteins with tandem mass spectra." *Bioinformatics.* 20: 1466–1467.

Craig, R. *et al.* (2004). "Open source system for analyzing, validating, and storing protein identification data." *J. Proteome Res.* 3: 1234–1242.

Cristianini, N. and Hahn, M. (2006). *Introduction to Computational Genomics.* Cambridge University Press.

Curtis Jamison. (2003). *Perl Programming for Biologists.* John Wiley & Sons, Inc. NJ.

Dandekar, T., Snel, B., Huynen, M. and Bork, P. (1998). "Conservation of gene order: a fingerprint of proteins that physically interact." *Trends Biochem. Sci.* (23): 324–328.

Dayhoff, M.O., Schwartz, R.M. and Orcutt, B.C. (1978). "A model of evolutionary change in proteins." In: *Atlas of Protein Sequence and Structure.* Dayhoff, M.O. (ed.). National Biomedical Research Foundation, Washington D.C. pp. 345–352.

Debe, D.A., Danzer, J.F., Goddard, W.A. and Poleksic, A. (2006). "STRUCTFAST: Protein sequence remote homology detection and alignment using novel dynamic programming and profile-profile scoring." *Proteins.* 64: 960–967.

Delcher, A., Harmon, D., Kasif, S., White, O. and Salzberg, S. (1999). "Improved microbial gene identification with GLIMMER." *Nucl'. Acids Res.* 27(23): 4636–4641.

Dietmann, S., Park, J., Notredame, C., Heger, A., Lappe, M. Holm, L. (2001). "A fully automatic evolutionary classification of protein folds: Dali domain dictionary version 3." *Nucleic Acids Res.* 29(1): 55–7.

Doolittle, R.F., Hunkapiller, M.W., Hood, L.E., Devare, S.G., Robbins, K.C., Aaronson, S.A. and Antoniades, H.N. (1983). "Simian Sarcoma *Onc* Gene, *v-sis,* is derived from the gene (or genes) encoding platelet derived growth factor." *Science.* 221: 275–277.

Durbin, R., Eddy, S., Krogh, A. and Mitchison, G. (1998). *Biological Sequence Analysis.* Cambridge University Press.

Edgar, R.C. (2004). "MUSCLE: multiple sequence alignment with high accuracy and high throughput." *Nucleic Acids Res.* 32: 1792–1797.

Edward, T. Maggio and Kal Ramnarayan. (2001). "Recent developments in computational proteomics." *Trends in Biotechnology.* 19(7): 266–272.

Enright, A.J., Iliopoulos, I., Kyripides, N.C. and Ouzounis, C.A. (1999). "Protein interaction maps for complete genomes based on gene fusion events." *Nature.* (402):86–90.

Etzold, T. and Argos, P. (1993). "SRS an indexing and retrieval tool for flat file data libraries." *Computer Applications of the Biosciences.* 9: 49–57.

Etzold, T., Ulyanov, A. and Argos, P. (1996). "SRS-Information-retrieval system for molecular biology databanks." *Methods in Enzymology.* 266: 114–128.

Falicov, A. and Cohen, F.E. (1996). "A surface of minimum area metric for the structural comparison of proteins." *J. Mol. Biol.* 258: 871–892.

Falkner, J. and Andrews, P. (2005). "Fast tandem mass spectra-based protein identification regardless of the number of spectra or potential modifications examined." *Bioinformatics.* 21: 2177–2184.

Fleischmann, R.D. *et al.* (1995). "Whole-genome random sequencing and assembly of *Haemophilus influenzae* Rd." *Science.* 269: 496–512.

Gardner, M.J., Hall, N. and Fung, E. *et al.* (2002). "Genome sequence of the human malaria parasite *Plasmodium falciparum.*" *Nature.* 419(6906): 498–511.

Gasteiger, E., Hoogland, C., Gattiker, A., Duvaud, S., Wilkins, M.R., Appel, R.D. Bairoch, A. (2005). "Protein identification and analysis tools on the ExPASy server." In: *The Proteomics Protocols Handbook.* John, M. Walker. (ed.). Humana Press. pp. 571–607.

George, D., Barker, W. and Hunt, L. (1986). "The Protein identification resource." *Nucleic Acids Research.* 14: 11–15.

George, P. Redei. (2003). *Encyclopedia Dictionary of Genetics, Genomics, and Proteomics,* 2nd edn. Wiley. p.1024.

Gibbs, A.J. and McIntyre, G.A. (1970). *Eur. J. Biochem.* 16: 1–11.

Gil Alterovitz, Jiwaji, A. and Ramoni, M.F. (2008). "Automated programming for bioinformatics algorithm deployment." *Bioinformatics.* 24(3): 450–451.

Gilbert, D. (2004). "Bioinformatics software resources." *Briefings in Bioinformatics.* 5(3): 300–304.

Goffeau, A. *et al.* "Life with 6000 genes." *Science.* 274: 546: 563–7.

Gray, J.J. (2006). "High-resolution protein-protein docking." *Current Opinion in Structural Biology.* 16: 183–193.

Green, E.D. (2001). "Strategies for systematic sequencing of complex organisms." *Nature Reviews (Genetics).* 2: 573–83.

Gregory, S.G., Barlow, K.F., McLay, K.E. *et al.* (2006). "The DNA sequence and biological annotation of human chromosome 1." *Nature.* 441(7091): 315–21.

Gross *et al.* "CONTRAST (2007): A discriminative, phylogeny-free approach to multiple informant de novo gene prediction." *Genome Biology.* 8 (12).

Hamm, G. and Cameron, G. (1986). "The EMBL Data Library." *Nucleic Acids Research.* 14: 5–9.

Herzberg, C., Weidinger L.A., Dörrbecker, B., Hübner, S., Stülke, J. and Commichau, F.M. (2007). "SPINE: A method for the rapid detection and analysis of protein-protein interactions *in vivo.*" *Proteomics.* 7(22): 4032–4035.

Hideo Matsumura and Stefanie Reich *et al.* (2003). "Gene expression analysis of plant host–pathogen interactions by SuperSAGE." *PNAS.* 100: 15718–15723.

Hochachka, P.W. and Somero, G.N. (1984). *Biochemical Adaptation.* Princeton University Press, Princeton, NJ.

Holm, L. and Sander, C. (1993). "Protein structure comparison by alignment of distance matrices." *J. Mol. Biol.* 233(1): 123–38.

Hooke, R. and Jeeves, T.A. (1961). "Direct search solution of numerical and statistical problems." *J. ACM.* 8: 212–229.

Human Genome Sequencing Consortium, International (2004). "Finishing the euchromatic sequence of the human genome." *Nature.* 431(7011): 931–45.

International Human Genome Sequencing Consortium. (2001). "Initial sequencing and analysis of the human genome." *Nature.* 409: 860–921.

Ion Korf, Paul Flecik, Daniel Daun and Brent, M.R. (2001). "Integrating genomic homology into gene structure prediction." *Bioinformatics.*12: 140–148.

Jaeger, J.A., Turner, D.H. and Zucker, M. (1990). *Methods Enzymol.* 183: 281–306.

Jansen, R., Yu, H., Greenbaum, D., Kluger, Y., Krogan, N.J., Chung, S., Emili, A., Snyder, M., Greenblatt, J.F. and Gerstein, M. (2003). "A Bayesian networks approach for predicting protein-protein interactions from genomic data." *Science.* 302(5644): 449–53.

Jin Xiong. (2006). *Essentials Bioinformatics.* Cambridge University Press, USA.

Julio Licinio and Ma-Li Wong. (eds.). (2002). *Pharmacogenomics:The Search for Individualized Therapies.* Wiley-VCH. p.600.

Kabsch, W. and Sander, C. (1983). "Dictionary of protein secondary structure: pattern recognition of hydrogen-bonded and geometrical features." *Biopolymers.* 22: 2577–2637.

Keedwell, E. (2005). *Intelligent Bioinformatics: The Application of Artificial Intelligence Techniques to Bioinformatics Problems.* Wiley.

Kohane *et al.* (2002). *Microarrays for an Integrative Genomics.* The MIT Press.

Lai, C.H. *et al.* (2000). "Identification of novel human genes evolutionarily conserved in *Caenorhabditis elegans* by comparative proteomics." *Genome Research.* 10(5): 703–713.

Laskowaski, R.A., Hutchinson, E.G., Michie, A.D., Wallace, A.C., Jones, D.T. and Thornton, J.M. (1997). "PDBsum -A web-based database summaries of and analysis of all PDB structures." *Trends in Biochemical Science.* 22: 488–490.

Le Novere, N. (2001). "MELTING, computing the melting temperature of nucleic acid duplex." *Bioinformatics.* 17: 1226–1227.

Lesk, A.M. (2001). *Introduction to Protein Architecture: The Structural Biology of Proteins.* Oxford University Press, Oxford.

Lesk, M. (1997). *Practical Digital Libraries: Books, Bytes and Bucks.* Morgan Kaufmann, San Francisco.

Lewin, B. (1999). *Genes VII.* Oxford University Press.

Lewis, J.P. and Ulrich Neumann (2007). *Performance of Java versus C++.* Computer Graphics and Immersive Technology Lab, University of Southern California.

Li, W.H. (1997). *Molecular Evolution,* Sinauer. Associates, Inc., Sunderland, MA, USA.

Li, W.H. and Graur, D. (1991). *Fundamentals of Molecular Evolution.* Sinauer Associates, Inc.,Sunderland, MA, USA.

Liebler, D.C. (2002). *Introduction to Proteomics: Tools for the New Biology.* Humana Press, Totowa, NJ.

Lindblad-Toh, K., Wade, C.M. and Mikkelsen, T.S. *et al.* (2005). "Genome sequence, comparative analysis and haplotype structure of the domestic dog." *Nature.* 438(7069): 803–19.

Lodish *et al.* (1995). *Molecular Cell Biology,* 3rd edn. Scientific American Books, Freeman and Company, New York.

Loftus, B., Anderson, I. and Davies, R. *et al.* (2005). "The genome of the protist parasite *Entamoeba histolytica.* " *Nature.* 433 (7028): 865–8.

Lohar, P.S. (2006). "Bioinformatics-*In silico* tools in modern biology." In the Proc. of UGC sponsored Natl. Sem. on Interdisciplinary Applications of Electronics. A.S.C. College, Chopda (Dist. Jalgaon), MS. India.

Lohar, P.S. (2008). *Cell and Molecular Biology,* MJP Publishers, Chennai.

Lund, O. *et al.* (2005). *Immunological Bioinformatics.* The MIT Press.

Majoros, W.H., Pertea, M. and Salzberg, S.L. "TigrScan and GlimmerHMM: two open-source *ab initio* eukaryotic gene-finders." *Bioinformatics.* 20: 2878–2879.

Makarov, V. (2002). "Computer programs for eukaryotic gene prediction." *Brief. Bioinform.* 3: 195–9.

Mangalam, H. (2002). "The Bio* toolkits – a brief overview." *Brief. Bioinform.* 3: 296–302.

Marcotte, E.M., Pellegrini, M., Ng, H.L., Rice, D.W., Yeates, T.O. and Eisenberg, D. (1999). "Detecting protein function and protein-protein interactions from genome sequences." *Science.* 285: 751–753 .

Martens, L. *et al.* (2005). "DBToolkit: processing protein databases for peptide-centric proteomics." *Bioinformatics.* 21: 3584–3585.

Mathews, D.H. *et al.* (1999). "Expanded sequence dependence of thermodynamic parameters improves prediction of RNA secondary structure." *J. Mol. Biol.* 288: 911–940.

Matthews, B.W. (1975). "Comparison of the predicted and observed secondary structure of T4 phage lysozyme." *Biochim. Biophys. Acta.* 405: 442–451.

May, R.M. (1988). "How many species are there on earth?" *Science.* 241: 1441–1450.

Meyer, E.F. (1997). "The first years of the protein data bank." *Protein Science.* 6: 1591–1597.

Michael, S. Waterman. (1995). *Introduction to Computational Biology: Sequences, Maps and Genomes.* CRC Press.

Møller, M.F. (1993). "A scaled conjugate gradient algorithm for fast supervised learning." *Neural Network.* 6: 525–533.

Mount David, W. (2002). *Bioinformatics: Sequence and Genome Analysis.* Spring Harbor Press.

Mouse Genome Sequencing Consortium. (2002). "Initial sequencing and comparative analysis of the mouse genome." *Nature.* 420: 520–562.

Murzin, A.G., Brenner, S.E., Hubbard, T. and Chothia, C. (1995). "SCOP -A Structural classification of proteins database for investigation of sequence and structures." *Journal of Molecular Biology.* 247: 536–540.

Nagaraj, Shivashankar, H., Gasser, Robin, B. and Ranganathan, Shoba. (2007). "A hitchhiker's guide to expressed sequence tag (EST) analysis." *Brief Bioinform.* 8: 6–21.

Nayeem, A., Sitkoff, D., Krystek, Jr. S. (2006). "A comparative study of available software for high-accuracy homology modeling: From sequence alignments to structural models." *Protein Sci.* 15: 808–824.

Needleman, S.B. and Wunsch, C.D. (1970). "General method applicable to the search for similarities in the amino acid sequence of two proteins." *J. Mol.Biology.* 48: 443–453.

Orengo, C.A., Michie, A.D., Jones, D.T., Swindells, M.B. and Thornton, J.M. (1997). "CATH-a hierachical classification of protein domain structure." *Structure.* 5 (8): 1093–1108.

Pachter, Lior and Sturmfels, Bernd. (2005). *Algebraic Statistics for Computational Biology.* Cambridge University Press.

Pazos, F. and Valencia, A. (2001). "Similarity of phylogenetic trees as indicator of protein-protein interaction." *Protein Engineering.* 96(14): 609–614.

Pellegrini, M., Marcotte, E.M., Thompson, M.J., Eisenberg, D. and Yeates, T.O. (1999). "Assigning protein functions by comparative genome analysis: protein phylogenetic profiles." *Proc. Natl. Acad. Sci. USA.* 96: 4285–8.

Penny, D., Hendy, M.D., Zimmer, E.A. and Hamby, R.K. (1990). "Trees from sequences: Panacea or Pandora's box?" *Australian Systematic Botany.* 3: 21–38.

Perkins, D.N. *et al.* (1999). "Probability-based protein identification by searching sequence databases using mass spectrometry data." *Electrophoresis.* 20: 3551–3567.

Pevzner and Pavel, A. (2000). *Computational Molecular Biology: An Algorithmic Approach.* The MIT Press.

Pocock, M.R. *et al.* (2000). "BioJava: open source components for bioinformatics." ACM SIGBIO Newsl. 20: 10–12.

Proteomics. (2000). "New tools for a new era." *Modern Drug Discovery.* 3(7): 35–44.

Rat Genome Sequencing Project Consortium. (2004). "Genome sequence of the Brown Norway rat yields insights into mammalian evolution." *Nature.* 428: 493–521.

Rosman, K.J.R. and Taylor, P.D.P. (1997). "Isotopic compositions of the elements." *Pure Appl. Chem.* 70: 217–235.

Rost, B. and Donoghue, S. (1997). "Sisyphus and prediction of protein structure." *Computer Application in the Biosciences.* 13(4): 345–356.

Rozen, S. and Skaletsky, H. (2000). "Primer3 on theWWW for general users and for biologist programmers." *Methods Mol. Biol.* 132: 365–386.

Sali, A. and Blundell, T.L. (1993). "Comparative protein modelling by satisfaction of spatial restraints." *J. Mol. Biol.* 234: 779–815.

Sankoff, D. and Kruskal, J. B. (1984). *Time Warps, String Edits and Macromolecules: The Theory and Practice of Sequence Comparison*, Addison-Wesley, Reading, MA, USA.

Schuler, G.D. (1998). "Electronic PCR: bridging the gap between genome mapping and genome sequencing." *Trends Biotechnol.* 16: 456–459.

Selbach, M. and Mann, M. (2006). "Protein interaction screening by quantitative immunoprecipitation combined with knockdown (QUICK)." *Nature Methods.* 3: 981–983.

Siew, N. *et al.* (2000). "MaxSub: an automated measure for the assessment of protein structure prediction quality." *Bioinformatics.* 16: 776–785.

Smith, T.F. (1999). "The art of matchmaking: sequence alignment methods and their structural implications." *Structure with Folding and Design.* 7: R7–R12.

Smith, T.F. and Waterman, M.S. (1981). "Identification of common molecular subsequences." *J. Mol. Biology.* 147: 195–197.

Staden, R. (1982). *Nucleic Acids Res.* 10: 2951–2961.

Stephen, R.M. and Schneider, T.D. (1992). "Features of spliceosome evolution and function inferred from an analysis of the information at human spliced sites." *J. Mol. Bio.* 228: 1124–1136.

Suchanek, M., Radzikowska, A. and Thiele, C. (2005). "Photo-leucine and photo-methionine allow identification of protein-protein interactions in living cells." *Nature Methods.* 2: 261–268.

Sussman, J.L., Lin, D., Jiang, J., Manning, N.O., Prilusky, J., Ritter, O. and Abola, E.E. (1998). "Protein data bank (PDB): a database of 3D structural information of biological macromolecules." *Acta Cryst.* D54: 1078–1084.

Tateno, Y., Fukami-Kobayashi, K., Miyazaki, S., Sugawara, H. and Gojobori, T. (1998). "DNA Data Bank of Japan at work on genome sequence data." *Nucleic Acids Res.* 26: 16–20.

Tisdall, James. (2001). *Beginning Perl for Bioinformatics.* O'Reilly.

Twyman, R.M. (2004). *Principles of Proteomics.* BIOS Scientific Publishers, New York.

Vasan, R.S. (2006). "Biomarkers of cardiovascular disease: molecular basis and practical considerations." *Circulation.* 113: 2335–2362.

Velculescu, V.E., Zhang, L., Vogelstein, B. and Kinzler, K.W. (1995). "Serial analysis of gene expression." *Science.* 270(5235): 484–7.

Voigt, C.A., Gordon, D.B. and Mayo, S.L. (2000). "Trading accuracy for speed: A quantitative comparison of search algorithms in protein sequence design." *J. Mol. Biol.* 299(3): 789–803.

Watson, J.D., Hopkins, N.H. and Roberts, J.W. *et al.* (1987). *Molecular Biology of the Gene*, 4th edn. Benjamin-Cummings, Menlo Park, CA.

Wu, C.H. and Yeh, L.S. *et al.* (2003). "The protein information resource." *Nucleic Acids Res.* 31(1): 345–7.

Zhang, Y. and Skolnick, J. (2005). "The protein structure prediction problem could be solved using the current PDB library." *Proc. Nat. Acad. Sci.* USA. 102(4): 1029–1034.

Websites

Deploying Applets in a Mixed-Browser Environment (The Java Tutorials > Deployment > Applets) http://java.sun.com/docs/books/tutorial/deployment/applet/mixedbrowser.html

Design Goals of the JavaTM Programming Language http://java.sun.com/docs/white/langenv/Intro.doc2.html

Ethical Issues in Human Gene Therapy - A Human Genome News article at http://www.ornl.gov/sci/techresources/Human_Genome/publicat/hgn/v10n1/16walter.shtml

Introduction to Antibodies (2006): Enzyme-Linked Immunosorbent Assay (ELISA). http://www.chemicon.com/resource/ANT101/a2C.asp.

Jon Byous, *Java technology* (2005)*: The early years.* Sun Developer Network, no date [ca. 1998]. Retrieved April 22,. http://java.sun.com/features/1998/05/birthday.html

Java Study Group; Why Java Was-Not-Standardized Twice; http://techupdate.zdnet.com/techupdate/stories/main/0,14179,2832719,00.html

Java Community Process website http://www.jcp.org/en/home/index

JAVAONE (open.itworld.com) Sun-The bulk of Java is open sourced at http://open.itworld.com/4915/070508opsjava/page_1.html

Kanehisa, M., Goto, S., Kawashima, S., Okuno, Y., Hattori, M., The KEGG resource for deciphering the genome, Nucleic Acids Res. 2004 Jan 1;32 (Database issue):(http://www.ncbi.nlm.nih.gov/entrez/query.fcgi? cmd =Retrieve&db =PubMed & list_uids=14681412&dopt=Abstract).

Lu, J.P., Beatty, L.K., Pinthus, J.H. (2008). "Dual expression recombinase based (DERB) single vector system for high throughput screening and verification of protein interactions in living cells." Nature Preceding <http://hdl.handle.net/10101/npre.2008.1550.2>.

National Academy of Sciences (2002). Defining the Mandate of Proteomics in the Post- Genomics Era, http://www.nap.edu/books/NI000479/html/R1.html

The Java Language Specification, 2nd edn. http://java.sun.com/docs/books/jls/second_edition/html/intro.doc.html#237601

U.S. Department of Energy Office of Science (http://www.er.doe.gov/), Office of Biological and Environmental Research (http://www.sc.doe.gov/ober/ober_top.html), Human Genome Program.

INDEX

Made in the USA
Monee, IL
07 July 2026

56552349R00351